ANNUAL REVIEW OF EARTH AND PLANETARY SCIENCES

ANNUAL REVIEW OF EARTH AND PLANETARY SCIENCES

FRED A. DONATH, *Editor*
University of Illinois—Urbana

FRANCIS G. STEHLI, *Associate Editor*
Case-Western Reserve University

GEORGE W. WETHERILL, *Associate Editor*
Carnegie Institution of Washington

VOLUME 4

1976

ANNUAL REVIEWS INC. 4139 EL CAMINO WAY PALO ALTO, CALIFORNIA 94306

ANNUAL REVIEWS INC.
Palo Alto, California, USA

International Standard Book Number 0-8243-2004-2
Library of Congress Catalog Card Number 72-82137

Annual Reviews Inc. and the Editors of its publications assume no responsibility for the statements expressed by the contributors to this Review.

REPRINTS

The conspicuous number aligned in the margin with the title of each article in this volume is a key for use in ordering reprints. Available reprints are priced at the uniform rate of $1 each postpaid. The minimum acceptable reprint order is 10 reprints and/or $10, prepaid. A quantity discount is available.

PRINTED AND BOUND IN THE UNITED STATES OF AMERICA

PREFACE

The *Annual Review of Earth and Planetary Sciences* is a title which subsumes a considerable proportion of human knowledge. When the Editorial Committee first met six years ago to plan Volume 1, there was some doubt as to whether so broad an area could indeed be covered in any adequate way. In the succeeding years, rapid expansion in both the earth and the planetary sciences bridged the gap between them. Each is built on a descriptive base extended by an experimental and quantitative superstructure. They are tied together by the threads of mathematics, physics, and chemistry. The present volume illustrates both the infusion of other sciences and the breadth of present inquiry. Its chapters range from the loss of hydrogen by terrestrial planets to consideration of the uranium decay series basic to quantitative dating, from generation of the Earth's magnetic field in the core to the ionosphere of Mars, from the evolution of lunar rocks to the diagenesis of clays, from mathematical models of lakes to manganese nodules on the deep-sea floor, from CO and CH_4 in the terrestrial atmosphere to diffusion in electrolytic solutions. Biology, as well as mathematics, physics, and chemistry, has become integrated with the earth sciences, and it is fitting that the first article in this volume contains the reminiscences of George Gaylord Simpson. Throughout Simpson's career, his keen intellect, enormous curiosity, and great talent for synthesis forged the links which bind the earth sciences and biology inextricably together with the chains of evolutionary theory.

THE EDITORS

SOME RELATED ARTICLES APPEARING IN OTHER ANNUAL REVIEWS

From the *Annual Review of Astronomy and Astrophysics*, Volume 13 (1975)

Young Stellar Objects and Dark Interstellar Clouds, S. E. Strom, K. M. Strom, and G. L. Grasdalen

Unseen Astrometric Companions of Stars, Peter van de Kamp

From the *Annual Review of Biochemistry*, Volume 44 (1975)

Pathways of Carbon Fixation in Green Plants, Israel Zelitch

Energy Capture in Photosynthesis: Photosystem II, Richard Radmer and Bessel Kok

From the *Annual Review of Ecology and Systematics*, Volume 6 (1975)

Structure and Climate in Tropical Rain Forest, Egbert Giles Leigh, Jr.

Spectral Analysis in Ecology, Trevor Platt and Kenneth L. Denman

The Population Biology of Coral Reef Fishes, Paul R. Ehrlich

Late Quaternary Climatic Change in Africa, D. A. Livingstone

Experimental Studies of the Niche, Robert K. Colwell and Eduardo R. Fuentes

Simulation Models of Ecosystems, Richard G. Wiegert

Modes of Animal Speciation, Guy L. Bush

From the *Annual Review of Fluid Mechanics*, Volume 8 (1976)

Mixing and Dispersion in Estuaries, Hugo B. Fischer

Multiphase Fluid Flow Through Porous Media, R. A. Woodling and H. J. Morel-Seytoux

Currents in Submarine Canyons: An Air-Sea-Land Interaction, Douglas L. Inman, Charles E. Nordstrom, and Reinhard E. Flick

From the *Annual Review of Materials Sciences*, Volume 5 (1975)

Dislocations and Disclinations in Material Structures: The Basic Topological Concepts, E. Kröner and K.-H. Anthony

The Structure of Surfaces, M. A. Chesters and G. A. Somorjai

From the *Annual Review of Microbiology*, Volume 29 (1975)

Ecological Implications of Metal Metabolism by Microorganisms, Arne Jernelöv and Ann-Louise Martin

Regulation and Genetics of Bacterial Nitrogen Fixation, Winston J. Brill

From the *Annual Review of Nuclear Science*, Volume 25 (1975)

Radiometric Chronology of the Early Solar System, George W. Wetherill

From the *Annual Review of Physical Chemistry*, Volume 26 (1975)

Isotope Effects in Chemical Kinetics, Fritz S. Klein

Pollution of the Stratosphere, Harold S. Johnston

CONTENTS

INDEXES

ANNUAL REVIEWS INC. is a nonprofit corporation established to promote the advancement of the sciences. Beginning in 1932 with the *Annual Review of Biochemistry*, the Company has pursued as its principal function the publication of high quality, reasonably priced Annual Review volumes. The volumes are organized by Editors and Editorial Committees who invite qualified authors to contribute critical articles reviewing significant developments within each major discipline.

Annual Reviews Inc. is administered by a Board of Directors whose members serve without compensation.

Annual Reviews are published in the following sciences: Anthropology, Astronomy and Astrophysics, Biochemistry, Biophysics and Bioengineering, Earth and Planetary Sciences, Ecology and Systematics, Energy, Entomology, Fluid Mechanics, Genetics, Materials Science, Medicine, Microbiology, Nuclear Science, Pharmacology and Toxicology, Physical Chemistry, Physiology, Phytopathology, Plant Physiology, Psychology, and Sociology. In addition, two special volumes have been published by Annual Reviews Inc.: *History of Entomology* (1973) and *The Excitement and Fascination of Science* (1965).

George Gaylord Simpson

THE COMPLEAT PALAEONTOLOGIST?

George Gaylord Simpson
Department of Geosciences, University of Arizona, and the Simroe Foundation, 5151 East Holmes Street, Tucson, Arizona 85711

INTRODUCTION

When I was asked to write this prefatory chapter I was told that it should not be a review in the usual sense, but rather should be autobiographical, philosophical, or preferably both. For someone who has already spent more than 50 years in professional activities, the task of dealing with the biographical and philosophical aspects of those years in a few pages is appalling. I am not conditioned, like some venerable professors, to be unable to say anything in either less or more than a 50-minute lecture period. Still it is true that I was once assigned the task of revising a small pamphlet on the evolution of horses, and this wound up as a new book of more than 260 pages plus 32 plates.

One trouble (and this is already both autobiography and philosophy) is that I have been unable to work steadily on one subject or even in a single field. I am primarily a vertebrate paleontologist specializing (but not exclusively) on mammals. The "Compleat Palaeontologist" (as Izaak Walton did not put it) is necessarily interdisciplinary, with one foot in geology and the other in biology. If he is contemplative, philosophical, or scatterbrained, he soon begins to need more than two feet. Omitting, for a start, such things as curating, administration, teaching and lecturing, and routine advising and informing, most of my other professional activities and studies have involved the following:

(*a*) Collecting fossils and making field stratigraphic studies.
(*b*) Identifying, classifying, and describing individual specimens and faunas.
(*c*) Devising descriptive, taxonomic, and various comparative methods.
(*d*) Studying the course, the principles or theories, and the significance of organic evolution.
(*e*) Working both in concrete particulars and in broader principles on the philosophy, methodology, and results of biological systematics.
(*f*) Studying the history of organisms in relationship to their geography and the history of the earth.
(*g*) Doing a variety of other things that do not fit well into even a loose definition

of paleontology and can be considered detours, self-indulgences, or hobbies—such as living with and writing a book about a tribe of Indians, penguin-watching, or venturing into linguistics and etymology. Perhaps being coauthor of a textbook of general biology belongs here. Surely also belonging here is my investigation of the chances of communication with extraterrestrial intelligence, with results that made me almost as unpopular with the exobiologists as I am with the antievolutionists. I should mention too writing a good many book reviews, which often required as much original thought as any other writing.

I hereby clean up this miniautobiography and miniphilosophy a bit by dropping topics *a, b,* and *g,* except for passing mention. I proceed to discuss *c, d, e,* and *f* at varying, but always too brief, length and in varying ways, but generally with a view to changes in these fields as they involved me and as I see their outcome up to now.

METHODOLOGY

Collecting, describing, and interpreting are all involved here. All have changed markedly in some ways during my adult life, and some are in a state of flux today.

The most important development in collecting methods has been the invention and increasing use of washing and screening techniques. These methods, although experimentally developed earlier, did not come into wide use by vertebrate paleontologists until the 1940s. That was long after I had started collecting and only a decade or so before an accident put an end to my most active participation in field work, although I have since resumed less strenuous field observations. These methods have enormously increased our knowledge of small fossil vertebrates. With their use, specimens from beds previously considered almost barren have been obtained, as have large samples of forms previously rare in collections. For example, these techniques are resulting in a revolution in the study of some faunas of Mesozoic mammals, which are among the rarest of fossils and have always interested me (my doctoral dissertation was on that subject).

Increase in sizes of available samples of fossil vertebrates, not only by screening but also by intensifying all methods of collection and greatly increasing personnel, has been accompanied by a change from typological approaches to population concepts in biological systematics in general. That meant changing from simple measurements of type specimens and subjective or rule-of-thumb interpretation of their taxonomic significance to statistical estimates of population parameters and of confidence intervals in comparisons of them. Now it seems almost incredible that well into the 1930s such methods were commonly unknown or, when known, were often opposed by both neo- and paleozoological systematists. I believe that one of the turning points in the change in attitudes and methods was a book by my wife and me (Simpson & Roe 1939). Since then, there has been a great increase in the variety and sophistication of relevant statistical methods. The development of electronic computers, still unknown in 1939, has also made it practical to follow much more complex programs and to include much more numerous measurements and more numerous estimates of parameters. I later mention the computer revolution again in connection with principles of systematics.

There are a number of other, less broadly applicable methods that were unusual when I began to use them, and it is possible that I was the inventor of some of them, although it is always foolish to try to establish priority or to argue over it. Among those that come to mind are (*a*) the use of fluorescence under ultraviolet radiation to examine obscure fossils (Simpson 1926) (dates here are those of publication, not of my actual first use of the methods); the illustration of fossil vertebrates by stereophotographs (1929); serial sectioning for study of detailed internal anatomy of fossil mammal skulls (1933) (this had been used on other fossils as early as 1903 but had not been generally adopted); ratio diagrams for comparing proportions of different individuals or species (1941); "Simpson's index" for measuring resemblance between biotas (1943); and the contour mapping of species densities (1964). These are examples of methods now widely used. Some of my other brainstorms subsided without, so to speak, leaving any precipitation. I see no reason to bother anyone with them now, but just as an example that such do, or rather did, indeed exist, I mention the concept and calculation of standard ranges.

Relevant methodology has indeed (fortunately) become almost incredibly more complex and sophisticated since 1924 when I first sat down to write a technical paper destined to be published. Among many other references, I may mention for those curious just Kummel & Raup (1965), Schopf (1972), Sokal & Rohlf (1969), or such journals as *Evolution, Systematic Zoology, Journal of Paleontology,* or *Journal of Mammalogy* (seriatim).

EVOLUTION

When Darwin's *Origin of Species* was published in 1859, paleontology was already a well-established science, but mostly it was pursued either by amateurs or by professionals whose primary interests were in other fields. Among the most eminent previous contributors to vertebrate paleontology were two zoologist-anatomists, Richard Owen in England and Louis Agassiz, who was of Swiss origin with an early French career but who was permanently established in the United States before 1859. They were almost contemporary with Darwin—Owen was then 55 years old and Agassiz 52 to Darwin's 50—but they could not grasp the Darwinian revolution and remained until their deaths (in 1892 and 1873, respectively) violent anti-evolutionists.

Nevertheless, within a few years practically all paleontologists, especially vertebrate paleontologists, were convinced evolutionists, and they have remained so ever since. Their work is with a historical record of life that cannot sensibly be interpreted in any other way. It is, however, one thing to be an evolutionist and another to be a Darwinian. The vertebrate paleontologists (until recently most other paleontologists were less inclined to theory and philosophy) remained almost to a man anti-Darwinian until well into the present century. They were, of course, not anti-Darwinian in the sense of rejecting evolution but rather in the sense of rejecting or relegating to a minor position the Darwinian theory of natural selection. Eminent anti-Darwinians in the United States included E. D. Cope (1840–1897), who was a convinced neo-Lamarckian, H. F. Osborn (1857–1935), who developed a completely idiosyncratic theory of evolution by inner forces working toward aristocratic ends, and W. B. Scott (1858–1947), who in his later years decided that evolution simply could not be explained, especially not by natural selection.

Oddly enough—or it seems odd in retrospect—the rediscovery of Mendelism in 1900 and the realization that it had a bearing on evolution (absent in Mendel's own work) at first obscured more than it clarified the search for explanation in this field. It took another generation to begin developing a new general theory, which incorporated many of Darwin's views, especially natural selection in a somewhat more sophisticated sense, but also bit by bit brought in the vastly greater knowledge now available in the increasingly specialized branches of the life sciences. I have called this still-developing theory the synthetic theory of evolution, to signalize its synthesis from many specialties and to distinguish it both from strict Darwinism, which lacked much that is now involved and included some factors now known to be incorrect, and from neo-Darwinism, which was on the right track as far as it went but which did not go much beyond concentrated faith in a rather naive concept of natural selection.

The synthetic theory has developed gradually over many years in studies by many hands and so does not have a clear founder or date of inception. Nevertheless the synthesis may be said, perhaps simplistically, to have begun by a reconciliation between twentieth-century genetics and neo-Darwinism. That reconciliation was signalized, if not begun, in work by R. A. Fisher (1890–1962) and J. B. S. Haldane (1892–1964) and appears in two of their books, written in quick succession [Fisher 1930, Haldane 1932 (based on earlier lectures)]. It should be mentioned that Sewall Wright (1889–) was also among those already deeply involved, although a full and connected publication of his many studies was not begun until much later.

In that inchoate epoch of the synthetic school, I was not long out of graduate school and I did not yet know Haldane and Wright personally (I never met Fisher), but I was aware of their studies and greatly interested by them. A number of my associates at the American Museum, where I was then employed, were non- or anti-Darwinian, although all were of course evolutionists. In that period, this was also true of almost all paleontologists who were interested in theoretical problems and not only in descriptive systematics. The greater part of my regular work was also perforce in descriptive systematics, but I was increasingly taken up with problems of evolutionary theory. That interest was intensified when the German invertebrate paleontologist Otto Schindewolf published a small book attempting a synthesis of evolutionary theory derived from genetics, embryology, and paleontology (Schindewolf 1936.) Unfortunately Schindewolf's attempted synthesis was based on mutationist, antiselectionist genetics, which had already been made obsolete by studies, mostly in English, that were little-known in Germany and unknown to most paleontologists anywhere.

The fact that Schindewolf's views did not agree with my own interpretations of the fossil record or with what I had so far gathered about evolutionary genetics made me resolve to go further into these matters, at least in what spare time could be devoted to them. I was further both inspired and aided by the appearance of the first of Dobzhansky's now numerous and invaluable books on genetics and evolution (Dobzhansky 1937). In due course, this led to my first book on evolution, begun in 1938 and completed in 1942, although not published until two years later (Simpson 1944). Because I was absent on active military service, I did not then have the advantage of consulting Ernst Mayr's book (1942), which was published before mine and which brought systematics into the growing synthesis.

Since then there has been a tremendous increase in information and improvement of theoretical formulations. I am keenly aware that much of *Tempo and Mode in Evolution* now seems primitive and that parts of it have been invalidated. Nevertheless, I am consoled by the conviction that it had some historic value, that it was a success at least to the extent that it did bring a new field of study and a new thesis into the development of the synthetic theory, and that its thesis has stood up well. That thesis, in briefest form, is that the history of life, as indicated by the available fossil record, is consistent with the evolutionary processes of genetic mutation and variation, guided toward adaptation of populations by natural selection, and furthermore that this approach can substantially enhance evolutionary theory, especially in such matters as rates of evolution, modes of adaptation, and histories of taxa, particularly at superspecific levels.

More recent contributions—even strictly paleontological contributions—in this field, some by me and many more by others, cannot be reviewed here, only briefly exemplified. Two recent paleontological examples, which both clarify and emphasize ideas nascent in previous studies of the synthetic theory, are Eldredge & Gould's (1972) "punctuated equilibria" (the quantum evolution of species in marginal populations and their subsequent expansion into the fossil record) and Stanley's (1975) "macroevolution" by "species selection" (the origination of higher categories by inter- rather than intraspecific selection).

Before leaving this subject, some notice must be taken of effects of the fairly recent revolution in molecular biology. This has been important also for the subject of evolution and has provided underpinning for genetic aspects of the synthetic theory. Genes and gene mutations, which were still abstractions or purely operational concepts in 1944, are now known as objectively definable physicochemical objects and events. Still, contributions from molecular biology to advancing the knowledge of evolution at organismal levels have so far been somewhat restricted.

The most important claimed contribution of molecular biology (or biochemistry) to evolutionary theory in general is the exposition by some molecular biologists and biochemists of what they usually, but confusingly, call non-Darwinian evolution. They maintain that a large amount of the genetic material (DNA) and hence of the amino acid sequences in proteins is selectively neutral but undergoes mutational changes, thus evolving biochemically, without relationship to natural selection. If true (or to the extent it is true), this would imply to paleontologists and some other organismal biologists that the "non-Darwinian" evolution of molecules has no effect on the somatic-phenetic evolution of organisms and hence may have little interest for understanding or explaining organic evolution. Some molecular biologists agree in part but still hold that molecular evolution can (*a*) aid in the reconstruction of phylogenies and (*b*) provide an evolutionary time scale. The first point is accepted by all evolutionists and systematists, at least to the extent that molecular evolution is clearly evidence on phylogeny even if not conclusive in itself. The second point rests on the belief that base substitutions in the relevant DNA proceed at constant rates. That has been disputed, however, and on the evidence presented it seems to me highly improbable. There is now a large, often polemic literature on this subject, and I do not intend to review it. A fairly recent, convenient symposium with contributions from all sides has been presented by Le Cam et al (1972). The Society for the Study of Evolution held a related symposium in 1975.

SYSTEMATICS

Possibly several hundred of my publications could be considered as in the field of systematics in a broad sense. Two, however, are most specifically and conspicuously in that field: an application of systematics to a large group of animals (the classification of all living and extinct mammals at generic and higher levels) (Simpson 1945) and a more general and theoretical study of the principles of animal taxonomy (Simpson 1961). The system that I have expounded and to which I still hold has been called phylogenetic, but that is confusing because recently this term has commonly been applied to a different, or narrower, approach better designated as cladistic. I prefer to call the more usual and more eclectic taxonomic system evolutionary taxonomy.

Taxonomy by any system is a complicated matter and obviously cannot be discussed adequately here. Reduced to the barest and quite inadequate summary, the procedures of evolutionary taxonomy include:

1. Obtaining data on the characteristics of individual sample organisms.
2. Inferences from those data by statistical procedures or principles on the populations represented by the observed samples.
3. Judgments and measures, preferably quantitative, of the resemblances and differences among populations.
4. Inferences from degrees of resemblance and difference and also from the nature of those differences (considered on evolutionary principles) as to the relationships of the populations in question.
5. Inferences as to the most probable phylogenetic pattern represented, taking into account geological sequence (as far as known); interpretation of resemblances as ancestral, parallel, or convergent; and differences as successive or divergent, with due consideration of the continuation, splitting, divergence, and convergence of lineages.
6. Devising a classification that is consistent with the inferred phylogeny, in a carefully defined sense of the concept consistency and with further criteria of convenience and stability of classifications as long as they are not demonstrably inconsistent.

Rather recently, there have been developed two other approaches to taxonomy. Although they have been considered quite distinct by their more extreme supporters, some of whom are combative to the point of fanaticism, each of these makes a definite contribution to a more general and eclectic system of evolutionary taxonomy. One has been called numerical taxonomy, a confusing term because any modern system of taxonomy must be extensively numerical. It is better and now more commonly called phenetic and has been expounded especially in Sokal & Sneath (1963) and Sneath & Sokal (1973). The main principle of this school—unduly but not unfairly simplified and compressed—is that classification should result directly from numerical expressions of degrees of resemblance, without evolutionary, phylogenetic, or genetic considerations.

Hull (1970), among others before and since, has pointed out that phenetic taxonomy excludes from classification the most important characteristics of organisms, namely, that they evolve and have phylogenetic and genetic relationships. Thus classification on this basis alone lacks informational content that evolutionary and, in general, organismal biologists usually seek and generally assume to be present in a classification. Others, such as Bock (1973), note that a satisfactory taxonomy must weight characters and must have clear evolutionary criteria of homology, both of which are absent in a purist version of phenetics. An argument in favor of phenetics that has sometimes weighed particularly with paleontologists has been that genetical affinity, in the strictest sense, cannot be directly observed in fossils and usually is not in recent organisms. In regard to classification, however, that argument is based on confusion between the definition of taxa, which may well be genetic, and evidence that the definition is met, which need not be and usually is not directly genetic. That is a variation of the twin fallacy, which is that two people are twins because they look alike, whereas in fact and logic the reverse is true—they look alike because they are twins.

Any usable taxonomic system must rely heavily on resemblances and differences among populations of organisms. The invention of increasingly sophisticated electronic computers has made possible the comparative quantitative handling of large numbers of characters in large numbers of taxa. The pheneticists have provided extremely useful and elegant means of achieving such results. These are seen more and more as a contribution to and one step toward evolutionary classification rather than as a sufficient end in themselves.

The other system (or philosophy) of taxonomy to be considered briefly here is cladistic, often misleadingly called phylogenetic and also often called Hennigian because in its most extreme development it was largely the work of Hennig (1966). Its leading principle is that classification should be based on dichotomies in phyletic lines and that hierarchic levels should be based on relative remoteness (or inversely on relative recency) of the relevant dichotomies. If applied strictly to classification in accordance with procedures laid down in detail by Hennig and some of his followers, the results are inadequate, omitting much truly phylogenetic information, often impractical, and sometimes absurd—as adequately demonstrated by Hull (1970), Bock (1973), and Mayr (1974), among others.

In any evolutionary classification, much depends on recognition of the origin of different clades (lines of descent) from common ancestries at different times in the past. It has long been recognized, to the point of being taken for granted by evolutionary biologists, that cladistic analysis requires distinguishing ancestral from derived characters. All that Hennig has contributed in that respect is an esoteric terminology. As Hennig's severest critic (Mayr 1974) adds, "Nevertheless, Hennig deserves great credit for having fully developed the principles of cladistic analysis. The clear recognition of the importance of [derived characters in common] for the reconstruction of branching sequences is Hennig's major contribution." (I would add that not all members of a monophyletic taxon necessarily share any derived characters.)

There is no objection to most of the Hennigian procedures for forming cladograms. On the contrary, they are another useful adjunct for proceeding to a fully evolutionary classification. They do not automatically provide acceptable classifications. That is especially true, I believe, of attempts to classify fossils. Here manifestly undesirable results of some Hennigian principles (of classification, not of the cladistic aspect of phylogeny) have led a few paleontologists to the peculiar conclusion that fossils should be classified as if they were all contemporaneous. Even that amusingly drastic remedy is insufficient.

The progress of knowledge of the systematics of mammals has been so great since 1942, when writing of Simpson (1945) was ended, that that classification is decidedly out of date. Changes in my circumstances during the 1950s prevented my planned revision. If I had been able to make the needed revision, I would have followed, in the main, the principles set forth in Simpson (1961), Mayr (1969), or indeed most of the work on paleontological and neontological systematics now appearing. If I were to revise Simpson (1961), I would retain essentially the same philosophy overall but would make extended use of the advanced numerical methods of the pheneticists and the advanced cladistic methods of the Hennigians.

BIOGEOGRAPHY

I now proceed to my final topic, the one probably of greatest interest to most readers of this volume: the plate tectonics revolution in geology and its effect on the study of biogeography, including paleobiogeography. I had planned to write a conventional review on this topic, with emphasis on the history of mammals, for the *Annual Review of Ecology and Systematics* (1975) but was prevented by personal circumstances. The present discussion cannot be even a summary of such a review and will continue as a cursory comment on my own involvement and present opinions.

I have been fascinated by biogeography and paleogeography ever since I first knew that such subjects exist. I was early exposed to the views of Charles Schuchert, who was retired but very much present at Yale while I was a graduate student there. He was a believer in transoceanic continents and mapped a Gondwanaland that became fragmented by the sinking of segments and not by continental drifting. However, I soon was drawn instead to the essentials of W. D. Matthew's views—my first professional job was as his summer field assistant in 1924, while I was a graduate student. He believed that the continents had existed as relatively permanent crustal segments, at least during the Mesozoic and Cenozoic. He rejected the theories of transoceanic continents and most of the many land bridges that had been postulated, particularly those directly connecting any two southern continents. He demonstrated that all then-available knowledge of fossil and recent mammals was consistent with the existence of only three land connections not now above water: between Alaska and Siberia, Canada and northern Europe, and Australia and southeastern Asia. His principal work on the subject (Matthew 1915) was written before the first edition of Wegener's book on continental drift (Wegener 1915), and in later years, until his death in 1930, he did not publish on drift. However in a note written sometime between 1927 and 1930 but not published until long after his death, he wrote that "pending [a really critical] examination one may view with an open mind the very ingenious and plausible theory of continental drift set forth by Professor Wegener and others (while not at all subscribing to his attempt to interpret the relations of land faunas, which is quite impossibly wrong)." He still held that if drift did occur, it had no effect on Cenozoic land faunas.

Although I came to differ with Matthew on some details and some other aspects of biogeography, extended study of Cenozoic mammals convinced me that he was essentially right on the points specified in the preceding paragraph (e.g. Simpson 1943). I did not deny the possibility of pre-Cenozoic continental drift, but until the 1960s I did consider the biogeographic evidence for it so scanty and equivocal that it remained an unconfirmed hypothesis. I held that the Cenozoic mammalian evidence did not support, but instead strongly opposed, the reality of any biogeographic effects of continental drift during that era.

All geologists know that the development and now nearly unanimous acceptance of the theory of plate tectonics has revolutionized the science of geology. There are already two lengthy histories of the origins and development of this revolution (Marvin 1973, Hallam 1973a). The theory of continental drift was strongly involved in that history, although I believe that its importance in its Wegenerian form has been overemphasized. Now one can say that it is acceptance of the theory of plate tectonics that has revolutionized concepts of continental drift and made them generally accepted.

Some explanation is needed for the fact that the Wegener, Du Toit, etc, views of continental drift were long questioned or rejected outright by most geologists, including almost all paleontologists (and including me), but now are accepted by almost all geologists, including paleontologists (and me). Hallam, especially, has repeatedly belabored this point (1973a,b, 1975). He maintains that the evidence against the sinking of transoceanic continents was already conclusive when Wegener wrote and that slowness to accept continental drift was due primarily to conservative prejudice, "stabilist" dogma, and the fact that Wegener did not belong to the geological establishment, and secondarily to the inadequacy of data and to what Wegener himself acknowledged as lack of a really plausible mechanism for drift.

Undoubtedly, there was an element of conservatism, as there always is in reluctance to accept ideas at a time when they are unorthodox. (I have mentioned the cases of Owen and Agassiz vs Darwin.) However, I do not believe that Hallam's is a fair judgment in this case, especially in reference to paleontologists and paleobiogeographers. I take space here to state three main refutations of that point of view.

First, during the years of dispute most of us (including me) had already abandoned the untenable hypothesis of transoceanic continents and accepted only a small number of land bridges, the reality of which is not disproved or disputed by anything in drift or plate theory.

Second, much or indeed most of the claimed biological, so-called (often incorrectly) paleontological, or biogeographic evidence for drift advanced by Wegener, Du Toit, and other early defenders of the theory was flatly wrong, demonstrably so even at that time. A paleontologist could hardly take Wegener seriously when he made such statements in his final (1929) revision as that in the Oligocene 31% of the mammals were "identical" in Europe and North America; that is among the less egregious paleobiogeographical misstatements of the drift school around the 1920s to 1940s.

Third, all crucial evidence for plate tectonics was acquired after the work of Wegener, Du Toit, and other early proponents of continental drift. It was this, not the hypotheses and arguments of the pioneers, that made continental drift first plausible and finally virtually incontrovertible.

From now on paleobiogeographers must work with paleogeographic patterns that involve continental drift. There is already a large literature based on that necessity. I cannot review it here, but will just cite a few prominent examples: Keast (1972), Hallam (1973b), Hughes (1973), Tarling & Runcorn (1973). Except for the cited chapter by Keast, these are all symposia involving a greater number of authors. Two special problems may be just mentioned: first, there is not yet close agreement on geographic configurations through long, relevant, dated spans of geological time, and, second, the juxtaposition of what were later two distinct continents does not exclude the possible presence of definite, even impassable barriers within or between them. Progress is being made on both these points.

My former belief that continental drift, if it occurred at all, had no effect on the geography of Cenozoic mammals has been changed in only a few ways by my present belief that continental drift did indeed occur. I believed at first that there had been an early Cenozoic land bridge between North America and Europe but later abandoned that view. I now believe that there was indeed a connection but that it was broken by the drifting apart of the continents around the end of the early Eocene, as indicated by McKenna (1972).

I formerly believed that marsupials originated in North America, spread thence on separate land bridges to Europe, Asia, and South America, and island-hopped from Asia to Australia. Continued failure to find fossil marsupials in Asia and probable plate tectonic configurations for the Cretaceous and early Cenozoic now make it highly improbable that marsupials reached Australia via Asia. Four or five mutually exclusive hypotheses all based on continental drift have now been proposed to explain the early presence of marsupials in Australia and South America. Lillegraven (1974) has excellently reviewed them and I need not do so. I agree with Lillegraven and with Tedford (1974) that spreading between Australia and South America was probably via Antarctica, without involving any other continent. It was probably in part by island-hopping and not by a continuous land route. Lillegraven believes that the marsupials originated in North America and spread to South America, then to Antarctica, and finally to Australia; Tedford believes that they originated in South America and spread separately to North America and Australia. I do not know where they originated or in which direction they spread. I am also now uncertain where the other elements in the extremely peculiar early Cenozoic South American fauna came from. There is no evidence at all that this was a Gondwana fauna and I consider that highly improbable.

Much of theoretical biogeography involves the ways in which taxa spread, especially when their subsequent distribution is disjunct. Drift now provides a ready mechanism for disjunction by the separation of continents once united, as seen in early Triassic reptilian faunas of Antarctica and Africa or early Eocene mammalian faunas of Europe and North America. Long ago I discussed the distinct problem of routes of spread in terms of (*a*) corridors, where no important barrier exists and where spread of essentially whole biotas may be deterministic; (*b*) filter bridges, where some taxa spread in a deterministic way but others spread stochastically or not at all; and (*c*) sweepstakes routes, where any spread is stochastic (Simpson 1940). Then, and thereafter, I discussed examples, characteristics, and criteria for those paths of spread. Recently McKenna (1973) has pointed out that there are two other means of spread that must almost certainly have occurred as results of continental drift, although so far no undoubted examples have been adduced: Noah's arks, crustal segments that drifted carrying their biotas on a one-way trip, and Viking funeral ships, similar segments containing fossils that may thereafter be found adjunct to a continent where the fossilized taxa never lived. Thus plate tectonics is leading and will surely lead further to enrichment both in knowledge and in the several principles of biogeography.

ENVOI: COMPLEAT?

I am at this writing still trying to learn with as much zest as when I began first grade, although not quite the same vigor. I have made a great many mistakes, but I have also learned much that was not mistaken—as far as I know. I have called this miniautobiography and miniphilosophy "The Compleat Palaeontologist?" and I am well aware that the word "compleat" is obsolete and that one of its meanings is "finished." I insist on the question mark.

Literature Cited[1]

Bock, W. J. 1973. Philosophical foundations of classical evolutionary classification. *Syst. Zool.* 22:375–92

Dobzhansky, T. 1937. *Genetics and the Origin of Species.* New York: Columbia Univ. Press

Eldredge, N., Gould, S. J. 1972. Punctuated equilibria: an alternative to phyletic gradualism. See Schopf (1972), pp. 82–115

Fisher, R. A. 1930. *The Genetical Theory of Natural Selection.* Oxford: Clarendon

Haldane, J. B. S. 1932. *The Causes of Evolution.* New York & London: Harper & Bros. 235 pp.

Hallam, A. 1973a. *A Revolution in the Earth Sciences. From Continental Drift to Plate Tectonics.* Oxford: Clarendon. 127 pp.

Hallam, A., ed. 1973b. *Atlas of Palaeobiogeography.* Amsterdam: Elsevier. 531 pp.

Hallam, A. 1975. Alfred Wegener and the hypothesis of continental drift. *Sci. Am.* 232(2):88–97

Hennig, W. 1966. *Phylogenetic Systematics.* Urbana, Chicago, London: Univ. Illinois Press. 263 pp.

Hughes, N. F., ed. 1973. *Organisms and Continents Through Time.* London: Palaeontological Assoc. 334 pp.

Hull, D. L. 1970. Contemporary systematic philosophies. *Ann. Rev. Ecol. Syst.* 1:19–54

Keast, A. 1972. Continental drift and the biota of the mammals on southern continents. In *Evolution, Mammals, and Southern Continents,* ed. A. Keast, F. C. Erk, B. Glass, 23–87. Albany, NY: State Univ. New York Press.

Kummel, B., Raup, D., eds. 1965. *Handbook of Paleontological Techniques.* San Francisco: Freeman. 852 pp.

Le Cam, L. M., Neyman, J., Scott, E. L., eds. 1972. *Darwinian, Neo-Darwinian, and Non-Darwinian Evolution, April 9–12, 1971. Proc. Berkeley Symp. Math. Stat. Probab.* 5:i-xvi, 1–309

Lillegraven, J. A. 1974. Biogeographic considerations of the marsupial-placental dichotomy. *Ann. Rev. Ecol. Syst.* 5:263–83

Marvin, U. B. 1973. *Continental Drift. The Evolution of a Concept.* Washington DC: Smithsonian Inst. Press. 239 pp.

Matthew, W. D. 1915. Climate and evolution. *Ann. NY Acad. Sci.* 24:171–318

Mayr, E. 1969. *Principles of Systematic Zoology.* New York: McGraw-Hill. 428 pp.

Mayr, E. 1942. *Systematics and the Origin of Species.* New York: Columbia Univ. Press. 334 pp.

Mayr, E. 1974. Cladistic analysis or cladistic classification? *Z. Syst. Evol.* 12(2):94–128

McKenna, M. C. 1972. Eocene final separation of the Eurasian and Greenland–North American landmasses. *Int. Geol. Congr., 24th,* Sect. 7, pp. 275–81

McKenna, M. C. 1973. Sweepstakes, filters, corridors, Noah's arks, and beached Viking funeral ships in palaeogeography. See Tarling & Runcorn (1973), pp. 295–308

Schopf, T. J. M., ed. 1972. *Models in Paleobiology.* San Francisco: Freeman, Cooper. 250 pp.

Schindewolf, O. H. 1936. *Paläontologie, Entwicklungslehre, und Genetik. Kritik und Synthese.* Berlin: Borntraeger. 108 pp.

Simpson, G. G. 1926. Mesozoic Mammalia. V. *Dromatherium* and *Microconodon. Am. J. Sci., Ser.* 5 12:87–108

Simpson, G. G. 1929. American Mesozoic Mammalia. *Mem. Peabody Mus. Yale Univ.* 3(1):I–XV, 1–171

Simpson, G. G. 1933. A simplified serial sectioning technique for the study of fossils. *Am. Mus. Novitates* 634:1–6

Simpson, G. G. 1940. Mammals and land bridges. *J. Wash. Acad. Sci.* 30:137–63

Simpson, G. G. 1941. Large Pleistocene felines of North America. *Am. Mus. Novitates* 1136:1–27

Simpson, G. G. 1943. Mammals and the nature of continents. *Am. J. Sci.* 241:1–31

Simpson, G. G. 1944. *Tempo and Mode in Evolution.* New York: Columbia Univ. Press. 237 pp.

Simpson, G. G. 1945. The principles of classification and a classification of mammals. *Bull. Am. Mus. Nat. Hist.* 85:I–XVI, 1–350

Simpson, G. G. 1961. *Principles of Animal Taxonomy.* New York: Columbia Univ. Press. 247 pp.

Simpson, G. G. 1964. Species density of North American Recent mammals. *Syst. Zool.* 13:57–73

Simpson, G. G., Roe, A. 1939. *Quantitative Zoology. Numerical Concepts and Methods in the Study of Recent and Fossil Animals.*

[1] Only publications explicitly cited in the text are here listed. For my own publications these are often not the most important on their subjects and are rarely the most recent. A nearly complete bibliography of my scientific publications up to 1971 was published in *Evol. Biol.* (1972) 6:7–29. A continuation for 1971–1975 can be obtained from me.

New York & London: McGraw-Hill. 414 pp.
Sneath, P. H. A., Sokal, R. R. 1973. *Numerical Taxonomy. The Principles and Practice of Numerical Classification.* San Francisco: Freeman. 573 pp.
Sokal, R. R., Rohlf, F. J. 1969. *Biometry, The Principle and Practice of Statistics in Biological Research.* San Francisco: Freeman. 776 pp.
Sokal, R. R., Sneath, P. H. A. 1963. *Principles of Numerical Taxonomy.* San Francisco & London: Freeman. 359 pp.
Stanley, S. M. 1975. A theory of evolution above the species level. *Proc. Nat. Acad. Sci. USA* 72:646–50
Tarling, D. H., Runcorn, S. K., eds. 1973. *Implications of Continental Drift to the Earth Sciences. Part 3. Palaeontological Implications.* Vol. 1, pp. 219–446. London & New York: Academic
Tedford, R. H. 1974. Marsupials and the new paleogeography. *Soc. Econ. Paleontol. Mineral. Spec. Publ.* 21:109–26
Wegener, A. 1915. *Die Enstehung der Kontinente und Ozeane.* Braunschweig: Vieweg. 94 pp.
Wegener, A. 1966. *The Origin of Continents and Oceans.* New York: Dover. 246 pp. (Transl. from German, 4th ed., 1962)

PETROLOGY OF LUNAR ROCKS AND IMPLICATION TO LUNAR EVOLUTION[1]

×10049

W. Ian Ridley

Lamont-Doherty Geological Observatory and Department of Geological Sciences, Columbia University, Palisades, New York 10964

INTRODUCTION

The lunar surface can be divided into three major lithologic units: regolith or comminuted rock, mare bedrock, and terra bedrock. Each of these units can be recognized by distinctive physical and chemical properties, the most obvious of which is the brightness or albedo of the individual units. In general, the terra regions are uniformly much brighter than the mare.

Wherever sampled, the mare basalts, which partially fill large impact basins, are overlain by a veneer of impact-comminuted rock. A physically similar, but chemically different, thicker veneer covers the terra regions. As a consequence of this surface blanket, no in situ rock sample has ever been returned from the Moon. Despite this, it has been possible to collect samples of local bedrock from both the mare and terra regions, because small scale meteorite impact penetrated the regolith, ejecting bedrock to the surface.

The geological aspects of the Apollo program were designed so that maximum sample coverage of major geologic units could be achieved within the framework of a limited number of missions. This goal was achieved by the sampling of four large maria (Mare Tranquillitatis, Oceanus Procellarum, Palus Putredinus, Mare Serenitatis) and four major terra units (Fra Mauro, Apennine Mountains, Cayley-Descartes Highlands, Taurus-Littrow Highlands). Additional coverage was possible through two Russian Luna missions, which sampled Mare Fecunditatis and Apollonius Mountains; extrapolation of local chemical data to a more regional scale was achieved through chemical data obtained from the orbiting command modules on Apollo 15 and 16.

Chemically and petrologically the lunar samples held many surprises and properties beyond our terrestrial experience. Probably one of the most important discoveries was the simple observation that the Apollo 11 basalts had textures similar

[1] Contribution No. 2237 from Lamont-Doherty Geological Observatory.

to terrestrial basalts, and had crystallized from silicate liquids. Clearly the Moon has not always been a cold planet. As data from further missions accumulated, it became evident that the Moon was a dynamic, hot planet in its youthful stages. In addition, its overall chemical composition is distinctly different from the Earth; silicate liquids produced within the Moon are uniquely lunar in composition, and the conditions under which lunar magmas originated and crystallized were alien to the terrestrial environment.

In order to understand the petrogenesis of the Moon, it is best to present the broad picture that has emerged since 1969, and later to follow up with more specific evidence. Although many important petrogenetic questions remain, the first-order evolutionary sequence has generally been accepted.

The Moon accreted homogeneously at ~4.65 AE in a short time period (< 100 yr) with the storing of enough accretionary energy to cause melting of the outer few hundred kilometers. Cooling of the molten outer shell occurred during a period of high meteorite flux, and the cooling crust was continually broken and stirred into the molten interior. Eventually the lithosphere thickened and became rigid enough to withstand large meteorite bombardments, and the underlying molten zone cooled quiescently and underwent fractional crystallization. Gravity sinking of high density ferromagnesian minerals (pyroxene, olivine, ilmenite) relative to lighter plagioclase caused a chemical/mineralogical layering into an ultrabasic pyroxenitic mantle and a plagioclase-rich crust. The rigid crust was impacted by large meteorites, which lead to the excavation of the large mare basins between 4.6 AE and 3.9 AE. At ~3.9 AE a decrease in meteorite flux occurred, although by this time the lunar surface was saturated with 50 km diameter craters.

Internal heat sources (K, U, Th) were concentrated towards the surface between 4.6 AE and 3.9 AE, so that partial melting of the lunar upper mantle occurred between 3.8 AE and 3.1 AE, generating the mare basalts. These magmas ascended into the large mare basins, presumably because these were regions of crustal weakness. There seems to be no temporal relationship between mare basin formation and mare basalt genesis. Gradual thickening of the lunar lithosphere, as radioactive heat generation failed to match surface heat loss, depressed the zone of partial melting deeper into the lunar mantle. At ~3.1 AE the lithosphere was too thick to allow ascent of large volumes of mare basalt to the surface, and surface activity essentially ceased.

At present the Moon is not a cold, dead planet, but relative to the Earth it is in a condition of extreme quiescence. Moonquakes at depths of 800–1000 km suggest this may represent the base of the lunar lithosphere; heat flow calculations indicate a partially molten zone may still be present within the lunar asthenosphere; seismic velocity data suggest that the Moon may have a molten core equivalent to approximately 2% of its volume; lunar transient events identify contemporary surface activity on the Moon.

In summary, the three most important dynamic processes operative during the initial 1.6 AE of lunar history were external impact bombardment, internal fractional crystallization and internal partial melting. The latter process is best exemplified by the mare basalts, which fortunately erupted after the rapid decrease in meteorite flux. They have therefore retained original textures and chemistry and are texturally similar, but are chemically unlike terrestrial basalts.

In contrast, the terra rocks are much more complex, reflecting the period between 4.6 AE and 3.9 AE, in which the lunar crust was continuously being fragmented, remelted, metamorphosed, mixed, and contaminated by large impacting meteorites. Most of the terra rocks are texturally complex, impact breccias or impact-produced melts, but a rare few are probably of pristine igneous origin derived from deep crustal levels. Studies of terra rocks are essentially studies of impact processes. These studies are fascinating, but the ultimate aim is to look behind this complicated curtain and evaluate the original stratigraphy of the lunar crust.

In the following sections, the petrochemistry of lunar rocks (mare basalts, terra assemblages, regoliths) are reviewed in the context of the generalized evolutionary framework already outlined. Mare and terra assemblages are treated separately because they form distinct temporal and chemical units.

Much lunar literature may be found within the impressive 15 volumes of the Proceedings of the Annual Lunar Science Conferences. Recent reviews (Taylor 1975) present a timely precis of the wealth of data amassed since 1969.

MARE BASALTS

Major Element Composition

Three major groups of mare basalts have been recognized; representatives of each are shown in Table 1. Titanium-rich basalts with 11–13% TiO_2 and 9% Al_2O_3

Table 1 Major element composition of mare samples[a]

	b	c	d	e	f	g	h	i	j
SiO_2	40.69	39.04	46.13	44.59	45.07	47.81	43.97	37.64	41.27
TiO_2	11.92	11.32	3.35	2.88	4.62	1.77	2.31	13.45	10.17
Al_2O_3	7.78	9.51	9.95	8.02	9.96	8.87	8.43	8.20	9.75
Cr_2O_3	—	—	0.46	0.55	0.26	—	—	0.57	0.27
FeO	19.49	19.40	20.70	22.03	20.25	19.97	22.58	18.78	18.24
MnO	0.28	0.27	0.28	0.29	0.28	0.28	0.33	0.28	0.29
MgO	7.51	7.73	8.07	12.66	7.21	9.01	11.14	9.49	6.84
CaO	10.76	11.28	10.89	9.05	11.45	10.32	9.40	10.29	12.30
Na_2O	0.51	0.36	0.26	0.20	0.28	0.28	0.21	0.40	0.44
K_2O	0.30	0.05	0.07	0.07	0.08	0.03	0.03	0.05	0.09

[a] All analyses by X-ray fluorescence (XRF).
[b] Apollo 11: high-K ilmenite basalt 10017 (Compston et al 1970).
[c] Apollo 11: low-K ilmenite basalt 10045 (Compston et al 1970).
[d] Apollo 12: pigeonite basalt 12052 (Compston et al 1971).
[e] Apollo 12: olivine basalt 12004 (Compston et al 1971).
[f] Apollo 12: ilmenite basalt 12051 (Compston et al 1971).
[g] Apollo 15: pyroxene-phyric basalt 15058 (Rhodes & Hubbard 1973).
[h] Apollo 15: olivine-phyric basalt 15016 (Rhodes & Hubbard 1973).
[i] Apollo 17: olivine porphyritic ilmenite basalt 75075 (Rhodes et al 1974).
[j] Apollo 17: "Apollo 11 low-K type" 75055 (Rhodes et al 1974).

were collected from Mare Tranquillitatis and Serenitatis, titanium-poor basalts with 2–4% TiO_2 and 9% Al_2O_3 occur in Oceanus Procellarum and Palus Putredinus, and aluminous basalts with 11–14% Al_2O_3 occur in Mare Fecunditatis and as rare samples at other sites. However, this does not imply that these maria are filled by basalts of these specific compositions, since spectral and photogeologic analysis (Bryan & Adams 1973, Howard et al 1973, Pieters et al 1975) indicate a variety of basaltic types in each mare. Figures 1 and 2 indicate that basalts from each mare can be distinguished by simple binary plots (Rhodes & Hubbard 1973, Rhodes et al 1974), and within each mare a number of subgroups, each with slightly different chemistry, may be recognized.

This leads to a classification of mare basalts based upon a combination of major element chemistry and mineralogy, as illustrated in Table 2. Textural classifications are also possible (Warner 1971) and are useful for cooling rate studies, but they have less petrogenetic significance.

At the Apollo 12, 15, and 17 sites, both quartz- and olivine-normative basalts have been analyzed (James & Wright 1972, Rhodes & Hubbard 1973, Shih et al 1975), but the relationships between each type are complex. For instance, at the Apollo 12 site the olivine- and quartz-normative basalts may be related by low pressure fractional crystallization (James & Wright 1972), but not at the Apollo 15 or 17 sites. In the latter cases, the two basaltic magmas are related to two different parental magmas (Rhodes & Hubbard 1973, Shih et al 1975).

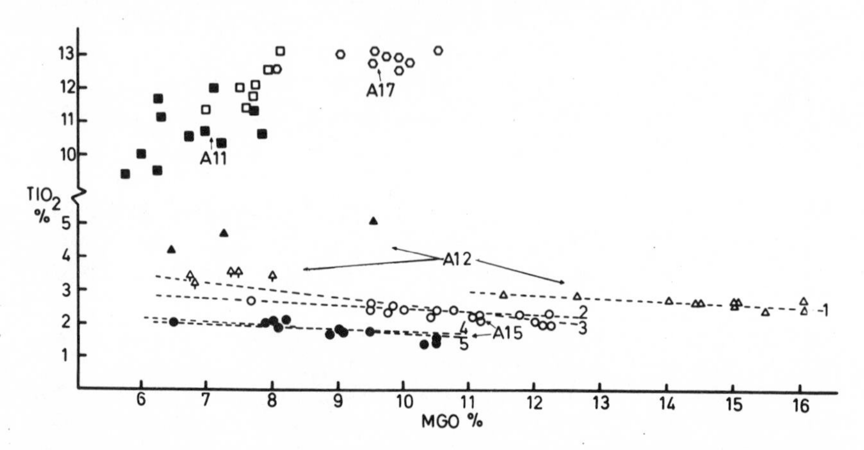

Figure 1 TiO_2-MgO relations for mare basalts. Each of the sampled mare may be separated using this diagram. Note the wide range in MgO content of Apollo 12 and 15 basalts. Lines *1*, *2*, and *4* are olivine fractionation control lines; lines *3* and *5* are pigeonite fractionation control lines. Apollo 11 (A11): high-K type (open squares) and low-K type (filled squares). Apollo 17 (A17): (open polygons). Apollo 12 (A12): ilmenite basalts (filled triangles), olivine basalts (open triangles), and pigeonite basalts (dashed triangles). Apollo 15 (A15): olivine-normative basalts (open circles) and quartz-normative basalts (filled circles). Data are from Compston et al (1970, 1971), Rose et al (1972, 1973), Rhodes & Hubbard (1973), Rhodes et al (1974).

Low pressure fractional crystallization is a viable method of producing within-group variation at nearly all mare sites, as demonstrated in Table 3. Only in some cases has testing of this process included trace element budgets (Shih et al 1975). Because mare magmas are about three times less viscous than terrestrial basalts (Weill et al 1970), gravitation settling should be an efficient process. However, most of the textural and mineral chemical features can be accounted for by a one-stage cooling history (Dowty et al 1974a) from liquids erupted essentially crystal-free. Hence the major fractional crystallization must occur during the cooling of lava at the lunar surface. Most fractionation schemes underscore the minor role of plagioclase in this process, indicating that the ubiquitous europium anomaly (see below) cannot be produced by fractionation of large volumes of plagioclase.

The spectrum of basalts observed in large hand specimens has been expanded by studies involving smaller and more numerous lithic fragments and glasses found in the lunar soil. Glass studies involve the recognition of preferred chemical compositions among impact-produced glasses (Reid et al 1972, Reid et al 1973c), as shown in Table 4. Green glass from Palus Putredinus and orange glass from the Littrow valley have aroused much interest because of their unusual compositions (Table 3) and remarkable chemical homogeneity (in contrast to the heterogeneous impact glass). Although there is no consensus of opinion on their origin, they are most likely volcanic glasses (Cadenhead 1973, Prinz et al 1973a, Reid et al 1973a),

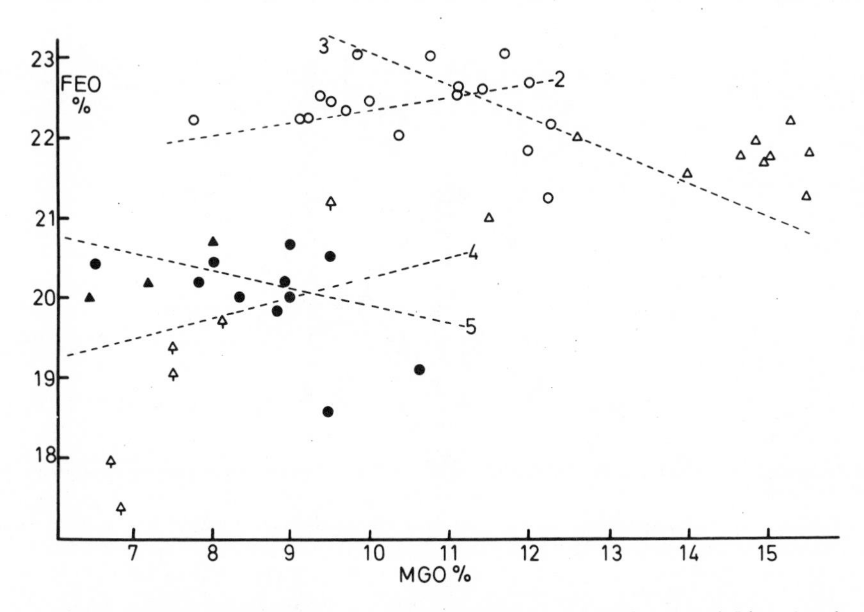

Figure 2 FeO-MgO relations for mare basalts to illustrate pigeonite control. Lines *2* and *4* are olivine control lines; lines *3* and *5* are pigeonite control lines. Symbols and data sources as in Figure 1.

Table 2 Classifications of mare basalts

Mission	Chemical/mineralogical classification	Reference	Textural classification	Reference
Apollo 11	High-K ilmenite basalt =	a, b, c	Type A, fine-grained	
	Group 1 basalt ($K_2O > 0.11\%$)	d	vesicular basalt =	
	Low-K ilmenite basalt =	a, b	intersertal ilmenite basalt	b, e
	Group 2 basalt ($K_2O < 0.11\%$)	d	Type **B**, medium-grained,	
			vuggy basalt = ophitic	
			ilmenite basalt	b, e
	No systematic relationship between classifications			
Apollo 12	Olivine basalt ($MgO > 11.5\%$)	b	Porphyritic basalt	e
	Pigeonite basalt ($MgO < 8.5\%$)	b	Ophitic basalt	e
	Ilmenite basalt ($TiO_2 > 4\%$)	b		
	Feldspathic basalt ($Al_2O_3 > 11.4\%$)	b		
	(Less complete classification given by: f, g, h)			
	No systematic relationship between classifications			
Apollo 15	Olivine-normative basalt ($SiO_2 < 46\%$)	i	Porphyritic clinopyroxene basalt	j
			= pyroxene-phyric basalt	k
			= mare basalt I	l
	Quartz-normative basalt ($SiO_2 > 47\%$)	i	Porphyritic clinopyroxene vitrophyres	j
			= pyroxene-phyric basalt	k
			= mare basalt II	l
			Porphyritic olivine basalt	j
			= olivine-phyric basalt	k
			= mare basalt III	l

Apollo 17	Olivine-normative basalt (TiO_2 13%) (OB)	m	Olivine porphyritic ilmenite basalt = OB	n
	Quartz-normative basalt (TiO_2 10%) (QB)	m	Plagioclase poikilitic ilmenite basalt = OB	n
	= Apollo 11 low-K type	n		
	= Type II basalt	o		

[a] Brown et al (1970).
[b] James & Wright (1972).
[c] Tera et al (1970).
[d] Compston et al (1970).
[e] Warner (1971).
[f] Brown et al (1971).
[g] Compston et al (1971).
[h] Gay et al (1971).
[i] Rhodes & Hubbard (1973).
[j] LSPET (1972).
[k] Dowty et al (1973).
[l] Brown et al (1972).
[m] LSPET (1973).
[n] Papike et al (1974).
[o] Brown et al (1974).

Table 3 Low-pressure fractional crystallization schemes for mare basalts

Site	Basalt group	Minerals removed	Relationship to natural sequence	Reference
Apollo 11	Low-K ilmenite basalts	Minor olivine and minor armalcolite	Consistent	a, b
	No relationship between low-K and high-K ilmenite basalts.			
Apollo 12	Olivine basalts	Up to 17% olivine, minor chrome-spinel. Rarely 5% pigeonite	Consistent	a
	Pigeonite basalts	6–10% pigeonite, minor olivine and chrome-spinel	Consistent	a
	Olivine basalts to pigeonite basalts	Up to 20% olivine, up to 6% pigeonite, minor chrome-spinel	Consistent	a
	Ilmenite basalts	21% olivine, 12–16% pyroxene, 8% plagioclase, 3% ilmenite	Inconsistent	a
Apollo 15	Olivine basalts	Up to 14% olivine (Fo 73), 1% spinel	Consistent	c
	Pigeonite	Up to 14% pigeonite, minor olivine (Fo 70), and spinel	Consistent	c
	Olivine basalt to pigeonite basalt	Olivine, pigeonite, plagioclase, spinel	Inconsistent	c
Apollo 17	All	Minor olivine, armalcolite	Consistent	b

[a] James & Wright (1972).
[b] Longhi et al (1974).
[c] Rhodes & Hubbard (1973).

formed either by lava fountaining or by meteorite impact into a molten lava lake (Roedder & Weiblen 1973). They are an enigma, chemically unlike the local mare basalts with which they are closely associated.

Comparison of the mare surface basalts with orbital X-ray fluorescence (Adler et al 1972a, b) and gamma-ray spectrographic data (Metzger et al 1973, 1974) allows a more regional view of the composition of mare surfaces (Figure 3). In general, the mare surfaces have higher Al/Si ratios than the mare basalts. This anomaly is conveniently explained—the orbital data reflect the composition of mare regoliths, i.e. a mixture of mare basalt fragments and aluminous terra rocks (Wood et al

Table 4 Mare basalt compositions deduced from glass studies

	a	b	c	d	e	f
SiO_2	45.62(0.55)	38.63(0.33)	48.22	40.2	45.48	44.55
TiO_2	0.41(0.04)	8.96(0.47)	1.09	7.2	2.77	3.79
Al_2O_3	7.74(0.39)	5.87(0.36)	12.93	13.2	10.86	11.77
Cr_2O_3	0.44(0.04)	0.55(0.03)	0.37	—	—	0.26
FeO	19.51(0.68)	22.00(0.32)	15.55	16.5	18.14	18.83
MgO	17.65(0.62)	14.79(1.15)	8.69	8.5	11.21	8.84
MnO	—	—	—	—	—	—
CaO	8.33(0.48)	7.31(0.40)	12.38	12.0	9.56	10.46
Na_2O	0.12(0.05)	0.37(0.13)	0.45	0.3	0.39	0.34
K_2O	0.02(0.02)	0.08(0.02)	0.02	0.10	0.32	0.13
Site	Apollo 15	Apollo 17	Apollo 17	Apollo 11	Apollo 14	Apollo 15

[a] Green glass in soil 15021.
[b] Orange glass in soil 74240 (Reid et al 1973b).
[c] Preferred composition in soil 74240 (Reid et al 1973b). Similar compositions occur in Apollo 15 soil.
[d] "Tranquillitatis-type" composition in Apollo 11 soil. Similar compositions occur in soil at Apollo 12, and 14 and Luna 16 sites.
[e] Mare-type glass in soil 14259.
[f] Mare-type glass in Apollo 15 soil.
Note: Preferred glass compositions similar to the local mare basalt occur in most soils.

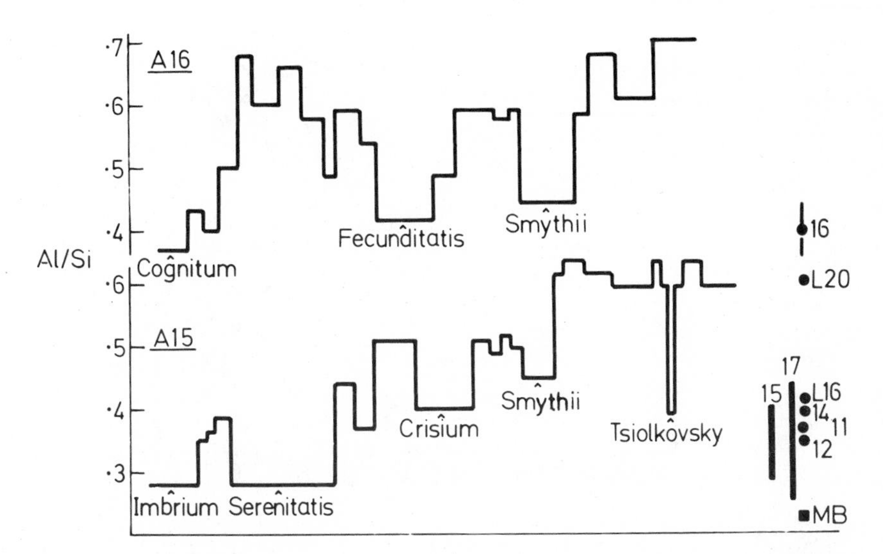

Figure 3 Al/Si atomic ratios for lunar regolith from orbital X-ray fluorescence data (Adler 1972a, b). Also shown are Al/Si ratios or ranges for samples of mare and terra soils. Note the low ratios for mare basalt (*MB*) and high ratios for mare geographically close to terra regions. This is almost certainly because of lateral mixing.

1970). The Al/Si ratios remain remarkably constant across the large mare basins, suggesting an efficient mixing of mare and terra materials, without significant compositional gradients towards the highlands. Because all the mare landing sites have been close to the highlands, it has not been possible to determine the extent of terra mixing with mare at sites distant from the highlands.

Trace Element Compositions

The large ion lithophile (LIL) trace elements generally show an affinity for the silicate melt phase during both melting and crystallization. Since crystal/melt distribution coefficients are reasonably well known for several of these trace elements in silicate phases [e.g. Rb, Sr, K, Ba, rare earth elements (REE)], their behavior during both melting and crystallization can be predicted with some confidence. Particular attention has been directed at the abundance patterns for mare basalts. The absolute abundances of LIL elements relative to chondritic abundances vary by an order of magnitude, the relative abundances less so (Figure 4). Some of the

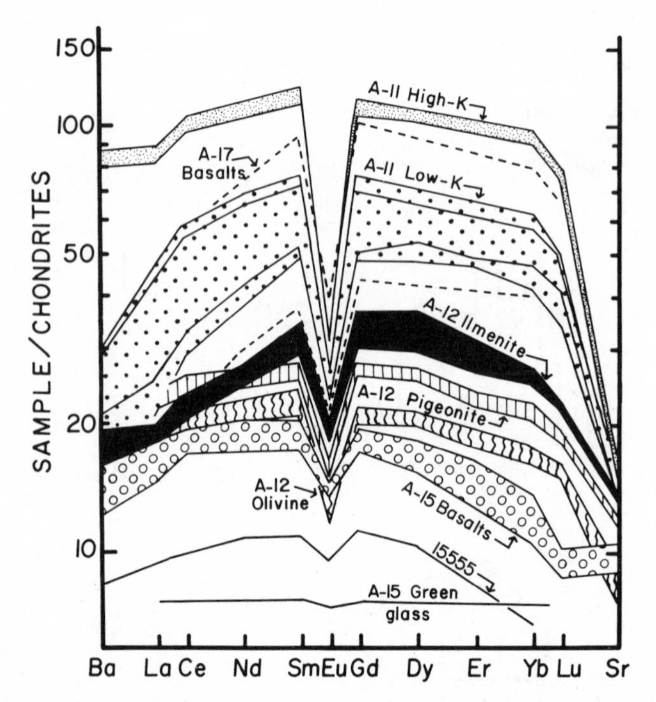

Figure 4 LIL trace element abundances for mare basalts and Apollo 15 green glass, normalized to chondritic abundances. Note the ubiquitous negative Eu anomaly and wide range in LIL elements. The Apollo 15 green glass has the least fractionated pattern but still maintains a small Eu anomaly. Note also the coherence of Eu and Sr and the cross-cutting relationship of some mare rare earth patterns. Data derived from Haskin et al (1970), Gast (1972), Rhodes & Hubbard (1973), Rhodes et al (1974).

more important observations regarding the LIL element distributions are (*a*) the ubiquity of a negative Eu anomaly, (*b*) correlation of Sm/Eu ratio with absolute abundance of LIL elements, (*c*) chemical coherence between Eu and Sr, (*d*) weak fractionation of light relative to heavy rare earths and tendency for fractionation to increase with increasing atomic number and with absolute abundances, (*e*) lack of close coherence between K and less volatile LIL elements, and (*f*) fundamental differences in abundances of LIL elements and rare earth patterns between mare and terrestrial basalts.

Large Sm/Eu ratios indicate the important role played by plagioclase (a sink for Eu^{2+} but not for trivalent REE) at some stage in the generation of mare basalts, although other complex schemes not involving plagioclase have been evolved (Ringwood 1974). Model calculations (Haskin et al 1970) indicate that prohibitive amounts of plagioclase would have to separate from unfractionated mare magmas in order to produce the observed Sm/Eu ratios. Such a process would also be incompatible with the delayed crystallization of plagioclase in mare basalts (Ringwood & Essene 1970, Green et al 1971, Biggar et al 1971, Bence & Papike 1972).

Observation (Anders et al 1971) has lead to the suggestion that the source for mare basalts would have Sm/Eu $\simeq 1$ and that gradually decreasing degrees of partial melting give gradually increasing Sm/Eu ratios. In detail, however, there are many exceptions: the Luna 16 basalts have lower Sm/Eu ratios than might be predicted from the REE abundances, the Apollo 12 ilmenite basalts have REE patterns quite different from other mare basalts, and so on. This suggests multiple source regions for mare basalts.

Variations in Ce/Sm ratio indicate that clinopyroxene played a role in mare basalt genesis (Shih et al 1975), but Ce/Sm and Gd/Lu ratios show no well-defined relationship (the Gd/Lu ratios are strangely high in Apollo 15 basalts).

The coherence of Eu and Sr in mare basalts provides some clue to the crystallo-chemical behavior of Eu under lunar conditions. Such coherence could only be maintained if Eu were largely present as Eu^{2+}, in which case both Sr and Eu would display diadochy with Ca. It has generally been accepted that the low redox conditions under which lunar magmas evolve would result in Eu^{2+}/Eu^{3+} ratios higher than observed in terrestrial basalts (Philpotts 1970, Drake et al 1974, Drake 1975). In this case, a large fraction of Eu as well as Sr would be locked in plagioclase, producing high Ce/Eu ratios in mare basalts equilibrated with plagioclase.

Fractionation of minerals at low pressure does account for some of the chemical variations in mare basalts (Table 3), but removal of reasonable amounts of olivine and chrome spinel cannot significantly perturb the relative LIL element distributions. Hence, it has generally been accepted that uniquely lunar patterns reflect processes operating within the lunar interior during the partial melting events giving rise to the mare basalts.

Important information has also been obtained from studies of siderophile and volatile elements in lunar rocks. Some of these elements, e.g. Ir, Re, Au, and Ni, are abundant and well characterized in meteorites, but are quite depleted in pristine lunar rocks such as mare basalts; therefore they allow an estimate of the contribution of extralunar meteoritic material to the mare regolith (Figure 5). A striking characteristic of mare basalts is the depletion in volatile elements, relative to their terrestrial and meteoritic abundances (Figure 5). Since these volatiles may be siderophile, chalcophile, or lithophile in their chemical behavior, one crystal-liquid process within the Moon is unlikely to have initially produced such uniform depletions. This conclusion points towards a Moon that was born already depleted in volatile components, either as a result of accretion at too high temperatures to condense volatile elements, or a solar nebula that was itself depleted in these elements.

Some skepticism has prevailed, however: some people argue that volatile loss occurred during magma eruption into the lunar vacuum. This seems highly unlikely (Ganapathy et al 1970) because both Rb-Sr and U-Th-Pb systematics demonstrate loss of either Rb or Pb not at the time of basalt crystallization but rather at a time approaching that of the Moon's genesis. In addition, the low vapor pressures of these elements require extremely efficient exposure of all parts of the magma body to the lunar atmosphere (Gibson et al 1972), such as might occur during fire-fountaining. Because both quenched basalts and slowly cooled

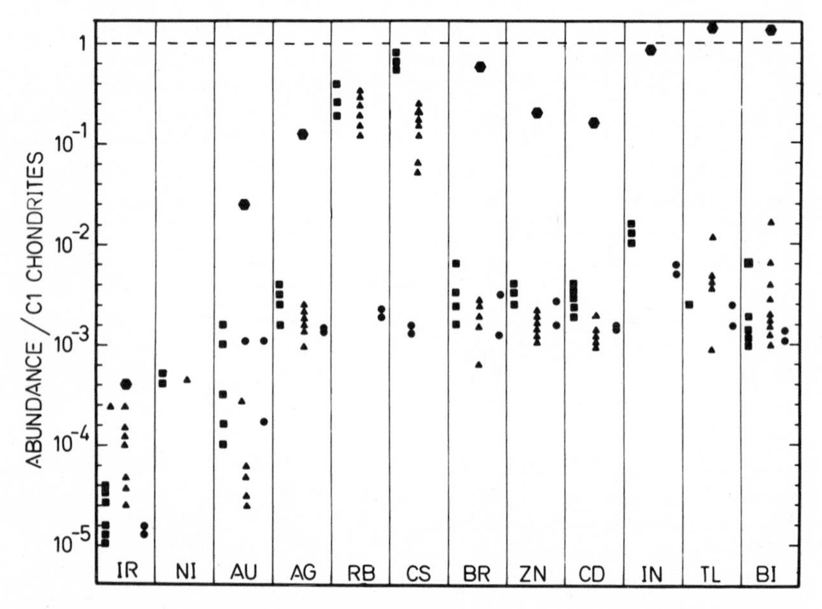

Figure 5 Abundance of siderophile elements (Ir, Ni, Ag, Au) and volatile elements in Apollo 11 (squares), Apollo 12 (triangles), and Apollo 15 (circles) mare basalts and terrestrial basalts (polygons), relative to Cl chondrites. Br and Bi may also be siderophile, Zn and Cd chalcophile, and Rb, Cs lithophile. Note the dramatic depletions in all these elements relative to chondrites and terrestrial basalts. Data from Ganapathy et al (1970), Anders et al (1971), Morgan et al (1972).

gabbros are equally devolatilized, the conclusion that this characteristic was inherited from the lunar interior seems inescapable.

The mare basalts are also drastically depleted in volatiles and siderophiles relative to terrestrial basalts; the latter are also much depleted relative to ordinary and hypersthene chondrites (Figures 5, 9). Some authors have used this difference to deduce that the Earth and the Moon are not genetically related (Singer & Bandermann 1970), but such differences can also be explained if the Moon were produced as an Earth satellite (Ganapathy et al 1970). Removal of siderophile elements in a period of metal extraction may produce the observed differences, if the four to six orders of magnitude oxygen fugacity difference between the Earth and Moon are taken into account. Siderophile elements may have been much more strongly fractionated into the metal rather than the silicate phase within the Moon (Anders et al 1971).

Texture and Mineralogy

The two most important parameters affecting texture are viscosity and undercooling, the latter being necessary to promote any crystallization at all. Both parameters determine crystal growth and nucleation. The texture of mare basalts are the result of crystallization from silicate liquids derived from partial melting of the lunar mantle rather than as impact melts, a conclusion borne out by numerous trace element and isotopic studies (Papanastassiou et al 1970, Papanastassiou & Wasserburg 1970, 1972, Gast 1972). Green and orange glass collected during Apollo 15 and 17, respectively, may represent marelike magmas that underwent extreme undercooling. Aside from these, all mare magmas crystallized extensively, even where rapid cooling rates can be demonstrated. This is in marked contrast to terrestrial basalts, and reflects the lower viscosity and variable but higher degree of supercooling of mare magmas.

Both empirical and experimental observations (Dowty et al 1974a, Lofgren et al 1975) demonstrate that many mare basalts, including those with phenocrysts many times larger than ground-mass crystals, are the result of a one-stage, approximately linear cooling-rate superimposed upon varying degrees of undercooling. Based on crystallochemical grounds, it was previously argued that some mare basalts had undergone a two-stage cooling history (Boyd & Smith 1971, Bence et al 1971, Bence & Papike 1972).

A voluminous literature exists on the mineral chemistry of mare basalts, providing clues to cooling rate, crystallization history, elemental fractionation, oxygen fugacity and petrogenesis. Such studies provide an essential part of the classification of mare basalts. Most attention has been focused upon the pyroxenes and spinels, whose chemical variations are very large compared to the equivalent terrestrial mineral groups. In addition, mare basalts contain assemblages considered incompatible on the Earth (e.g. Mg-olivine and tridymite), assemblages unstable on Earth (e.g. Fe-Ni metal and troilite), and some uniquely lunar minerals (e.g. armalcolite and tranquillityite). Glossaries of lunar minerals may be found in Brown (1970), Levinson & Taylor (1971), Frondel (1972), and Taylor (1975).

Idealized elemental variations in mare pyroxenes are shown in Figure 6. Mare pyroxenes characteristically cover much of the pyroxene quadrilateral, frequently plotting within the "forbidden zone" for terrestrial pyroxenes. Zoning is ubiquitous, complex, and extreme, involving large changes in Ca, Fe, Mg, Ti, Al, and Cr. Several authors have used zonal trends to separate different basalt classes, and in this respect, subtle changes in relative abundances of Al, Ti, and Cr may be successfully utilized.

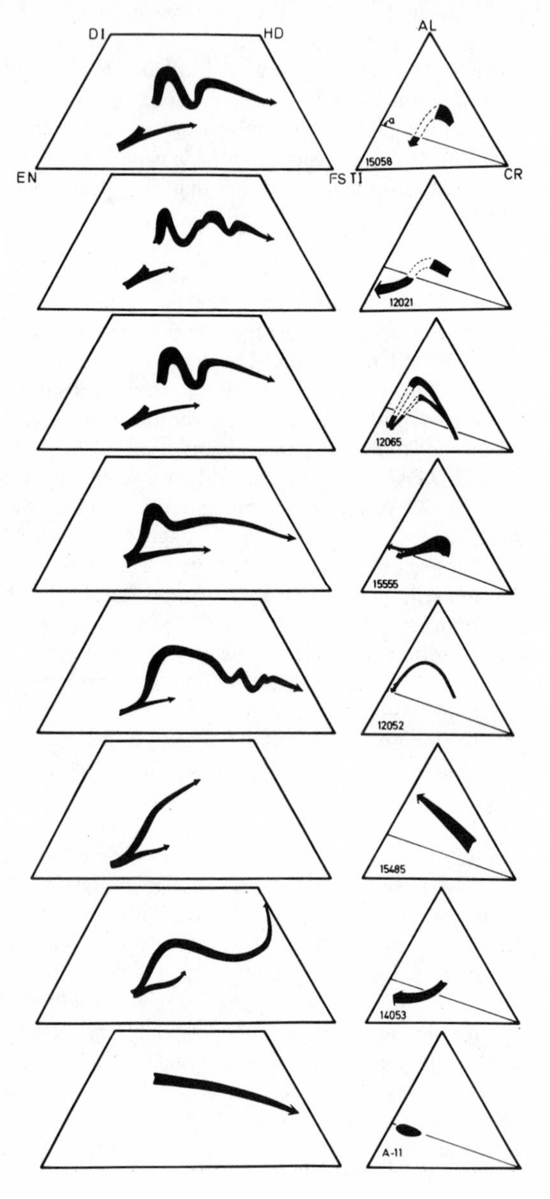

Generally, pyroxene chemistry is strongly determined by magma bulk chemistry, for example, the much larger absolute concentrations of Ti (and hence Al) in Apollo 11 and 17 pyroxenes. Al/Ti ratios are determined initially by bulk chemistry but later by the buffering effect of crystallization of other minerals, e.g. ilmenite and plagioclase. In some cases, the relative abundances of Al, Ti, Cr in late-stage, iron-rich pyroxenes can only be accounted for by assuming the presence of Ti^{3+} and/or Cr^{2+}.

In most mare basalts, with the exception of Apollo 11 and 17, the initial pyroxene to crystallize was Mg-pigeonite, or rarely hypersthene (Hollister et al 1971). The core low-Ca pyroxene may be sharply rimmed by, or gradationally pass into, a calcium-rich augite as a consequence of epitaxial growth. Further zoning involves progressive iron enrichment and possibly several excursions into the subcalcic, ferroaugite field. The ultimate crystallization is either Mg-free triclinic, pyroxenoid, pyroxferroite, or ferrohedenbergite. Complications arise if the pyroxene undergoes sector zoning (Ross et al 1970, Hollister et al 1971, Bence & Papike 1972, Dowty et al 1974a) in which [110] and [010] sectors of different composition grow. Note in Figure 6 the divergence of pyroxene trends in some cases is caused by sector zoning.

A number of circumstances may have given rise to the remarkable variations in pyroxene compositions (Brown et al 1970). These are (*a*) undercooling towards the cotectic in the subcalcic augite field (this does not explain the strong iron enrichment), (*b*) near-equilibrium crystallization of augite, largely in the absence of olivine, pigeonite or orthopyroxene, so that augite was no longer restricted to a solvus relationship, (*c*) availability of abundant Fe^{2+} in the melt relative to Mg, Ca allowing strong iron enrichment, (*d*) removal of Ca from the melt by crystallization of bytownite-anorthite (Bence & Papike 1972).

Plagioclase is more calcic than observed in terrestrial basalts, reflecting the

---

← *Figure 6* (*opposite*) Idealized compositional trends in a variety of lunar pyroxenes. The pyroxene quadrilateral represents ideal end-members $CaMgSi_2O_6$ (*DI* = diopside), $CaFeSi_2O_6$ (*HD* = hedenbergite), $MgSiO_3$ (*EN* = enstatite), and $FeSiO_3$ (*FS* = ferrosilite). Two trends in some diagrams generally represent an 010 pigeonite trend and 110 augite trend in sector-zoned crystals. Note the tendency for late-stage pyroxenes to approach the magnesium-free join, resulting in the crystallization of the pyroxenoid, pyroxferroite.

Triangles represent variations in minor elements Al, Ti, and Cr with crystallization. Note the Ti and Cr apices are at 50% Ti and 50% Cr, respectively. Detailed discussion to the use of this diagram may be found in Hollister et al (1971) and variations in Bence & Papike (1972). Line *a* represents an Al/Ti = 2/1. Note most pyroxenes approach or cross this line to Al/Ti < 2/1 at some stage in their cooling history. *12038* to *12052* approximately represents a sequence in which plagioclase appears progressively later in the crystallization sequence (Bence & Papike 1972); in *15485* plagioclase does not appear at all. Note the unique trend for *14053*, which crystallized at lower P_{O_2} than most other mare basalts. Pyroxenes in Apollo 11 ilmenite basalts have Al/Ti $\simeq$ 2 throughout their crystallization period. Note also that the position at which line *a* is intersected by the pyroxene trend is a rough guide to the beginning of plagioclase crystallization, except for *14053* and Apollo 11 pyroxenes.

very low alkali content of mare magmas. Zoning of more than 15 mol % anorthite is rare. Mare plagioclases also contain substantial ferrous iron and magnesian (Smith 1971), the Fe/Mg ratio increasing with fractionation (Crawford 1973). This relationship is potentially useful in distinguishing xenocrystic crystals and as a monitor of alkali loss during crystallization. Nonstoichiometry is frequently observed in mare plagioclases, which can only be partly explained by Fe^{2+} in tetrahedral sites. Rapid cooling may be an important factor (Weill et al 1970), but the observation that stoichiometric deviation correlates with bulk composition (Storey 1974), suggests that the structure of the mare melts may influence plagioclase stoichiometry.

Olivine crystallized in varying amounts in most mare basalts. In some Apollo 17 basalts, olivine seems to have had a reaction relationship with the melt, since rims of augite on olivine are frequently observed (Papike et al 1974). In Apollo 15 olivine basalts, much of the variation in bulk chemistry is due to variations in olivine content, probably as a result of low-pressure olivine settling. Zoning of olivine phenocrysts to about Fo_{40} has been observed, with a distinct compositional break to small amounts of pure fayalite in the mesostasis that suggests again a reaction relationship may hold. Unlike terrestrial basalts, forsterite and cristobalite crystallized simultaneously from mare basalts, although this may be a disequilibrium assemblage (Brown 1970).

The chemical variations and cationic substitutions in mare opaque phases is at least as complex as observed for mare pyroxenes, and much more variable than for opaques in terrestrial basalts. Probably the major nonterrestrial feature is the absence of magnetite or hematite solid solutions, which can be attributed to the lack of ferric iron. Instead, the major solid solutions involve complex substitutions of Fe-Mg, Al-Cr, Ti-Zr, so that ilmenite, members of the ulvospinel (Fe_2TiO_4)-picrochromite ($MgCr_2O_4$)-chromite ($FeCr_2O_4$)-hercynite ($FeAl_2O_4$), and members of the armalcolite series ($MgTi_2O_5$) have been described in mare basalts. Other opaque phases include Fe-Ni alloys and troilite, and several exotic Zr-rich minerals have also been identified.

Spinel compositional data in mare basalts have been summarized in the Luna 16 volume (Haggerty 1972b) and are shown in Figure 7. Chrome-rich spinels frequently crystallized first, followed by ulvospinel and finally ilmenite, representing different stable compositions as oxygen fugacity and temperature decreased. High temperature reduction of spinels has also been described (Haggerty 1972a), representing the antithesis of high temperature oxidation frequently observed in terrestrial igneous spinels (Watkins & Haggerty 1967).

Titania-rich basalts frequently contain armalcolite (an Mg-rich member of the pseudobrookite group), presumably in response to high Ti activity in the melt. Both observation and experiment support a limited stability range for armalcolite, which reacts with melt to produce ilmenite with decreasing temperature (Green et al 1975). Complex solid state reductions of armalcolite have also been reported (Haggerty 1973, Ridley & Brett 1973) involving the stabilization of ilmenite, ulvospinel, Al-spinel, and Fe metal.

A number of mesostasis minerals have been described from mare basalts, of which the most common is tranquillityite, a lunar mineral rich in zirconium. However, this is just one of a complex solid solution series of minerals containing substantial Zr and rare earth elements that have no terrestrial counterparts.

Ages and Isotopic Compositions

Mare basalts have crystallization ages from 3.1 AE to 3.8 AE, as determined by Rb-Sr, U-Th-Pb and ^{39}Ar–^{40}Ar methods. The titania-rich basalts from Tranquillity

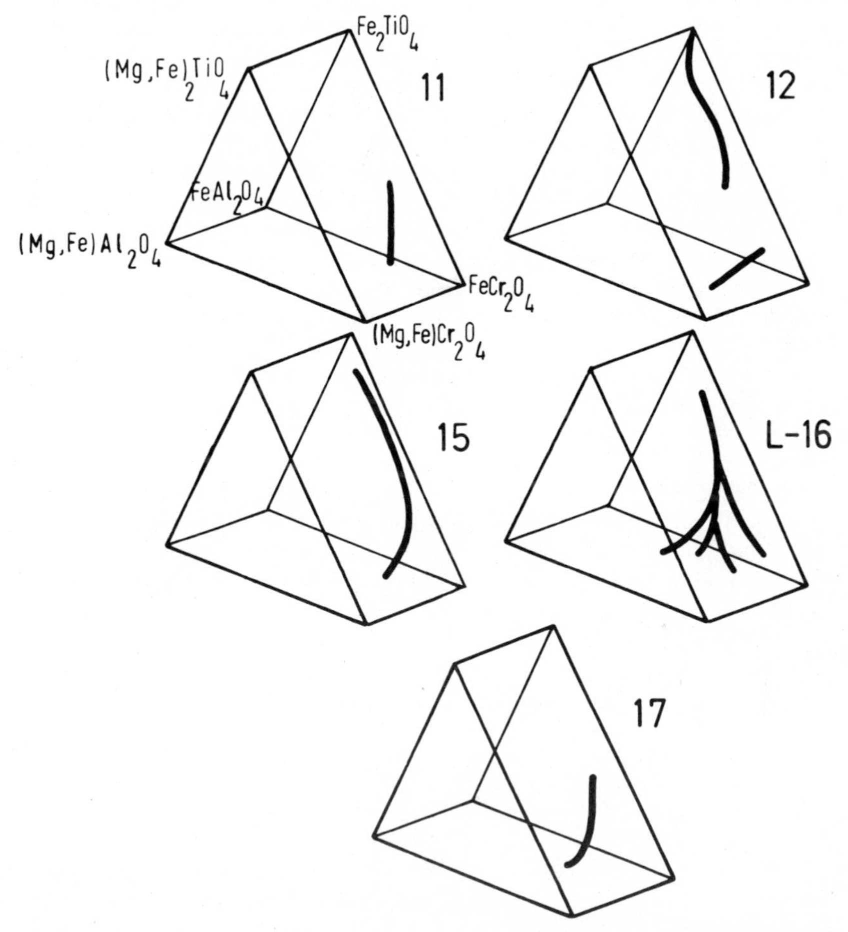

Figure 7 Idealized trends of spinel variation in mare basalts plotted in a modified Johnston prism (after Haggerty 1971). The end-members are described in the text. Note the similarity between Apollo 11 and 17 spinel trends, and Apollo 15 and 12 trends, except for a discontinuity in the latter and a greater variability in the chromite-picrochromite solid solution. Luna 16 spinels are similar but show greater hercynite-spinel solid solution and more variability in the ulvospinel-poor members. Data from Haggerty (1971, 1972b), El Goresy et al (1974).

and Serenitatis have consistently older ages (3.6–3.8 AE) than the low-titania basalts from Procellarum and Imbrium (3.2–3.4 AE). Basalt from Fecunditatis has been dated at 3.4–3.5 AE (Papanastassiou & Wasserburg 1972). Temporal changes in chemistry are indicated (Papike et al 1974), but the areal sampling is statistically poor and warrants cautious interpretation. Large scale volcanism must have ceased at about 3 AE, judging from crater densities on mare surface.

Rb-Sr systematics are rather straightforward, at least compared to terra samples, and suggest that the systems have remained undisturbed since the lavas cooled (Papanastassiou & Wasserburg 1972). The low Rb/Sr ratios, together with crystallization ages, indicate an origin of mare basalts by partial melting of a low Rb/Sr mantle (0.003–0.008) evolved essentially since 4.6 AE (Figure 8). This in turn indicates that low Rb/Sr ratios as a result of depletion in Rb were a characteristic of the Moon's mantle at approximately 4.6 AE.

Most mare basalts give discordant U-Pb ages (Tera & Wasserburg 1974); Concordant basalts (Nunes et al 1974, Tera & Wasserburg 1974) with intersections at ~4.4 AE, are rarely observed, which implies no fractionation of Pb and U in these basalts relative to their source. Tera & Wasserburg (1974) discuss several ways such an unlikely situation might arise. The evolutionary history of the discordant mare basalts must be complex and highly dependent on the multistage model chosen. Preliminary U-Pb internal isochron determinations (Tera & Wasserburg 1974), which are model determined, again indicate an important event at ~4.4 AE in the evolution of the lunar mantle.

Experimental Petrology

Much of the rhetoric and argument on mare basalts has centered upon the interpretation of high-pressure and temperature-melting experiments. Such heated debate reflects the inherent importance of this discipline in determining the origin of mare basalts and the mineralogy of the lunar mantle. Reduced to its simplest

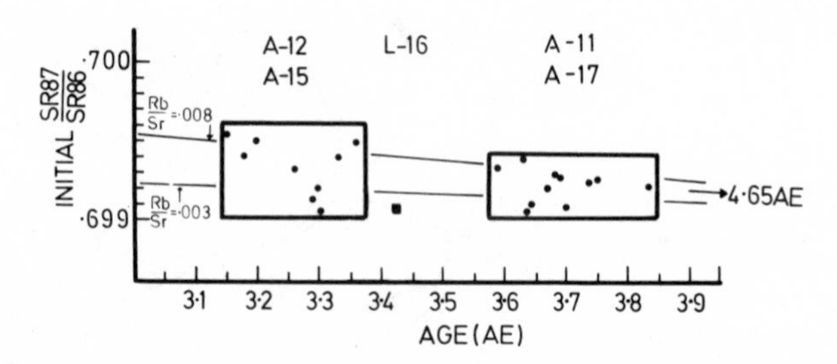

Figure 8 Rb-Sr systematics for mare basalts. Note the two groups corresponding to overall similarities in bulk chemistry. Two lines of Rb/Sr = 0.003 and Rb/Sr = 0.008 roughly bracket the Rb/Sr ratio of the mare source regions that could have evolved from $^{87}Sr/^{86}Sr$ = 0.69898 (initial $^{87}Sr/^{86}Sr$ ratio of basaltic achondrite meteorites since 4.65 AE and produced mare basalts at the intervals shown without previous fractionation of the Rb/Sr ratio. Data are from Compston et al (1970, 1971), Papanastassiou et al (1970), Papanastassiou & Wasserburg (1970, 1972), Nyquist et al (1972), Evensen et al (1973).

terms, and ignoring the esoteric complexities of "run conditions," the arguments revolve around the degree to which mare basalt chemistry has been perturbed by low pressure fractional crystallization. If this process has operated in a major way, then experiments that simulate mantle pressures cannot be used to evaluate phase relations in the Moon's mantle; conversely, if the basalts are pristine and largely unfractionated melts, such experiments provide valuable information.

Much of the information previously discussed indicates that some low pressure fractional crystallization should occur in lavas of such low viscosity. This process is likely to have operated most effectively in the high-titanium basalts that crystallized abundant, dense iron-titanium oxide. However, both textural evidence and low pressure cooling experiments clearly demonstrate that the presence of abundant phenocrysts cannot be an unequivocal test for a high degree of crystal settling. At least some of the mare basalts arrived at the lunar surface as liquids and justifiably can be used for high pressure experiments (Grove et al 1973, Longhi et al 1974).

High-titanium basalts are multiply saturated, or nearly so, with olivine + clinopyroxene + ilmenite + spinel at pressures corresponding to 100–150 km (Ringwood & Essene 1970, Longhi et al 1974). Some experimentalists (Longhi et al 1974) accept this as the depth range in which mare basalts are produced. Others (Ringwood & Essene 1970) argue that heat flow considerations require the zone of partial melting to be deeper (200–400 km), which according to experimental data would indicate a mantle dominated by low-calcium pyroxene (pyroxenite).

In contrast, experimental data for low-titanium, picritic basalt, carefully selected for its primitive nature (Grove et al 1973), indicate a source mineralogically unlike the source of high-titanium basalts. Extended liquidus crystallization of olivine at low pressure and low-calcium pyroxene at higher pressure is unlike the multiply saturated liquidi noted above.

The Sources of Mare Basalts

Assuming that some mare basalts are primitive enough to render high pressure experiments relevant, there are two model-dependent sources for mare basalts. According to the first model, mare basalts were produced by partial melting of a pyroxenitic mantle at depths between 200–400 km, determined by the fluctuating position of the melting zone with time (Wood 1972, Toksoz 1974). In this model (Ringwood & Essene 1970, Green et al 1971) the sources of the high-titanium and low-titanium basalts are very similar and do not require the presence of substantial ilmenite. Production of the LIL element abundances and the ubiquitous europium anomaly require disequilibrium melting but do not require the presence of plagioclase. In the second model, mare basalts were produced at less than 150 km depth by partial melting of cumulates containing substantial pyroxene, ilmenite, and olivine. Heat sources to remelt such a refractory assemblage have not been satisfactorily located. The cumulates are the product of an early lunar differentiation and have a negative Eu anomaly, which is then inherited by the mare melts. By this scheme the low-titanium and high-titanium basalts cannot have the same source. Ilmenite plays an essential part in generating high-titanium melts.

Other conclusions that are not model-determined can be drawn. The mare source was depleted in volatile trace elements by two orders of magnitude over chondritic abundances (Figure 9). Radiometric considerations indicate this was a primitive feature developed at ~4.6 AE, which corresponds to the Moon's inception as a planet. The mare sources are also depleted in siderophile and chalcophile elements, some of which are also volatile, suggesting early removal in metal and sulphide phases. For reasonable degrees of partial melting (1–10%) the source would be enriched 5–10 times over chondritic abundances in refractory LIL elements, perhaps reflecting an earlier period of enrichment during mantle formation. For a cumulate source, the LIL elements may have resided in intercumulus material.

TERRA ASSEMBLAGES

With few exceptions, samples collected from the lunar terra have clastic and/or metamorphic textures. These textures are believed to be a consequence of intense meteorite bombardment, which fragmented, shocked, annealed, partially melted and recrystallized the terra rocks. During this process, pristine, igneous textures were modified extensively, sometimes up to the point of obliteration. In this sense, the terra rocks contrast vividly with the mare basalts, which still retain pristine igneous textures and mineral assemblages. Our limited knowledge of physical and chemical changes attendant upon meteorite impact has led to a renewed interest in terrestrial

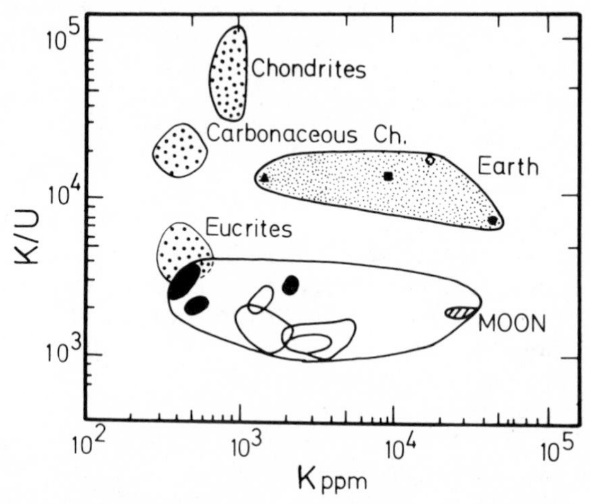

Figure 9 K concentrations normalized to a constant Si concentration of 18.5, and K/U ratios for lunar, meteoritic, and terrestrial samples. Range of values for mare basalts shown as solid areas; other areas are mare soils and breccias, except for the shaded area (12013). Specific rock types identified amongst terrestrial rocks are oceanic tholeiite (triangle), continental tholeiites (square), andesite (open circle), and granite (filled circle). Modified after Gast (1972).

Note that the low K/U of the Moon is not characteristic of all meteorites; ordinary and enstatite chondrites have significantly higher ratios.

impact craters (Chao 1974, Grieve et al 1974) in the hope that such studies may have application to terra breccias.

Rare, but nonetheless important, coarse-grained rocks occur in terra assemblages. These show no evidence for extensive meteoritic reworking and are thought to represent deeper parts of the lunar crust that were occasionally excavated during particularly intense bombardment. In the next section these rocks are discussed separately from the petrochemistry of terra breccias. Space considerations require us to deal cursorily with the terra soils that form a thick veneer over the terra bedrock. Figure 10 indicates the physiochemical processes operating to produce and modify lunar soils.

Breccias and Soils

Petrographic and chemical data indicate that impact may crush, shock, partly or wholly remelt, partly volatilize, and metamorphose bedrock. A minimal effect would be mild thermal induration of disaggregated material, e.g. regolith to produce a rather friable "rock," commonly called a regolith or soil breccia. At the other extreme, total melting may have occurred so that the rock acquired a completely new texture.

Two research fields have developed in breccia studies. Firstly, there are studies toward understanding the process of brecciation and how breccias evolve at or near the lunar surface. In these cases particular emphasis is placed upon matrix textures. Secondly, there are studies using breccias as carriers of previous lithologies; particular emphasis is placed upon vitric, crystalline, and lithic clasts within the breccias.

Breccias are commonly the product of many impact events and are polymict

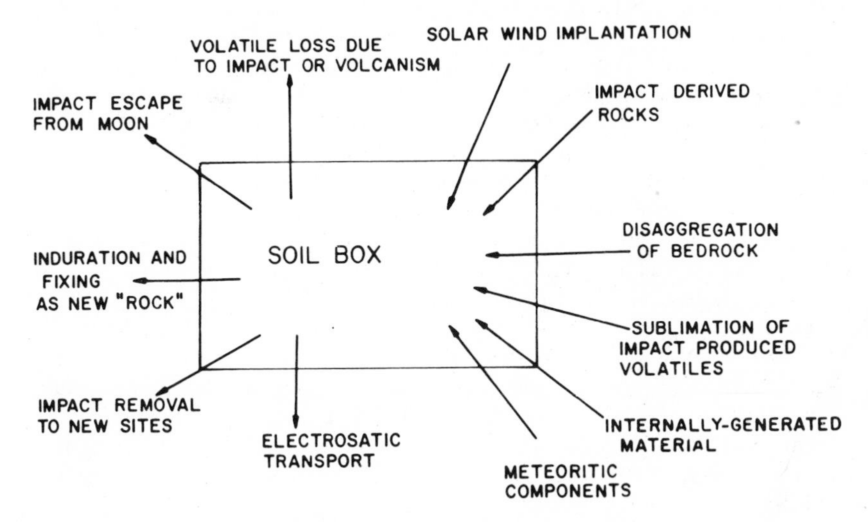

Figure 10 Idealized diagram illustrating various chemical and physical inputs to and outputs from the lunar "soil box."

rocks with several generations of brecciated clasts, best exemplified at the Fra Mauro site. Monomict breccias, representing one or a few minor impacts into a homogeneous target are rather rare. Useful criteria for recognizing mature breccias are high siderophile element content as a result of admixture of meteorite components; high metallic iron content as a result of long exposure to the reducing solar wind, high solar wind components; and high fission track density. Clastic matrix and overall inequigranular texture also help to separate breccias from primary igneous rocks. All of the criteria are most easily applied to low-grade breccias that have not been subjected to temperatures above about 700°C over long periods of time. Outgassing of breccias, annealing of fission tracks, volatile loss, and resorption of clasts into matrix take place at more elevated temperatures, obscuring the prehistory of the breccia (Tera et al 1970). If carried to extremes, breccia textures approach those of pristine igneous rocks, ultimately leading to debate as to the origin of high-grade metamorphic rocks (somewhat reminiscent of the granite controversy).

Regolith-or-soil-breccias have heterogeneous matrices, shock-produced matrix glass, and a notable porosity (Christie et al 1973). Increased shock may lead to plastic deformation, heating, loss of porosity and formation of a recrystallized, cryptocrystalline, isotropic matrix. These lithification processes probably occur in situ, involving propagation of a shock wave through loose soil, and do not develop in thick impact ejecta blankets.

The most successful and widely applicable breccia classifications are based upon matrix textures (Warner 1971, 1972, Warner et al 1973, Wilshire et al 1973, James 1974, Simonds et al 1974), because increasing breccia grade involves a continuous variation in matrix texture and mineralogy.

Much controversy has centered on the origin of high-grade breccias with granulitic, poikilitic, subophitic, and ophitic textures. Such breccias are abundant at the Cayley-Descartes Mountains and Taurus-Littrow Valley and to a lesser extent at all other terra sites. Arguments center upon the presence and amount of a melt phase and environment of formation. In the poikilitic textures are largely the result of solid state metamorphic recrystallization, then this might occur in a thermally insulated, thick ejecta blanket. Conversely, the presence of a substantial melt phase (> 10%) would indicate origin as an impact melt. Evidence from terrestrial craters suggests that these textures are associated with 20–60 km diameter craters (Grieve et al 1974) and are a function of stratigraphic position within the impact melt.

Cataclastic textures are typically developed in bedrock lining impact craters. Cataclastic anorthosites, anorthositic gabbros, norites, and dunites—displaying varying degrees of recrystallization, brecciation and deformation—have all been found at terra sites. More complex breccias indicating injection of fragment-laden impact melt may represent fallout or fallback breccias associated with major impacts.

Chemically, terra breccias are very variable, as shown in Figure 11. The Fra Mauro breccias display the least variability, being dominated by an aluminous basaltic component rich in LIL trace elements and designated KREEP basalt. Exceptions are the white breccias excavated from Cone Crater, which are more aluminous and contain no significant KREEP component. Brown-glass matrix breccias dominate the stratigraphy of the Apennine Front (Phinney et al 1972) and also have a basaltic composition, but only have about one third of the LIL elements of KREEP basalt.

The most feldspathic terra breccias occur in the Cayley-Descartes highlands, including cataclastic anorthosites and light-matrix breccias, chemically equivalent

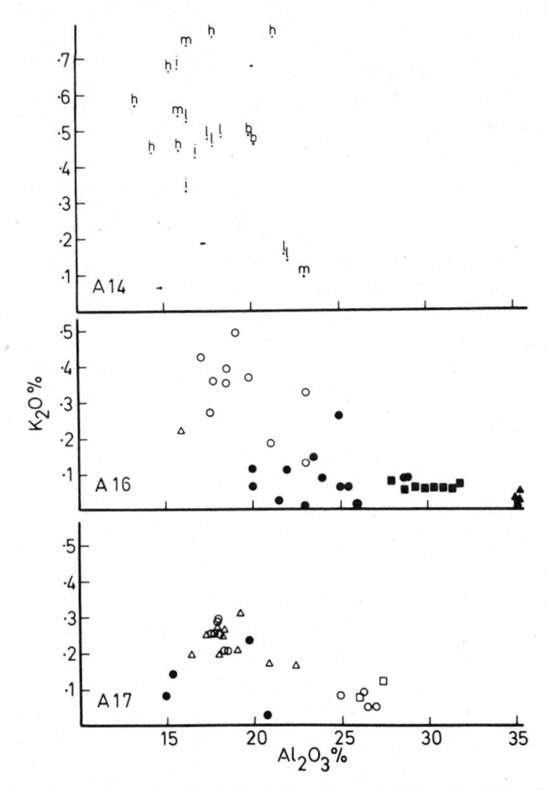

Figure 11 K_2O-Al_2O_3 plot for terra breccias from Apollo 14, 16, and 17. Al_2O_3 is a reflection of the feldspathic component and K_2O of the KREEP component (Simonds et al 1974). Apollo 14: high (*h*), medium (*m*), and low (*l*) grade breccias and basalts (*b*); data from Hubbard et al (1972), Laul et al (1972), Rose et al (1972), Taylor et al (1972), Willis et al (1972). Apollo 16: mesostasis-rich breccias and basalts (open circles), poikilitic breccias (filled circles), light-matrix breccias (filled squares), cataclastic anorthosite (filled triangles); data from Duncan et al (1973), Hubbard et al (1973), Laul & Schmitt (1973), Rose et al (1973). Apollo 17: subophitic breccias (open triangles), others as above; data from Simonds et al (1974). Breccia descriptions usually refer to matrix texture.

Note: (*a*) the overwhelming frequency of KREEP breccias at the Apollo 14 site, (*b*) a distinct division between poikilitic and mesostasis-rich breccias at the Apollo 16 site and abundant, highly aluminous ($>30\%$ Al_2O_3) breccias, and (*c*) general lack of KREEP-like breccias at the Apollo 17 site and different compositions for poikilitic rocks compared to the Apollo 16 rocks.

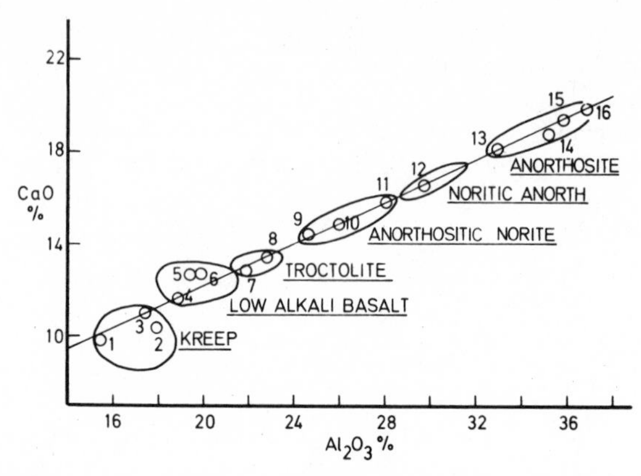

Figure 12 Major terra rock types plotted in terms of CaO and Al_2O_3. Individual rock type averages (open circles) are as follows: (*1*) Apollo 15 high-K KREEP glass, (*2*) Luna 20 KREEP lithics, (*3*) Apollo 15 medium-K KREEP glass, (*4*) Apollo 15 glass, (*5, 6, 7*) lithics, glass, and rocks in Luna 20, (*8*) Luna 20 lithics (*9, 10*) lithics and glass in Luna 20, (*11*) Apollo 16 glass, (*12*) Luna 20 glass, (*13*) Luna 20 lithics, (*14*) Apollo 14 glass, (*15*) rock 15415, (*16*) anorthite. Data from Reid et al (1972, 1973b, c), Prinz et al (1973c).

to gabbroic anorthosite. In general, however, there seems little relationship between matrix texture and chemistry if Apollo 16 and 17 breccias are compared. The variations in chemistry can be accounted for by physical mixing of a few major rock types, including KREEP basalt, anorthosite, troctolite, and dunite.

Table 5 Suggested fundamental terra rock types

	a	b	c	d	e	f	g
SiO_2	48.01	45.84	44.57	43.98	44.29	43.06	39.93
TiO_2	2.02	0.78	0.37	0.33	0.05	0.17	0.03
Al_2O_3	17.12	21.16	25.84	29.75	34.60	23.66	1.53
Cr_2O_3	—	0.17	—	0.05	0.08	0.30	—
FeO	10.56	8.34	5.22	3.70	0.77	4.48	11.34
MgO	8.72	10.41	7.38	4.97	0.85	14.50	43.61
MnO	—	0.12	—	0.05	0.01	0.09	0.13
CaO	10.77	12.77	14.84	16.81	18.89	12.95	1.14
Na_2O	0.71	0.39	0.41	0.33	0.50	0.32	—
K_2O	0.60	0.11	0.09	0.05	0.02	0.05	—

[a] KREEP basalt, Fra Mauro basalt. Large rocks, lithic fragments, glasses.

[b] Low-alkali, high-alumina basalt; low-K Fra Mauro basalt. Occur in glass populations and lithic fragments.

[c] Anorthositic norite, anorthositic gabbro, highland basalt, feldspathic basalt, ANT norite. Occur as large rocks, lithic fragments and glasses.

[d] Gabbroic anorthosite, noritic anorthosite. Large rocks, lithic fragments, glasses.

[e] Anorthosite. Large rocks, lithic fragments, glasses.

[f] Troctolite. Lithic fragments.

[g] Dunite. Lithic fragments.

Fundamental Terra Rock Types

In evaluating fundamental terra rock types, emphasis has been placed upon rare terra clasts with pristine igneous textures, and the recurrence of rocks with similar chemistry at several highland sites (Figure 12). Consistently similar rock types emerge from analyses of large rocks, lithic fragments, and impact glasses, as shown in Table 5 and Figure 13. Only some of these can be shown to have unequivocal igneous textures.

Terra rock types can be divided into those with fundamentally basaltic chemistry, i.e. KREEP basalt and low-alkali, high-alumina basalt, which are considered to have originated as melts, and those with compositions that deviate considerably from any reasonable liquid line of descent. These latter rocks are considered to be cumulates from basaltic magmas and include troctolites, spinel troctolites, norites, noritic anorthosites, anorthosites, and dunites.

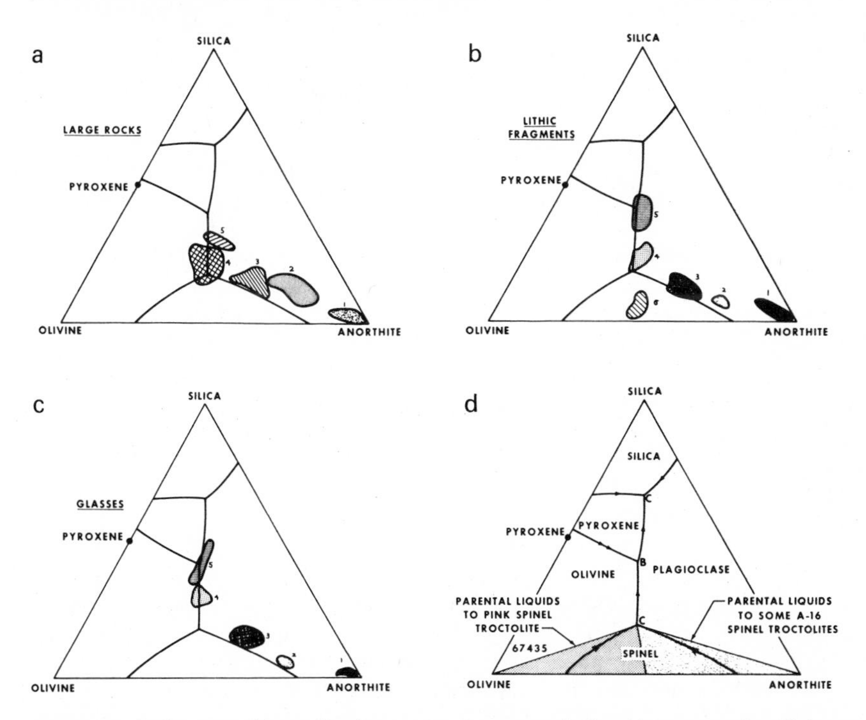

Figure 13 Bulk compositions of major terra rock types recalculated into normative olivine, anorthite, and quartz, and plotted in the pseudoternary system olivine-anorthite-silica. Phase boundaries are after Walker et al (1972). Individual fields are (*1*) anorthosite, (*2*) noritic anorthosite, (*3*) anorthositic norite, (*4*) low-alkali, high-alumina basalt, (*5*) KREEP, (*6*) (spinel) troctolite. Average compositions from large rocks (*a*), lithic fragments (*b*), and glasses (*c*). In (*d*), shaded areas represent possible parental liquids to some A-16 troctolites, assuming they are cumulates and based upon the cumulus and intercumulus phases. Note most troctolites could be derived, according to this diagram, from liquids lying close to peritectic point C.

KREEP Basalt with well-developed igneous texture has been sampled at the Apollo 14, 15, and 16 sites. However, the high siderophile and chalcophile element contents of some Apollo 14 KREEP indicates it was locally formed by impact melting [Morgan et al (1972), but see Crawford & Hollister (1974) for an opposing view]. Most KREEP basalt clasts have very low contents of meteoritic elements and are undoubtedly endogeneous to the Moon. As the acronym implies, this basalt type is rich in potassium, rare earth elements, and phosphorus. It is generally very enriched in LIL trace elements, including Th and U and presents a unique signature to the orbital gamma-ray data (Metzger et al 1973, Metzger et al 1974), indicating KREEP has a distinct provinciality to the Imbrium-Procellarum region. This is consistent with the dominance of KREEP basalt at the Fra Mauro site compared to at all other highland regions. Impact processes have tended to distribute KREEP in small amounts throughout the terra regions. KREEP basalt can be distinguished from mare basalts by high LIL element abundances, distinctive rare earth pattern (Figure 14), and high-alumina and low-titania content. KREEP basalts crystallize calcic plagioclase (An 94-58) as liquidus phase, followed by magnesian orthopyroxene containing about 3% Al_2O_3. Orthopyroxenes are zoned continuously to pigeonite and ferropigeonite with only rare metastable pyroxene compositions (cf mare pyroxenes). In some slowly cooled KREEP gabbros, microscopic exsolution of augite and ferroaugite from pigeonite may be observed (Figure 13). Spinels are much less abundant than in mare basalts, and they include ilmenite and ulvospinel-rich phases (Haggerty 1972b) that occasionally have undergone subsolidus reduction to ilmenite and metallic iron.

Low-alkali, high-alumina basalt has been recognized as lithic fragments and preferred glass compositions, particularly in the Apollonius region (Prinz et al 1973c, Reid et al 1973c), but also at the Apollo 15 and 16 sites. Even in lithic fragments, the textures are commonly high-grade metamorphic textures, and it cannot easily be demonstrated that these rocks were initially magmas. At the Apollo 15 site some low-alkali basalts are polymict breccias (Phinney et al 1972). This basaltic composition is essentially similar to KREEP but with higher TiO_2, Mg/Fe ratio, about one third of the LIL element concentrations, and a similar REE pattern (Figure 14).

Anorthosites were first recognized as lithic fragments in Mare Tranquillitatis soil (Wood 1972) and surmised at that time to be derived from the terra regions. They have been recognized dominantly as lithic fragments in soils (Taylor 1972, Dowty et al 1974b)—rarely as large samples (James 1972)—and form a small population in impact glasses (Reid et al 1972). Many anorthosites are believed to be genetically related to terra norites and troctolites and form part of the "ANT" suite of terra assemblages. Calcic plagioclase forms 85–95% of the rock and has a very restricted composition (An 98-94). Ferromagnesian minerals include olivine (Fo 90-55) and calcium-rich and calcium-poor pyroxene (Taylor 1972, Dowty et al 1974b, Drake et al 1974). Textures are either cataclastic or granoblastic, and the origin of these rocks as essentially monomineralic cumulates is based upon their extreme chemistry and the abnormally high temperatures required to generate them as liquids. Anorthosites have distinctive trace element abundances, being low in siderophile and chalcophile elements, but some are unusually enriched in volatile elements such as lead

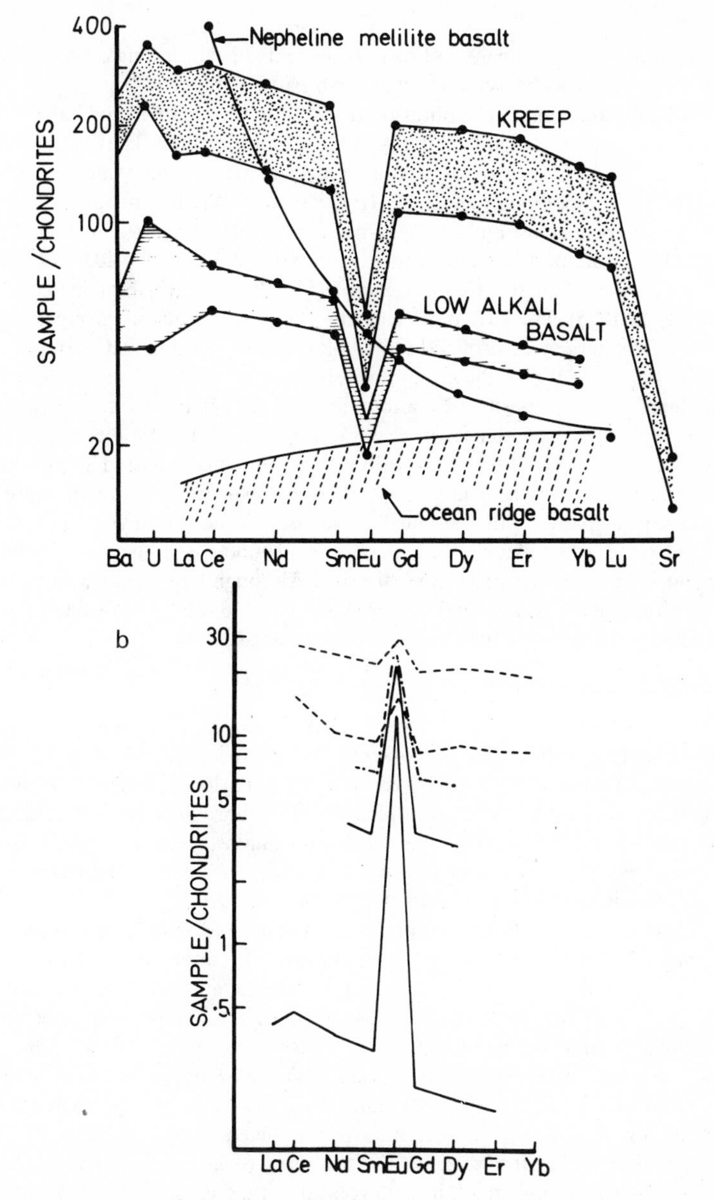

Figure 14 Lithophile trace element abundances in (*a*) terra basaltic rocks and (*b*) terra cumulate rocks. Abundances are normalized to chondritic abundances. In (*a*), patterns for terrestrial nepheline-melilite basalt and ocean ridge basalt are shown for comparison. In (*b*) are shown range for anorthosites (continuous lines), norite (broken lines), and the lower limit for troctolites (dash-dot line). Other troctolites have large negative Eu anomalies and up to 100 times the chondritic abundances of rare earths. Data from Hubbard et al (1972, 1973), Drake et al (1974).

and thallium. They also contain low LIL element concentrations except for enrichment in Eu giving a characteristic rare earth pattern (Figure 14).

Troctolites and spinel troctolites occur as lithic fragments in soils at most terra sites and as larger samples at the Apollo 16 and 17 sites. They are principally composed of olivine with plagioclase, spinel, minor pyroxene, and metallic iron (Prinz et al 1973b, Gooley et al 1974). Those with olivine and pink spinel have intercumulus plagioclase and have well-defined cumulate textures (Prinz et al 1973b). On the basis of mineralogy and texture, some troctolites are considered to be parts of deep crustal cumulates (Gooley et al 1974), although this remains contentious (Albee et al 1975). The LIL trace element composition of this group is not well defined, being largely controlled by the volume of intercumulus mesostasis present (Drake et al 1974). Where the groundmass is negligible, the rocks have low absolute LIL abundances and positive Eu anomalies (Figure 14), but with increasing mesostasis the absolute abundances increase and negative Eu anomalies appear.

Norites have been observed as lithic fragments and preferred groups in terra impact glasses, but rarely as large rocks. Metamorphic textures predominate. These rocks are principally composed of calcic plagioclase and orthopyroxene with minor clinopyroxene and olivine. With increasing plagioclase content they grade into anorthositic norites and noritic anorthosites. Although large rocks with these latter compositions have been found at the Apollo 15, 16, and 17 sites, their textures have been extremely perturbated by impact processes.

Petrogenesis of Terra Rocks

Numerous schemes have been evolved to account for the mineralogy and chemistry of terra samples within a framework of lunar crust and mantle evolution (Wood et al 1970, Smith et al 1970, Taylor & Jakes 1974, Taylor & Bence 1975). Although these models have become more sophisticated as more data has accumulated, the basic mechanism for crust-mantle evolution remains the operation of extremely efficient crystal fractionation within a moon-wide silicate melt. This was first proposed following the Apollo 11 mission (Smith et al 1970).

In a gross sense the formation of the lunar crust and mantle as complementary cumulates is suggested by (*a*) the inferred negative Eu anomaly for the lunar upper mantle (Nava & Philpotts 1973) and positive Eu anomaly for the terra crust (Taylor et al 1973); (*b*) the high Fe/Mg ratio of the mare basalts (derived from the upper mantle) compared to the low Fe/Mg ratio of terra rocks; (*c*) the high density of upper mantle minerals (olivine, orthopyroxene, clinopyroxene, ilmenite) compared to plagioclase that dominates the crust; and (*d*) the terra rocks and mare basalts that lie on opposite sides of the meteoritic Ca/Al line.

In detail, however, it is evident that the terra rocks, particularly the cumulate rocks, represent several stages in the evolution of the crust. Those with magnesian ferromagnesian minerals may be related to early crustal formation when plagioclase was separating with ferromagnesian minerals having a high Mg/Fe ratio. The implication here would be that the initial ferromagnesian minerals that largely sank should have produced an upper mantle whose deep regions had a high Mg/Fe ratio. However the extreme range observed in Fe/Mg ratios of olivines and pyroxenes in terra cumulates strongly suggests some cumulates developed from highly fractionated silicate melts. Separation of early magnesian olivine and orthopyroxene from the primitive melted outer shell would result in residual melts with increasing Fe/Mg ratio (Taylor & Jakes 1974) but would also correspond to a drastic decrease in the amount of plagioclase precipitating and accumulating. It seems most unlikely that the more iron-rich cumulate members of the anorthosite-norite-troctolite suite could form by this process.

Once the crust had developed, the meteorite flux would still be very substantial, producing very extensive areas in which crustal remelting occurred. Some of the presently observed crustal cumulates may have developed in these "secondary" magma systems, basically unrelated to the development of the protocrust. Such magmas, if in equilibrium with plagioclase would have large negative Eu anomalies (Haskin et al 1970) and may account for the presence of terra rocks with large Eu depletions relative to trivalent REE. It is also clear that the source of KREEP magmas need not contain substantial plagioclase ($<10\%$) (Weill et al 1974) and may represent partial melting of olivine and orthopyroxene cumulates, with minor plagioclase near the base of the lunar crust.

EPILOGUE

Lunar petrology is a very large and complex subject, involving many interrelated disciplines. We have probably accumulated more petrologic information on lunar rocks than any one type of terrestrial rock, and certainly more disciplines have been brought to bear on lunar rocks. One might surmise that similar broad-based approaches to terrestrial petrologic problems might lead to the surprising conclusion that terrestrial igneous and metamorphic rocks are much more complex than we presently consider them to be. In the lunar case, the broad brush scenario of lunar evolution has been constructed. The added problem of dealing with meteorite bombardment and the huge physicochemical perturbations this causes to rocks has insured that no physicochemical evolutionary scheme, satisfying in detail, has yet been put together. It remains a challenging intellectual exercise.

Acknowledgments

Mary-Linda Adams and Michael Perfit painstakingly checked the manuscript and bibliography and substantially improved the final version. The work was carried out under NASA grant NGR-33-008-199. Lamont-Doherty Geological Observatory contribution number 2237.

Literature Cited

Adler, I., Trombka, J., Gerard, J., Lowman, P., Schmadebeck, R., Blodgett, H., Eller, E., Yin, L., Lamothe, R., Osswald, R., Gorenstein, P., Bjorkholm, P., Gursky, H., Harris, B., Golub, L., Harnden, F. R. 1972a. Apollo 16 X-ray fluorescence experiment. *Apollo 16 Prelim. Sci. Rep. NASA SP-315*, Sect. 19, pp. 1–14

Adler, I., Trombka, J., Gerard, J., Lowman, P., Schmadebeck, R., Blodgett, H., Eller, E., Yin, L., Lamothe, R., Gorenstein, P., Bjorkholm, P. 1972b. Apollo 15 geo-

chemical X-ray fluorescence experiment: preliminary report. *Science* 175:436–40
Albee, A. L., Dymek, R. F., DePaolo, D. J. 1975. Spinel symplectites: High-pressure solid-state reaction or late-stage magmatic crystallization. *Lunar Science VI*, 1–3. Houston: Lunar Sci. Inst.
Anders, E., Ganapathy, R., Keays, R. R., Laul, J. C., Morgan, J. W. 1971. Volatile and siderophile elements in lunar rocks: Comparison with terrestrial and meteoritic basalts. *Proc. Lunar Sci. Conf., 2nd* 2:1021–36. Cambridge: M.I.T. Press
Bence, A. E., Papike, J. J., Lindsley, D. H. 1971. Crystallization histories of clinopyroxenes in two porphyritic rocks from Oceanus Procellarum. *Proc. Lunar Sci. Conf., 2nd* 1:559–74. Cambridge: M.I.T. Press
Bence, A. E., Papike, J. J. 1972. Pyroxenes as recorders of lunar basalt petrogenesis: Chemical trends due to crystal-liquid interaction. *Proc. Lunar Sci. Conf., 3rd. Geochim. Cosmochim. Acta* 1: Suppl. 3, pp. 431–69
Biggar, G. M., O'Hara, M. J., Peckett, A., Humphries, D. J. 1971. Lunar lavas and the achondrites: Petrogenesis of protohypersthene basalts in the maria lava lakes. *Proc. Lunar Sci. Conf., 2nd. Geochim. Cosmochim. Acta* 1: Suppl. 2, pp. 617–43
Boyd, F. R., Smith, D. 1971. Compositional zoning in pyroxenes from lunar rock 12021, Oceanus Procellarum. *J. Petrol.* 12:439–64
Brown, G. M. 1970. Petrology, mineralogy and genesis of lunar crystalline igneous rocks. *J. Geophys. Res.* 75:6480–96
Brown, G. M., Emeleus, C. H., Holland, J. G., Phillips, R. 1970. Mineralogical, chemical, and petrological features of Apollo 11 rocks and their relationship to igneous processes. *Proc. Apollo 11 Lunar Sci. Conf. Geochim. Cosmochim. Acta* 1: Suppl. 1, pp. 195–219
Brown, G. M., Emeleus, C. H., Holland, J. G., Peckett, A., Phillips, R. 1971. Picrite basalts, ferrobasalts, feldspathic norites, and rhyolites in a strongly fractionated lunar crust. *Proc. Lunar Sci. Conf., 2nd* 1: 583–600. Cambridge: M.I.T. Press
Brown, G. M., Emeleus, C. H., Holland, J. G., Peckett, A., Phillips, R. 1972. Petrology, mineralogy, and classification of Apollo 15 mare basalts. *The Apollo 15 Lunar Samples*, 40–44. Houston: Lunar Sci. Inst.
Brown, G. M., Peckett, A., Emeleus, C. H. Phillips, R. 1974. Mineral-chemical properties of Apollo 17 mare basalts and terra fragments. *Lunar Science V*, 89–91. Houston: Lunar Sci. Inst.
Bryan, W. B., Adams, M. L. 1973. Some volcanic and structural features of Mare Serenitatis. In *Apollo 17 Prelim. Sci. Rep., NASA SP-330*, Sect. 29, pp. 26–28
Cadenhead, D. A. 1973. Lunar volcanic glasses and cinder formation. *EOS. Trans. Am. Geophys. Union* 54:582
Chao, E. C. T. 1974. Impact cratering models and their application to lunar studies—a geologists view. *Proc. Lunar Sci. Conf., 5th. Geochim. Cosmochim. Acta* 1: Suppl. 5, pp. 35–52
Christie, J. M., Griggs, D. T., Heuer, A. H., Nord, G. L., Radcliffe, R. V., Lally, J. S., Fisher, R. M. 1973. Electron petrography of Apollo 14 and 15 breccias and shock-produced analogs. *Proc. Lunar Sci. Conf., 4th. Geochim. Cosmochim. Acta* 1: Suppl. 4, pp. 365–82
Compston, W., Berry, H., Vernon, M. J., Chappell, B. W., Kaye, M. J. 1971. Rubidium-strontium chronology and chemistry of lunar material from the Ocean of Storms. *Proc. Lunar Sci. Conf., 2nd.* 2:1471–85. Cambridge: M.I.T. Press
Compston, W., Chappell, B. W., Arriens, P. A., Vernon, M. J. 1970. The chemistry and age of Apollo 11 lunar materials. *Proc. Apollo 11 Lunar Sci. Conf. Geochim. Cosmochim. Acta* 2: Suppl. 1, pp. 1007–27
Crawford, M. L. 1973. Crystallization of plagioclase in mare basalts. *Proc. Lunar Sci. Conf., 3rd. Geochim. Cosmochim. Acta* 1: Suppl. 3, pp. 705–17
Crawford, M. L., Hollister, L. S. 1974. KREEP basalt: A possible partial melt from the lunar interior. *Proc. Lunar Sci. Conf., 5th. Geochim. Cosmochim. Acta* 1: Suppl. 5, pp. 399–419
Dowty, E., Klaus, K., Prinz, M. 1974a. Lunar pyroxene-phyric basalts: Crystallization under supercooled conditions. *J. Petrol.* 15:419–53
Dowty, E., Prinz, M., Keil, K. 1973. Composition, mineralogy, and petrology of 28 mare basalts from Apollo 15 rake samples. *Proc. Lunar Sci. Conf., 4th. Geochim. Cosmochim. Acta* 1: Suppl. 4, pp. 423–44
Dowty, E., Prinz, M., Keil, K. 1974b. Ferroan anorthosite: a widespread and distinctive lunar rock type. *Earth Planet. Sci. Lett.* 24:15–25
Drake, M. J. 1975. The crystallization of plagioclase feldspar from silicate melt. *Lunar Science VI*, 199–201. Houston: Lunar Sci. Inst.
Drake, M. J., Taylor, G. J., Goles, G. G. 1974. Descartes Mountains and Cayley Plains: composition and provenance. *Proc. Lunar Sci. Conf., 5th. Geochim. Cosmochim. Acta* 2: Suppl. 5, pp. 991–1008

Duncan, A. R., Erlank, A. J., Willia, J. P., Ahrens, L. H. 1973. Composition and inter-relationships of some Apollo 16 samples. *Proc. Lunar Sci. Conf., 4th. Geochim. Cosmochim. Acta* 2: Suppl. 4, pp. 1097–1113

El Goresy, A., Ramdohr, P., Medenbach, O., Bernhardt, H. 1974. Taurus-Littrow TiO_2-rich basalts: Opaque mineralogy and geochemistry. *Proc. Lunar Sci. Conf., 5th. Geochim. Cosmochim. Acta* 1: Suppl. 5, pp. 627–52

Evensen, N., Murthy, V. R., Coscio, M. R. 1973. Rb-Sr ages of some mare basalts and the isotopic and trace element systematics in lunar fines. *Proc. Lunar. Sci. Conf., 4th. Geochim. Cosmochim. Acta* 2: Suppl. 4, pp. 1707–24

Frondel, J. W. 1972. Harvard University preprint

Ganapathy, R., Keays, R. R., Laul, J. C., Anders, E. 1970. Trace elements in Apollo 11 lunar rocks: Implications for meteorite influx and origin of moon. *Proc. Apollo 11 Lunar Sci. Conf. Geochim. Cosmochim. Acta* 2: Suppl. 1, pp. 1117–42

Gast, P. W. 1972. The chemical composition and structure of the moon. *Moon* 4:630–57

Gay, P., Bown, M. G., Muir, I. D., Bancroft, G. M., Williams, P. G. L. 1971. Mineralogy and petrographic investigation of some Apollo 12 samples. *Proc. Lunar Sci. Conf., 2nd.* 1: 377–92. Cambridge: M.I.T. Press

Gibson, E. K., Hubbard, N. J., Wiesmann, H., Bansal, B. M., Moore, G. W. 1972. How to lose Rb, K, and change the K/Rb ratio: an experimental study. *Proc. Lunar Sci. Conf., 3rd. Geochim. Cosmochim. Acta* 2: Suppl. 3, pp. 1263–73

Gooley, R., Brett, R., Warner, J. L., Smyth, J. R. 1974. A lunar rock of deep crustal origin: sample 76535. *Geochim. Cosmochim. Acta* 38: 1329–39

Green, D. H., Ringwood, A. E., Ware, N. G., Hibberson, W. O., Major, A., Kiss, E. 1971. Experimental petrology and petrogenesis of Apollo 12 basalts. *Proc. Lunar Sci. Conf., 2nd. Geochim. Cosmochim. Acta* 1: Suppl. 2, pp. 601–15

Green, D. H., Ringwood, A. E., Ware, N. G., Hibberson, W. O. 1975. Experimental petrology and petrogenesis of Apollo 17 mare basalts. *Lunar Science VI*, 311–13. Houston: Lunar Sci. Inst.

Grieve, R. A. F., Plant, A. G., Dence, M. R. 1974. Lunar impact melts and terrestrial analogs: Their characteristics, formation, and implications for lunar crustal evolution. *Proc. Lunar Sci. Conf., 5th. Geochim. Cosmochim. Acta* 1: Suppl. 5, pp. 261–73

Grove, T. L., Walker, D., Longhi, J., Stolper, E. M., Hays, J. F. 1973. Petrology of rock 12002 from Oceanus Procellarum. *Proc. Lunar Sci. Conf., 4th. Geochim. Cosmochim. Acta* 1: Suppl. 4, pp. 995–1011

Haggerty, S. E. 1971. Compositional variations in lunar spinels. *Nature* 233: 156–60

Haggerty, S. E. 1972a. Subsolidus reduction of lunar spinels. *Nature* 234: 113–17

Haggerty, S. E. 1972b. Luna 16: An opaque mineral study and a systematic examination of compositional variations of spinels from Mare Fecunditatis. *Earth Planet. Sci. Lett.* 13: 328–52

Haggerty, S. E. 1973. Armalcolite and genetically associated opaque minerals in the lunar samples. *Proc. Lunar Sci. Conf., 4th. Geochim. Cosmochim. Acta* 1: Suppl. 4, pp. 777–97

Haskin, L. A., Allen, R. O., Helmke, P. A., Paster, T. P., Anderson, M. R., Korotev, R. L., Zweifel, K. A. 1970. Rare earths and other trace elements in Apollo 11 lunar samples. *Proc. Apollo 11 Lunar Sci. Conf. Geochim. Cosmochim. Acta* 2: Suppl. 1, pp. 1213–31

Hollister, L. S., Trzcienski, W. E., Hargraves, R. B., Kulick, C. G. 1971. Petrogenetic significance of pyroxenes in two Apollo 12 samples. *Proc. Lunar Sci. Conf. 2nd* 1: 529–57. Cambridge: M.I.T. Press

Howard, K. A., Carr, M. H., Muehlberger, W. R. 1973. Basalt stratigraphy of southern Mare Serenitatis. In *Apollo 17 Prelim. Science Report, NASA SP-330,* Sect. 29, pp. 1–12

Hubbard, N. J., Gast, P. W., Rhodes, J. M., Bansal, B. M., Wiesmann, H., Church, S. E. 1972. Nonmare basalts: Part II. *Proc. Lunar Sci. Conf., 3rd. Geochim. Cosmochim. Acta* 2: Suppl. 3, pp. 1161–79

Hubbard, N. J., Rhodes, J. M., Gast, P. W., Bansal, B. M., Shih, C., Wiesmann, H., Nyquist, L. E. 1973. Lunar rock types: The role of plagioclase in non-mare and highland rock types. *Proc. Lunar Sci. Conf., 4th. Geochim. Cosmochim. Acta* 2: Suppl. 4, pp. 1297–1312

James, O. B. 1972. Lunar anorthosite 15415: texture, mineralogy, and metamorphic history. *Science* 175: 432–34

James, O. B. 1974. Lunar highlands breccias generated by major impacts. *Proc. Sov. Am. Conf. Cosmochem. Moon Planets* (preprint)

James, O. B., Wright, T. L. 1972. Apollo 11 and 12 mare basalts and gabbros: classification, compositional variations, and possible petrogenetic relations. *Geol. Soc. Am. Bull.* 83: 2357–82

Laul, J. C., Schmitt, R. A. 1973. Chemical

composition of Apollo 15, 16 and 17 samples. *Proc. Lunar Sci. Conf., 4th. Geochim. Cosmochim. Acta* 2: Suppl. 4, pp. 1349–67

Laul, J. C., Wakita, H., Showalter, D. L., Boynton, W. V., Schmitt, R. A. 1972. Bulk, rare earth, and other trace elements in Apollo 14 and 15 and Luna 16 samples. *Proc. Lunar Sci. Conf., 3rd. Geochim. Cosmochim. Acta* 2: Suppl. 3, pp. 1181–1200

Levinson, A. A., Taylor, S. R. 1971. *Moon Rocks and Minerals.* Oxford: Pergamon

Lofgren, G. E., Usselman, T. M., Donaldson, C. H. 1975. Cooling history of Apollo 15 quartz normative basalts determined from cooling rate experiments. *Lunar Science VI,* 515–17. Houston: Lunar Sci. Inst.

Longhi, J., Walker, D., Grove, T. L., Stolper, E. M., Hays, J. F. 1974. The petrology of the Apollo 17 mare basalts. *Proc. Lunar Sci. Conf., 5th. Geochim. Cosmochim. Acta* 1: Suppl. 5, pp. 447–69

LSPET (Lunar Sample Preliminary Examination Team). 1972. The Apollo 15 lunar samples: A preliminary description. *Science* 175:363–75

LSPET (Lunar Sample Preliminary Examination Team). 1973. Preliminary examination of lunar samples. *Apollo 17 Preliminary Science Report, NASA SP-330,* Sect. 7, pp. 1–46

Metzger, A. E., Trombka, J. I., Peterson, L. E., Reedy, R. C., Arnold, J. R. 1973. Lunar surface radioactivity: preliminary results of the Apollo 15 and Apollo 16 gamma ray spectrometer experiments. *Science* 179:800

Metzger, A. E., Trombka, J. I., Reedy, R. C., Arnold, J. R. 1974. Elemental concentrations from lunar orbital gamma-ray measurements. *Proc. Lunar Sci. Conf., 5th. Geochim. Cosmochim. Acta* 2: Suppl. 5, pp. 1067–78

Morgan, J. W., Laul, J. C., Krahenbuhl, U., Ganapathy, R., Anders, E. 1972. Major impacts on the moon: characterization from trace elements in Apollo 12 and 14 samples. *Proc. Lunar Sci. Conf., 3rd. Geochim. Cosmochim. Acta* 2: Suppl. 3, pp. 1377–95

Nava, D. F., Philpotts, J. A. 1973. A lunar differentiation model in light of new chemical data on Luna 20 and Apollo 16 soils. *Geochim. Cosmochim. Acta* 37: 963–73

Nunes, P. D., Tatsumoto, M., Unruh, D. M. 1974. U-Th-Pb systematics of some Apollo 17 samples. *Lunar Science V,* 562–64. Houston: Lunar Sci. Inst.

Nyquist, L. E., Gast, P. W., Church, S. E., Wiesmann, H., Bansal, B. 1972. Rb-Sr systematics for chemically defined Apollo 15 materials. *The Apollo 15 Lunar Samples* 380–84. Houston: Lunar Sci. Inst.

Papanastassiou, D. A., Wasserburg, G. J. 1970. Rb-Sr ages from the Ocean of Storms. *Earth Planet. Sci. Lett.* 8:269–78

Papanastassiou, D. A., Wasserburg, G. J., Burnett, D. S. 1970. Rb-Sr ages of lunar rocks from the Sea of Tranquillity. *Earth Planet. Sci. Lett.* 8:1–19

Papanastassiou, D. A., Wasserburg, G. J. 1972. Rb-Sr age of a Luna 16 basalt and the model age of lunar soils. *Earth Planet. Sci. Lett.* 13:368–74

Papike, J. J., Bence, A. E., Lindsley, D. H. 1974. Mare basalts from the Taurus-Littrow region of the moon. *Proc. Lunar Sci. Conf., 5th. Geochim. Cosmochim. Acta* 1: Suppl. 5, pp. 471–504

Philpotts, J. A. 1970. Redox estimation from a calculation of Eu^{2+} and Eu^{3+} concentrations in natural phases. *Earth Planet. Sci. Lett.* 9:257–68

Phinney, W. C., Warner, J. L., Simonds, C. H., Lofgren, G. E. 1972. Classification and distribution of rock types at Spur Crater. *The Apollo 15 Lunar Samples,* 149–53. Houston: Lunar Sci. Inst.

Pieters, C., Head, J. W., McCord, T. B., Adams, J. B., Zisk, S. 1975. Geological and geochemical units of Mare Humorum: Further definition using remote sensing and lunar sample information. *Lunar Science VI,* 637–39. Houston: Lunar Sci. Inst.

Prinz, M., Dowty, E., Keil, K. 1973a. A model for the origin of orange and green glasses and the filling of mare basins. *EOS. Trans. Am. Geophys. Union* 54: 605–6

Prinz, M., Dowty, E., Keil, K., Bunch, T. E. 1973b. Spinel troctolite and anorthosite in Apollo 16 samples. *Science* 179:74–76

Prinz, M., Dowty, E., Keil, K., Bunch, T. E. 1973c. Mineralogy, petrology and chemistry of lithic fragments from Luna 20 fines: origin of the cumulate ANT suite and its relationship to high-alumina and mare basalts. *Geochim. Cosmochim. Acta* 37:979–1006

Reid, A. M., Lofgren, G. E., Heiken, G. H., Brown, R. W., Moreland, G. 1973a. Apollo 17 orange glass, Apollo 15 green glass and Hawaiian lava fountain glass. *EOS. Trans. Am. Geophys. Union* 54: 606–7

Reid, A. M., Ridley, W. I., Donaldson, C., Brown, R. W. 1973b. Glass compositions in the orange and gray soils from Shorty Crater, Apollo 17. *EOS. Trans. Am. Geophys. Union* 54:607–8

Reid, A. M., Warner, J., Ridley, W. I., Johnston, D. A., Harmon, R. S., Jakes, P., Brown, R. W. 1972. The major element compositions of lunar rocks as inferred from glass compositions in the lunar soils. *Proc. Lunar Sci. Conf., 3rd. Geochim. Cosmochim. Acta* 1: Suppl. 3, pp. 363–78

Reid, A. M., Warner, J. L., Ridley, W. I., Brown, R. W. 1973c. Luna 20 soil: abundance and composition of phases in the 45-125 micron fraction. *Geochim. Cosmochim. Acta* 37:1011–1030

Rhodes, J. M., Hubbard, N. J. 1973. Chemistry, classification, and petrogenesis of Apollo 15 mare basalts. *Proc. Lunar Sci. Conf., 4th. Geochim. Cosmochim. Acta* 2: Suppl. 4, pp. 1127–48

Rhodes, J. M., Rodgers, K. V., Shih, C., Bansal, B. M., Nyquist, L. E., Wiesmann, H., Hubbard, N. J. 1974. The relationship between geology and soil chemistry at the Apollo 17 landing site. *Proc. Lunar Sci. Conf., 5th. Geochim. Cosmochim. Acta* 2: Suppl. 5, pp. 1097–1117

Ridley, W. I., Brett, R. 1973. Petrogenesis of basalt 70035: a multi-stage cooling history. *EOS. Trans. Am. Geophys. Union* 54:611–12

Ringwood, A. E. 1974. Petrogenesis of maria basalts and composition of lunar interior. *EOS. Trans. Am. Geophys. Union* 55:325 (Abstr.)

Ringwood, A. E., Essene, E. 1970. Petrogenesis of Apollo 11 basalts, internal constitution and origin of the moon. *Proc. Apollo 11 Lunar Sci. Conf. Geochim. Cosmochim. Acta:* Suppl. 1, pp. 769–99

Roedder, E., Weiblen, P. W. 1973. Origin of orange glass spherules in Apollo 17 sample 74220. *EOS. Trans. Am. Geophys. Union* 54:612–13

Rose, H. J., Cuttitta, F., Annell, C. S., Carron, M. K., Christian, R. P., Dwornik, E. J., Greenland, L. P., Ligon, D. T. 1972. Compositional data for twenty-one Fra Mauro lunar materials. *Proc. Lunar Sci. Conf., 3rd. Geochim. Cosmochim. Acta* 2: Suppl. 3, pp. 1215–29

Rose, H. J., Cuttitta, F., Berman, S., Carron, M. K., Christian, R. P., Dwornik, E. J., Greenland, L. P., Ligon, D. T. 1973. Compositional data for twenty-two Apollo 16 samples. *Proc. Lunar Sci. Conf., 4th. Geochim. Cosmochim. Acta* 2: Suppl. 4, pp. 1149–58

Ross, M., Bence, A. E., Dwornik, E. J., Clark, J. R., Papike, J. J. 1970. Mineralogy of the lunar pyroxenes, augite and pigeonite. *Proc. Apollo 11 Lunar Sci. Conf. Geochim. Cosmochim. Acta* 1: Suppl. 1, pp. 839–48

Shih, C., Wiesmann, H., Haskin, L. A. 1975. On the origin of high-Ti mare basalts. *Lunar Science VI*, 735–37. Houston: Lunar Sci. Inst.

Simonds, C. H., Phinney, W. C., Warner, J. L. 1974. Petrography and classification of Apollo 17 non-mare rocks with emphasis on samples from the Station 6 boulder. *Proc. Lunar Sci. Conf., 5th. Geochim. Cosmochim. Acta* 1: Suppl. 5, pp. 337–53

Singer, S. F., Bandermann, L. W. 1970. Where was the moon formed? *Science* 170:438–39

Smith, J. V. 1971. Minor elements in Apollo 11 and Apollo 12 olivine and plagioclase. *Proc. Lunar Sci. Conf., 2nd.* 1: 143–50. Cambridge: M.I.T. Press

Smith, J. V., Anderson, A. T., Newton, R. C., Olsen, E. J., Wyllie, P. J., Crewe, A. V., Isaacson, M. S., Johnson, D. 1970. Petrologic history of the moon inferred from petrography, mineralogy, and petrogenesis of Apollo 11 rocks. *Proc. Apollo 11 Lunar Sci. Conf. Geochim. Cosmochim. Acta* 1: Suppl. 1, pp. 897–925

Storey, W. C. 1974. Anomalous lunar plagioclase composition and the composition of coexisting melt. *Nature* 251: 403–5

Taylor, G. J. 1972. The composition of the lunar highlands: evidence from modal and normative plagioclase contents in anorthositic lithic fragments and glass. *Earth Planet. Sci. Lett.* 16: 263–68

Taylor, S. R. 1975. Lunar sciences, a post-Apollo view. Oxford: Pergamon

Taylor, S. R., Bence, A. E. 1975. Petrogenesis of the lunar highland crust. *Lunar Science VI,* 804–6. Houston: Lunar Sci. Inst.

Taylor, S. R., Gorton, M. P., Muir, P., Nance, W. B., Rudowski, R., Ware, N. 1973. Composition of the Descartes region, lunar highlands. *Geochim. Cosmochim. Acta* 37:2665–83

Taylor, S. R., Jakes, P. 1974. The geochemical evolution of the moon. *Proc. Lunar Sci. Conf., 5th. Geochim. Cosmochim. Acta* 2: Suppl. 5, pp. 1287–1305

Taylor, S. R., Kaye, M., Muir, P., Nance, W., Rudowski, R., Ware, N. 1972. Composition of the lunar uplands: chemistry of Apollo 14 samples from Fra Mauro. *Proc. Lunar Sci. Conf., 3rd. Geochim. Cosmochim. Acta* 2: Suppl. 3, pp. 1231–49

Tera, F., Eugster, O., Burnett, D. S., Wasserburg, G. J. 1970. Comparative study of Li, Na, K, Rb, Cs, Ca, Sr and Ba abundances in achondrites and in Apollo 11 lunar samples. *Proc. Apollo 11 Lunar Sci. Conf. Geochim. Cosmochim. Acta* 2: Suppl. 1, pp. 1637–57

Tera, F., Wasserburg, G. J. 1974. U-Th-Pb systematics on lunar rocks and inferences about lunar evolution and the age of the moon. *Proc. Lunar Sci. Conf., 5th. Geochim. Cosmochim. Acta* 2: Suppl. 5, pp. 1571–99

Toksoz, M. N. 1974. Geophysical data and the interior of the moon. *Ann. Rev. Earth Planet. Sci.* 2:151–77

Walker, D., Longhi, J., Hays, J. F. 1972. Experimental petrology and origin of Fra Mauro rocks and soil. *Proc. Lunar Sci. Conf., 4th. Geochim. Cosmochim. Acta* 1: Suppl. 4, pp. 797–817

Warner, J. L. 1971. Lunar crystalline rocks: petrology and geology. *Proc. Lunar Sci. Conf., 2nd.* 1: 469–80. Cambridge: M.I.T. Press

Warner, J. L. 1972. Metamorphism of Apollo 14 breccias. *Proc. Lunar Sci. Conf., 3rd. Geochim. Cosmochim. Acta* 1: Suppl. 3, pp. 623–43

Warner, J. L., Simonds, C. H., Phinney, W. C. 1973. Apollo 16 rocks: classification and petrogenetic model. *Proc. Lunar Sci. Conf., 4th. Geochim. Cosmochim. Acta* 1: Suppl. 4, pp. 481–504

Watkins, N. D., Haggerty, S. E. 1967. Primary oxidation variation and petrogenesis in a single lava. *Contrib. Mineral. Petrol.* 15:251–71

Weill, D. F., McCallum, I. S., Bottinga, Y., Drake, M. J., McKay, G. A. 1970. Mineralogy and petrology of some Apollo 11 igneous rocks. *Proc. Apollo 11 Lunar Sci. Conf. Geochim. Cosmochim. Acta* 1: Suppl. 1, pp. 937–55

Weill, D. F., McKay, G. A., Kridelbaugh, S. J., Grutzeck, M. 1974. Modelling the evolution of Sm and Eu abundances during lunar igneous differentiation. *Proc. Lunar Sci. Conf., 5th. Geochim. Cosmochim. Acta* 2: Suppl. 5, pp. 1337–52

Williams, R. J. 1972. The lithification and metamorphism of lunar breccias. *Earth Planet. Sci. Lett.* 16:250–56

Willis, J. P., Erlank, A. J., Gurney, J. J., Theil, R. H., Ahrens, L. H. 1972. Major, minor, and trace element data for some Apollo 11, 12, 14 and 15 samples. *Proc. Lunar Sci. Conf., 3rd. Geochim. Cosmochim. Acta* 2: Suppl. 3, pp. 1269–73

Wilshire, H. G., Stuart-Alexander, D. E., Jackson, E. D. 1973. Petrology and classification of the Apollo 16 samples. *Lunar Science IV,* 784–86. Houston: Lunar Sci. Inst.

Wood, J. A. 1972. Thermal history and early magmatism in the moon. *Icarus* 16:229–40

Wood, J. A., Dickey, J. S., Marvin, U. B., Powell, B. N. 1970. Lunar anorthosites and a geophysical model of the moon. *Proc. Apollo 11 Lunar Sci. Conf. Geochim. Cosmochim. Acta* 1: Suppl. 1, pp. 965–88

NUMERICAL MODELING OF LAKE CURRENTS

Wilbert Lick
Department of Earth Sciences, Case Western Reserve University, Cleveland, Ohio 44106

INTRODUCTION

In recent years, there has been increased interest in the numerical modeling of various aspects of aquatic systems. Examples of problems of interest are (*a*) the dispersion of contaminants, including the resuspension of sediments; (*b*) eutrophic modeling, for instance, population dynamics of planktonic organisms; and (*c*) modeling of strictly hydrodynamic problems, for example, the temperature distribution in thermal plumes from power plants or lake water levels during a storm. The major interest in this work has resulted from potential practical uses of the models. More recently, however, there has been increased interest in using these models for scientific purposes, in order to explore specific aquatic physical, chemical, and biological processes of a more fundamental nature.

A knowledge of the physical processes of dispersion is essential to understanding the above problems. A large amount of field, laboratory, and theoretical work has been and is being done to acquire this knowledge. Numerical modeling has been a basic and useful tool in the theoretical work. In the last few years, numerical models of lake currents have become quite detailed and realistic and have contributed greatly to our understanding of physical dispersion processes. The present article attempts to review some of the more recent and significant contributions on numerical modeling of lake currents.

When considering the currents in a lake, it is immediately evident that a large variation exists in the length and time scales describing these currents. Length scales vary from the size of the basin (up to several hundred kilometers) to the vertical dimensions of the microstructure of stratified flows (as little as a few centimeters), and time scales vary from many years to a few seconds. Although in principle the equations of fluid dynamics can describe motions that include all of these length and time scales, practical difficulties prohibit the use of the full equations of motion for problems involving large length and time scales. Therefore, considerable effort and ingenuity have been expended to approximate these equations, in order to obtain simpler equations and methods of solution.

The result is that many different numerical models of lake currents presently exist. The differences among these models can usually be related to the different length and time scales that the investigator believes to be significant for the specific problem. For example, if the details of the flow in the vertical are not thought to be pertinent, one may use a vertically averaged model, either steady state or time-dependent, based on whether the time variation is considered significant. In general, the choice of the numerical grid sizes in both space and time is determined by the physical length and time scales of the problem, because for accuracy the numerical grid sizes must be much smaller than the significant physical scales of the problem.

The present review, instead of completely surveying the numerous numerical models that have been developed, describes various representative models in somewhat more detail than would otherwise be possible. The emphasis is on a comparison of the various types of models and their uses, advantages, and limitations. Most of the discussion is limited to hydrodynamic models of the Great Lakes. These bodies of water are extremely important, have been extensively studied by aquatic scientists, and have also received most of the attention of lake modelers. The models developed for the Great Lakes can generally be applied with ease to other bodies of water, and have been in some cases.

In the past, most of the modeling has been concerned with the overall circulation in lakes. More recently, increased emphasis has been given to understanding and predicting the currents and related phenomena in the nearshore regions of lakes. These regions are important because (*a*) the nearshore regions are where contaminants are generally introduced, and therefore their concentrations and effects are generally greater here than in the offshore regions, and (*b*) the nearshore regions are of more particular interest to us for such uses as recreation, water supplies, and fishing. Both overall circulation and nearshore models are discussed in the present review.

In the first section, a brief discussion of some general characteristics of currents in lakes is followed by a presentation of the three-dimensional, time-dependent equations, boundary conditions, and parameters basic to all the models.

Vertically averaged models are the topic of the next section. These models result from an averaging of the three-dimensional equations over the depth. A bottom friction parameter, which is difficult to determine a priori, must also be introduced and related to the mean flow. The result of these manipulations is a reduced two-dimensional set of equations, which is comparatively easy to analyze and requires relatively little computer time, but does not give details of the vertical variation of the flow. This vertical detail is necessary for the complete understanding of the flow characteristics and for an accurate description of dependent problems, such as the dispersion of a contaminant. However, for some problems this detail is not necessary, and a two-dimensional model is adequate. The two-dimensional model is also useful for preliminary qualitative investigations of flows, especially parametric studies. The model is also the basis for the three-dimensional layered models discussed later.

Models of steady state currents are discussed next. The most useful of these is a constant-density model. Although limited by the assumptions of constant density and steady state, the model has been shown (by comparison with field observations) to give good results for periods when the winds were relatively constant. In this model, a major assumption is that the vertical eddy diffusivity is either constant or a relatively simple function of depth. This assumption, along with other, more minor ones, allows vertical, analytical integration of the equations of motion (without introducing a new parameter such as the bottom-stress relation used in the two-dimensional, vertically averaged models mentioned above). This procedure also reduces the governing equations to two dimensions. Further manipulations lead to a comparatively simple set of equations that can be solved relatively quickly on the computer but that still give vertical details of the flow. The steady state model can also be considered as a time-averaged model over periods when the winds are not constant. The fluctuations in the currents are then treated as random phenomena and are parameterized by means of an enhanced horizontal eddy diffusivity. Applications of these models are described and representative results are shown.

After the steady state models, time-dependent models are discussed. The more sophisticated of these models also include variable density effects. Several of these models have been extensively developed, applied, and at least partially verified by comparison with field observations. Although realistic, detailed, and capable of modeling quite complicated phenomena, these models are difficult to program and consume large amounts of computer time. Their very complexity makes it sometimes difficult to interpret the results of the calculations. Representative applications of these models to overall lake circulations and to thermal plumes from power plants are presented.

GENERAL CONSIDERATIONS IN MODELING OF CURRENTS

Characteristics of Lake Currents

The currents in the Great Lakes are primarily driven by the wind. Currents caused by through flows from incoming rivers moving to a single outflowing river are comparatively small except locally near the mouths of the rivers. In addition to these two causes of currents, which are present more or less continuously, temperature and hence density gradients cause currents and also modify existing currents. These density gradients may have a relatively large influence on currents and must be considered during the late spring, summer, and early fall, when stratification occurs and these gradients are large. The Great Lakes are large enough (horizontal dimensions of 100–300 km) so that Coriolis forces are important.

In shallow water, usually near shore, the effects of waves become important. The direct effect of waves on the water is to cause an oscillatory motion of the water particles. Because the motion is oscillatory, it is generally unimportant in causing transport (except in a narrow layer near the bottom in relatively shallow water), although it may be significant in causing a bottom shear stress and hence re-suspension of bottom sediments. The major effect of waves on transport is indirectly through longshore currents caused by the breaking of waves and the resulting dissipation of energy and momentum. Wave effects are not discussed here.

The currents due to the combined actions of winds, temperature gradients, and through flows and as modified by the basin geometry are quite complex and involve many different length and time scales. A thorough description of the dynamics of a lake is not given here, but some of the more significant motions and the associated length and time scales are briefly discussed because of the relevance of these matters to numerical modeling. An excellent and more thorough discussion of the significant physical processes occurring in lakes is given by Boyce (1974).

It is convenient to separate the currents into quasi-steady motions and time-dependent motions, although in practice this is difficult to do. Steady state currents are discussed first. When the winds act on the surface of a deep, constant-density body of water, a circulation is set up which in the steady state consists approximately of (*a*) top and bottom boundary layers (Ekman layers), in which vertical turbulent mixing is important; (*b*) horizontal boundary layers near shore, in which horizontal turbulent mixing is important; and (*c*) a geostrophic, inviscid core. In shallow lakes or in shallow, nearshore areas of deep lakes, this description is no longer valid. In these shallow regions, the Ekman layers are merged, and vertical turbulent diffusion is important throughout the water column.

Throughout the oceans, an Ekman layer is typically on the order of 100 m thick and hence is quite small compared to the depth of the ocean. In most of the Great Lakes, where the thickness of an Ekman layer d is on the order of 20–40 m, the depth of the lake h is generally several times as great as d. In Lake Erie, the average depth of the lake is only about 20 m and is about the same as the thickness of an Ekman layer. The thickness of the boundary layers near shore in which horizontal mixing is important is generally on the order of a few kilometers or less, and is therefore quite small by comparison with the horizontal dimensions L of a lake.

Heating and the subsequent stratification introduce additional phenomena. During the summer, a thermocline or region of large vertical temperature gradients generally develops. The depth of this thermocline is generally 20–100 m, whereas its thickness may be as little as a few meters. Again because of its shallow depth, Lake Erie is atypical in that the thermocline is quite often very near the bottom, sometimes only a few meters from the bottom.

Time dependence introduces even more complexities. Consider first the constant density flow in a lake that is initially still but at time zero is acted on by a uniform and constant wind stress. The time t_s to reach a new steady state, which, of course, is determined by this wind stress, depends on the parameter $\beta = gh/L^2f^2$ (Haq & Lick 1975), where $f = 10^{-4}$ sec^{-1} and is the Coriolis parameter. For the Great Lakes, this parameter can be assumed to be large, and as a first approximation the result for t_s is independent of β. For a shallow lake where $h/d \ll 1$ or $0(1)$, t_s is then of order h^2/A_v, where A_v is the vertical eddy viscosity. That is, t_s is a viscous diffusion time. For a deep lake where $h/d \gg 1$, the time t_s is of order $2\pi h/df$. In this case, t_s is the usual spin-up time for a rotating container with a rigid lid (Greenspan 1968). For the Great Lakes, the time to reach steady state varies from one or two days (Lake Erie) to probably one or two weeks (Lake Superior). This is to be compared with the average period of storm cycles in the Great Lakes region of two to seven days (Oort & Taylor 1969). On a smaller time scale are the surface oscillations, or seiches, of lakes. For shallow wide lakes, these have a period of oscillation close to the inertial period $2\pi/f$, which is about 17.5 hours. For deep narrow lakes, they have a period close to that of gravitational waves, i.e. $L/(gh)^{1/2}$.

If the lake is stratified, internal oscillations analogous to the free surface seiches may be present. The period of these waves is generally close to the inertial period. Also, the time to reach a steady state in stratified conditions may be considerably longer than that in nonstratified conditions because of the effect of the stratification on turbulent mixing.

These are just a few of the many different physical phenomena occurring in lakes and the length and time scales associated with them. No model is capable of describing all of these phenomena. Therefore the modeler must decide which phenomena are important to his problem and then choose or develop his model accordingly.

Basic Equations and Boundary Conditions

The basic equations used in the modeling of lake currents are the usual hydrodynamic equations for conservation of mass, momentum, and energy plus an equation of state. In sufficiently general form for most lake modeling, these equations are

$$\frac{\partial u}{\partial x} + \frac{\partial v}{\partial y} + \frac{\partial w}{\partial z} = 0 \tag{1}$$

$$\frac{\partial u}{\partial t} + \frac{\partial u^2}{\partial x} + \frac{\partial uv}{\partial y} + \frac{\partial uw}{\partial z} - fv = -\frac{1}{\rho_r}\frac{\partial p}{\partial x} + \frac{\partial}{\partial x}\left(A_H \frac{\partial u}{\partial x}\right) + \frac{\partial}{\partial y}\left(A_H \frac{\partial u}{\partial y}\right) + \frac{\partial}{\partial z}\left(A_v \frac{\partial u}{\partial z}\right) \tag{2}$$

$$\frac{\partial v}{\partial t} + \frac{\partial uv}{\partial x} + \frac{\partial v^2}{\partial y} + \frac{\partial vw}{\partial z} + fu = -\frac{1}{\rho_r}\frac{\partial p}{\partial y} + \frac{\partial}{\partial x}\left(A_H \frac{\partial v}{\partial x}\right) + \frac{\partial}{\partial y}\left(A_H \frac{\partial v}{\partial y}\right) + \frac{\partial}{\partial z}\left(A_v \frac{\partial v}{\partial z}\right) \tag{3}$$

$$\frac{\partial p}{\partial z} = -\rho g \tag{4}$$

$$\frac{\partial T}{\partial t} + \frac{\partial uT}{\partial x} + \frac{\partial vT}{\partial y} + \frac{\partial wT}{\partial z} = \frac{\partial}{\partial x}\left(K_H \frac{\partial T}{\partial x}\right) + \frac{\partial}{\partial y}\left(K_H \frac{\partial T}{\partial y}\right) + \frac{\partial}{\partial z}\left(K_v \frac{\partial T}{\partial z}\right) + S \tag{5}$$

$$\rho = \rho(T) \tag{6}$$

where u, v, and w are the fluid velocities in the x, y, and z directions, respectively; t is the time; f is the Coriolis parameter, which is assumed constant; p is the pressure; ρ is the density; ρ_r is the ambient or reference density; A_H is the horizontal eddy viscosity and A_v is the vertical eddy viscosity; K_H is the horizontal eddy conductivity and K_v is the vertical eddy conductivity; g is the acceleration due to gravity; T is the temperature; and S is a heat source term.

Several approximations are implicit in these equations. These are (*a*) the pressure is assumed to vary hydrostatically; (*b*) the Boussinesq approximation (which assumes that density variations are small and can be neglected compared to other terms except in the hydrostatic equation) is valid; and (*c*) eddy coefficients are used to account for turbulent diffusion effects in both the momentum and energy equations.

The appropriate boundary conditions are dependent on the particular problem to be solved. At the free surface, $z = \zeta$, usual conditions are: (*a*) the specification of a stress due to the wind,

$$\rho A_v \frac{\partial u}{\partial z} = \tau_x, \quad \rho A_v \frac{\partial v}{\partial z} = \tau_y \tag{7}$$

where τ_x, τ_y are the specified wind stresses in the x and y directions, respectively; (*b*) a kinematic condition on the free surface,

$$\frac{\partial \zeta}{\partial t} + u\frac{\partial \zeta}{\partial x} + v\frac{\partial \zeta}{\partial y} - w = 0; \tag{8}$$

(*c*) the pressure is continuous across the water-air interface and therefore the fluid pressure at the surface equals atmospheric pressure p_a,

$$p(x, y, \zeta, t) = p_a; \tag{9}$$

and (*d*) a specification of the heat flux at the surface,

$$q = -\rho K_v \frac{\partial T}{\partial z} = H(T - T_a) \tag{10}$$

where q is the energy flux, H is the surface heat transfer coefficient, and T_a is the air temperature.

At the bottom, the conditions are (*a*) those of no fluid motion or a specification of stress in terms of either integrated mass flux or bottom velocity and (*b*) either a specification of temperature or a specification of heat flux.

Variations on these boundary conditions and other boundary conditions are discussed in the following sections.

Eddy Coefficients

The numerical grid sizes for a problem are usually determined from considerations of the physical detail desired and computer limitations. Once the grid size is chosen, it is implicitly assumed that all physical processes smaller than this can either be neglected or approximately described by random processes, i.e. turbulent fluctuations. It can be shown that random turbulent fluctuations manifest themselves in an apparent increase in the viscous stresses of the basic flow. These additional stresses are known as Reynolds stresses. The total stress is the sum of the Reynolds stress and the usual molecular viscous stress, but because in turbulent flow the latter is comparatively small, it may be neglected in many cases. Analogous to the coefficients of molecular viscosity, an eddy viscosity coefficient can be introduced (as has been done in the equations above) so that the shear stress is proportional to a velocity gradient. Similarly, an eddy conductivity coefficient can be introduced so that the heat flux is proportional to a temperature gradient. In turbulent flow, these coefficients are not properties of the fluid as in laminar flow but depend on the flow itself, i.e. on the processes generating the turbulence. The determination of these turbulent eddy coefficients in terms of mean flow variables is a major problem in hydrodynamic modeling.

Although considerable work has been done on the theory of turbulence, in practice one must resort to experiments and semiempirical theories for realistic values of the eddy coefficients. Since the scale and intensity of the vertical and horizontal components of turbulence are generally quite different, it is convenient to consider these effects separately as has been done in the equations presented above. The vertical eddy viscosity A_v in general should vary throughout the lake. Some of the more important generating processes of this vertical turbulence and causes for its variation are (*a*) the direct action of the wind stress on the lake surface, (*b*) the presence of vertical shear in currents due to horizontal pressure gradients, (*c*) the presence of internal waves, and (*d*) the effect of bottom irregularities and friction on currents.

In addition, if the density of the water changes with depth, stability effects will change the intensity of the turbulence. The stability effect is dependent on the Richardson number, defined by

$$Ri = -\frac{g\dfrac{\partial \rho}{\partial z}}{\rho\left(\dfrac{\partial \bar{u}}{\partial z}\right)^2} \tag{11}$$

where $\bar{u}$ is the mean horizontal velocity. Various empirical values have been developed relating the variation of the eddy viscosity coefficient with the Richardson number [e.g. see Koh & Chang (1973) or Sundarem & Rehm (1970) for summaries]. A typical relation is that developed by Munk & Anderson (1948):

$$A_v = A_{vo}(1 + 10Ri)^{-1/2} \tag{12}$$

where A_{vo} is the value of A_v in a nonstratified flow. Typical values for A_{vo} in the Great Lakes are on the order of 1–50 cm^2/sec.

Horizontal viscosity coefficients are generally much greater than the vertical coefficients. It is found from experiments that the values of the horizontal viscosity coefficient increase with the scale l of the turbulent eddies, i.e.

$$A_H = a\varepsilon^{1/3} l^{4/3} \tag{13}$$

where a is a proportionality constant and ε is the rate of energy dissipation (Stommel 1949; also see Orlob 1959, Okubo 1971, Csanady 1973). Observations indicate values of 10^4 to 10^5 for A_H for the overall circulation in the Great Lakes (Hamblin 1971), with smaller values indicated in the nearshore regions.

In a nonstratified flow, it is believed that the eddy conductivity is approximately equal to the eddy viscosity. However, for a stratified flow, the mechanisms of turbulent transfer of momentum and heat are somewhat different, and this leads to different dependences of these coefficients on the Richardson number. For example, a semiempirical relation (Munk & Anderson 1948) similar to equation 12 suggests

$$K_v = K_{vo}(1+3.33Ri)^{-3/2} \tag{14}$$

where K_{vo} is the value of the vertical eddy conductivity in a nonstratified flow.

Wind Stress

In the modeling of the wind-driven circulation in a lake, it is necessary to know the horizontal shear stress imposed as a boundary condition at the surface of the lake. This stress is caused by the interaction of the turbulent air and water. The relation of this stress to the wind speed is very difficult to determine from theoretical considerations and its value is usually based on semiempirical formulas and on observations. A common relation assumed between these quantities is

$$\tau = \rho_a C_d W_a^{n-1} \mathbf{W}_a \tag{15}$$

where C_d is a drag coefficient, ρ_a is the density of the air, $\mathbf{W}_a$ is the wind velocity at 10 m above the water surface, and n is an empirically determined exponent not necessarily integer.

Wilson (1960) has analyzed data from many different sources and has given a best fit to the data. For $\mathbf{W}_a$ in units of cm/sec, ρ_a in units of gm/cm^3, and τ in units of dynes/cm^2, Wilson suggests a value of $n = 2$ and $C_d = 0.00237$ for strong winds and 0.00166 for light winds. From a comparison of field data and results of numerical models, Simons (1974) suggests values of $n = 2$ and $C_d = 0.003$.

An additional problem in determining the wind stress is that $\mathbf{W}_a$ in the above formula is the wind velocity at the specified location over the lake. Unfortunately, wind velocities are generally measured on shore and not at the over-the-lake location desired. Over-the-lake winds tend to be higher than the land values by as much as a factor of 1.5 (Gedney & Lick 1972). This phenomena is not well understood. Some investigation of this problem has been done (Simons 1974, Donelin, Elder & Hamblin 1974), but more is needed before this effect can be determined accurately.

Numerical Stability

In the numerical calculation of space- and time-dependent lake problems, one would like for efficiency to use as large space and time steps as are consistent with accuracy and physical detail desired. However, other restrictions on the allowable space and time steps are dictated by the stability of the calculation procedure. These restrictions of course depend on the particular numerical scheme used, but when present can usually be related to the physical space and time scales of the problem.

For example, consider an explicit, forward-time, central-space scheme. Simple theory indicates that limits on the time step Δt and space steps Δz in the vertical and Δx in the horizontal are approximately given by the following:

(*a*) $\Delta t < \Delta x/(gh)^{1/2}$, a restriction indicating that the numerical time step must be less than the time it takes a surface gravity wave [speed of $(gh)^{1/2}$] to travel the horizontal distance between two grid points Δx;

(*b*) $\Delta t < \Delta x/u$, i.e. the time step must be less than the time it takes a fluid particle to be convected horizontally a distance Δx;
(*c*) $\Delta t < (\Delta z)^2/2A_v$, i.e. the time step must be less than the time for diffusion between two grid points in the vertical;
(*d*) $\Delta t < (\Delta x)^2/2A_H$, i.e. the same argument as (*c*) but applied to horizontal diffusion;
(*e*) $\Delta t < 2\pi/f$, i.e. the time step must be less than the inertial period;
(*f*) $\Delta t < \Delta x/u_i$, where u_i is the speed of an internal wave—an argument similar to *a* above but for internal waves.

All of these restrictions may be eliminated by various numerical procedures. However, each numerical method has its own difficulties, and which procedure is most advantageous depends on the particular problem being studied.

VERTICALLY AVERAGED MODELS

Vertically averaged models have proved to be quite useful, especially in the prediction of lake levels during storms and of longtime contaminant transport. The model is relatively simple conceptually and requires little computer time compared to three-dimensional models. The model does not predict the vertical variation of the flow field but can give vertically averaged velocities and surface elevations correctly in many cases. An important limitation of the model is that it does not consider the effects of density variations.

The basic equations for this model are obtained by integrating the equations of motion, Equations 1–5, from the bottom $z = -h(x, y)$ to the surface $z = \zeta(x, y)$. The linear terms can then be expressed in terms of the integrated velocities, U and V, and the surface displacement ζ, where $U \equiv \int_{-h}^{\zeta} u dz \doteq \int_{-h}^{0} u dz$ and $V \equiv \int_{-h}^{\zeta} v\, dz \doteq \int_{-h}^{0} v dz$. However, nonlinear terms of the form $\int_{-h}^{\zeta} (\partial u^2/\partial x)\, dz$ are also obtained and can not be expressed in terms of U and V or their derivatives without a further assumption. A convenient assumption is that the velocity profiles are similar, i.e. $u(x, y, z) = UF(\sigma)/h$ and $v(x, y, z) = VF(\sigma)/h$ where $\sigma = z/h$ and $F(\sigma)$ is a shape factor normalized such that $\int_{-1}^{0} F(\sigma)\, d\sigma = 1$.

The resulting equations can be written as (Paul 1975):

$$\frac{\partial \zeta}{\partial t} + \frac{\partial U}{\partial x} + \frac{\partial V}{\partial y} = 0 \tag{16}$$

$$\frac{\partial U}{\partial t} + \beta\left[\frac{\partial(U^2/h)}{\partial x} + \frac{\partial(UV/h)}{\partial y}\right] - fV = -\frac{h}{\rho}\frac{\partial P_a}{\partial x} - \frac{gh\,\partial \zeta}{\partial x} + A_H\left(\frac{\partial^2 U}{\partial x^2} + \frac{\partial^2 U}{\partial y^2}\right) + \frac{1}{\rho}(\tau_x - \tau_x^B) \tag{17}$$

$$\frac{\partial V}{\partial t} + \beta\left[\frac{\partial(UV/h)}{\partial x} + \frac{\partial(V^2/h)}{\partial y}\right] + fU = -\frac{h}{\rho}\frac{\partial P_a}{\partial y} - \frac{gh\,\partial \zeta}{\partial y} + A_H\left(\frac{\partial^2 V}{\partial x^2} + \frac{\partial^2 V}{\partial y^2}\right) + \frac{1}{\rho}(\tau_y - \tau_y^B) \tag{18}$$

where τ_x, τ_y are the surface shear stresses due to the wind in the x, y directions, τ_x^B, τ_y^B are the bottom shear stresses in the x, y directions, and β is equal to $\int_{-1}^{0} F^2(\sigma)\, d\sigma$. A reasonable guess must be made for the functional relation $F(\sigma)$ in order to evaluate β. A constant or quadratic $F(\sigma)$ is generally assumed.

The wind shear stresses are related to the wind over the lake and are specified parameters. However, the bottom shear stresses are dependent on the flow and in vertically averaged models must be related somehow to U and V. It is usually assumed that $\tau_x^B = \rho\alpha U^n/h^m$ and $\tau_y^B = \rho\alpha V^n/h^m$, where α is a coefficient of proportionality and n and m are constants, usually integers. If sufficient field data are available, these parameters can be determined from the best fit between the calculated results and observational data.

Alternatively, theoretical arguments can be presented relating τ to U and V, at least for limiting cases. If one considers flow in a shallow basin and assumes similar velocity profiles as above, then one can show that $\tau_x^B = \rho\alpha U/h^2$ and $\tau_y^B = \rho\alpha V/h^2$, where $\alpha = A_v\, dF/d\sigma$. If one considers flow in a deep lake with an interior geostrophic flow of magnitude U and assumes the presence of a bottom Ekman layer, then one can show that $\tau_x^B \sim U/h$ and $\tau_y^B \sim U/h$. In addition, if one postulates that the eddy viscosity is proportional to the wind stress and hence to the integrated velocities, then it follows that the bottom shear stress is proportional to the square of the integrated velocities. This is a common assumption.

The form of the above equations is such that the most obvious method of numerical integration is some sort of explicit method, of which a variety exists (Roache 1972). Numerous applications of the vertically averaged equations exist in which an explicit procedure has been used. For example, the flow in Lake Ontario has been studied by Rao & Murthy (1970), Simons (1971), and Paskausky (1971), and Lake Erie has been studied by Haq, Lick & Sheng (1975) and Simons (1976).

An alternative numerical procedure is the use of the rigid-lid assumption (Berdahl 1968, Bryan 1969), i.e. $w(z = 0) = 0$. This approximation eliminates surface gravity waves and the small time scales associated with them, greatly increasing the maximum time step possible in the numerical computations. In this approximation, the high frequency surface variations associated with gravity waves are neglected, whereas the steady state results are calculated correctly and are the same as for the free-surface case. This procedure, together with the integrated equations above, has been used for the calculation of river discharges into a lake (Paul & Prahl 1971).

Another promising numerical method is the alternating-direction-implicit (ADI) method (Peaceman & Rachford 1955, Douglas 1955). The method is usually stable for much larger numerical time steps than are allowed by explicit methods. Considerable savings in computer time may be possible.

The vertically averaged equations and the ADI method have been used to solve for the time-dependent flow in Saginaw Bay (Allender 1975). In addition to the use of the ADI method, this analysis is interesting because of an important problem that it illustrates. Saginaw Bay is a large bay (30 km by 70 km) opening into Lake Huron. The average depth of the bay is only 9 m. Surface wind stresses and forcing of the bay waters by seiches in Lake Huron are the main driving mechanisms of the currents in the bay. In the Allender analysis, seiche activity in Lake Huron was simulated by specifying a time-dependent surface elevation along the open boundary between Saginaw Bay and Lake Huron. It is known (Reid & Bodine 1968, Wurtelle, Paegle & Sielecki 1971) that specifying the flow or surface elevation at an open boundary will partially block the flow at the boundary and reflected waves will result. Less restrictive conditions at an open boundary have been investigated by Chen & Miyakoda (1974) and satisfactory results have been obtained by using numerical smoothing techniques near the boundary.

An alternative approach to specifying conditions at the open boundary is to allow full interaction between the solutions in the region being investigated and the surrounding region. Using a fine grid size in the surrounding region usually requires considerable computer capacity so that a coarser grid is usually used there. However, it is known (Matsuno 1966, Browning, Kreiss & Oliger 1973) that wave motions in two unequal meshes have different phase speeds because of the truncation error. As a result, numerical difficulties may develop in this method also.

As indicated previously, a vertically averaged model does not give details of the vertical variation of the flow. In particular it should be noted that the maximum horizontal velocity can be many times as great as the average horizontal velocity. Indeed, for a constant depth basin with a constant wind, the vertically averaged, steady state velocity is zero everywhere, whereas in general the actual steady state velocity at any depth is nonzero and proportional to the wind velocity. Because of this, contaminants may be transported to much greater distances and in different directions than indicated by the mean flow (Sheng & Lick 1975). To include the approximate effect of vertical velocity gradients in the vertically averaged models, one must use an effective eddy coefficient greater than the eddy coefficient used in a three-dimensional model. This procedure is valid when the subgrid scale convection can be approximately treated as random, as implied by the use of an eddy coefficient (Csanady 1973, Galloway & Vakil 1975).

STEADY STATE MODELS

Although vertically averaged models are simple and quite useful, vertical details of the flow are often needed and therefore three-dimensional models are required. Time-dependent effects can be very important, of course, and three-dimensional, time-dependent models are discussed in the following section. However, there are periods when currents in lakes are essentially steady or can be treated in a quasi-steady or time-averaged manner, and this makes a steady state analysis applicable.

In much of the modeling of steady state currents, variations of Welander's shallow lake model (Welander 1957) are used. In this model, in addition to the steady state assumption, it is assumed that the density is constant and the nonlinear convection and horizontal diffusion terms can be neglected. The validity of these latter two approximations can be seen by looking at the momentum equations. The relative magnitudes of the nonlinear convection, horizontal diffusion, and vertical diffusion terms with respect to the Coriolis force are characterized by the Rossby number, $R_B = U_R/fL$, the horizontal Ekman number, $E_H = A_H/fL^2$, and the vertical Ekman number, $E_V = A_V/fh^2$, where U_R, h, and L are a characteristic velocity, a characteristic depth, and a characteristic horizontal length scale, respectively. In Lake Erie, typical values of these parameters are $U_R = 10$ cm/sec, $h = 20$ m, $L = 100$ km, $A_V = 20$ cm^2/sec, and $A_H = 10^5$ cm^2/sec. This gives $R_B = 0.01$, $E_H = 10^{-5}$, and $E_V = 0.05$. This indicates that, throughout most of the lake, nonlinear convection and horizontal diffusion can be neglected. Vertical diffusion is important especially in shallow areas and, as described before, in top and bottom boundary layers in deeper waters. Numerical calculations substantiate these arguments not only for Lake Erie but also for the other Great Lakes.

In the near shore, horizontal boundary layers are present in which the horizontal diffusion and nonlinear convection terms are important. These coastal boundary layers have been studied analytically by Janowitz (1970, 1972), who neglected the nonlinear convection terms, and numerically by Sheng & Lick (1975), who included nonlinear convection as well as horizontal diffusion. From these analyses, it can be shown that, for reasonable values of A_H of 10^4 to 10^5 cm^2/sec, these boundary layers are less than 1 km wide. Because of this, these narrow layers are generally neglected, at least in the models discussed in this section.

The above assumptions, along with the assumption that the vertical eddy viscosity is constant or a simple function of depth, allows one to integrate the equations of motion in the vertical direction analytically. This procedure reduces the governing equations to two dimensions. Introduction of an integrated stream function ψ, defined by $U = \partial\psi/\partial y$ and $V = -\partial\psi/\partial x$, reduces the equations to the single equation (Gedney & Lick 1970)

$$\nabla^2\psi + \gamma_1 \frac{\partial\psi}{\partial x} + \gamma_2 \frac{\partial\psi}{\partial y} = \gamma_3 \tag{19}$$

where γ_1, γ_2, and γ_3 are functions of the local depth and bottom slopes and γ_3 is also a function of the applied wind stress. The appropriate boundary condition is that of no normal flux at the shore, or ψ is constant along shore. Once this equation has been solved for ψ, all three components of the velocity as continuous functions of depth, as well as the surface elevation, can be calculated.

The present model has been used to describe the overall circulation in Lake Erie (Gedney & Lick 1970, 1972) and Lake Ontario and the Rochester Embayment (Bonham-Carter, Thomas & Lockner 1973, Bonham-Carter & Thomas 1973). A few details of the Lake Erie calculation will be given to illustrate the calculation procedure, the results, and the advantages and limitations of the model.

The region of Lake Erie for which the above stream function equation must be solved is multiply connected because of the presence of relatively large islands in the western part of the lake (see Figure 1). The value of the stream function on the mainland shore is determined by the river inflows and outflows. The values of the stream function on the island boundaries are not known a priori. These values are determined by the condition that the surface elevation be continuous around each island, i.e. $\oint (\partial\zeta/\partial s)\,ds = 0$, where the integration path is around each island. In the calculations presented here, three islands were incorporated.

For purposes of the computation, Lake Erie was divided into two regions. One was a region approximately 80 km by 64 km, surrounding the islands, in which a 0.805 km (0.5 mile) square grid was used. The second region was composed of the remainder of Lake Erie, and in this region a 3.22 km (2 mile) square grid was used. A total of 5050 grid points was used. The 0.805 km grid size was found necessary in order to obtain consistent and accurate ζ line integrals around each island and also to accurately represent the island boundaries. A combination of successive over-relaxation by points and lines was used to solve the system of finite-difference equations. Numerical coupling between the solutions in the two regions was allowed by iterating between the two solutions.

For a constant eddy viscosity, numerical solutions for the stream function and velocities were obtained for a variety of wind directions and magnitudes. For the calculation presented here, the wind was assumed to be uniform over the entire lake surface with a magnitude of 10.1 m/sec and direction W 50° S. This uniform wind condition was found to be a valid approximation for the period for which the calculations and field observations were compared. A friction depth d of 27.4 m was used because it provided the best agreement between the numerical results and current-meter measurements. This value of d corresponds to an eddy viscosity A_V of 38.0 cm^2/sec. Calculations were also made for other wind magnitudes. For best agreement between the calculations and observations, it was found that A_V must be taken to increase as the wind magnitude increases, with either a linear or quadratic relationship [also see Simons (1974) for a similar conclusion]. The results presented

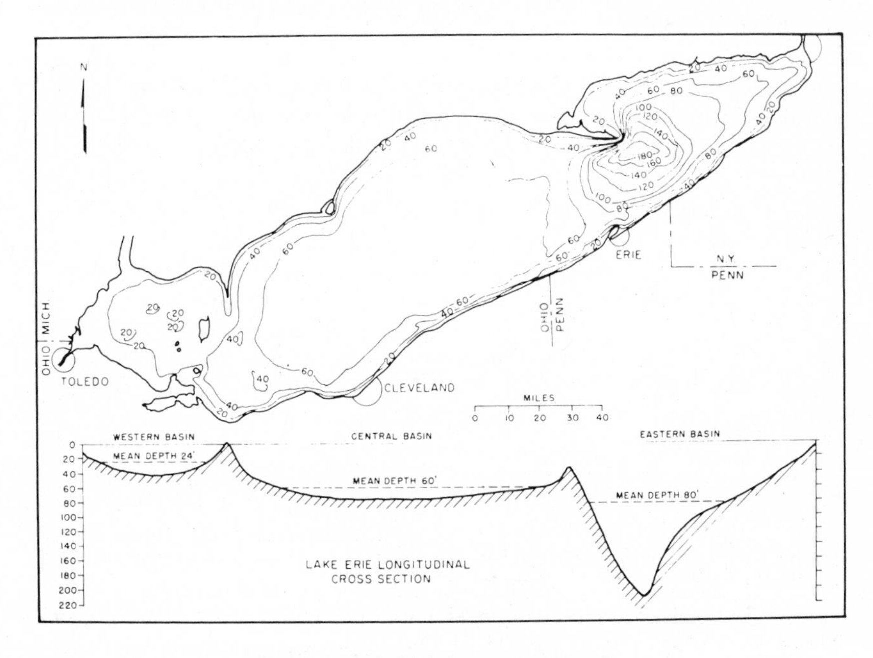

Figure 1 Lake Erie bottom topography

here include a Detroit River inflow of 5380 m^3/sec and an equal outflow via the Niagara River.

Lake velocity plots are shown in Figures 2 and 3. In these figures, the beginning of the arrow represents the actual location of the current represented by the arrow. The magnitude of the velocity can be determined from the velocity scale indicated on the figure. Figure 2 shows that a top surface mass flux is being transported toward the eastern and southern boundaries, primarily in the direction of the wind but deflected to the right by the Coriolis force. As shown in Figure 3, a subsurface current (driven by a pressure gradient due to the slope of the free surface) returns the surface mass flux in the opposite direction. In the central and eastern basins, surface currents are in general smaller in the center of the lake than near the shore. This effect is essentially caused by the relatively large subsurface return current down the center of the lake, which is opposite in direction to the surface current and subtracts from it.

The U.S. Environmental Protection Agency (EPA) established a system of automatic current-metering stations in Lake Erie in 1964. For days on which the observations and calculations were to be compared, this EPA data was vectorially averaged over 24 hours. The 24-hour average wind, as measured at shore stations and with its magnitude increased by 1.48, was used to determine the shear stress at the water surface. The 1.48 factor was determined by comparing shore data with over-the-lake wind data taken by the EPA.

On May 24, 1964, the resultant over-the-lake wind was determined to be 10.1 m/sec with direction W 50° S. The resultant winds for the prior two days had been within 20° of this direction and at somewhat less magnitude. The current-meter data for May 24, 1964, as measured at 10 m below the surface, is shown in Figure 3. Note that the positions of the measurements are different from those of the calculated currents. The agreement is markedly good in both magnitude and direction. The

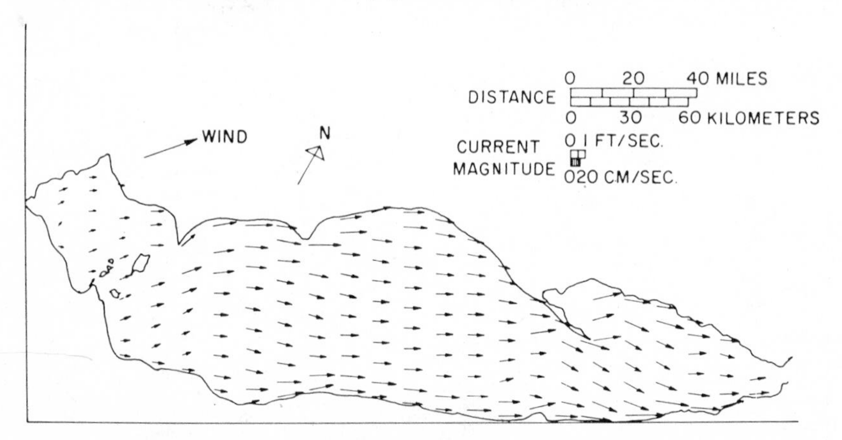

Figure 2 Horizontal velocities at a constant 0.4 m (1.5 ft) from surface. Wind direction, W 50 S; wind magnitude, 10.1 m/sec (22.7 mph); friction depth, 27.4 m (90.0 ft). Rivers: Detroit, Niagara.

discrepancy between the magnitudes of the measurements and calculations at point *A* at 10 m is believed to be a measurement error since this measurement became erratic at a later date. The magnitudes of the measurements in the region of point *B* at first appear to be considerably different from the calculated values. However, the agreement is believed to be satisfactory when one considers that the currents are changing rapidly with distance in the point *B* area. Also in Figure 3 are plotted the meter measurements for a W 43° S wind at a velocity of 8.6 m/sec, taken on October 25, 1969. Again the agreement is quite good.

The present model has also been applied to Lake Ontario (Bonham-Carter, Thomas & Lockner 1973, Bonham-Carter & Thomas 1973). Lake Ontario (maximum depth of 220 m) is much deeper than Lake Erie (maximum depth of 70 m) and its depth is generally much greater than the thickness of an Ekman layer. However, extensive nearshore regions, such as the Rochester Embayment, are shallow, and therefore shallow-lake theory is essential.

In addition, in order to predict the hydrodynamics and especially the dispersion of contaminants in the nearshore with adequate accuracy, a much smaller grid is needed in the nearshore than that usually used in overall lake circulation models. Because of this, calculations were made using a large grid (2.5 km) for most of the lake and a finer grid (0.625 km) for the Rochester Embayment, where more detail was needed. Coupling between the two regions was allowed as in the Lake Erie calculation above.

General comparison of the calculated results with field observations (D. J. Casey, unpublished report, 1965) were made for both Lake Ontario and the Rochester Embayment. General agreement was obtained. The calculated results were also compared with the results from a discrete four-layer model (Simons 1972). There was excellent agreement between the two.

The same basic model as described above has also been applied to a nearshore region of Lake Erie (Sheng & Lick 1975). The purpose was to analyze in more

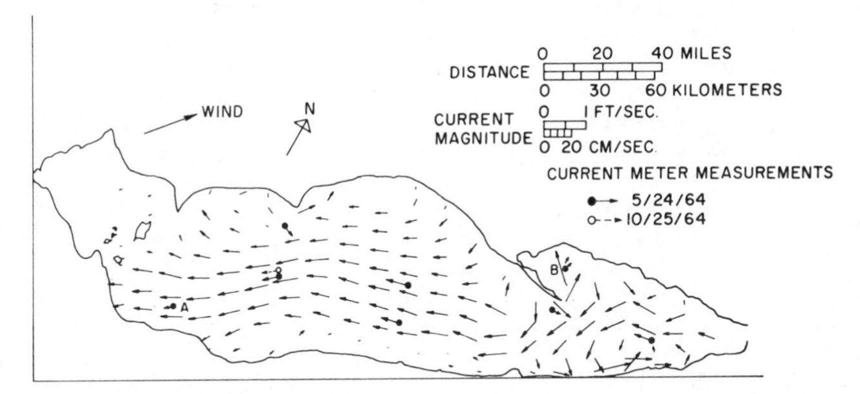

Figure 3 Horizontal velocities at a constant 9.9 m (32.8 ft) from surface. Wind direction, W 50 S; wind magnitude, 10.1 m/sec (22.7 mph); friction depth, 27.4 m (90.0 ft). Rivers: Detroit, Niagara.

detail the currents in the nearshore Cleveland area in the absence of and including a proposed Lake Erie jetport. Advocates of this airport have proposed that it be situated on a large manmade island in the lake approximately eight kilometers offshore, in waters approximately 15 m deep.

The steady state currents were calculated for various jetport configurations and for different wind velocities. In these calculations, a 0.4 km grid was used nearshore and a 3.2 km grid was used in the open lake. Two jetport configurations were studied in detail, an island approximately 4.8 km by 3.2 km, and this same island with an extension and 0.4 km causeway to shore.

An interesting result in the cases studied, which were for a southwesterly wind, was that the island configuration did not modify the flow appreciably, but the island-with-extension did. The reason for this can be understood as follows. From the hydrodynamic calculations, it can be shown that in the absence of the jetport (although there are strong currents near the surface at any horizontal location), there are generally also strong opposing currents near the bottom, with the result that the vertically integrated mass flux in most locations near the island is approximately zero. This is not true near shore, where the vertically integrated mass flux is moderately large. Because of this, the island does not appreciably block the flow; however, an island that extends to shore does block the flow and extensively modifies the nearshore flow field. The extent of the effect of the jetport depends on the jetport configuration and location and on the wind direction.

In general, the coupling between the nearshore region and the open lake was treated by an iteration process as described above. A procedure that is simpler but approximate is to first calculate the overall circulation in the lake, and then calculate the flow field in the nearshore region assuming that conditions at the boundary of the nearshore region remain fixed. No further iteration is required if the nearshore region is taken to be sufficiently large so that the effect of the jetport is not felt appreciably outside of the nearshore region. In the above calculations, the nearshore region was taken to be approximately 25.6 km by 22.4 km. This was sufficiently large so that the island configuration could be modeled without iteration (this was shown by comparison of solutions with and without iteration), but for the case of the island with extension to shore, iteration was necessary.

The above models have assumed a vertical eddy viscosity independent of depth. In reality, the eddy coefficients must vary over the depth depending on several different factors. A calculation using the same basic model as above but including a depth-dependent eddy viscosity has been made (Witten & Thomas 1975) and applied to Lake Ontario. An eddy viscosity of the form $A_V = A_{VO} \exp(-\alpha z)$ was assumed where A_{VO} and α are constants and z is the depth from the free surface. It was found that the vertically integrated mass flux was fairly insensitive to changes in A_{VO} and α, but that the three-dimensional current pattern was sensitive. Differences in the velocity field were especially apparent near shore.

The Great Lakes are all or partially ice-covered during the winter. A preliminary investigation of the steady state currents in a partially ice-covered lake has been made (Sheng & Lick 1973). The basic model used was the same as that described above, except that in the region where the lake was ice-covered the boundary condition at the ice-water interface was modified so as to allow no horizontal motions. The ice cover was assumed to have negligible thickness and negligible rigidity.

The basic equations can be vertically integrated as before and a similar equation for ψ, but with slightly different coefficients, is obtained. Results were obtained for Lake Erie with an ice-covered eastern or else western basin. It was found that horizontal velocities under the ice-covered portion of the lake but within approximately 45 km from the interface were comparable in magnitude to currents in the ice-free portion of the lake. This is because of the pressure gradient caused by the wind stress and the subsequent tilting of the free surface.

An interesting extension of.the above basic model is to a two-layered model of a stratified lake (Gedney, Lick & Molls 1972, 1973a,b). In this model, it was assumed that the flow consisted of two homogeneous layers, each with a different density and different eddy diffusivity and with the interface between the two layers being impermeable and of negligible thickness. The basic assumptions for the flow within each layer of fluid were identical to those used in the models described above, except for the boundary conditions at the interface. At the interface, the conditions connecting the two layers were assumed to be that the horizontal shear stress and the horizontal velocities were continuous across the interface and that there was no flow normal to the interface.

The latter assumption is particularly objectionable. However, the model does give an indication of the vertical structure of the velocity in a strongly stratified flow. Results have been calculated that approximately show the effects on a strongly stratified flow of variations in the eddy viscosity, the wind, and bottom topography.

TIME-DEPENDENT MODELS

Several three-dimensional, time-dependent models of currents in lakes have been developed. The more sophisticated of these include variable density effects. Even without variable density effects (which require the energy equation), the numerical computations are lengthy because numerical integration in three space dimensions and one time dimension is required. For efficiency in numerical computations, one would like to use large space and time steps. However, the space grid is usually chosen so that it is smaller than the significant physical length scales of the problem. Numerical stability considerations then restrict the maximum allowable time step. For the usual explicit, forward or central time, central space scheme, this limiting condition is $\Delta t < \Delta x/(gh)^{1/2}$, a condition that states that the numerical time step must be less than the time it takes a surface gravity wave to travel the distance Δx between two horizontal grid points.

Time-dependent computations of currents in lakes have been made using numerical schemes with this type of time limitation. Although correct, the procedure is not very efficient. Two alternate and useful procedures have been developed for modeling of three-dimensional, time-dependent lake currents. Both are much more efficient than the method indicated above and both have been used previously in atmospheric and oceanic modeling. The first method to be discussed is that developed for lakes by Simons (1971, 1972, 1974, 1975). In this method, the external (vertically averaged) and internal modes of the flow field are treated separately. The time step used in the calculation of the external mode is limited by the stability requirements indicated above. However, the calculation of the internal structure of the flow is not limited by these requirements and a much larger time step can be taken for these calculations. The second method (Paul & Lick 1973, 1974) also involves separating the calculations into external and internal modes. However, the external mode is treated by using a rigid-lid assumption, an assumption which effectively eliminates free-surface gravity waves. More details of the methods are given below.

Free-Surface Models

The most realistic and detailed applications of this type of model have been to Lake Ontario by Simons (1974, 1975). Extensive numerical computations have been made and the results compared in detail with observations for two physical events. For these calculations, the horizontal grid spacing was 5 km and the vertical resolution consisted of four layers, separated by horizontal levels at 10, 20, and 40 m below the surface, to approximate the location of current meters used in the field studies.

As mentioned above, the calculations for the external and internal modes were treated differently. This is done as follows. For each of the four layers, the equations of motion are written in conservative form similar to Equations 1–6. By summing these equations over the four layers (equivalent to vertically averaging over the depth), one obtains the equations for the external mode. By subtracting adjacent layer equations, one obtains the equations for the internal mode. In this manner, it is found that much larger time steps can be taken for the internal mode computations than those for the external mode computations. For Lake Ontario, a time step of 100 sec was used for the external mode and a time step of 15 min was used for the internal mode.

The model has been applied to two physical events. The first was a calculation of the time-dependent flow during tropical storm Agnes in June 1972, when Lake Ontario was only weakly stratified (Simons 1974). The second was a calculation of the flow in August 1972, when Lake Ontario was strongly stratified (Simons 1975).

In the first calculation, the flow was essentially at constant density. The model calculations were made starting on June 18, five days before the storm maximum and continuing until June 27, approximately three days after the storm subsided. Initial conditions for the calculation were determined from field measurements, as were the wind velocities during this period. The wind stress was taken proportional to the square of the wind velocity, and the bottom stress was taken proportional to the square of the water velocity in the lowest model layer. The vertical eddy viscosity was taken to be proportional to the wind stress and of the form

$$A_v = A_{vo} + K\frac{\tau}{\rho} \tag{20}$$

where A_{vo} and K are constants and A_{vo} is the eddy viscosity in the absence of wind. Good agreement with measurements was obtained by taking $A_{vo} = 25$ cm^2/sec and $K = 100$ sec.

Wind velocities were obtained from buoys stationed in the lake. The wind stresses at any particular grid point were then obtained by calculating the wind stress at the observation point and then interpolating to the grid point assuming a weighting of each stress inversely proportional to the square of the distance from the observation point to the grid point.

By this procedure, the surface elevations as well as the water velocities throughout the lake were calculated. It was found that for Lake Ontario, storm surges are quite small, and the wind-induced setup is of the same order as the effect from atmospheric pressure changes. In general, the agreement between the computed and observed data was good, with one exception. This disagreement was attributed to either incorrect observations or smaller-scale effects not resolved by the model. A characteristic result, also obtained in the steady state, constant-density model discussed in the previous section, was that nearshore currents were driven with the wind whereas offshore currents, especially at depth, flowed counter to the wind and were driven by the wind-induced pressure gradient. Coriolis forces were found to be significant. Especially notable at certain stations was the rotation of the current vector through 360° in about 17 hr, approximately the inertial period equal to $2\pi/f$.

During the second event to be monitored and modeled (in August 1972), Lake Ontario was strongly stratified. The purpose of the modeling effort was to obtain a hydrodynamic model sufficiently valid to serve as a basis for contaminant transport models. No effort was made to investigate problems such as thermocline formation or decay, problems requiring greater vertical resolution than was employed. The modeling was accomplished in conjunction with a field monitoring program.

The simulation experiment was for the period August 2 to August 15. Ship cruises were made on August 2 and August 15 to specify the initial temperature structure of the lake and to verify the model results, respectively. Water currents, surface elevations, wind velocities, and surface heat input were monitored continuously during this period.

Turbulent diffusion of momentum was treated in the same manner as for the constant density problem described above, with no dependence of the eddy viscosity on temperature gradient. Turbulent diffusion of heat was not included in the calculations but was determined by a comparison of the predicted and observed results. In general, the predictions of surface water levels and currents were quite good. Stratification was shown to have an appreciable effect on the temporal variations of currents. Phenomena such as internal waves were not simulated in a satisfactory manner because of the choice of model parameters. Temperature predictions showed general agreement with large-scale averaged observations. However, because of the neglect of turbulent heat transfer, the detailed temperature distribution was not given adequately.

From the differences in the calculated and observed temperature distributions, a lake-wide diffusive heat flux was calculated. From this, an effective eddy conductivity was determined. For this event, effective eddy conductivities were about

1 cm^2/sec. For an earlier period during storm Agnes, effective eddy conductivities were determined to be approximately 2–4 cm^2/sec, which indicates the effect of higher wind stresses on producing turbulence.

A free-surface model has also been developed for Lake Huron (Freeman, Hale & Danard 1972). The separation into external and internal modes was not made and so the computation was not very efficient. However, a stretched vertical coordinate proportional to the local depth was introduced. With this transformation, the same number of vertical grid points are present in the shallow as in the deeper parts of the basin. This ensures that in the shallow areas there is no loss of accuracy in the computations because of lack of vertical resolution, a significant factor in near-shore calculations. There is the additional advantage that in the new coordinate system the free and bottom surfaces are defined by constant values of the vertical coordinate, thus simplifying the definition of the location of the boundaries and the application of boundary conditions.

The vertically stretched coordinate is defined by $\sigma = Z/H$, where $Z = Z(x, y, z, t)$ is the position of the fluid element relative to the free surface and $H = H(x, y, t)$ is the depth also relative to the free surface. Because of this transformation, the form of the equations of motion is somewhat modified.

For Lake Huron (including Saginaw Bay), two calculations were made. These were for (*a*) constant depth but variable density, and (*b*) variable depth but constant density. No comparison with field data was made.

The vertically stretched coordinate system has also been used for a three-dimensional, time-dependent calculation of Lake Erie (Haq, Lick & Sheng 1975). No separation into external and internal modes was made. Analyses were made of (*a*) simple, rectangular basins for different values of Ekman number $E_v = A_v/fh^2$ and $\beta = gh/L^2f^2$, (*b*) Lake Erie under a uniform wind stress started impulsively, and (*c*) Lake Erie for conditions during Storm Agnes in June 1972, using measured wind conditions. In the latter case, good agreement between observations and calculated results were obtained for surface elevation, a quantity relatively insensitive to the details of the flow. No current measurements were made during the period.

The flow in the near shore of Lake Erie has also been studied (Sheng & Lick 1975) using the same model as above but with two different size grids, a small grid (1.6 km) near shore and a coarse grid (6.4 km) in the open lake. The solutions in the two grids were completely coupled dynamically such that mass and momentum were conserved between the two grids. In numerical experiments with simpler basins, the results obtained were consistent with calculations using a single size space grid and a single size time step over the entire domain. The results were smooth, showing no irregularities across the boundaries.

A free-surface model has also been applied to the calculation of thermal discharges from power plants (Waldrop & Farmer 1974a, b). No separation of the calculations into external and internal modes was made and no vertical coordinate stretching was used. Variable grid spacing in the horizontal was used to increase resolution near the discharge. Effects of inertia of the discharge and the ambient flow (both significant in their applications) and buoyancy were considered. Applications were made to the thermal discharge from power plants into a river and to the discharge from the Mississippi River.

Rigid-Lid Models

The rigid-lid model has been applied to the important nearshore problems of predicting the flow field and temperature distribution in river discharges and thermal plumes from power plants (Paul & Lick 1973, 1974, Lick 1975). Calculations have been made for simple river-lake flows, with parameters typical of the Cuyahoga River entering Lake Erie and of the discharge from the Point Beach power plant on Lake Michigan. Representative results of these latter calculations are discussed.

The direct application of the rigid-lid condition $[w(z=0)=0]$ is difficult to implement in a numerical solution of Equations 1–6. To get around this problem, an additional equation for the pressure containing the rigid-lid condition can be derived. This is accomplished by taking the divergence of the vertically integrated horizontal momentum equations and using the vertically integrated continuity and hydrostatic pressure equations. The resulting equation then has the general form

$$\frac{\partial}{\partial x}\left(\frac{h\,\partial P_s}{\partial x}\right)+\frac{\partial}{\partial y}\left(\frac{h\,\partial P_s}{\partial y}\right)=F(u,v,w,T) \tag{21}$$

The term P_s is the integration constant resulting from the vertical integration of the hydrostatic pressure equation and is also the surface pressure at the lid $(z=0)$. This surface pressure can be interpreted as a pressure due to a height of water above or below the surface $z=0$. In this way, surface displacements (neglecting the transient motion due to surface gravity waves) can be calculated.

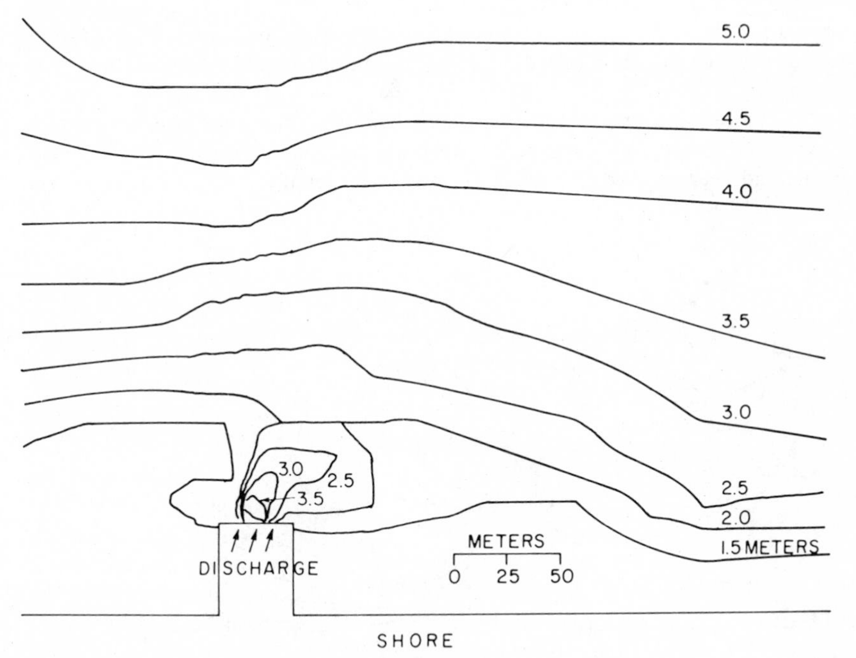

Figure 4 Bottom topography for the Point Beach power plant.

The calculation to be described includes realistic geometry, buoyancy effects, wind stresses, and cross-flows in the lake. A stretched vertical coordinate proportional to the depth was used. A variable horizontal grid was used for increased resolution near the discharge.

For the Point Beach area, the bottom topography is shown in Figure 4. The outfall extends into the lake and the discharge forms a 60° angle with the shore. Relevant parameters for the calculation were as follows: flow rate, 24.7 m^3/sec; outfall width, 10.8 m; outfall depth, 4.2 m; ambient lake temperature, 9.5°C; discharge temperature, 18°C; maximum discharge velocity, 0.9 m/sec.

The vertical coefficient was taken as dependent on the local vertical temperature gradient and assumed to be given by

$$A_v = \alpha - \beta \frac{\partial T}{\partial z} \tag{22}$$

where α and β are constants depending on local conditions. The constant α is equal to the vertical eddy diffusivity under vertically stable conditions and was taken to be 50 cm^2/sec and β was 200 cm^3/°C sec.

The boundary conditions were as follows. The inlet velocity profile was specified as a smoothed average of that measured at the outfall. The inlet temperature was taken as the constant value measured. A surface heat transfer proportional to the difference in temperature between the surface water and the air was assumed with the heat transfer coefficient determined from the work of Edinger & Geyer (1965). The stress acting on the water surface caused by the wind (measured) was calculated by the formulas developed by Wilson (1960). At the outer boundaries, the conditions used were such that the normal derivatives of the velocities and the temperatures were zero.

For the above boundary conditions and parameters, time-dependent calculations were made, starting with an initial guess of the flow and proceeding until a steady state had been reached. A calculation was made for the case when a cross-flow

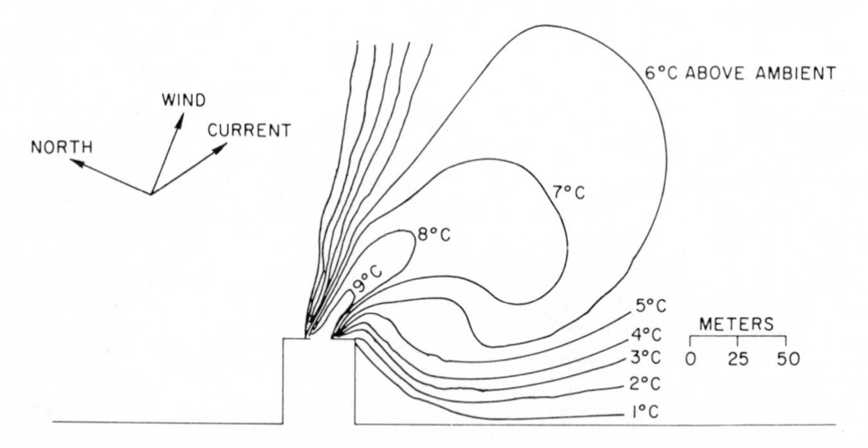

Figure 5 Surface temperature distribution with wind and cross-flow.

(current of 9.1 cm/sec at an angle of 125°) and a wind (approximately 5 m/sec at an angle of 270°) were present. In this case, the discharge is physically swept in the direction of the cross-flow. The temperatures (see Figure 5) are displaced in the direction of the cross-flow and toward the shore.

A comparison of the calculated results with field observations (Frigo, Frye & Tokar 1974) was made. Good agreement was obtained for both the temperature decay along the centerline and for the surface isotherm areas for various temperatures. Additional field data is available from which one can determine more details of the flow field. However, because of its turbulent nature [see Csanady (1973) for a general description of the turbulent diffusion and nature of plumes], the flow is highly variable both in space and time. Continuous field measurements over a sufficiently long period of time to average out these variations must be made before more general comparisons can be made between the above calculations and observations. This has not been done yet. Nonetheless, the above calculated results, although limited, seem more than adequate at this point and do instill confidence in the numerical model.

The rigid-lid model has also been applied to a calculation of the wind and thermally driven flow in Lake Erie [see Lick (1975) for preliminary results]. The purpose here was to investigate the thermocline formation and decay and the accompanying changes in the flow velocities. A vertically stretched coordinate was used as in the calculations above. A vertical eddy viscosity and conductivity dependent on temperature were used. As the dependence of the eddy coefficients on temperature was changed, major changes in the temperature structure occurred, indicating the importance of this parameter in determining the flow field. Numerical experiments with this model are continuing.

SUMMARY AND CONCLUSIONS

As one can see from the previous pages, many different numerical models of currents in lakes have been developed. The present review is mainly limited to models of the Great Lakes; even then, all the existing models have not been surveyed. Only a few representative models have been discussed in any detail.

For convenience of discussion, classification of the models has been made into the categories of vertically averaged models, three-dimensional steady state models, and three-dimensional time-dependent models. Each class of models has its advantages and limitations.

Hydrodynamic models are becoming quite detailed. Information on their use, sufficient to place confidence in their predictions is becoming available. In particular, these models are accurate enough to be used as bases for models of sediment resuspension and transport, phytoplankton and zooplankton growth, community succession, etc, and should be so used.

Although the models presented are sufficiently general (by themselves or with slight extensions) for almost all lake circulation problems, a great deal of further numerical experimentation using these models needs to be done. The purposes of these computations should be as follows:

(*a*) To understand the general characteristics of flows in lakes, especially in the near-

shore region (where contaminants are introduced) and under stratified conditions, e.g. upwelling and mass flux through the thermocline;
(*b*) To calculate details of specific flows and verify these by means of field observations, again especially in the nearshore region. A combination of field observations, numerical experiments, and laboratory work is required to determine more accurately the magnitude and functional dependence of the vertical and horizontal eddy coefficients, as well as the wind stress–wind velocity relation;
(*c*) To understand the numerical accuracy and characteristics of the various models and to further improve the numerical efficiency of these models.

The models discussed in this review do not consider the effects of waves, effects increasingly important as the depth of water decreases. To be able to predict adequately the dispersion of contaminants in the nearshore region, one must be able to model wave effects and also to combine these wave models with models of currents similar to those above.

Acknowledgments

This work was partially supported by the U.S. Environmental Protection Agency. Mr. William R. Richardson served as Grant Project Officer.

Literature Cited

Allender, J. H., 1975. *Numerical simulation of circulation and advection diffusion processes in Saginaw Bay, Michigan.* PhD thesis. Univ. Michigan, Ann Arbor

Berdahl, P. 1968. *Oceanic Rossby Waves, A Numerical Rigid-Lid Model.* ITD-4500, UC-34, Lawrence Radiation Lab., Univ. Calif., Livermore

Bonham-Carter, G., Thomas, J. H., Lockner, D. 1973. *A Numerical Model of Steady Wind-Driven Currents in Lake Ontario and the Rochester Embayment Based on Shallow-Lake Theory,* Rep. No 1, Univ. Rochester, NY

Bonham-Carter, G., Thomas, J. H. 1973. Numerical calculation of steady wind-driven currents in Lake Ontario and the Rochester Embayment. *Proc. Conf. Great Lakes Res., 16th,* pp. 640–62. Int. Assoc. Great Lakes Res.

Boyce, F. M. 1974. Some aspects of Great Lakes physics of importance to biological and chemical processes. *J. Fish. Res. Board Can.* 31:689–730

Browning, G., Kreiss, H., Oliger, J. 1973. Mesh refinement. *J. Math. Comp.* 27:29–39

Bryan, K. 1969. A numerical method for the study of the world ocean. *J. Comput. Phys.* 4:347–76

Chen, J. H., Miyakoda, K. 1974. A nested grid computation for the barotropic free surface atmosphere. *Mon. Weather Rev.* 102:181–90

Csanady, G. T. 1973. *Turbulent Diffusion in the Environment.* Boston: Reidel

Donelan, M. A., Elder, F. C., Hamblin, P. F. 1974. Wind stress from water set-up. *Proc. Conf. Great Lakes Res., 17th.* Int. Assoc. Great Lakes Res.

Douglas, J. 1955. On the numerical integration of $\partial^2 u/\partial x^2 + \partial^2 u/\partial y^2 = \partial u/\partial t$ by implicit methods. *J. Soc. Ind. Appl. Math* 3:42–65

Edinger, J. E., Geyer, J. C. 1965. Heat Exchange in the Environment. Publ. No. 65-902, Edison Electric Inst., New York

Freeman, N. G., Hale, A. M., Danard, M. B. 1972. A modified sigma equations approach to the numerical modeling of Great Lakes hydrodynamics. *J. Geophys. Res.* 77:1050–60

Frigo, A. A., Frye, D. E., Tokar, J. V. 1974. *Field Investigations of Heated Discharges from Nuclear Power Plants on Lake Michigan.* ANL/ES-32, Argonne Nat. Lab. Argonne, Illinois

Galloway, F. M., Vakil, S. J. 1975. *Criteria for the use of vertical averaging in Great Lakes dispersion models. Abstr. Conf. Great Lakes Res., 18th,* p. 23. Int. Assoc. Great Lakes Res.

Gedney, R. T., Lick, W. 1970. *Numerical calculations of the steady-state, wind-driven currents in Lake Erie. Proc. Conf., Great Lakes Res., 13th.* Int. Assoc. Great Lakes Res.

Gedney, R. T., Lick, W. 1972. Wind-driven currents in Lake Erie. *J. Geophys. Res.* 77:2714–23

Gedney, R. T., Lick, W., Molls, F. B. 1972. *Effect of Eddy Diffusivity on Wind-Driven Currents in a Two-Layer Stratified Lake. NASA TN D-6841.* Washington DC: NASA

Gedney, R. T., Lick, W., Molls, F. B. 1973a. *Effect of Bottom Topography, Eddy Diffusivity, and Wind Variation on Circulation in a Two-Layer Stratified Lake. NASA TN D-7235*

Gedney, R. T., Lick, W., Molls, F. B. 1973b. A Simplified Stratified Lake Model for Determining Effects of Wind Variation and Eddy Diffusivity, *Proc. Conf. Great Lakes Res. 16th,* pp. 710–22. Int. Assoc. Great Lakes Res.

Greenspan, H. P. 1968. *The Theory of Rotating Fluids.* London: Cambridge Univ. Press

Hamblin, P. F. 1971. An investigation of horizontal diffusion in Lake Ontario, *Proc. Conf. Great Lakes Res., 14th,* pp. 570–77. Int. Assoc. Great Lakes Res.

Haq, A., Lick, W. 1975. On the time-dependent flow in a lake. *J. Geophys. Res.* 80: 431–37

Haq, A., Lick, W., Sheng, Y. P. 1975. *The Time-Dependent Flow in Large Lakes with Application to Lake Erie.* Case Western Reserve Univ. Rep., Cleveland, Ohio

Janowitz, G. S. 1970. The coastal boundary layers of a lake when horizontal and vertical Ekman numbers are of different orders of magnitude. *Tellus* 22:585–96

Janowitz, G. S. 1972. The effect of finite vertical Ekman number on the coastal boundary layers of a lake. *Tellus* 24: 416–20

Koh, R. C. Y., Chang, Y. P. 1973. *Mathematical Model for Barged Ocean Disposal of Wastes,* EPA-660/2-73-029, Environ. Prot. Agency

Lick, W. 1975. *Numerical Models of Lake Currents.* Environ. Prot. Agency Rep.

Matsuno, T. 1966. Numerical integrations of the primitive equations by a simulated backward difference method. *J. Meteorol. Soc. Jpn.* 2:76–84

Munk, W. H., Anderson, E. R. 1948. Notes on the theory of the thermocline. *J. Mar. Res.* 1:276–95

Okubo, A. 1971. Oceanic diffusion diagrams. *Deep-Sea Res.* 18:789–802

Oort, A. H., Taylor, A. 1969. On the kinetic energy spectrum near the ground. *Mon. Weather Rev.* 97:623–36

Orlob, G. T. 1959. Eddy diffusion in homogeneous turbulence. *J. Hydrol. Div., Proc. ASCE, HY9,* pp. 75–101

Paskausky, D. F. 1971. Winter circulation in Lake Ontario. *Proc. Conf. Great Lakes Res., 14th,* pp. 593–606. Int. Assoc. Great Lakes Res.

Paul, J. F., Prahl, J. 1971. *Investigations of a Constant Temperature Rectangular Jet.* Case Western Reserve Univ. Rep., Cleveland, Ohio

Paul, J. F., Lick, W. 1973. *A Numerical Model for a Three-Dimensional, Variable-Density Jet.* Tech. Rep. Case Western Reserve Univ., Cleveland, Ohio

Paul, J. F., Lick, W. 1974. A Numerical Model for Thermal Plumes and River Discharges. *Proc. Conf. Great Lakes Res., 17th.,* pp. 445–55. Int. Assoc. Great Lakes Res.

Paul, J. F. 1975. A Note on the Vertically-Averaged Equations of Motion. Case Western Reserve Univ. Rep. Cleveland, Ohio.

Peaceman, D. W., Rachford, H. H. 1955. The numerical solution of parabolic and elliptic differential equations. *J. Soc. Ind. Appl. Math.* 3:28–41

Rao, D. B., Murthy, T. S. 1970. Calculations of the steady state wind-driven circulation in Lake Ontario. *Arch. Meteorol. Geophys. Bioklimatol. A19,* pp. 195–210

Reid, R. O., Bodine, B. R. 1968. Numerical model for storm surges in Galveston Bay. *J. Waterways Harbors, Div. Amer. Soc. Civil Eng.* 94:33–57

Roache, P. J. 1972. *Computational Fluid Dynamics.* Albuquerque, New Mexico: Hermosa Publ.

Sheng, Y. P., Lick, W. 1973. Wind-driven currents in a partially ice covered lake. *Proc. Conf. Great Lakes Res., 16th,* pp. 1001–8. Int. Assoc. Great Lakes Res.

Sheng, Y. P., Lick, W. 1975. The Wind-Driven Currents and Contaminant Dispersion in the Near-Shore of Large Lakes. Case Western Reserve Univ. Rep. Cleveland, Ohio.

Simons, T. J. 1971. Development of numerical models of Lake Ontario. *Proc. Conf. Great Lakes Res., 14th,* pp. 654–69. Int. Assoc. Great Lakes Res.

Simons, T. J. 1972. Development of numerical models of Lake Ontario. *Proc. Conf. Great Lakes Res., 15th,* pp. 655–72. Int. Assoc. Great Lakes Res.

Simons, T. J. 1974. Verification of numerical models of Lake Ontario. Part I, Circulation in spring and early summer. *J. Phys. Oceanogr.* 4:507–23

Simons, T. J. 1975. Verification of numerical models of Lake Ontario. II, Stratified circulations and temperature changes. *J. Phys. Oceanogr.* 5:98–110

Simons, T. J. 1976. Continuous dynamical calculations of water transports in Lake Erie in 1970. *J. Fish. Res. Board Can.*

Stommel, H. 1949. Horizontal diffusion due to oceanic turbulence. *J. Mar. Res.* 8(3): 199–225

Sundaram, T. R., Rehm, R. G. 1970. *Formation and Maintenance of Thermoclines in Stratified Lakes Including the Effects of Power-Plant Thermal Discharges.* AIAA Paper No. 70-238

Waldrop, W. R., Farmer, R. C. 1974a. Three-dimensional computation of buoyant plumes. *J. Geophys. Res.* 79:1269–76

Waldrop, W. R., Farmer, R. C. 1974b. *Thermal Plumes from Industrial Cooling Water. Proc. Heat Transfer and Fluid Mechanics Inst. 1974.* Stanford Calif.: Stanford Univ. Press

Welander, P. 1957. Wind action on a shallow sea. *Tellus* 9:47–52

Wilson, B. W. 1960. Note on surface wind stress over water at low and high speeds. *J. Geophys. Res.* 65:3377–82

Witten, A. J., Thomas, J. H. 1975. Calculation of steady wind-driven currents in Lake Ontario with a spatially variable eddy viscosity. *Abstr. Proc. Conf. Great Lakes Res., 18th.* Int. Assoc. Great Lakes Res.

Wurtelle, M. G., Paegle, J., Sielecki, A. 1971. The use of open boundary conditions with the storm-surge equations. *Mon. Weather Rev.* 99:537–44

FEATURES INDICATIVE OF PERMAFROST

Robert F. Black
Department of Geology, University of Connecticut, Storrs, Connecticut 06268

INTRODUCTION

Permafrost is synonymous with perennially frozen ground. It is a euphonic term coined during World War II to attract the attention of American policy makers to construction problems in the frozen northland (Muller 1945). The term was defined originally as ". . . a thickness of soil or other superficial deposit, or even of bed-rock, at a variable depth beneath the surface of the Earth in which a temperature below freezing has existed continually for a long time (from two to tens of thousands of years)" (Muller 1945, p. 3). The definition is based exclusively on temperature, irrespective of texture, degree of induration, water content, or lithologic character. Problems arise in the strict application of the definition to ground that is precisely at the freezing point of fresh water (rather than below) or to material containing salts which lower the freezing point below that of fresh water. Hence, a more restricted definition has been introduced: "The thermal condition in soil or rock of having temperatures below 0°C persist over at least two consecutive winters and the intervening summer" (Brown & Kupsch 1974). (Soil is used in both these definitions in the broad geologic or engineering sense of unconsolidated earth materials rather than in a pedologic sense.) Ice does not have to be present.

Permafrost extends over several million square kilometers in North America alone. With the recent advent of the Alaska pipeline and related construction problems, special priority has been given again, as it was during World War II (Black 1954, Terzaghi 1952), to the recognition, character, and effects of permafrost. A recent comprehensive review of permafrost is available (National Academy of Sciences 1973), along with several excellent books (Bird 1967, Brown 1970, Embleton & King 1968, Washburn 1973). Reference to these works or to the literature cited therein is axiomatic for most statements in this paper. Relatively few other references need be cited here.

A host of adjectives is applied to permafrost to connote special meanings, most of which are self-explanatory, such as contemporary, relic, continuous, discontinuous, sporadic, dry, ice-rich, marginal, submarine, and syngenetic. The permafrost table, or surface, is the upper boundary of permafrost. It is the lower limit of the active layer, the ground over permafrost that freezes and thaws each year or, more precisely, in which the temperature fluctuates above and below 0°C during each year. Because normally we don't see permafrost, except in recent exposures, we use surface characteristics and features in the active layer to infer the presence and character of the underlying permafrost. The annual freezing and thawing of the active layer promote frost heaving, frost slumping, frost riving, frost stirring, and other geomorphic processes because water is often trapped on the permafrost surface, where it generates much seasonal ice. These processes in turn markedly affect the plant and animal communities and the pedologic processes. The effects need to be seen to be appreciated fully.

A negative temperature by itself has little effect geomorphically in solid rock and dry soil and must be measured with instruments for it to be detected. A negative temperature in ground containing fresh water generates ice in many forms. These forms may have surface manifestations and may be diagnostic of permafrost. The annual temperature changes in the upper 10 m or so of permafrost induce contraction cracking, which leads to other forms that are commonly reflected at the surface as polygons.

Permafrost is a progeny of climate. Surface and ground characteristics are second-order effects that in most instances modulate the impact of climate on heat exchange at the earth's surface. Hence, a knowledge of regional climate and microclimate has highest priority in predicting permafrost. However, present-day permafrost in a number of places was first formed many thousands of years ago when the climate of these localities was different. It is relic and not necessarily in equilibrium with its environment. Temperature profiles and, rarely, other characteristics, such as

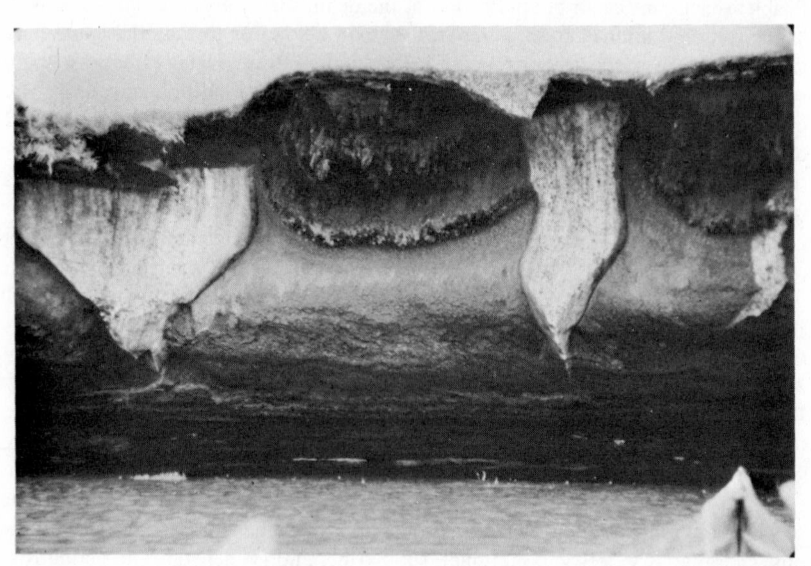

Figure 1 Ice wedges of three generations in a marine cut bank 5 m high in northern Alaska.

buried ice or relic surface forms, may indicate these former climates. Permafrost under the ocean may be inherited from a terrestrial origin by transgressing seas or may be forming today in a freshwater environment beneath the negative mean annual temperatures of saline waters above. Thus, permafrost is a complicated phenomenon, affecting nature and man in many ways.

The objectives of this summary are to provide some insight for the uninitiated reader into those surface forms and conditions that indicate the presence of permafrost today and something of its characteristics. Mention is made of those surface forms and deposits that indicate the former occurrence of permafrost in the present temperature regions of the United States. As this summary cannot be all-inclusive, the more diagnostic and striking forms are emphasized. Many have similar counterparts that are produced by other processes in areas that have never had permafrost climates. Hence, for recognition of ancient permafrost, few forms are diagnostic, and their origins are hard to prove. Mostly, indirect evidence and several converging lines of suggestive evidence, rather than proof positive, are all that remain of the record of the former periglacial regions. This is one of the most intriguing aspects of permafrost. The reconstruction of the distribution and character of former permafrost can lead to a better understanding of past climate changes and in turn to prediction of future climates.

FEATURES INDICATIVE OF PERMAFROST

Certain kinds of patterned ground, such as ice- and sand-wedge polygons, and some large mounds, such as pingos, grow within permafrost and are diagnostic of it. The problem is to ascertain whether the genetic terms are correctly applied, because other patterns and mounds not produced in permafrost may be superficially similar. Thermokarst, the collapse of the surface from thaw of underground ice, overlaps with the widely recognized formation of kettles from glacial ice. Creep and gelifluction features on permafrost are difficult to distinguish from creep and solifluction without permafrost, unless caught in the act. Some soil structures and distributional patterns of vegetation are also suggestive of permafrost. Each of these phenomena is discussed in turn.

Patterned Ground

Wedges of ice (Figure 1) and wedges of sand and rubble are two specific end members of components of one kind of patterned ground (Figures 2, 3). They represent the humid and arid polar climates respectively (Black 1974a,b, 1976, Mackay 1974). They are diagnostic of permafrost. Freezing and thawing, although commonly present, are not necessary for their formation. Thermal contraction cracking of permafrost on an annual basis, in polygons 5–30 m in plan diameter, opens vertical cracks in which ice or soil are added. The process repeated perennially produces wedge-shaped forms oriented vertically with apices downward. The wedges range from a few millimeters to 6 m wide and are 4–6 m high. Sand wedges require snow-free conditions during at least a part of each winter. Moisture is lost by sublimation to the air from contraction cracks in permafrost, and sand falls in. Sand wedges and composite wedges of ice and soil, including sand and rubble, are actively growing in the dry valley regions of Antarctica today; ice wedges occur on the humid coasts in Antarctica and in much of the Arctic. Depending upon local microclimatic conditions, either ice or soil may be added to contraction cracks during any part of one or more years, the presence or absence of snow cover changing from time to time. Moisture for ice wedges comes from the atmosphere in the form of snow, hoar frost, and spring meltwater that freezes in the voids. Wedges grow in width by a fraction of a millimeter to several millimeters per year.

Troughs over wedges that outline the polygons widen as wedges grow. Centers of polygons are flat, high-centered, or low-centered, in a continuum in which relief generally is several decimeters to one meter. Wedge widths and particularly surface processes determine the superficial appearance of the underlying wedges, which are separated from the surface by the active layer. Surface runoff, slope, vegetation, moisture content of the active layer, soil texture, and other factors are involved in the surface processes. In humid regions marshes and ponds fill the low centers of polygons (Figure 2).

The perennial addition of material to the wedges poses space problems, which can only be relieved by surfaceward flow and shear of the permafrost, adjacent to or within the wedges. The amount of movement is predicated on the amplitude of the annual temperature cycle in the upper part of permafrost and the coefficient of thermal expansion of the materials. Ice content is especially important, because ice

Figure 2 Ice-wedge polygons with double raised rims and medial trough over the wedges. Polygons are 5–20 m across. Arctic Coastal Plain of northern Alaska.

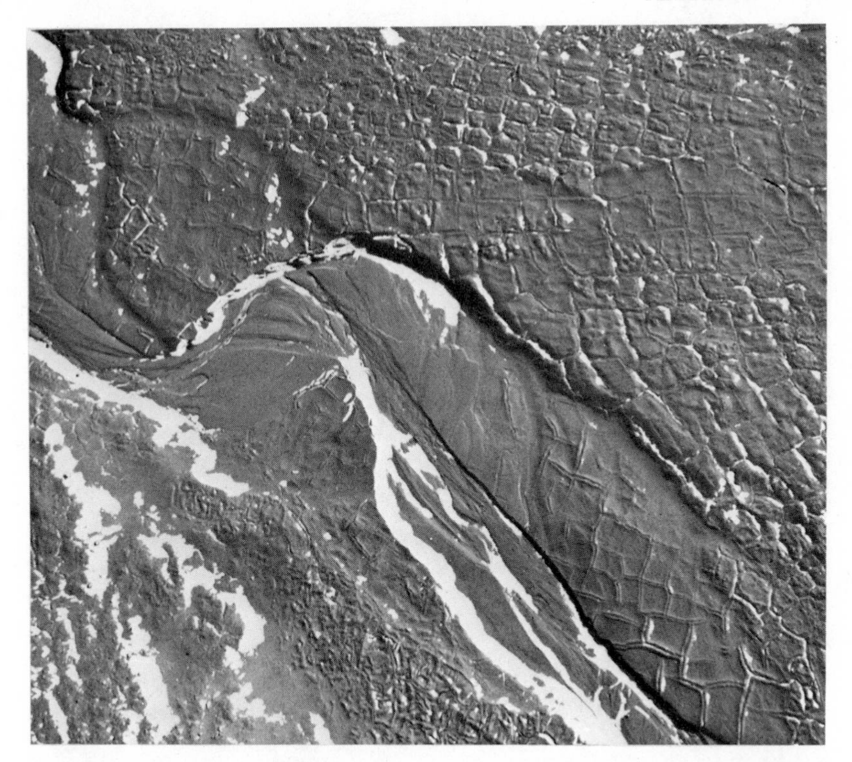

Figure 3 Sand-wedge polygons of several generations in Victoria Land, Antarctica.

has a thermal coefficient of expansion about five times that of most rock-forming minerals.

If much ice is involved, concurrent surface thaw, slump, and modification by geomorphic processes may disperse the material squeezed to the surface. No rims are produced adjacent to the wedges. In the tundra, material brought to the surface by wedge growth commonly has fresh minerals, high pH, and readily usable plant food, aiding vegetal growth adjacent to wedges. These areas show less change between the active layer and permafrost than the older centers of the polygons wherein horizontally stratified materials are being laid down. Thus, soil profiles are dependent on position with respect to an actively growing wedge. Through time, soil-forming processes may alter fabric, organic and mineral content, and pH in the active layer so that it becomes markedly different from the underlying permafrost (Figure 4).

As wedges grow, the ground surface often becomes more irregular (Figure 5). This leads to differences in vegetation, active layer thickness, snow cover, etc. A point may be reached at which the thermal regime of an initial wedge is changed—the upper part of the permafrost cooling more slowly in early winter than adjacent

Figure 4 Permafrost table represented by dark concentration of iron and manganese oxides below leached and frost stirred active layer. Barrow, Alaska.

Figure 5 Older tundra near Barrow, Alaska, showing surface morphology indicative of large (4–6 m wide) ice wedges under troughs and small (10–50 cm wide) ice wedges under narrow cracks across high areas of peat laid down in former low centers of polygons generated by the large wedges.

areas. The wedge is subsequently abandoned as a locus of thermal contraction cracking and a new crack is initiated elsewhere. Original polygons may be subdivided into halves, thirds, and quarters. Still more complexity is produced in second- or third-generation wedges. Peat laid down in original low-centered polygons may end up capping higher areas, because the tops of ice wedges are thawed at irregular intervals during unusually warm dry periods, or because material is squeezed toward the centers of polygons.

Because ice comes from the atmosphere, much or all the ground in the upper few meters of permafrost may be replaced by ice. The mineral and organic materials left may be a suspension in ice. The turf and soil of the active layer float over ground ice. Thus, the coefficient of thermal expansion of the ground may increase with time also. Where the ground responds like ice, a sudden change of only 4°C in the upper part of permafrost is sufficient to generate contraction cracks. Where ground temperatures change less rapidly or permafrost characteristics are less controlled by ice, differences of 8–10°C are required to initiate the cracks. Obviously rapid heat flow is not accomplished under thick snow cover or a thick active layer.

The environment of active ice and sand wedges is reasonably well understood. In general, only regions of continuous permafrost, where temperatures at the zero annual change level are below −5°C, have the thin active layer necessary to permit rapid autumn freeze-up and rapid major temperature drops in the upper permafrost in early winter. Microclimates in local areas of discontinuous permafrost may also induce thermal contraction cracking and growth of wedges, but growth is slower and more aperiodic. Sporadic permafrost rarely seems to permit growth of ice or sand wedges. Warm permafrost temperatures and very thick active layers that

Figure 6 Sorted polygons of coarse rubble with fines in the center. Colorado Front Range.

require most of the winter to refreeze to the permafrost effectively reduce heat loss from the ground.

Both the rapidity and magnitude of temperature change in the upper 5–10 m of permafrost and the ice content determine cracking potential, making correlations with regional climates difficult. The average annual air temperature differs by 1–6°C from the ground temperature at the zero annual change level. This is influenced by a host of meterologic factors in addition to the snow cover, vegetation, and conditions of the active layer. The mean annual air temperature of a region, though a poor index for correlation with patterned ground, is the one most widely used.

Other kinds of patterned ground are commonplace in areas of frost action (Goldthwait 1976), and some occur in desert and tropical regions as well. Sorted patterns produced within the active layer over permafrost are of many kinds. Some have coarse rubble surrounding cores of finer material (Figure 6); others, in soil lacking very large clasts, show active cells of frost stirring, surrounded by vegetated troughs (Figure 7). Generally in permafrost areas, the diameters of sorted polygons, circles, and cells in nets are proportional to the thickness of the active layer and to the size of the material being sorted. The size range of material is also important. Abundant water and numerous freeze-thaw cycles may produce striking forms in only a few years, but most are centuries or millenia in the forming.

Normally the larger and better-developed forms of sorted patterned ground are produced in the active layer over permafrost where trapping of water on the permafrost table leads to ideal growth conditions. However, they are not confined to the permafrost environment and as indicators of former permafrost they must be used in combination with other lines of evidence.

Figure 7 Frost stirred centers surrounded by vegetated troughs in Colorado Front Range.

Mounds

A perplexing variety of polygenetic mounds are often found in permafrost environments. Large conical mounds, called pingos from an abbreviated Eskimo term, are common in the continuous and discontinuous zones of permafrost in the Arctic (Figure 8) and are characterized by a core of massive ice. The term pingo is applied mostly to mounds that are 10–50 m high and 50–200 m in diameter, but smaller mounds of similar origin are included. Other terms are commonplace.

There are two main types of pingos, distinguished because of their environmental implications and different origins. Closed-system pingos are indigenous to thaw basins in continuous permafrost, where permafrost is aggrading. Water is trapped between the downward freezing surface and the permafrost below and on the sides. Pressure from growing ice crystals in the saturated trap bows the freezing surface, and water is injected to form the resulting massive ice core. However, water also produces segregated ice to help lift the surface directly. Such pingos cease growing when the basin is refrozen, which takes from a few years to several decades.

Open-system pingos originate in thin discontinuous or sporadic permafrost on or at the base of slopes where artesian pressure of water below the permafrost bows it and injects water to form the ice core. These pingos tend to be smaller than closed-system pingos and are markedly different in distribution and climatic inference. Open-system pingos cluster and interfere with each other, because new ones are produced aperiodically after older ones have ceased growing. The process can continue irregularly for centuries or millenia.

Palsas (anglicized plural of a Fennoscandian term for round or elongate mounds

Figure 8 Pingo in alas or thermokarst depression in eastern Siberia.

Figure 9 Palsas in northern British Columbia.

Figure 10 Segregated ice-soil-peat mound formed at base of the active layer, northern Alaska.

Figure 11 Plano-convex clear ice lens formed in the active layer, northern Alaska.

of peat, soil, and segregated ice) (Brown & Kupsch 1974) are common in the Subarctic. They may be as much as 10 m high and some tens of meters across, but most are smaller (Figure 9). For the most part they are isolated occurrences of sporadic permafrost in unfrozen bogs. Elevation of the palsa is attributed to segregated ice. In places, the individual palsas have merged to form peat plateaus and complexes. Long cold winters with little snow cover are required for their information. In a single bog it is typical to see palsas or parts of one palsa in different stages of growth and decay. Microhabitats seem capable of tipping the balance either way.

Mounds similar to small palsas are produced in the active layer above permafrost and may be related genetically by production of segregated ice (Figure 10). Others are also produced only in the active layer in wet places but are caused by bowing of the seasonally freezing surface and injection of water to form ice cores, as in pingos (Figure 11).

Thermokarst

Karst is a well-known term for the irregular topography produced by solution of underlying soluble rocks. Thermokarst is a term employed for somewhat similar topography in the permafrost regions wherein melting of ground ice promotes settlement and collapse of the surface. Thus, it is a thermal effect rather than a chemical one. Buried ice in mounds loses its protective cover of vegetation and soil quickly by gravity movements and, once exposed, melts rapidly during summer. Mineral matter moves to the perimeter of the mound, leaving a small rim and the center of the mound eventually turns into a pit or pond (Figure 12). The more

Figure 12 Pond 2 m deep resulting from the melting of an open-system pingo in northern Sweden.

Figure 13 Irregular topography produced by the thaw of ice wedges after stripping of the forest near Fairbanks, Alaska.

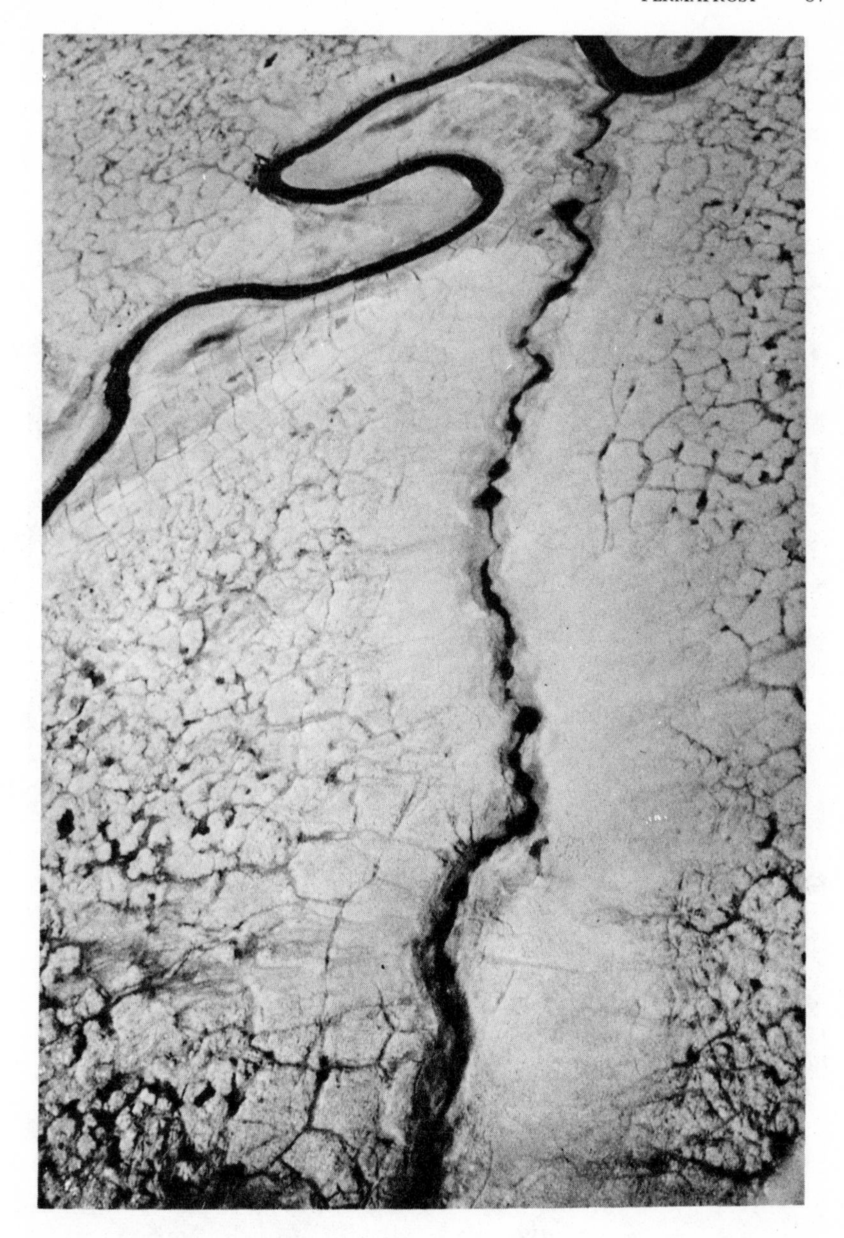

Figure 14 Beaded stream in which rounded ponds result from thawing at the intersections of ice wedges. Northern Alaska.

vulnerable position of ice-cored mounds makes them ephemeral features of permafrost landscapes, but some pingos are known to have survived many thousands of years. Small ice-cored mounds may disappear in a single summer. Deposits from inferred former pingos are identified from many parts of the world and tentatively from the United States (Flemal 1976).

Figure 15 Ice-wedge cast of sand that replaced a former ice wedge in west-central Wisconsin. Irregular dark bands are ground-water stains.

Disturbance of the vegetation over ice-wedge polygons and massive beds of segregation ice induces thaw. Thaw following ice wedges leaves a characteristic pattern of polygonal ground in which conical highs mark the centers of polygons (Figure 13). A drainage line produces a beaded stream (Figure 14). Fire is particularly effective in removing vegetation and may disrupt permafrost for decades. Flooding produces similar, but less drastic and shorter-lived effects. Thawing of individual ice wedges under favorable circumstances permits the replacement of the ice with sand or other overlying material to produce an ice-wedge cast (Figure 15). Such casts have been identified, often erroneously (Black 1976), in many parts of the temperate regions of the United States and elsewhere in the world.

Where ground contains much more ice than mineral matter, thawing enlarges initial pits into lakes (Figure 16) and large basins many kilometers across. Ice-wedge growth followed by deep thaw goes on cyclically in the Arctic, with an aperiodic time span of many thousands of years. In northern Alaska, this process of addition of ice to the ground and eventual thaw has generated tens of thousands of lakes apparently oriented by present prevailing winds (Figure 17). The large thaw basins of Siberia are the best agricultural areas despite underlying aggrading permafrost and pingos (Figure 8).

It is clear that thermokarst landscapes can coexist in the same general area as aggrading permafrost. The thermal balance of permafrost is easily upset by whims of nature or the works of man. The Alaska pipeline, designed to carry oil at a temperature of perhaps 75°C across Alaska, will have to be elevated where large quantities of ice exist in the ground in order to avoid melting the permafrost. Local

Figure 16 Large thermokarst lake in eastern Siberia that is enlarging 0.5 to 1 m per year.

Figure 17 Oriented thermokarst lakes of northern Alaska. The dark centers have water depths greater than 2 m. Wind streaks on the water surface are normal to the long axes of the lakes.

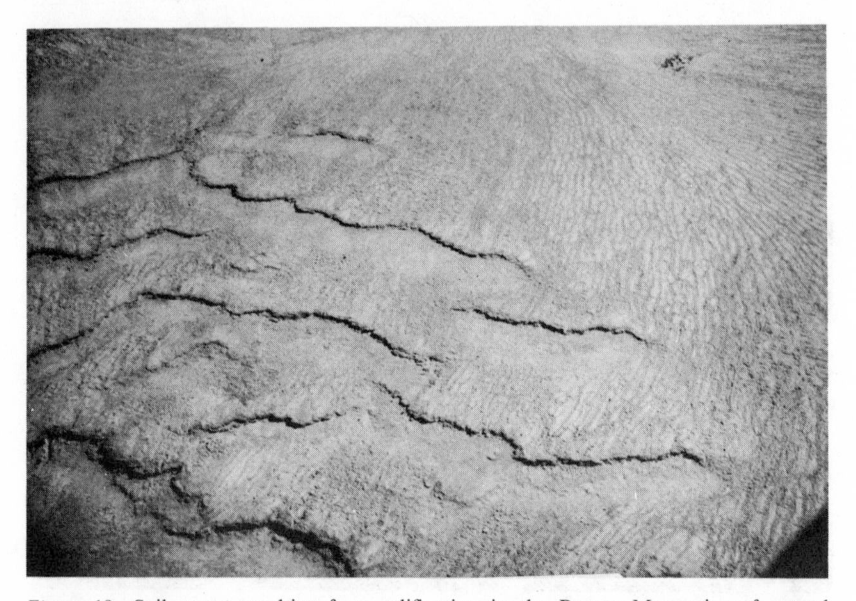

Figure 18 Soil waves resulting from gelifluction in the Beaver Mountains of central Alaska. Small elongate polygonal patterns result from interaction of vegetation, frost stirring, and gravity in the active layer. Fronts of the waves are 1–4 m high.

artificial refrigeration will be necessary to prevent thermokarst development and resulting pipe breakage. Only about half the length of the pipe can be buried where permafrost does not exist or where its destruction will not cause collapse.

Frost Creep and Gelifluction Features

Frost creep is the rachetlike, downslope movement of particles produced by freezing and thawing. These processes lift particles normal to slope and drop them vertically for a net downslope movement (Benedict 1976, Carson & Kirkby 1972, Washburn 1973). Wetting and drying may produce creep of particles on slopes to the same degree as frost creep (Black & Hamilton 1972). Creep as a general process may be exceedingly complicated. Gelifluction is the flow of soil down slopes where frozen ground is present. Gelifluction is a form of solifluction, soil flow, which is not confined to areas of frost action. The results of frost creep and gelifluction processes are difficult to distinguish even with detailed instrumentation of slope processes. Hence, we speak commonly of slope processes as a general part of mass wasting of the landscape. Morphometric terms are used to describe many features produced by the combined processes.

Where the active layer retains much water, downslope movements are rapid; a rate of several centimeters per year is common. Lobes and waves of soil may be seen (Figure 18). Sheets, benches, streams, terraces, and other terms are used to describe the form of the deposits. Vegetation plays an active role by providing a flexible canopy to retain fines. The flow of soil downslope turns the vegetal mat and part of the soil profile upside down. Sorted patterns of the active layer are strung out in stripes (Figure 19).

Figure 19 Stone stripes at the foot of a snow patch in the mountains of northern British Columbia. Permafrost is 50–75 cm below the surface.

With rare exceptions, the more striking and better-developed slope features produced by creep and soil flow are found in the permafrost regions. In many places the slope movements are so rapid that stream channels become choked with debris, which is brought to them from the hillsides faster than the flowing water can carry it away. A cycle of erosion characterizes the humid permafrost regions in which mass wasting is dominant.

Soils

Soil-forming processes in permafrost regions are not particularly different although slower than in more temperate regions. Chemical reactions go on, and especially where vegetation is present, may proceed rapidly. Embryonic podzols form. Organic materials accumulate in low-centered polygon ponds in the tundra regions. The drier polar deserts without much vegetation exhibit low temperature calcification and alkaline reactions or saline, as in Antarctica. Moisture and vegetation are apparently more critical than temperature. The permafrost table provides an impervious layer on which moisture and descending fine clastics and chemical substances can be trapped (Figure 4). Ice-wedge growth brings fresh material to the active layer in the tundra where organic acids leach the solubles readily. Thus, strong microgradients of pH both horizontally and vertically can be measured in some areas. The composition of the parent material plays a major pole.

Fabrics of the ground in the upper part of permafrost may reflect its genesis, for instance, in showing it to be a marine deposit. In the active layer inherited fabric is quickly lost by frost action which involves stirring, heaving, etc. In wet places, flow and injection of mud occur; in dry areas desiccation hollows develop. The permafrost table is a major transition zone or break in the pedogenic processes.

Vegetation

Cold ground temperatures and moisture are major controls on the types of vegetation that may grow in the Arctic. Shallow active layers permit only tundra; progressively deeper permafrost allows a progression through black spruce and white spruce, to birch and poplar. Such horizontal zonation of vegetation typically parallels meandering streams in interior Alaska and Canada. Partly it is a function of age of the surface, the older areas having the thinner active layers; partly it is a function of drainage and direct influence of the warm water of the stream.

Vertical zonation of vegetation follows the relief generated by ice-wedge polygons and mounds. Horizontal zonation accompanies sorted patterned ground. In most permafrost areas, vegetation is the primary indicator used to suggest the depth to permafrost and the texture and moisture of the active layer. Vegetation in effect has the same role in regulating ground temperature as human clothing does in regulating body temperature. A coniferous forest canopy intercepts more of the sun's heat than tundra species; they have different albedo (reflectivity).

FEATURES INDICATIVE OF FORMER PERMAFROST

Hamelin & Cook (1967) discussed periglacial phenomena under these general headings: ground ice, frost shattering (congelifraction), nivation, floating ice forms, fluvial features, eolian forms, patterned ground, solifluction, and congeliturbation (frost churning). They covered 51 subheadings in their classification, which cannot be reviewed here. Black (1969) reviewed the following general categories for their significance as indicators of former climate in the north-central United States: patterned ground, solifluction (gelifluction), frost riving and frost heaving, asymmetrical valleys, eolian phenomena, thermokarst, ice mounds, icings, and congeliturbation.

Of all those fossilized phenomena, perhaps the only truly diagnostic forms reflecting the temperature at time of formation are ice wedges. Even with them, it is difficult to ascertain the thermal coefficients of expansion of the ground as it existed at the time of wedge growth, because the coefficient is dependent on the amount of ice present. Furthermore, the widths of the wedges may be either a function of age or of growth rate per year. Nonetheless, by analogy with areas of existing ice wedges, those and other parameters can be approximated within certain limits, providing a better control on their former environment than the other features.

Pingos are confined to areas of permafrost, but fossilized pingos are difficult to identify positively in the present temperate regions. Geomorphic processes modify the small pit or pond too quickly, and the form alone is not diagnostic. Other lines of evidence must corroborate the interpretation. Most of those identified in the present temperate regions are of the open-system type suggestive of thin sporadic permafrost. Palsas and other small mounds are even harder to identify in fossil form.

Lakes, some oriented as the Carolina Bays of the Atlantic seaboard, are characteristic of the coastal plains of many parts of the world. They are not confined to the humid Arctic. Identification of their origins has been difficult, and differences of opinion are extant. Recognition of thermokarst pits and lakes as relic features of the landscape of thousands of years ago is frought with difficulty. Corroborative evidence is mandatory.

Gelifluction and frost-stirring phenomena are almost impossible to associate directly with permafrost, even though they are generally best developed there. Frost and gravity processes work effectively outside the permafrost borders. Hence, such phenomena can only be suggestive. Fossilization of soil profiles, from thousands of years ago, that still show evidence of a permafrost table is most difficult. I know of no incontrovertible example.

CONCLUSIONS

Many features in the cold polar regions and some in the high alpine regions result from frost action associated with permafrost. Many have surface manifestations that permit the experienced observer to infer their origins and something of the soil texture and ice content of the active layer and upper permafrost. Experience and a knowledge of microclimates are needed to make best use of such features.

Means of rapid and accurate mapping of the distribution of permafrost and its characteristics, including that beneath the ocean margins, are needed (National Academy of Sciences 1974). Unfortunately no remote sensing method yet devised can cover large areas rapidly and detect and delineate accurately all bodies of permafrost beneath the surface of the ground, let alone determine their characteristics. Detailed surface and underground studies must be done for on-site construction problems. Costs could be reduced considerably if best sites for pipelines, roads, and other structures could be selected by less painstaking methods. Aerial photographic interpretation techniques have been used widely and, in the hands of the experienced scientists, remain the most fruitful. Infrared thermal imagery and several multifrequency systems are being experimented with. Low-frequency airborne electromagnetic sensors, to map the electrical resistivity of frozen ground, appear promising for some areas. Multispectral bands from satellites record some surface characteristics. All these techniques require considerable "ground truth," and they are not yet panaceas.

Identification of fossilized or relic features in areas where permafrost no longer exists is even more difficult. The burden of proof must be placed on the investigator. No single feature or group of like features is proof positive, because of the many possible ways to produce similar counterparts. Several lines of suggestive evidence must lead to the same conclusion before the interpretation of former permafrost should be made with all its climatic implications.

Literature Cited

Benedict, J. B. 1976. Frost creep and gelifluction: a review. *Quat. Res.* 6: In press

Bird, J. B. 1967. *The Physiography of Arctic Canada.* Baltimore, Md: Johns Hopkins Press. 336 pp.

Black, R. F. 1954. Permafrost—a review. *Geol. Soc. Am. Bull.* 66:839–55

Black, R. F. 1969. Climatically significant fossil periglacial phenomena in north-central United States. *Biul. Peryglacjalny* 20:225–38

Black, R. F. 1974a. Ice-wedge polygons of northern Alaska. In *Glacial Geomorphology,* ed. D. R. Coates, Chap. 9, pp. 247–75. Binghamton, NY: Publ. Geomorphology, SUNY. 398 pp.

Black, R. F. 1974b. Cryomorphic processes and micro-relief features, Victoria Land, Antarctica. In *Research in Polar and Alpine Geomorphology,* ed. B. D. Fahey, R. D. Thompson, 11–24. Norwich, England Geo Abstracts. 206 pp.

Black, R. F. 1976. Periglacial features indicative of permafrost: ice and soil wedges. *Quat. Res.* 6: In press

Black, R. F., Hamilton, T. D. 1972. Mass-movement studies near Madison, Wisconsin. In *Quantitative Geomorphology: some aspects and applications,* ed. M. Morisawa, Chap. 5, pp. 121–79. Binghamton, NY: Dep. Geomorphology, SUNY. 315 pp.

Brown, R. J. E. 1970. *Permafrost in Canada.* Toronto: Univ. Toronto Press. 234 pp.

Brown, R. J. E., Kupsch, W. O. 1974. *Permafrost Terminology. Nat. Res. Counc. Can. Tech. Mem. 111.* 62 pp.

Carson, M. A., Kirkby, M. J. 1972. *Hillslope Form and Process.* London: Cambridge Univ. Press. 475 pp.

Embleton, C., King, C. A. M. 1968. *Glacial and Periglacial Geomorphology.* New York: St. Martin's Press. 608 pp.

Flemal, R. C. 1976. Pingos and pingo scars: their characteristics, distribution, and utility in reconstructing former permafrost environments. *Quat. Res.* 6: In press

Goldthwait, R. P. 1976. Frost-sorted patterned ground: a review. *Quat. Res.* 6: In press

Hamelin, L. E., Cook, F. A. 1967. *Le Périglaciaire par L'Image (Illustrated Glossary of Periglacial Phenomena).* Quebec: Presses Univ. Laval. 237 pp.

Mackay, J. R. 1974. Ice-wedge cracks, Garry Island, Northwest Territories. *Can. J. Earth Sci.* 11:1366–83

Muller, S. W. 1945. *Permafrost or Permanently Frozen Ground and Related Engineering Problems.* Washington, DC: Off. Chief Engineers, US Army. 231 pp. (Lithoprinted 1947. Ann Arbor, Mich: Edwards Bros.)

National Academy of Sciences 1973. *Permafrost, North American Contribution to Second International Conference.* Washington, DC: Nat. Acad. Sci. 783 pp.

National Academy of Sciences 1974. *Priorities for Basic Research on Permafrost.* Washington, DC: Nat. Acad. Sci. 54 pp. Reprinted 1975. *Quat. Res.* 5:125–50

Terzaghi, K. 1952. Permafrost. *J. Boston Soc. Civ. Eng.* 39:1–50

Washburn, A. L. 1973. *Periglacial Processes and Environments.* New York: St. Martin's Press. 318 pp.

MULTICOMPONENT ELECTROLYTE DIFFUSION

David E. Anderson and Donald L. Graf
Department of Geology, University of Illinois at Urbana-Champaign,
Urbana, Illinois 61801

INTRODUCTION

The interest of geochemists, petrologists, structural geologists, and geophysicists in diffusion is indicated by the growing publication rate for that subject in recent years. Diffusion, for example, is considered a significant process in sediment piles undergoing diagenesis (Fanning & Pilson 1974, Lasaga & Holland 1974, Michard et al 1974, Berner 1975, Graf & Anderson 1975), and in lacustrine and oceanic water masses (Ben-Yaakov 1972, Lerman & Jones 1973, Li & Gregory 1974; also D. E. Anderson and D. L. Graf, "Ionic diffusion in naturally-occurring electrolyte solutions," in preparation). Some of the phenomena considered for igneous and metamorphic rocks include exsolution, partitioning, and zoning of elements (Anderson & Buckley 1973, Buening & Buseck 1973, Fisher & Elliott 1974, Anderson 1975); the volume edited by Hofmann et al (1974) contains additional papers on these subjects. The roles that diffusion plays in pressure solution (Paterson 1973, Elliott 1973), geochronology (Giletti 1974), and stable isotope redistribution (Yund & Anderson 1974, Muehlenbachs & Kushiro 1974) have also been examined. The importance of diffusion in groundwater systems is less evident, because diffusion is typically lumped in a chemical dispersion term that also includes the mixing contributed by pore tortuosity and viscous flow (Bear 1972), and because practical considerations result in the greatest attention being paid to aquifers with high flow rates, in which diffusive transport may indeed be minor compared to convective transport.

In all these diverse geological environments, the point of interest is the simultaneous diffusion of two or more components in a partially ionized liquid or solid solution. The body of theory upon which these studies rely is scattered among the literatures of several fields, and we believe that reviewing this area will be of greater value than expanding to a full coverage the limited sampling of publications given above.

The treatment of diffusion in systems of two components grew independently of that for multicomponent diffusion, which involves three or more components. In retrospect, however, there is a great advantage in treating binary diffusion as a special case of multicomponent diffusion. As Cooper (1974) has pointed out, the theoretical problems of multicomponent diffusion are neither more nor less complicated than those of binary diffusion.

The analytical scheme for diffusion derived by de Groot & Mazur (1962), although somewhat abstract, is by far the best available. When combined with the methods of Miller (1966, 1967a, 1967b) for imposing electrical neutrality, it provides a complete foundation for the kinetic and macroscopic treatment of diffusion in electrolyte solutions. Some further development of this scheme is required, however, to treat diffusion phenomena in (*a*) crystalline and molten silicates if there is no neutral solvent (Anderson 1975), and (*b*) in aqueous electrolytes if there is more than one anionic species (D. E. Anderson and D. L. Graf, in preparation).

Multicomponent diffusion has another important aspect: the concepts of nonequilibrium thermodynamics have been worked out and the consequences utilized with significant success in this area. Nonequilibrium thermodynamics, an extension of the macroscopic theory of thermodynamics to natural, i.e. irreversible, processes will undoubtedly become an important tool for geologists. It is concerned with the history or evolution of a system, in contrast to equilibrium thermodynamics, which can only predict the terminal state of a system for a given set of constraints. Although nonequilibrium thermodynamics may in principle be made completely macroscopic, it is like equilibrium thermodynamics in having greater analytical and heuristic value if combined with kinetics and statistical mechanics.

This article considers diffusion resulting from the presence of concentration gradients and excludes systems in which there is chemical reaction or in which thermal, electromagnetic, or pressure gradients exist. By chemical reaction we mean the transfer of material across phase boundaries, e.g. by dissolution, precipitation, or exchange between a liquid and a solid phase. There are circumstances in which exchange between two phases may closely approximate diffusion in a single, nonhomogeneous phase. Specifically, this situation results when the kinetic description of each transfer process at the interface is what would be expected for a smooth transition between the kinetic descriptions of diffusion in the phases. The construction of experimental diffusion couples is limited by the requirement that there not be an interface process, i.e. a process for which there is a discrete jump in kinetics at the interface.

We hope by focusing on such a restricted and relatively simple phenomenon to demonstrate clearly the way in which nonequilibrium thermodynamics works and the source of problems in the theory. Hopefully, our discussion serves as an introduction to more comprehensive and necessarily more complex treatments (Fitts 1962, de Groot & Mazur 1962, Katchalsky & Curran 1967, Haase 1969).

Different authors have often used different symbols for the same physical quantity; we prefer the system of notation of Prigogine & Defay (1954) for thermodynamic quantities and attempt to adhere to it. For simplicity of notation, all equations containing vectors are written in one-dimensional form. Subscripts attached to the component of a vector quantity refer to the chemical components of the solution and not to the Cartesian components of the vector.

TREATMENT OF MULTICOMPONENT DIFFUSION BY NONEQUILIBRIUM THERMODYNAMICS

Although some aspects of nonequilibrium thermodynamics had been treated earlier, the first unified theory was the work of Onsager (1931a, 1931b). Not until the last two decades did it become clear that nonequilibrium thermodynamics provides information that is neither trivial nor derivable from older, established theories. An appreciation of the contentious history of the theory may be gleaned by reading Truesdell (1969).

Abstract though the concepts of nonequilibrium thermodynamics may appear, the object is very simple: the analytical construction of a set of differential equations that describe the evolution of a system. There are simple limiting cases in which the application of the theory is trivial and the results may be won using only mass- and energy-balance equations. The flow of pure water through a nonreactive, porous medium is one such example.

Nonequilibrium thermodynamics utilizes the second law of thermodynamics in addition to mass- and energy-balance equations. Thermodynamic equilibrium is attained in a system when the state variables that characterize the system are independent of position and time. In a system in which gradients are maintained, it is assumed that local volume elements can be defined on some scale such that thermodynamic equilibrium is closely approximated at each instant in each element. Furthermore, it is assumed that each element passes through a series of local equilibrium states as the system evolves. If these local equilibrium states can be described with sufficient accuracy using a principle other than the second law of thermodynamics, the system as a whole may be treated without recourse to nonequilibrium thermodynamics. The assumption of mechanical equilibrium is a frequently used alternate principle.

Nonequilibrium thermodynamics usually becomes important when there is cross-coupling between two or more phenomena. The transport of heat by diffusional flow of matter (Dufour Effect) and the diffusion of matter on a temperature gradient (thermal diffusion) are coupled phenomena. Another example of cross-coupling, the diffusion of one component on the concentration gradient of another, was vividly demonstrated by Darken (1949, 1951) for Fe containing C and Si as interstitial components. Diffusion of carbon in this material is coupled to the concentration gradient of Si (Figure 1). The process by which the carbon is redistributed is sometimes referred to as "uphill" diffusion. Studies of K_2O-SrO-SiO_2 glasses by A. K. Varshneya and A. R. Cooper (cited in Cooper 1974) have shown that each component in that system may be made to diffuse uphill by varying the initial compositions of the diffusion couples.

Onsager (1945) suggested that Fick's first law be written for a system of n components as

$$J_i = -\sum_{k=1}^{n} D_{ik} \operatorname{grad} y_k \qquad (i = 1, 2, \ldots, n), \tag{1}$$

where J_i is the flux of component i, y_k is a compositional parameter (the concentration, mole fraction, chemical potential, etc, of component k), and D_{ik} is an $n \times n$ matrix of diffusion coefficients. The on-diagonal or direct diffusion coefficients, D_{ik} $(i = k)$, relate the flux of i to its own concentration gradient, whereas the off-diagonal or cross-coefficients, D_{ik} $(i \neq k)$, allow for the diffusion of i on the concentration gradient of another component. The existence of nonzero off-diagonal coefficients leads to the coupling of fluxes.

In some systems, the coupling of fluxes is limited. Thus, in the ternary system KCl-NaCl-H_2O, the flux of KCl is almost wholly independent of the gradient of NaCl, and the flux of KCl may be approximated quite accurately in most circumstances without reference to cross-coupling. On the other hand, Wendt & Shamim (1970) have shown that, for equal concentration gradients of $MgCl_2$ and NaCl in the system $MgCl_2$-NaCl-H_2O, 35% of the flux of NaCl results from the concentration gradient of $MgCl_2$. Neglect of cross-coupling of this magnitude leads to significant systematic error in the size of calculated fluxes. More importantly, the directions of one or more of the calculated fluxes may be reversed (Figure 1).

Even a casual reading of the literature on multicomponent diffusion reveals the difficulty of setting up the proper nonequilibrium formulation to treat a particular evolving system. The theory of nonequilibrium thermodynamics is a subtle and complex tool that inherits many of the problems of equilibrium thermodynamics. It proceeds through a series of steps that may be considered as postulates:

1. The entropy balance and consequently the entropy production may be computed from Gibbs' equation (Equation 29, below).

2. The entropy production is non-negative and can be written as a sum of products of fluxes and thermodynamic forces (the grad y_k of Equation 1 are examples of thermodynamic forces).

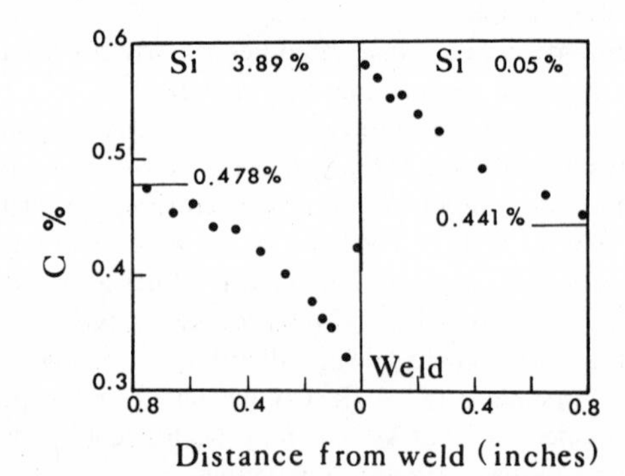

Figure 1 Distribution of carbon after annealing of a diffusion couple formed by welding two iron rods with nearly equal initial concentrations of carbon (0.478% and 0.441%), but different silicon contents. The annealing was carried out for 13 days at 1050°C. By permission from *Atom Movements*, copyright American Society for Metals, 1951 (Darken 1951).

3. Each flux in the expression for the entropy production is a linear combination of partial fluxes engendered by each of the thermodynamic forces.

4. From the expression for the entropy production it is possible to introduce a set of phenomenological equations (Equation 1 is an example of a phenomenological equation).

5. If the expression for the entropy production is constructed correctly, the matrix of coefficients in the phenomenological equations will be symmetric.

The first three postulates are extensions of the second law of thermodynamics and follow naturally from the assumption of local equilibrium. The fourth postulate is an extension of known linear laws such as Fick's law for diffusion, Ohm's law for electrical conduction, and Fourier's law of heat conduction. The final postulate (Onsager 1931a, 1931b) differs from the others in that it predicts a relationship that can be tested directly by experiment.

Symmetry relations similar to those predicted by Onsager (1931a,b) had in fact been verified previously for heat conduction in anisotropic crystals (Soret 1894, Voigt 1903) and for some thermo-electric phenomena. Onsager's treatment, however, was the first successful attempt to provide an explanation for the origin of the symmetry relations. An excellent discussion of the experiments of Soret and Voigt, the symmetry of the thermal conductivity tensor, and Onsager's derivation appears in Chapter 11 of Nye (1957). Experimental tests of the validity of Onsager's theorem for isothermal, isobaric diffusion in ternary systems have been reviewed by Miller (1967a) and for other phenomena by Miller (1960, 1969). These tests verify the theorem and, by implication, the rest of nonequilibrium theory for a range of phenomena, including diffusion in multicomponent solutions.

One of the cornerstones of nonequilibrium thermodynamics is the invariance of σ, the entropy production per unit volume per unit time. The great significance accorded to entropy derives from two observations: (*a*) that many systems, if left undisturbed for a time, proceed to a simple state—equilibrium—in which the thermodynamic parameters are constant over space and time, and (*b*) that natural processes, proceeding from one equilibrium state to another, are irreversible in time. Callen (1960) shows that these two facts alone are sufficient to construct the second law of thermodynamics and, thus, the entropy function. The foundations of Callen's axiomatic treatment may be detected in the opening pages of Gibbs (1961). The entropy is defined as a unique, monotonic, and continuously differentiable function of the internal energy, volume, and mole numbers of the system. With the assumption of local equilibrium, each volume element of a spontaneously changing system passes through a sequence of equilibrium states. Thus the entropy production in each volume element must be independent of any particular scheme of description. Otherwise, by choosing different schemes, we could produce different values of entropy for a particular equilibrium state. The invariance of the entropy production enters the analysis of multicomponent diffusion in a specific and fundamental way: it is used to relate diffusion fluxes and diffusion coefficients measured in different coordinate systems (reference frames).

REFERENCE FRAMES

A flux or, strictly speaking, each component of the vector representing a flux, is the amount of material crossing a plane of unit area per unit time. The plane to which the flux is referred may be chosen in a variety of ways. The significance of this simple fact first became apparent in connection with experiments performed by Hartley (1946) and by Smigelskas & Kirkendall (1947). In separate but similar treatments, Darken (1948) and Hartley & Crank (1949) [the bulk of this latter paper is repeated verbatim in Chapter 11 of Crank (1956)] realized that the experiments posed a basic question: what part of a flow of matter is to be regarded as diffusion?

The experiment of Smigelskas & Kirkendall (1947) is illustrated in Figure 2. The results initially created considerable controversy (e.g. see the discussion following Darken 1948). The motion of the markers implied that the amount of Cu diffusing into the brass was less than the outward flux of Zn. However, it appeared intuitively that in these crystalline materials the flux of Zn should match that of Cu. We see presently that this assumption is not necessarily true. Initial doubts about the validity of the experiment were dispelled by the thorough tests of Correa da Silva & Mehl (1951) and by the work of numerous later workers (e.g. see the review article of Lazarus 1960).

Diffusion coefficients are commonly obtained by solving Fick's second law in one dimension, e.g. parallel to x,

$$\partial c/\partial t = \partial/\partial x(D\, \partial c/\partial x), \tag{2}$$

using the data of a concentration vs distance curve. If D is a function of concentration but not of time, the solution of Equation 2 imposes an arbitrary coordinate system on the diffusion couple. The position of the zero for x is determined by the requirement that

$$\int_{c=c_1}^{c=c_2} x\, dc = 0 \tag{3}$$

(Crank 1956, pp. 232–40; Darken & Gurry 1953, pp. 453–58). Solutions of Equation

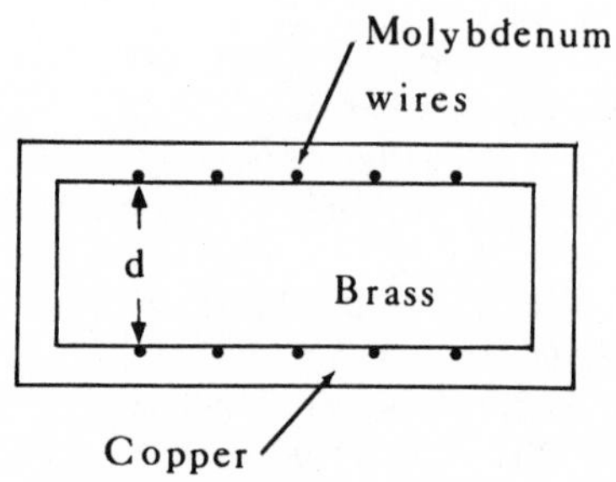

Figure 2 Schematic cross-section of the copper-brass diffusion couple used by Smigelskas & Kirkendall (1947).

2 for $D = f(c)$ (for the boundary conditions usually employed) were first given by Matano (1933, after Boltzmann 1894).

For liquid or solid solutions that are ideal with respect to volume—that is, those in which there is no volume change on mixing—the zero for x lies in a plane (perpendicular to the x-axis) across which there is no net transfer of volume. Therefore, the solution of Equation 2 yields a diffusion coefficient in a volume-fixed reference frame (Kirkwood et al 1960). These authors also proved that the Ds obtained from Equation 2 for volume-ideal systems are uniquely volume-fixed coefficients, except in binary systems, for which volume-fixed and molecular-fixed coefficients are numerically equal (Trimble et al 1965). (The equivalence of diffusion coefficients in the two frames is only valid in ternary solutions if the partial molar volumes of all components are equal. There is no such restriction in binary solutions.) The molecular (number-fixed) reference frame is defined in terms of the average velocity of all particles in the system, and therefore appears automatically in most kinetic calculations.

For experiments carried out in solutions that are not volume-ideal, Kirkwood et al (1960) have shown that the error in considering the measured diffusion coefficients to be volume-fixed is proportional to $(\Delta\rho_i)^2$, where $\Delta\rho_i$ is the difference between the partial densities of i at time $t = 0$ on the two sides of a diffusion couple. By keeping the $\Delta\rho_i$ small, errors can be minimized.

If there is no volume change on mixing, diffusion in a binary solution is described by a single volume-fixed diffusion coefficient (Crank 1956, p. 219):

$$D^v = D_1^v = D_2^v, \tag{4}$$

where the superscript defines the reference frame, and 1 and 2 are components.

Volume changes accompany diffusion in many natural and experimental systems. These changes in volume are compensated for by mass flow of the whole solution, i.e. bulk movement, relative to a fixed, external coordinate system. The movement of the markers in the Cu-brass system discussed above is caused by mass flow resulting from volume changes associated with (*a*) the solution of Cu in brass and (*b*) the solution of vacancies in brass (Shewmon 1963, Girifalco 1964, Manning 1967). The markers lie in a plane across which there is no flow of mass and thus describe a mass-fixed reference frame. Similarly, the volume changes that accompany diffusion in nonideal liquid couples formed in an open column may be monitored by following the position of the meniscus at the top of the column.

Diffusion signifies the motion of molecules or atoms of one kind relative to the rest of the solution, in contrast to mass flow, which is the simultaneous transport of atoms and molecules of all kinds in the same direction. The rate of diffusion may be measured in solids by following the movement of a radioactive isotope through a homogeneous sample. In a typical experiment, a radioactive isotope diffuses in from a surface layer deposited on the crystal or from a well-stirred liquid reservoir in contact with the crystal. The diffusion coefficient for the isotope may be calculated from a plot of specific activity vs depth from the surface. Because there are no chemical concentration gradients in the sample (the isotopes of an element to a good approximation are chemically identical), there is no mass flow.

However, in an experiment in which there are chemical concentration gradients and in which mixing is nonideal with respect to volume, the flux of matter across a volume-fixed plane results in part from diffusion and in part from mass flow. Thus the diffusion coefficient D^v is "unnecessarily complicated by the presence of mass flow" (Hartley & Crank 1949, p. 813).

An observer located on an inert marker can unambiguously distinguish between diffusion and mass flow. New diffusion coefficients D_1^m and D_2^m, not complicated by mass flow, can be determined in the mass-fixed reference frame defined by the marker. Except when diffusion is by direct interchange of atoms in a crystal, D_1^m need not equal D_2^m. It is particularly easy to visualize two species migrating at different rates in a binary liquid, but the same process occurs in crystals if diffusion proceeds by a vacancy mechanism (Shewmon 1963, Girifalco 1964), the common mechanism in most metals and oxides, or if the diffusing species are dilute, interstitial components in the crystal. Vacancies or other defects that participate in diffusion must be treated as chemical species, to which chemical potentials and other thermodynamic properties are assigned (Swalin 1972, van Gool 1966). Therefore, a crystal through which vacancies and two kinds of atom diffuse is strictly speaking a dilute ternary system.

Darken (1948) and Hartley & Crank (1949) showed independently that, for a binary solution,

$$D^v = x_2 D_1^m + x_1 D_2^m, \tag{5}$$

where x_1 is the mole fraction of 1, and the term *intrinsic diffusion coefficients* of the latter authors has often been applied to the D_i^m. The derivation of Equation 5 for a vacancy mechanism has been examined from the point of view of kinetics and nonequilibrium thermodynamics by Bardeen & Herring (1951) and Manning (1967, 1968a, 1968b). Clearly, D_1^m and D_2^m cannot be obtained directly by substituting a measured value of D^v into Equation 5. General relations that allow mass-fixed diffusion coefficients to be calculated from measurements in the volume-fixed frame are cited later in this paper.

There are other possible reference frames in addition to the three already mentioned. Reference frames defined by a plane moving with the local velocity of a particular species are especially important because these frames allow flux contributions from mass flow and diffusion to be distinguished. Any component may be chosen as the dependent one, but some choices are especially convenient. In liquids, the undissociated solvent is frequently chosen (solvent-fixed reference frame), in crystalline materials, a species that is more or less rigidly bound to the lattice (lattice-fixed reference frame) (Haase 1969, D. E. Anderson 1975). Vacancies are sometimes chosen in metals.

There was some concern in the early metallurgical literature on diffusion about whether gradients of chemical potential or of concentration were the fundamental driving forces for diffusion. The answer is that Equation 1 is expressed in an equally fundamental sense if y_k is taken to be μ_k or c_k. However, the diffusion coefficients that relate J_k to y_k have very different properties in the two cases. To emphasize the difference, Kirkwood et al (1960) have suggested the terms theoretical and practical diffusion coefficient for $y_k = \mu_k$ and $y_k = c_k$, respectively.[1] The chemical potential of a component is defined analytically in the development of thermodynamics (Callen 1960), so that the initial application of thermodynamics to diffusion involves chemical potentials. But concentrations, and not chemical potentials, are the measurable quantities. No analytical relationship between the chemical potential and the concentration of a component of a solution is prescribed by thermodynamics; the two quantities are related by assuming an equation of state. The usual quantities for nonideal solutions at constant temperature, T, and pressure, P, are

$$\mu_i = \mu_i^o + RT \ln \gamma_i x_i \qquad (i = 1, 2, \ldots, n), \tag{6}$$

where x_i is the mole fraction of i, but molarity, molality, etc, could equally well be used if the activity coefficient γ_i had the appropriate numerical value for the other concentration unit, and μ_i^o is an integration constant to be evaluated at some selected standard state (Denbigh 1971, Chaps. 8 and 9).

Formally, diffusion fluxes are defined in a particular reference frame a ($a = m$, v, etc) by

$$J_k^a = c_k(v_k - v^a) \qquad (k = 1, 2, \ldots, n), \tag{7}$$

where v_k is the average velocity of k, v^a is a reference velocity, and c_k is the molarity of k. The reference velocity is given by

$$v^a = \sum_{k=1}^{n} a_k v_k, \tag{8}$$

where the a_k are weighting factors, subject to the normalization condition

$$\sum_{k=1}^{n} a_k = 1. \tag{9}$$

Multiplication of Equation 7 by a_k/c_k, summation over the n components, and substitution for v^a from Equation 8 yields

$$\sum_{i=1}^{n} a_i J_i^a/c_i = \sum_{i=1}^{n} a_i J_i^a/x_i = 0. \tag{10}$$

Equation 10 may be taken as the definition of reference frame a.

Note that by using c_k in Equation 7, we have defined molar fluxes. If partial densities ρ_k are chosen instead, we obtain mass fluxes. Relations for converting mass to molar fluxes may be found in de Groot & Mazur (1962, p. 242). The weighting factors and reference velocities for various reference frames are given in de Groot & Mazur (1962, p. 240) for mass fluxes and in Haase (1969, p. 218) for molar fluxes. For mass fluxes, ρ_i and ω_i (mass fraction) replace c_i and x_i respectively in Equation 10.

[1] Other measurable concentration parameters such as mass fractions, partial densities, mole fractions, etc may be used in place of c_k. The numerical value of the practical diffusion coefficient will depend on the particular choice.

The relationship between fluxes in two reference frames a and b, with weighting factors a_k and b_k, respectively, may be derived from

$$J_k^a = J_k^b + c_k u \qquad (k = 1, 2, \ldots, n), \tag{11}$$

where u is the velocity of frame a relative to frame b. Equation 11 holds for two coordinate systems moving without rotation at a uniform velocity with respect to each other. The absence of rotation is sustained for diffusion in single crystals with orthogonal crystal axes and in isotropic media such as gases, liquids, and polycrystalline aggregates that have random orientation.

The condition of uniform velocity turns up in a different guise later in the development of the theory. Because the relative motion of a coordinate system results from mass flow, the condition of uniform velocity reduces to the requirement that $dv^m/dt = 0$, which corresponds to mechanical equilibrium in the system. Prigogine (1955) first realized that the description of irreversible processes is considerably simplified at mechanical equilibrium. In experiments where the $\Delta\rho_i$ terms are kept small, mechanical equilibrium is probably approximated closely. However, in natural systems with large concentration gradients, the transformation matrices that are derived from Equation 11 and from Prigogine's theorem for systems at mechanical equilibrium may break down.

Multiplying Equation 11 by a_k/c_k, summing over the components, and utilizing Equations 9 and 10,

$$u = -\sum_{k=1}^{n} a_k J_k^b/c_k. \tag{12}$$

Substituting this expression for u back into Equation 11 yields

$$J_k^a = J_k^b + c_k \sum_{i=1}^{n} a_i J_i^b/c_i \qquad (k = 1, 2, \ldots, n). \tag{13}$$

The change of indices in the last term follows from the presence of the summation in Equation 12.

By virtue of Equation 10, only $n-1$ of the fluxes are independent in any reference frame. Taking the nth component to be the dependent one,

$$J_n^b = -(c_n/b_n) \sum_{i=1}^{n-1} b_i J_i^b/c_i, \tag{14}$$

where the b_i are weighting factors in reference frame b. Substituting this expression for J_n^b back into Equation 13 gives

$$J_k^a = J_k^b + c_k \sum_{i=1}^{n-1} J_i^b[(a_i/c_i) - (a_n b_i/b_n c_i)] \qquad (k = 1, 2, \ldots, n-1) \tag{15}$$

or, more compactly,

$$J_k^a = \sum_{i=1}^{n-1} B_{ki} J_i^b \qquad (k = 1, 2, \ldots, n-1), \tag{16}$$

where B_{ki} is an $(n-1)$-dimensional matrix with elements

$$B_{ki} = \delta_{ki} + [(a_n b_i/b_n) - a_i c_k/c_i] \qquad (i, k = 1, 2, \ldots, n-1), \tag{17}$$

and δ_{ki} is the Kronecker delta.

Because of Equation 11, B_{ki} is a matrix that linearly transforms fluxes from one reference frame to another. This transformation is independent of nonequilibrium thermodynamics: it merely relates the descriptions of a vector in two different coordinate systems. To find a scheme for transforming theoretical diffusion coefficients, we turn to nonequilibrium thermodynamics.

BALANCE EQUATIONS

The change in the entropy of a local volume element of a thermodynamic system may be written as

$$dS = d_e S + d_i S, \tag{18}$$

where $d_e S$ is the entropy supplied to the system by its surroundings and $d_i S$ is the entropy produced inside the system, e.g. by chemical reactions, diffusion, heat flow, etc. The second law of thermodynamics requires that $d_i S > 0$ for all spontaneous processes in the system. Noting that entropy, like any other extensive thermodynamic quantity, may be visualized as a nonconserved fluid, it is possible to recast Equation 18 in the form of a balance equation, analogous to hydrodynamic balance equations for mass, momentum, and energy. We have, for any volume element dV

$$\begin{bmatrix}\text{Rate of increase of}\\ \text{entropy inside } dV\end{bmatrix} = \begin{bmatrix}\text{Outward flux of}\\ \text{entropy from } dV\end{bmatrix} + \begin{bmatrix}\text{Entropy production}\\ \text{inside } dV\end{bmatrix}$$

or

$$dS/dt = d_e S/dt + d_i S/dt. \tag{19}$$

Equation 19 may also be written in the form

$$\partial(\rho \underline{s})/\partial t = -\operatorname{div} J_{\underline{s}(\text{total})} + \sigma, \tag{20}$$

with

$$d_i S/dt = \int^{V} \sigma \, dV, \tag{21}$$

where ρ is density and $\underline{s}$ is the specific entropy of the solution. The conversion of Equation 19 to Equation 20 is a standard manipulation covered in many texts, e.g. Bird et al (1960), Haase (1969), and Bear (1972). Equation 20 may be applied to volume elements of any size or to the total system itself.

The operator $\partial/\partial t$ implies differentiation at a fixed position. It is related to the substantial derivative d/dt, taken at the center of mass of a volume element moving with a velocity v^m, by

$$d/dt = \partial/(\partial t) + v^m \text{ grad}. \tag{22}$$

Application of Equation 22 to 20 gives

$$(\mathrm{d}\underline{s}/\mathrm{d}t) = -\mathrm{div}\, J_{\underline{s}} + \sigma, \tag{23}$$

with

$$J_{\underline{s}} = J_{\underline{s}(\text{total})} - \rho \underline{s} v^{m}. \tag{24}$$

Obtaining explicit expressions for the entropy production σ in particular systems is a major task of nonequilibrium thermodynamics.

From the postulate cited above that entropy is a unique, monotonic function of the internal energy U, the volume V, and the mole numbers n_i of a system, it follows that entropy will be transported across the boundary of a volume element or of an open system by fluxes of both heat and matter. There is no single, unambiguous way in which $J_{\underline{s}}$ may be divided into these two components; the choice ultimately effected is always a matter of convenience. Even if one chooses to ignore explicit cross-coupling between thermal and diffusive phenomena, great care must be taken in defining a heat flow term to be added to the diffusion equations to find the entropy production. For the isothermal systems considered in this paper, of course, the calculation of entropy production takes a simpler form.

A balance equation similar to Equation 19 may be introduced for the conservation of mass. In the absence of chemical reactions, the source term on the right-hand side of Equation 19 disappears and we are left with

$$\partial \rho_i / \partial t = -\mathrm{div}\, \rho_i v_i \tag{25}$$

(de Groot & Mazur 1962, Chap. 2; Fitts 1962, Chap. 2). Alternatively,

$$\rho(\mathrm{d}\omega_i/\mathrm{d}t) = -\mathrm{div}\, \bar{J}_i^m, \tag{26}$$

where ω_i is the mass fraction of i, the diffusion fluxes are required to be in a mass-fixed reference frame because of the use of the substantial derivative (recall Equation 22), and the bar over the $\bar{J}_i^m$ indicates that mass fluxes rather than molar fluxes are being considered.

A balance equation for energy in an isobaric system may be obtained either from Equation 19 or from the conservation condition expressed by the first law of thermodynamics, differentiated with respect to time,

$$\mathrm{d}\underline{u}/\mathrm{d}t = \mathrm{d}q/\mathrm{d}t - P(\mathrm{d}\underline{v}/\mathrm{d}t), \tag{27}$$

where $\mathrm{d}q$ is the "heat" added per unit mass, and $\underline{u}$ and $\underline{v}$ are the specific internal energy and the specific volume, respectively.

The total differential of the second law of thermodynamics,

$$\underline{u} = \underline{u}(\underline{s}, \underline{v}, \omega_i), \tag{28}$$

is (Callen 1960)

$$\mathrm{d}\underline{u} = T\,\mathrm{d}\underline{s} - P\,\mathrm{d}\underline{v} + \sum_{i=1}^{n} \underline{\mu}_i\, \mathrm{d}\omega_i. \tag{29}$$

Differentiation of Equation 29 with respect to time, followed by substitution from Equation 27 for $\mathrm{d}\underline{u}/\mathrm{d}t$ and from Equation 22 for $\mathrm{d}\omega_i/\mathrm{d}t$, gives

$$\rho(\mathrm{d}\underline{s}/\mathrm{d}t) = -(1/T)\,\mathrm{div}\,J_q - \sum_{i=1}^{n} (\underline{\mu_i}/T)\,\mathrm{div}\,\overline{J_i^m}, \tag{30}$$

where the heat flux J_q is defined by

$$-\mathrm{div}\,J_q = \rho(\mathrm{d}q/\mathrm{d}t). \tag{31}$$

Each of the terms on the righthand side of Equation 30 may be split by using the relation

$$\mathrm{div}\,(AB) = B\,\mathrm{div}\,A + A\cdot\mathrm{grad}\,B, \tag{32}$$

where A is a vector and B is a scalar, and both are functions of position. Thus, Equation 30 becomes

$$\rho(\mathrm{d}\underline{s}/\mathrm{d}t) = -\mathrm{div}\left(J_q - \sum_{i=1}^{n} \mu_i \overline{J_i^m}\right)\Big/ T - (J_q\cdot\mathrm{grad}\,T)/T^2 - \left[\sum_{i=1}^{n} T\overline{J_i^m}\cdot\mathrm{grad}\,(\mu_i/T)\right]\Big/ T. \tag{33}$$

Comparison with Equation 23 suggests

$$J_{\underline{s}} = \left(J_q - \sum_{i=1}^{n} \underline{\mu_i}\,\overline{J_i^m}\right)\Big/ T, \tag{34}$$

and

$$\sigma = -(J_q\cdot\mathrm{grad}\,T)/T^2 - \left[\sum_{i=1}^{n} J_i\cdot T\,\mathrm{grad}\,(\mu_i/T)\right]\Big/ T. \tag{35}$$

By splitting the grad (μ_i/T) term and using Equation 34, Equation 35 can be rewritten as

$$T\sigma = -J_{\underline{s}}\cdot\mathrm{grad}\,T - \sum_{i=1}^{n} J_i\cdot\mathrm{grad}\,\mu \tag{36}$$

(de Groot & Mazur 1962, p. 27). It is evident that a more general way of writing Equation 36 is

$$T\sigma = \sum_{i=1}^{n} J_i\cdot X_i, \tag{37}$$

where X_i, the thermodynamic forces that produce the fluxes, are gradients except in the case of chemical reactions. Additional terms may be added to Equation 37 to treat systems in which, in addition to heat flow and diffusion, there are other phenomena such as electrical conduction, viscous flow, and chemical reactions. It is always possible to construct the entropy production for each new phenomenon as a combination of fluxes and forces similar to the right-hand side of Equation 37.

ONSAGER'S DERIVATION

From Equation 37, we infer a set of linear equations, the so-called phenomenological equations,

$$J_i = -\sum_{k=1}^{n-1} L_{ik} X_k \qquad (i = 1, 2, \ldots, n). \tag{38}$$

This step is taken by analogy with the empirical, linear laws of Fick, Fourier, and Ohm for diffusion, heat flow, and electrical conduction, respectively. Each of these laws depends for its validity on the occurrence of one phenomenon at a time, whereas Equation 38 allows for the mixing of different phenomena. Clearly, the three empirical laws are special cases of Equation 38.

Onsager's derivation asserts that if Equation 37 contains independent forces and fluxes and if the linear laws expressed in Equation 38 are valid, then the phenomenological coefficients L_{ik} are—in the absence of magnetic fields—symmetric:

$$L_{ik} = L_{ki} \qquad (i, k = 1, 2, \ldots, n). \tag{39}$$

Stated differently,

$$(\partial J_i/\partial X_k)_{X_{j \neq k}} = (\partial J_k/\partial X_i)_{X_{j \neq i}}, \tag{40}$$

that is, an increase in the flux J_i caused by unit increase in the force X_k (all other forces remaining constant) is equal to the increase in J_k because of unit increase in X_i. The equality expressed by Equation 39 is usually known as the Onsager reciprocity relation.

Because of Equations 20, 22, and 25, the expression for entropy production derived in the preceding section, Equation 30, is correct only for diffusion fluxes in a mass-fixed reference frame. At mechanical equilibrium, however, the velocity v^m in the definition of J_i^m, Equation 7, may be replaced by an arbitrary velocity v^a without affecting the expression for the entropy production (Prigogine 1955). Thus, any of the reference frames related by linear transformations may be introduced into Equation 37. If Equations 37 and 38 are written for two different reference frames a and b, then because of Equation 16 and the invariance of the entropy production, the thermodynamic forces transform according to

$$X_k^b = \sum_{i=1}^{n-1} B_{ik} X_i^a, \qquad (k = 1, 2, \ldots, n-1), \tag{41}$$

and the L coefficients according to

$$L_{ij} = \sum_{k=1}^{n-1} \sum_{l=1}^{n-1} B_{ik} L_{kl} B_{jl} \qquad (i, j = 1, 2, \ldots, n-1) \tag{42}$$

(Hooyman 1956, de Groot & Mazur 1962).

In an isothermal system, grad $T = 0$, and independent fluxes and forces may be introduced into Equation 37 by using the Gibbs-Duhem equation for an isothermal, isobaric system,

$$\sum_{i=1}^{n} \omega_i \, d\mu_i = 0, \tag{43}$$

together with Equation 10 and expressions for J_o^a and grad μ_o, to give

$$T\sigma = -\sum_{i=1}^{n-1} J_i^a \left[\text{grad } \mu_i + \sum_{k=1}^{n-1} (\omega_k a_i/\omega_i a_o) \text{ grad } \mu_k \right]. \tag{44}$$

Putting

$$A_{ik}^a = \delta_{ik} + \omega_k a_i/\omega_i a_o \qquad (i, k = 1, 2, \ldots, n-1), \tag{45}$$

and

$$X_i^a = -\sum_{k=1}^{n-1} A_{ik}^a \text{ grad } \mu_k/T \qquad (i, k = 1, 2, \ldots, n-1), \tag{46}$$

Equation 37 becomes

$$T\sigma = \sum_{i=1}^{n-1} J_i^a \cdot X_i^a, \tag{47}$$

and Equation 39 becomes

$$L_{ik} = L_{ki} \qquad (i = 1, 2, \ldots, n-1). \tag{48}$$

Some restrictions exist on the elements of the matrix of phenomenological coefficients. If, for example, a phenomenon may be described by

$$\begin{aligned} J_1 &= L_{11}X_1 + L_{12}X_2 \\ J_2 &= L_{21}X_2 + L_{22}X_2, \end{aligned} \tag{49}$$

then substitution for J_i in Equation 37 yields

$$\sigma = L_{11}X^2 \ + (L_{12} + L_{21})X_1X_2 + L_{22}X_2^2. \tag{50}$$

The entropy production is zero at equilibrium and positive for all spontaneous processes. These conditions require that

$$L_{11} > 0, \qquad L_{22} > 0, \tag{51}$$

and that

$$(L_{12} + L_{21})^2 > 4(L_{11}L_{22}). \tag{52}$$

The on-diagonal coefficients are always positive, whereas the off-diagonal ones may be either positive or negative, with their magnitudes subject only to Equation 52.

Onsager's (1931a,b) derivation extends the concepts of microscopic reversibility, fluctuation, and regression of fluctuations from the microscopic to the macroscopic domain. In the derivation, the selection of fluxes and forces to be used in an expression for the entropy production is dictated by microscopic considerations. Once this expression is established, the symmetry relations appear as a consequence of the principle of microscopic reversibility. In contrast, the flows and forces that enter into Equation 37 are usually formulated from purely macroscopic arguments and, as experience has shown, it is often possible to choose them in a variety of ways. Nevertheless, Onsager's theorem says that if an expression (Equation 37) can be obtained for a particular system by macroscopic means and if the linear laws (Equation 38) are valid for that system, then the symmetry relations should follow automatically. The construction of Equation 37 should make it possible to avoid a detailed microscopic analysis.

Unfortunately, the fact that a particular set of independent fluxes and forces satisfies Equations 37 and 38 is not sufficient to guarantee the Onsager reciprocal relation; additional microscopic conditions must also be satisfied. It is this blending of microscopic and macroscopic arguments that has created many of the conceptual difficulties associated with nonequilibrium thermodynamics.

In matrix notation, Equation 38 is

$$\mathbf{J}^a = \mathbf{L}^a \cdot \mathbf{X}^a \tag{53}$$

and, substituting from Equation 38 into Equation 37,

$$\sigma = \tilde{\mathbf{X}}^a \cdot \mathbf{L}^a \cdot \mathbf{X}^a, \tag{54}$$

where $\tilde{\mathbf{X}}^a$ is the transpose of $\mathbf{X}^a$. We may define a new matrix

$$(\mathbf{L}^a)' = \mathbf{L}^a + \mathbf{N}, \tag{55}$$

where the matrix $\mathbf{N}$ is formed by multiplying the mass fractions ω_j by arbitrary constants A_i:

$$N_{ij} = A_i \omega_j \qquad (i, j = 1, 2, \ldots, n). \tag{56}$$

The constants A_i may always be chosen such that the matrix $\mathbf{N}$ is nonzero and antisymmetric. Then

$$(\mathbf{L}^a)' \cdot \mathbf{X}^a = \mathbf{L}^a \cdot \mathbf{X}^a + \mathbf{N} \cdot \mathbf{X}^a. \tag{57}$$

But the term $\mathbf{N} \cdot \mathbf{X}^a$ is equivalent to the Gibbs-Duhem equation (Equation 43), and is therefore equal to zero. Thus either $\mathbf{L}^a$ or $(\mathbf{L}^a)'$ satisfies Equations 52 and 54, but $(\mathbf{L}^a)'$ is antisymmetric. Unless $\mathbf{N}$ is zero, the matrix of phenomenological coefficients is neither unique nor necessarily symmetric.

In an extension of the above arguments, Davies (1952; see also Hooyman et al 1955) proved that the Onsager reciprocal relation may also be destroyed by linear transformations, including, of course, the transformation that relates diffusion coefficients in different reference frames. Arguments analogous to those of the preceding paragraph, utilizing Equations 37 and 38 and assuming independent fluxes and forces, allow

$$\mathbf{L}^a = \mathbf{B}^{ab} \cdot (\mathbf{L}^b + \mathbf{M}) \cdot \mathbf{B}^{ab}, \tag{58}$$

where $\mathbf{M}$ is again a nonzero, antisymmetric matrix. Although the derivation of Equation 57 as given appeals explicitly to the Gibbs-Duhem equation, the derivations of both Equations 57 and 58 rely only on the fact that any matrix may be split into a symmetric and an antisymmetric part. The Onsager reciprocal relation follows from Equations 37 and 38 only if it can be shown on physical grounds that the antisymmetric matrix, **N** for chemical diffusion and **M** in the general case, is always zero.

A Macroscopic Approach

Equilibrium thermodynamics can be constructed from simple macroscopic observations and concepts, as is particularly evident in the axiomatic development of the theory by Callen (1960). The deduction from statistical mechanics of an expression for entropy, the Boltzmann equation, reinforces and elucidates the validity of the second law of thermodynamics but is not strictly necessary for its development. There appears to be significant advantage in retaining a purely macroscopic approach in extending thermodynamics to nonequilibrium systems. Thus, Fitts (1962), among others, has suggested that Equation 39 be treated as a plausible postulate to be tested against experiment, with the expectation that accumulated experience will make the proper forms of Equations 37 and 38 evident. This approach is at present proving fruitful in understanding isothermal diffusion, for which the body of experimental measurements is greater than for many other phenomena.

When applied to isothermal diffusion, Equations 37 and 38 yield the Onsager reciprocal relation if (*a*) the thermodynamic forces are taken to be chemical potential gradients and (*b*) either the set of fluxes or the set of forces is a mutually independent set, or if both are. Using these two selection rules for fluxes and forces, we may deduce from Equations 37 and 38 a matrix of L-coefficients that is symmetric within the limits of experimental accuracy (Miller 1967a). A more general set of rules for selecting fluxes and forces for other phenomena has been given by de Groot (1952, pp. 5–7).

Not only is this approach sufficient for the study of isothermal diffusion, but it may in practice be applied to systems in which there is diffusion of both heat and matter. Even though the heat flux and the diffusion fluxes as defined in Equations 26 and 31 are not entirely consistent with the dictates of Onsager's microscopic analysis (discussed below), experience has shown that a symmetric matrix results if the selection rules of de Groot (1952) are obeyed.

In many geological problems, phenomena other than the diffusion of heat and matter must be included to complete the analysis of the evolution of the system. For example, a description of the behavior of the fluid phase in a compacting sediment pile requires the addition to Equation 37 of terms for growth or dissolution of solid phases, for chemical reactions, and perhaps for viscous flow. There is relatively little accumulated experimental information from which to construct selection rules for the forces and fluxes associated with these phenomena. Microscopic analysis does provide a general set of selection rules, but, as we shall see, there are difficulties in projecting them from the microscopic to the macroscopic world. The source of the problem is easily brought out by examining the essence of Onsager's derivation.

Microscopic Analysis

The microscopic properties of a system may be described in terms of local fluctuations about the equilibrium state. Such fluctuations are generally too small, too local, and too transient to be detected by macroscopic instruments. Nevertheless fluctuations play an important role in the kinetic and statistical-mechanical treatment of transport processes and stability. The kinetic treatment of diffusion, for example, supposes that atoms jump continuously in a random fashion from position to position in a crystal (Manning 1968b, Shewmon 1963, Girifalco 1964, 1973; Flynn 1972). Even in the absence of microscopic concentration gradients, the jumping permits a continual rearrangement of atoms in a crystalline solution. The result is that there are local fluctuations of density, volume, energy, etc, as different atoms appear in local combinations. The fluctuations form and decay continuously, and it is only in the presence of concentration gradients imposed initially on the system that the random jumping leads to measurable change in composition at a point.

The principle of microscopic reversibility asserts that the microscopic equations of motion are invariant under a reversal of time. That is, if $-t$ is substituted for $+t$, the equations of motion will run backward through the sequence of states traversed previously in the positive time direction. The principle applies equally well to microscopic flucutations about an equilibrium state, for which it states that fluctuations have the same frequency to and from equilibrium in a system at macroscopic equilibrium. One consequence is that at equilibrium the probability of occurrence of any microscopic process must be equal to the probability of occurrence of the reverse process; this statement may be recognized as the familiar principle of detailed balance, which was used by Onsager (1931a) to derive his reciprocal relation. A broader derivation directly from microscopic reversibility followed in the second paper (Onsager 1931b).

If a_i is a particular value of a state variable (U, V, T, etc) whose equilibrium value is A_i^o, then the fluctuations may be described by

$$a_i = A_i - A_i^o \qquad (i = 1, 2, \ldots, r). \tag{59}$$

The local variation in entropy with a particular fluctuation, ΔS, may be written as a Taylor's series expansion in the a_i. A good approximation (Onsager 1931b, Reif 1965, Yourgrau et al 1966) gives

$$\Delta S = -(1/2) \sum_{i=1}^{r} \sum_{j=1}^{r} (\partial^2 S/\partial a_i \, \partial a_j)/a_i \, a_j, \tag{60}$$

where S is the local value of entropy at equilibrium. Each fluctuation creates a force proportional to the magnitude of the deviation from equilibrium and a corresponding flux that tends to restore the equilibrium state. Onsager (1931b) defined the fluxes as

$$J_i = \mathrm{d}a_i/\mathrm{d}t, \tag{61}$$

and the forces as

$$X_i = -\sum_{j=1}^{r} (\partial^2 S/\partial a_i \, \partial a_j) a_j. \tag{62}$$

Differentiating Equation 60 with respect to time and substituting J_i and X_i, respectively, for the expressions on the righthand sides of Equations 61 and 62, the local entropy production assumes the form of Equation 37.

Microscopic reversibility provides a symmetric correlation in time between the deviations in state variables,

$$\langle a_i(t)a_j(t+\Upsilon)\rangle = \langle a_i(t+\Upsilon)a_j(t)\rangle, \tag{63}$$

where Υ is an interval that is long relative to the characteristic time of molecular motion, e.g. the time for individual jumps in diffusion [in metals near their melting temperature, each atom changes position approximately 10^8 times per sec (Shewmon 1963)] but short relative to the time of decay towards equilibrium. The angular brackets in Equation 63 denote a time average over an extended period of time for a volume element of the system (or, alternatively, an average at one time over the whole ensemble that models the system). Onsager further postulated that the rate of macroscopic irreversible processes is the same as the average rate of decay, or regression, of a fluctuation from its extreme value toward equilibrium. Equivalently, the postulate asserts that at any point in a system we cannot distinguish between a fluctuation about an equilibrium state and a state with similar coordinates (A_i) that results from a nonequilibrium process. Provided that the fluxes and forces are linearly related, the system will relax to the same equilibrium state regardless of the source of the state with coordinates A_i. This statement is best regarded as a new, but reasonable postulate at the microscopic level (Casimir 1945).

The important point is that the fluxes must be time derivatives of the state variables, i.e. the variables treated in Equation 63, if Onsager's derivation is to work in a straightforward fashion. For diffusion fluxes, the pertinent state variables are concentration factors such as ω_k or c_k. Most diffusion experiments are arranged to produce strictly one-dimensional diffusion, either in an isotropic or pseudoisotropic medium, or along one of the principal axes in an anisotropic medium. Many natural systems can also be reduced to one-dimensional problems. As is evident from Equation 26, which applies to a volume element of any size, monitoring the change of composition with time on either side of a plane perpendicular to the axis of diffusion, x, measures not the flux J_x but rather one component of the divergence of this flux, $\partial J_x/\partial x$. The magnitude of the flux may be calculated from Fick's first law if we know the diffusion coefficient and the concentration gradient parallel to x at that point.[2] Nevertheless, it is the divergence of the flux that is the directly measurable quantity (Nye 1957, Chap. 11). The same statements may be made about the heat flux or, in fact, about any vector flux (Hooyman et al 1955), and clearly they remain true for the three-dimensional case, although the calculation of the

[2] It is worth noting that, except at infinite dilution, the diffusion coefficient will ordinarily be a function of concentration and therefore, implicitly, of x. The dependence of D on concentration is often exponential, but the exact form may be a function of temperature and pressure. Because the concentration of defects that effects diffusion in a crystal is determined by equilibration with other phases, the partial pressure of oxygen, sulfur, etc, may be a very important factor in diffusion in oxides, sulfides, etc (van Gool 1966, Swalin 1972, Buening & Buseck 1973, Stevenson 1973, Wagner 1973, Birchenall 1974).

flux at a point requires the appropriate components of the second-rank tensor of diffusion coefficients.

An inspection of Equation 26 also reveals that it is the divergence of the diffusion flux rather than the flux that is a time derivative of a state variable. Consequently, vector fluxes cannot be made consistent with Equation 63. Hooyman et al (1955) first proved that vector fluxes may be transformed in a way that preserves the Onsager relations; their proof depends on the invariance of the deviation of the entropy production from its equilibrium value rather than the invariance of the entropy production itself. In the absence of a magnetic field, the antisymmetric matrix in Equation 58 becomes zero if the thermodynamic forces are transformed by $\mathbf{P}^{-1}$, where $\mathbf{P}$ is the matrix that produces a linear transformation of the fluxes (de Groot & Mazur 1962). The matrix $\mathbf{B}$ for diffusion that occurs in Equations 16 and 41 is evidently a particular example of $\mathbf{P}$.

In a separate proof, de Groot & Mazur (1962) demonstrated that the L_{ik} connecting vector or tensor quantities may also be symmetric provided they satisfy Equations 37 and 38. Unfortunately, this proof tends to void the generality of Onsager's derivation and the selection rules obtained from it. Moreover, a certain arbitrariness remains in the choice of fluxes and forces. Arguments surrounding the selection of fluxes and forces have tended to enhance the speculation that the origin of symmetric coefficients is rooted in some more fundamental condition only partially elucidated by Onsager's derivation.

IONIC FLUXES

Diffusion in most geological environments takes place in solutions that are initially electrically neutral and that remain so. Consequently, diffusion may be most easily treated by choosing electrically neutral species, e.g. almandine, grossular and spessartine in garnets, or NaCl and KCl in aqueous solutions of these salts. Haase (1969) has discussed the selection of neutral species and their relationship to thermodynamic components. However, we must return to the migration of individual ions if we wish to estimate diffusion coefficients and obtain a fundamental understanding of diffusion.

Because of the labor involved in directly measuring the D_{ik} of Equation 1, especially in systems of four or more components (Duda & Vrentas 1965), there have been extensive efforts to utilize D_{ik} values obtained from simpler systems or calculated from other kinds of measurements. In particular, the D_{ik} can be related to ion mobilities determined from limiting ionic conductances (Wendt 1965, Pikal 1971a,b), or from radiotracer measurements. A parallel development of theory has included the generalization of Darken's equations from binary to multicomponent solutions (Cooper 1965, Ziebold & Cooper 1965, Cooper & Heasley 1966) and the formulation of the kinetic models of Lane & Kirkaldy (1964, 1965, 1966). Although the models of Lane & Kirkaldy are initially kinetic in form, the parameters they contain are necessarily approximated from tracer diffusion coefficients or limiting ionic conductances rather than being calculated from interionic forces. In their ultimate application, the models are in fact almost devoid of kinetic information.

However, they are extremely useful for keeping track of relationships, and they give an intuitive feel for diffusion processes that is lacking in purely macroscopic models. Calculations based upon interionic forces have been made for aqueous electrolytes (Onsager & Fuoss 1932, Onsager & Kim 1957, Pikal 1971a, 1971b), but are valid only for solutions less concentrated than 0.1 molal.

Relations of greater generality between diffusion coefficients and ionic conductances have been formulated by Miller (1966, 1967a, 1967b). Means of extrapolating from the limiting binaries into a ternary system have been explored by Kirkaldy (1959) and by Miller (1966, 1967a, 1967b). The methods developed by Miller for aqueous electrolytes have proved to be extremely accurate, but they require much more extensive thermodynamic and conductance data than do the simpler models of the previous paragraph. Miller (1967a) has compared the accuracies of the various models for aqueous electrolytes. D. E. Anderson & D. L. Graf (in preparation) have attempted to simplify the Lane-Kirkaldy model for aqueous electrolytes and to improve its accuracy.

The theoretical basis at the microscopic level for coupling of ionic fluxes is not clearly understood (Flynn 1972). Suggested explanations, stated in terms of macroscopic principles, include: (*a*) maintenance of local electrical neutrality, (*b*) mass-balance or stoichiometry requirements, (*c*) coupling of thermodynamic forces through the Gibbs-Duhem equation, and (*d*) relationships introduced by transformations among reference frames.

Undoubtedly, the motions of the ions K^+, Na^+, and Cl^- in an aqueous solution of NaCl and KCl are coupled by the necessity of preserving local electrical neutrality. In contrast, both stoichiometry and local electrical neutrality were invoked by Howard & Lidiard (1964) in their analysis of the coupling of fluxes in crystals. The total number of atomic sites in a crystal is conserved except in the vicinity of dislocations and internal surfaces. In more complicated crystals, diffusion is considered to occur separately on a number of sublattices. Within a particular sublattice, maintenance of stoichiometry results in coupling of the migrations of the ions and the vacancies that share the set of atomic sites constituting the sublattice. If ions of the same charge diffuse by one-to-one exchange, the constraints of stoichiometry and local charge balance coincide. Typically, the ions diffusing through a sublattice have the same charge, but their movement also involves charged or uncharged vacancies. In theory, it is possible to calculate the number of vacancies within the sublattice of a crystal at thermal equilibrium. Thus, stoichiometry might be employed to investigate coupling of fluxes. However, it is usually simpler to appeal to the preservation of electrical neutrality.

The third of the explanations listed above was suggested by Lane & Kirkaldy (1964). In their models, certainly, coupling arises in part from application of the Gibbs-Duhem equation and in part from the electrical neutrality requirement. However, the fluxes of individual ions are not coupled through the Gibbs-Duhem equation in the treatment of Miller (1967a); this form of coupling is evidently peculiar to the particular kinetic model employed by Lane & Kirkaldy. Their calculations may be rearranged so that the cross-terms originate entirely from transformations between the molecular frame, in which the kinetic equations are formulated, and other reference frames. Ziebold & Cooper (1965) also found that cross-coupling may result from transformation between reference frames.

Clearly, in the absence of detailed microscopic models, we may choose as a matter of convenience among a variety of macroscopic methods for handling cross-coupling of fluxes. For most problems involving ionic fluxes, the preservation of local electrical neutrality provides the simplest and most convenient method.

In the absence of an applied electrical field, any migration of ions that destroys local electrical neutrality will create an internal electrical field. Reciprocally, this virtual field will act to counter any migration of ions that does not preserve local electrical neutrality. Thus we may add to the thermodynamic force X_k in Equations 37 and 38 an additional term for the virtual electrical field (ϕ), to give

$$j_i^o = -\sum_{k=1}^{n-1} l_{ik}^o(\partial\mu_k/\partial x - z_k F\,\partial\phi/\partial x) \qquad (i = 1, 2, \ldots, n-1), \tag{64}$$

where z_k is the valence of the kth ion and F is the Faraday. The symbols j_i^o and l_{ik}^o denote the fluxes and phenomenological coefficients of ionic components. By demanding in the solution of Equation 64 that the total electrical current $\sum_k z_k F(\partial\phi/\partial x)$ always be zero, we automatically insure local electrical neutrality.

Equations 64 are written for the solvent-fixed reference frame. As has been shown by de Groot & Mazur (1962, Chap. 13), this equation is strictly valid only in reference frames for which the velocity v^a may be taken as zero. Only the laboratory-fixed and solvent-fixed (lattice-fixed) reference frames satisfy this condition. The restriction on suitable reference frames arises from the fact that the complete equation for the entropy production (de Groot & Mazur 1962) contains a term $(1/c)v_k \times B$, in which c is the velocity of light and B, the magnetic field. The symbol $\times$ indicates the cross-product of the vector v_k and the axial vector B. At mechanical equilibrium, we can by a simple manipulation replace v_k by v^a. If we are in a reference frame in which v^a is zero, this term may be eliminated, and it will not be necessary to employ the modified form of the Onsager reciprocity relation that obtains in the presence of a magnetic field (Casimir 1945). Calculations suggest, as perhaps might have been anticipated, that for diffusion in aqueous electrolytes less concentrated than 3 molar, imposing the electrical neutrality condition in other reference frames causes no significant error.

The chemical potentials in Equation 64 are those of ionic components. Although such potentials are well-defined physical quantities, they are unmeasurable (Denbigh 1971). They can be specified only in relation to a neutral component, according to

$$\mu_{ia} = r_{ic}\,\mu_i + r_{ia}\,\mu_a, \tag{65}$$

where the r_{ic} and r_{ia} are the stoichiometric coefficients of the cation (C) and anion (A), respectively, in the dissociation equation

$$C_{r_{ic}}A_{r_{ia}} = r_{ic}C^{z_i} + r_{ia}A^{z_a} \tag{66}$$

In the solution of Equation 64 for zero electric current, the chemical potentials of the ions appear in combinations consistent with Equation 65 and may be replaced with the chemical potentials of the neutral components, which are measurable. This result is hardly surprising, because an electrically neutral liquid or solid solution must contain ionic species only in combinations that are compatible with Equations 65 and 66. Solutions of Equation 64 are given by Miller (1967a) for electrolyte solutions with a neutral solvent and a single anion, by D. E. Anderson and D. L. Graf (in preparation) for electrolyte solutions with a neutral solvent and two anions, and by D. E. Anderson (1975) for certain classes of solutions in which there is no neutral species, solvent or otherwise.

PHENOMENOLOGICAL AND PRACTICAL DIFFUSION COEFFICIENTS

Suppose that Equation 1 is written separately for two reference frames a and b, the set of compositional parameters y_k being the same in both frames. The diffusion coefficients then transform between a and b according to the same scheme as the fluxes in Equation 16, i.e.

$$D_{ik}^{a} = \sum_{j=1}^{n-1} B_{ij} D_{jk}^{b}. \tag{67}$$

Thus, diffusion coefficients measured in the volume-fixed reference frame may be transformed into any other reference frame required for the solution of a particular problem. All possible transformations between mass-, volume-, solvent-, and molecular-fixed frames may be carried out if we know the partial densities and partial specific volumes of the $(n-1)$ components of the solution.

New compositional parameters y_k' may be substituted in Equation 1 for the y_k by obtaining solutions to

$$\partial y_k'/\partial x = \sum_{j=1}^{n-1} (\partial y_k'/\partial y_j)(\partial y_j/\partial x) \qquad (k = 1, 2, \ldots, n-1), \tag{68}$$

in which temperature, pressure, and all concentrations except those of k and j are held constant. Numerous solutions of Equation 68 may be found in de Groot & Mazur (1962).

By a derivation that exactly parallels the usual derivation of Fick's second law for a one-dimensional system (Crank 1956), a set of differential equations for multicomponent solutions may be derived from Equation 1:

$$\partial c_i/\partial t = \sum_{k=1}^{n-1} \partial/\partial x(D_{ik}^{a}\, \partial c_k/\partial x) \qquad (i = 1, 2, \ldots, n-1). \tag{69}$$

Thus, given measured or estimated values of the D_{ik}^{a}, all manipulations needed to solve practical problems may be completed without recourse to nonequilibrium thermodynamics. The fact that the matrices D_{ik}^{a} are not in general symmetrical is only a minor handicap in numerical calculations. It is always possible to find symmetrical L_{ik} matrices, but their appearance does not reduce the number of experiments needed to obtain D_{ik}^{a} for a particular system. From a practical point of view, the most important feature is the derivation of the matrix B_{ik} defined by

Equation 17. One may then reasonably ask, what is the value of nonequilibrium thermodynamics in the analyses of multicomponent diffusion?

Nonequilibrium thermodynamics has played only a minor role in the study of diffusion in metals and ceramics, at least in part because of the special interest of metallurgists in binary systems and in dilute ternary systems that approximate binary behavior. The use of nonequilibrium thermodynamics to interpret experimental measurements in ternary aqueous electrolyte solutions has arisen, not from a concern with the chemistry of the particular systems, but rather from a peripheral interest in testing the Onsager reciprocity relation. Only in the theoretical attempts to obtain expressions for diffusion coefficients from the fundamental properties of ions in aqueous solutions (Onsager & Fuoss 1932; Onsager & Kim 1957; Pikal 1971a, 1971b) has it played a central role.

Duda & Vrentas (1965) described the difficulty, if not the impossibility, of conducting diffusion experiments in systems of four or more components. Yet electrolyte solutions of this complexity are commonly encountered in geological problems. For geologists, therefore, the theoretical estimation procedures outlined above assume a vital importance. Existing models of diffusion in metals, simple oxides, and dilute electrolytes will have to be modified to deal with the problems of diffusion in silicates, silicate melts, and concentrated electrolyte solutions. For example, the kinetic models of Lane & Kirkaldy provide reasonable estimates of diffusion coefficients in aqueous solutions up to 1 or 2 molar concentration. The decreased accuracy of prediction toward the upper end of this concentration range suggests that the models will have to be modified if still more concentrated solutions are to be treated. Correlation and vacancy-wind effects (Manning 1968a, 1968b) have proved to be very important in iron oxides and may well be equally important in some iron-bearing silicates. It is an open question whether these effects are adequately treated in the Lane-Kirkaldy models, but the need for some modification might reasonably be expected. Important problems connected with the conservation of electrical neutrality in silicates remain unsolved; they have no counterpart in metals, halides, or simple oxides.

It is in the construction of theoretical models and the theoretical analysis of multicomponent diffusion that nonequilibrium thermodynamics becomes important. The manipulation of practical diffusion coefficients in Equations 1, 67, and 69 is purely phenomenological; nothing is predicted that can be tested, beyond the behavior of the particular system modeled. In contrast, nonequilibrium thermodynamics predicts a general relationship, Equation 39, that is independent of any particular system and that may be confirmed or denied by experiment, by trial numerical calculation, or by inspection of the form of the governing equations. At least in the study of isothermal, isobaric diffusion, the failure of a theoretical model or estimation procedure to produce a symmetric matrix of phenomenological coefficients L_{ik}^{a} is clear proof of the failure of the model. The successful prediction of the Onsager reciprocity relation is a focal point in the analysis of multicomponent diffusion.

The appearance of symmetry relations in isotropic ternary liquids and crystalline solutions in which vacancy or interstitial mechanisms prevail is something of a surprise. Any theoretical analysis of diffusion must ultimately account for these relations. Nonequilibrium thermodynamics, with all its inherent weaknesses, at least allows a coordinated macroscopic analysis of symmetrical phenomena.

Methods for calculating L_{ik}^{a} from measured D_{ik}^{a} may be found in Kirkwood et al (1960), Fitts (1962), de Groot & Mazur (1962), and Haase (1969). A specific example is worked out in detail in Fitts (1962).

Literature Cited

Anderson, D. E. 1976. Diffusion in metamorphic tectonites; lattice-fixed reference frames. *Phil. Trans. Roy. Soc. A:* In press

Anderson, D. E., Buckley, G. R. 1973. Zoning in garnets-diffusion models. *Contrib. Mineral. Petrol.* 40:87–104

Bardeen, J., Herring, C. 1951. Diffusion in alloys and the Kirkendall effect. In *Atom Movements,* 87–111. Cleveland, Ohio: Am. Soc. Metals

Bear, Jacob 1972. *Dynamics of Fluids in Porous Media.* New York: Elsevier. 764 pp.

Ben-Yaakov, S. 1972. Diffusion of sea water ions—I. Diffusion of sea water into dilute solution. *Geochim. Cosmochim. Acta* 36: 1395–1406

Berner, R. A. 1975. Diagenetic models of dissolved species in the interstitial waters of compacting sediments. *Am. J. Sci.* 275:88–96

Birchenall, C. E. 1974. Diffusion in sulfides. In Hofmann et al 1974, pp. 53–59

Bird, R. B., Stewart, W. E., Lightfoot, E. N. 1960. *Transport Phenomena.* New York: Wiley. 780 pp.

Boltzmann, L. 1894. Zur Integration der Diffusionsgleiching bei variabeln Diffusionscoefficienten. *Ann. Phys. Chem. N.F.* 53:959–64

Buening, D. K., Buseck, P. R. 1973. Fe-Mg lattice diffusion in olivine. *J. Geophys. Res.* 78:6852–62

Callen, H. B. 1960. *Thermodynamics.* New York: Wiley. 376 pp.

Casimir, H. B. G. 1945. On Onsager's principle of microscopic reversibility. *Rev. Mod. Phys.* 17:343–50

Cooper, A. R. Jr. 1965. Model for multicomponent diffusion. *Phys. Chem. Glasses* 6:55–61

Cooper, A. R. Jr. 1974. Vector space treatment of multicomponent diffusion. In Hofmann et al 1974, pp. 15–30

Cooper, A. R. Jr., Heasley, J. H. 1966. Extension of Darken's equation to binary diffusion in ceramics. *J. Am. Ceram. Soc.* 49:280–83

Correa da Silva, L. C., Mehl, R. F. 1951. Interface and marker movements in diffusion in solids and metals. *Trans. Am. Inst. Min. Metallurg. Pet. Eng.* 191:155–73

Crank, J. 1956. *The Mathematics of Diffusion.* Oxford: Oxford Univ. Press. 347 pp.

Darken, L. S. 1948. Diffusion, mobility, and their interrelation through free energy in binary metallic systems. *Trans. Am. Inst. Min. Metallurg. Pet. Eng.* 175:189–201

Darken, L. S. 1949. Diffusion of carbon in austenite with a discontinuity in composition. *Trans. Am. Inst. Min. Metallurg. Pet. Eng.* 180:430–38

Darken, L. S. 1951. Formal basis of diffusion theory. In *Atom Movements,* pp. 1–25. Cleveland: *Am. Soc. Metals*

Darken, L. S., Gurry, R. W. 1953. *Physical Chemistry of Metals.* New York: McGraw-Hill. 535 pp.

Davies, R. O. 1952. Transformation properties of the Onsager relations. *Physica* 18:182

de Groot, S. R. 1952. *Thermodynamics of Irreversible Processes.* Amsterdam: North Holland. 242 pp.

de Groot, S. R., Mazur, P. 1962. *Nonequilibrium Thermodynamics.* Amsterdam: North Holland. 510 pp.

Denbigh, Kenneth 1971. *The Principles of Chemical Equilibrium.* Cambridge, Engl: Cambridge Univ. Press. 494 pp.

Duda, J. L., Vrentas, J. S. 1965. Mathematical analysis of multicomponent free-diffusion experiments. *J. Phys. Chem.* 69:3305–13

Elliott, D. 1973. Diffusion flow laws in metamorphic rocks. *Geol. Soc. Am. Bull.* 84:2645–64

Fanning, K. A., Pilson, M. E. Q. 1974. The diffusion of dissolved silica out of deep-sea sediments. *J. Geophys. Res.* 79:1293–97

Fisher, G. W., Elliott, D. 1974. Criteria for quasi-steady diffusion and local equilibrium in metamorphism. In Hofmann et al 1974, pp. 231–41

Fitts, D. D. 1962. *Nonequilibrium Thermodynamics.* New York: McGraw-Hill. 173 pp.

Flynn, P. C. 1972. *Point Defects and Diffusion.* Oxford: Oxford Univ. Press. 826 pp.

Gibbs, J. W. 1961. *The Scientific Papers of*

J. Willard Gibbs, Volume 1, Thermodynamics. New York: Dover. 434 pp.
Giletti, B. J. 1974. Diffusion related to geochronology. In Hofmann et al 1974, pp. 61–76
Girifalco, L. A. 1964. *Atomic Migration in Crystals.* New York: Blaisdell. 162 pp.
Girifalco, L. A. 1973. *Statistical Physics of Materials.* New York: Wiley. 346 pp.
Graf, D. L., Anderson, D. E. 1975. Properties of one-dimensional diagenetic models. *J. Geol.* 83:331–48
Haase, Rolf 1969. *Thermodynamics of Irreversible Processes.* Reading, Mass: Addison-Wesley. 509 pp.
Hartley, G. S. 1946. Diffusion and swelling of high polymers Part I. The swelling and solution of a high polymer solid considered as a diffusion process. *Trans. Faraday Soc.* 42B:6–11
Hartley, G. S., Crank, J. 1949. Some fundamental definitions and concepts in diffusion processes. *Trans. Faraday Soc.* 45:801–18
Hofmann, A. W., Giletti, B. J., Yoder, H. S. Jr., Yund, R. A., eds. 1974. *Geochemical Transport and Kinetics.* Publ. No. 634. Washington DC: Carnegie Inst. Washington. 353 pp.
Hooyman, G. J. 1956. Thermodynamics of diffusion in multicomponent systems. *Physica* 22:751–59
Hooyman, G. J., de Groot, S. R., Mazur, P. 1955. Transformation properties of the Onsager relations. *Physica* 21:360–66
Howard, R. E., Lidiard, A. B. 1964. Matter transport in solids. *Rep. Prog. Phys.* 27: 161–240
Katchalsky, A., Curran, P. F. 1967. *Nonequilibrium Thermodynamics in Biophysics.* Cambridge, Mass: Harvard Univ. Press. 248 pp.
Kirkaldy, J. S. 1959. Diffusion in multicomponent metallic systems IV. A general theorem for construction of multicomponent solutions from solutions of the binary diffusion equations. *Can. J. Phys.* 37:30–34
Kirkwood, J. G., Baldwin, R. L., Dunlop, P. J., Gosting, L. G., Kergeles, G. 1960. Flow equations and frames of reference for isothermal diffusion in liquids. *J. Chem. Phys.* 33:1505–13
Lane, J. E., Kirkaldy, J. S. 1964. Diffusion in multicomponent systems. VIII. A kinetic calculation of the Onsager L coefficients in substitutional solid solutions. *Can. J. Phys.* 42:1643–57
Lane, J. E., Kirkaldy, J. S. 1965. A quasicrystalline model of diffusion in ternary liquid systems. *Can. J. Chem.* 43:1812–28
Lane, J. E., Kirkaldy, J. S. 1966. Diffusion in multicomponent aqueous systems. *Can. J. Chem.* 44:477–85
Lasaga, A. C., Holland, H. D. 1974. The mathematics of non-steady state diagenesis. *Trans. Am. Geophys. Union* 55: 696
Lazarus, D. 1960. Diffusion in metals. In *Solid State Physics,* ed. F. Seitz, D. Turnbull, 10:71–126. New York: Academic
Lerman, A., Jones, B. F. 1973. Transient and steady-state salt transport between sediments and brine in closed lakes. *Limnol. Oceanogr.* 18:72–85
Li, Y.-H., Gregory, S. 1974. Diffusion of ions in sea water and in deep-sea sediments. *Geochim. Cosmochim. Acta* 38: 703–14
Manning, J. R. 1967. Diffusion and the Kirkendall shift in binary alloys. *Acta Metallurg.* 15:817–26
Manning, J. R. 1968a. Vacancy-wind effect in diffusion and deviation from thermodynamic equilibrium conditions. *Can. J. Phys.* 46:2633–43
Manning, J. R. 1968b. *Diffusion Kinetics for Atoms in Crystals.* Princeton, NJ: Van Nostrand. 257 pp.
Matano, C. 1933. On the relation between the diffusion-coefficients and concentrations of solid metals (the nickel-copper system). *J. Phys. (Jpn)* 8:109–12
Michard, G., Church, T. M., Bernat, M. 1974. The pore water chemistry of Recent sediments in the Western Mediterranean Basin. *J. Geophys. Res.* 79:817–24
Miller, D. G. 1960. Thermodynamics of irreversible processes, the experimental verification of the Onsager reciprocal relations. *Chem. Rev.* 60:15–37
Miller, D. G. 1966. Application of irreversible thermodynamics to electrolyte solutions. I. Determination of ionic transport coefficients l_{ij} for isothermal vector transport processes in binary electrolyte systems. *J. Phys. Chem.* 70: 2639–59
Miller, D. G. 1967a. Application of irreversible thermodynamics to electrolyte solutions. II. Ionic coefficients l_{ij} for isothermal vector transport processes in ternary systems. *J. Phys. Chem.* 71:616–32
Miller, D. G. 1967b. Application of irreversible thermodynamics to electrolyte solutions. III. Equations for isothermal vector transport processes in n-component systems. *J. Phys. Chem.* 71:3588–92
Miller, D. G. 1969. The experimental verifications of the Onsager reciprocal relations. In *Transport Phenomena in Fluids,* ed

H. J. M. Hanley, 377–432. New York: Dekker

Muehlenbachs, K., Kushiro, I. 1974. Oxygen isotope exchange and equilibrium of silicates with CO_2 or O_2. In *Carnegie Inst. Washington. Yearb.* 73:232–36

Nye, J. F. 1957. *Physical Properties of Crystals.* Oxford: Oxford Univ. Press. 322 pp.

Onsager, L. 1931a. Reciprocal relations in irreversible process I. *Phys. Rev.* 37:405–26

Onsager, L. 1931b. Reciprocal relations in irreversible process II. *Phys. Rev.* 38: 2265–79

Onsager, L. 1945. Theories and problems of liquid diffusion. *Ann. NY Acad. Sci.* 46: 241–65

Onsager, L., Fuoss, R. M. 1932. Irreversible processes in electrolytes. Diffusion, conductance, and viscous flow in arbitrary mixtures of strong electrolytes. *J. Phys. Chem.* 36:2689–2778

Onsager, L., Kim, S. K. 1957. The relaxation effects in mixed strong electrolytes. *J. Phys. Chem.* 61:215–29

Paterson, M. S. 1973. Nonhydrostatic thermodynamics and its geologic applications. *Rev. Geophys. Space Phys.* 11: 355–89

Pikal, M. J. 1971a. Ion-pair formation and the theory of mutual diffusion in a binary electrolyte. *J. Phys. Chem.* 75:663–75

Pikal, M. J. 1971b. Theory of the Onsager transport coefficients l_{ij} and R_{ij} for electrolyte solutions. *J. Phys. Chem.* 75: 3124–34

Prigogine, I. 1955. *Introduction to Thermodynamics of Irreversible Processes.* Springfield, Ill: Thomas. 115 pp.

Prigogine, I., Defay, R. 1954. *Chemical Thermodynamics.* London: Longmans, Green. 543 pp.

Reif, F. 1965. *Fundamentals of Statistical and Thermal Physics.* New York: McGraw-Hill. 651 pp.

Shewmon, P. G. 1963. *Diffusion in Solids.* New York: McGraw-Hill. 203 pp.

Smigelskas, A. D., Kirkendall, E. O. 1947. Zinc diffusion in alpha brass. *Trans. Am. Inst. Min. Metallurg. Pet. Eng.* 171:130–42

Soret, Charles 1894. Coefficients rotatoires de conductibilité thermique dans les cristaux. *Arch. Sci. Phys. et Natur. (Genève)* 32:631–33

Stevenson, D. A. 1973. Diffusion in the chalcogenides of Zn, Cd, and Pb. In *Atomic Diffusion in Semiconductors,* ed. D. Shaw, 431–541. New York: Plenum

Swalin, R. A. 1972. *Thermodynamics of Solids.* New York: Wiley. 387 pp.

Trimble, L. E., Finn, D., Cosgarea, A. Jr. 1965. Mathematical analysis of diffusion coefficients in binary systems. *Acta Metallurg.* 13:501–7

Truesdell, Clifford 1969. *Rational Thermodynamics.* New York: McGraw-Hill. 208 pp.

van Gool, W. 1966. *Principles of Defect Chemistry of Crystalline Solids.* New York: Academic. 148 pp.

Voigt, W. 1903. Fragen der Kristallphysik, I. *Nachr. Ges. Wiss. Göttingen* 3:87–89

Wagner, J. B. Jr. 1973. Diffusion in oxide semiconductors. In *Atomic Diffusion in Semiconductors,* ed. D. Shaw, 543–600. New York: Plenum

Wendt, R. P. 1965. The estimation of diffusion coefficients for ternary systems of strong and weak electrolytes. *J. Phys. Chem.* 69:1227–37

Wendt, R. P., Shamim, Mohammed 1970. Isothermal diffusion in the system water-magnesium chloride-sodium chloride as studied with the rotating diaphragm cell. *J. Phys. Chem.* 74:2770–83

Yourgrau, W., van der Merwe, A., Raw, G. 1966. *Treatise on Irreversible and Statistical Thermophysics.* New York: Macmillan. 268 pp.

Yund, R. A., Anderson, T. F. 1974. Oxygen isotope exchange between potassium feldspar and KCl solution. In Hofmann et al 1974, pp. 99–105

Ziebold, T. O., Cooper, A. R. Jr. 1965. Atomic mobilities and multicomponent diffusion. *Acta Metallurg.* 13:465–70

RADAR IMAGERY IN DEFINING REGIONAL TECTONIC STRUCTURE

Terry A. Grant and Lloyd S. Cluff[1]
Woodward-Clyde Consultants, Two Embarcadero Center, San Francisco, California 94111

INTRODUCTION

Side-Looking Airborne Radar (SLAR) is a relatively new remote sensing tool used increasingly in the past few years for geologic interpretations of regional tectonic structures. Use of SLAR imagery still is not extensive, however, because of its limited availability and relatively high cost compared to other types of imagery. SLAR systems produce strip images of moderate to low resolution (SLAR systems can discriminate between reflecting targets that have a minimum spacing of 15–30 m) and small scales (1:250,000 to 1:1,000,000). These strip images usually are assembled as a mosaic to create images that cover relatively large areas.

SLAR imagery can be very useful in regional tectonics studies because the imagery affords a synoptic view of large regional structures. A single SLAR mosaic may cover several thousand square kilometers, whereas standard aerial photographs usually cover 100–200 km^2. Regional patterns and structures become evident on the SLAR mosaic because the entire region may be viewed on one image. Because individual aerial photographs cover only small areas, large regional structures that may extend across many photographs may go undetected on individual photographs. The shadowing characteristics of SLAR enhance topographic detail and emphasize linear features such as faults and regional joint and fracture patterns. Although commercial SLAR systems do not penetrate vegetation, their low resolution tends to suppress vegetative detail and emphasize topographic detail that reflects the underlying geology. SLAR reconnaissance has been used in all types of terrain and has been particularly successful in tropical areas because of SLAR's unique ability to penetrate clouds. SLAR is an active system that provides its own energy or illumination. Thus, SLAR imagery may be flown at any time of the day or night, and the direction and amount of shadowing can be controlled. The radar wavelength of commercial systems ranges between 0.75 cm and 3.75 cm. Two types

[1] The authors acknowledge with appreciation the technical reviews and helpful suggestions provided by Janet L. Born, Ronald H. Gelnett, William R. Hansen, and Charles L. Taylor.

of SLAR imaging systems are in use today: real aperture systems and synthetic aperture systems. The imagery produced by these systems is quite similar and is interpreted in the same manner.

This paper briefly reviews the types and characteristics of SLAR imagery and compares SLAR systems with other photographic systems. These discussions are followed by descriptions of the techniques of SLAR imagery interpretation and by examples of various applications of SLAR imagery in defining regional tectonic structure.

Real Aperture System

Figure 1 diagrams the operation of a real aperture system. A long rodlike antenna (*A*) is mounted beneath the aircraft with its long axis parallel to the direction of flight (V_a). Short pulses of radio frequency energy (*B*) are emitted at the velocity of light (3×10^8 m/sec). The configuration of the antenna confines the pulse to a narrow beam oriented perpendicular to the aircraft. After a pulse is emitted, the antenna switches to a receiving mode and listens for energy reradiated toward the antenna from the ground (MacDonald 1969).

Several factors affect the return amplitude and consequently the tonal variation on the radar image. Incidence angle is the most important factor in determining return amplitude. Low incidence angles, such as would occur on slopes facing the aircraft, produce a higher backscattering return to the antenna than do high incid-

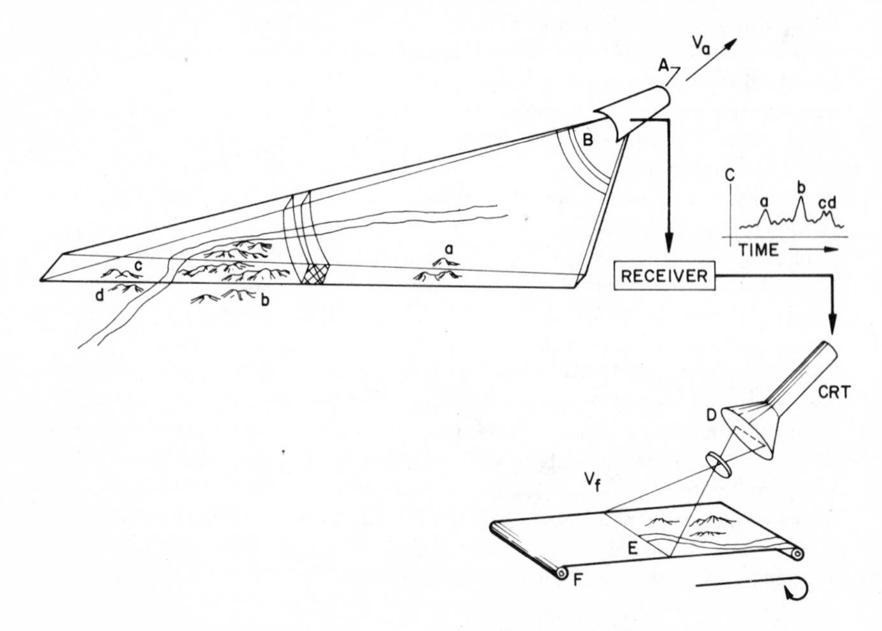

Figure 1 Diagram of the operation of a real aperture SLAR system (from MacDonald 1969).

ence angles, such as would occur on flat ground or slopes facing away from the antenna. The second most important factor in determining return amplitude is terrain roughness. Smooth surfaces, such as bodies of water, tend to reflect the radar pulse away from the antenna at an angle equal to the angle of incidence (Snell's Law) so that little energy is reflected back to the aircraft. These areas appear dark on the radar image. If the ground surface or vegetation have an appreciable surface roughness in terms of radar wavelength, diffuse reflection will occur and considerable energy will be reradiated back toward the aircraft, causing such areas to appear lighter-toned on the radar image. A third factor affecting the return is the complex dielectric constant, which includes the microwave reflectivity and conductivity of the terrain surface. The complex dielectric constant is greatly influenced by moisture content (Dellwig, MacDonald & Waite 1973). A relatively high soil moisture content will cause an increase in the dielectric constant and the reflectivity of the soil.

Figure 1 shows that hills *a, b, c,* and *d* have reradiated a higher amplitude return (*C*) than the surrounding terrain. The distance from the aircraft to a specific point is determined by the travel time of a pulse and its return. The return from a pulse is recorded at the receiver and displayed as a single scan line on a cathode ray tube (CRT). The scan line is passed through a lens and exposed on photographic film (*F*) as a single line at *E*. The film moves past the CRT display line at a velocity (V_f) proportional to the velocity of the aircraft (V_a), so that a continuous strip image is built up on the film by sequential radar pulses as the aircraft moves down the flight line (MacDonald 1969). Recording directly onto photographic film affords a greater dynamic range (more shades of gray) on the imagery than does synthetic aperture radar.

Synthetic Aperture System

In a real aperture system, the length of the antenna determines the horizontal width of the beam generated and the distance from the aircraft at which usable results can be gained. To obtain very narrow beam widths and operate at great distances from the aircraft, a very long antenna is required. A physical antenna of this length would be difficult or impossible to carry in the aircraft and keep stabilized. To avoid these difficulties, a synthetic aperture system is used. The synthetic aperture system uses a short antenna that produces a broadly lobed beam. The reflected return from a single object is recorded by the system, along with the Doppler shift of the radar frequency at several points along the flight path. This has the effect of creating a very long synthetic antenna. The Doppler shift is used to determine the angles at which the energy is returned. The end product of the succession of images is a one-dimensional "zone plate" recorded on photographic film that acts as a lens to focus laser light to a point that represents a reflecting point on the radar imagery. At the end of a SLAR mission, the zone plate photographic film is processed and fed into an optical correlator. Coherent laser light is passed through the film and optically focused on a second strip of photographic film on which the final SLAR image is produced. This system produces better resolution at long range than real aperture systems.

Radar Shadow

Because the radar beam in a SLAR system is directed to the side of the aircraft, the radar images the terrain at relatively high angles of incidence. At these angles, terrain having even moderate relief will block the radar signal from reaching the slopes facing away from the radar. These areas of no return appear as shadows on the radar image (Figure 2*a*). Shadowing emphasizes terrain detail, particularly linear geologic features, and provides a three-dimensional effect when viewing the imagery. Shadow lengths are not uniform across the image, but increase with distance from the radar receiver as the angle of incidence increases.

Look-Direction

The look-direction of the radar is of great importance in detecting linear features of the terrain that may be the expressions of joint systems and faults. Look-direction is the direction perpendicular to the ground track of the aircraft in which the radar scans. MacDonald (1969) shows several examples of linear joint, fault, and dike systems that appear prominently on radar images with look-directions perpendicular to the strike of the features and are nearly undetectable on radar images with look-directions parallel to the strike of the features. MacDonald (1969) believes that for geologic reconnaissance in relatively unmapped areas the specific region should be imaged from four orthogonal look-directions and, in more detailed studies where the regional geology is known, a configuration of two opposing look-directions should be used.

Radar Foreshortening and Layover

Radar foreshortening is a distortion inherent in all imaging radar. Figure 2*b* illustrates foreshortening. A radar pulse from the aircraft (*A*) images the slope *BC* in a time interval equivalent to distance *D*, and slope *CE* in a time interval equivalent to distance *F*. Since the radar measures distance by using travel time, slope *BC* will appear shortened with respect to slope *CE*. Figure 2*a* illustrates an extreme case of foreshortening called layover. A radar pulse from the aircraft (*A*) strikes points *C* and *D* at the same time so that these points are imaged together at the same point. Points *F* and *E* are encountered at a later time and are imaged together at a point farther from the aircraft than *C* and *D*. In this case, the top of the hill is displaced toward the aircraft on the image so that the positions of the top and the slope are reversed. Foreshortening and layover usually occur in the near range of a radar image where high relief is present.

COMPARISON OF SLAR WITH OTHER SYSTEMS

The principal advantage of SLAR imagery in studies of regional tectonics is the synoptic view that allows the eye to integrate information from a large area and recognize regional structures that might not be apparent on conventional aerial photography, wherein each frame covers a small area. The SLAR shadowing enhances this regional view by emphasizing topography. Other photographic systems, such as high-altitude low-sun-angle photography and Earth Resources Technology Satellite (ERTS) imagery, offer similar advantages. These two photographic systems are discussed in the following sections, and the results obtained from them are compared with the results obtained from SLAR.

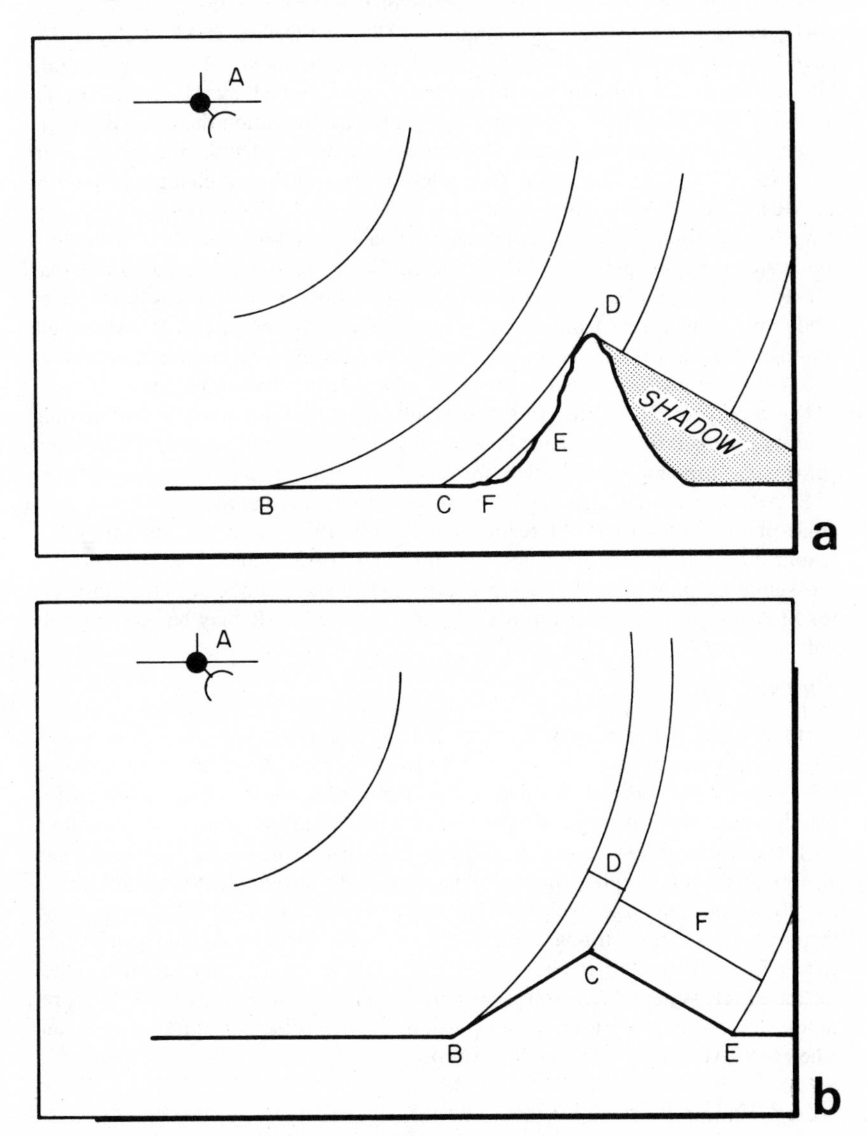

Figure 2 Diagrams illustrating shadowing, foreshortening, and layover (from Dellwig, MacDonald & Waite 1973).

High-Altitude Low-Sun-Angle Photography

Low-sun-angle aerial photography utilizes the low angle (less than 25° above the horizon) of sun illumination at different times of the day and year to obtain optimum enhancement of the irregularities of the surface of the earth. M. Clark (1971, 1973), Lyon, Mercado & Campbell (1970), and Dellwig & McCauley (1973) have discussed the relative advantages and disadvantages of SLAR and high-altitude low-sun-angle photography. In the desert region studied by M. Clark (1971), photography was thought to exhibit more geologic information than SLAR images because of its greater resolution, tonal range, geometric fidelity, and easier use in stereoscopic viewing. Dellwig & McCauley (1973) replied that changes in 1969 in SLAR imaging format, which went from a slant-range presentation to a ground-range presentation, and other improvements have increased geometric fidelity and made stereo SLAR practical. Dellwig & McCauley (1973) also pointed out that SLAR studies could be made from any look-direction and that it was desirable to study unexplored areas from several look-directions. In the case of low-sun-angle photography, the orientation of solar illumination is fixed by the time of the year, and some illumination directions are not available at any time of the year.

The ability of SLAR to penetrate clouds may be fundamental to obtaining imagery over certain areas, and the advantages of suppressed vegetative detail and emphasized topography in heavily forested areas as a result of the low resolution of SLAR imagery may also be important factors in choosing SLAR over low-sun-angle photography. Lyon, Mercado & Campbell (1970) have used special photographic processes to create images very similar to SLAR images from high-altitude low-sun-angle photography. Such pseudo-SLAR images can be a useful substitute for SLAR, especially for covering small areas where SLAR may be economically unfeasible.

ERTS Imagery

Earth Resources Technology Satellite (ERTS) imagery is very similar to SLAR imagery in the synoptic view that it affords of large areas of the earth. Because ERTS imagery is acquired by the federal government, it has the advantages of being readily available for most regions of the earth and relatively inexpensive to obtain. As is the case with low-sun-angle photography, shadowing on ERTS imagery can also highlight topography when solar lighting is at a low angle. SLAR imagery is usually obtained at larger scales and with better resolution than ERTS imagery and therefore shows more topographic detail. In many areas, the cloud penetration capability and the ability to choose illumination direction are important advantages for the SLAR system. The two systems are actually complimentary in the interpretation process in that ERTS senses portions of the visible and infrared spectrum whereas SLAR senses in the microwave spectrum.

SLAR INTERPRETATION

SLAR imagery is interpreted in much the same manner as other kinds of imagery, once the interpreter is cognizant of radar returns. Topographic, drainage, and tonal anomalies are key factors in recognizing gross lithologic boundaries and geologic structural features on SLAR imagery. The small-scale, synoptic view, and shadowing characteristics of SLAR imagery make it most useful for detecting linear terrain features (Wing & Dellwig 1970, Reeves 1969). Methods of detecting and interpreting such lineaments on SLAR imagery are similar to those described by Lattman (1958) and Tator (1960) for interpreting aerial photograph mosaics.

Elder, Jeran & Keck (1974) have demonstrated the use of the Ronchi grating to emphasize parallel sets of lineaments on SLAR mosaics. The Ronchi grating is a glass plate that has been engraved with a series of closely spaced parallel lines. When the imagery is viewed through the grating, lineaments oriented perpendicular to the lines on the grating are enhanced, and lineaments and other terrain features oriented in other directions are suppressed. The Ronchi grating is useful for detecting lineaments that are probably related to regional joint or fracture systems, but the lineaments detected by this method frequently are discontinuous, poorly developed, or not perfectly aligned; hence they must be carefully evaluated when applying this technique in fault investigations.

Recent investigations by MacDonald (1969) and Elder, Jeran & Keck (1974) have followed Lattman's (1958) terminology in calling linear terrain and vegetation features shorter than one mile in length "fracture traces" and those longer than one mile "lineaments." In these studies, fracture traces are commonly equated with joints, and lineaments with faults. Experience has shown, however, that only a very small percentage of lineaments are faults, regardless of length. Determination of the cause of a topographic or vegetational lineament is not usually possible from the SLAR imagery, and the best practice in the interpretation of SLAR imagery is not to attempt to classify lineaments as to origin. Lineaments should remain undefined until their origin can be determined by a field check. A thorough field check of lineaments detected on SLAR imagery is the only satisfactory method for classifying the lineaments. Unfortunately, this most important aspect of the SLAR interpretation process frequently receives the least amount of time and effort.

Joints and Fractures

In many areas, the most conspicuous feature on SLAR imagery is one or more sets of parallel lineaments. Some of these lineaments may be quite pronounced and lengthy; however, ground truth studies may indicate that they are not faults but rather are related to regional joint or fracture systems. The recognition of these regional lineament systems is not new (e.g. see Fisk 1944), but SLAR imagery's regional synoptic view and shadowing effect tend to emphasize these patterns. Because not all long linear features are necessarily faults, the lineament systems create an interpretation problem in differentiating between faults and regional joints. The north fork of the Stanislaus River, California (Figure 3), forms a striking linear feature approximately 65 km long that is not fault-related (Strand & Koenig 1965, L. Clark 1964). Although such long lineaments may be related to the regional jointing or fracture pattern, in most cases their origin cannot be determined without supporting ground truth data unless there is very clear geomorphic evidence. The following studies illustrate selected interpretations.

WASHINGTON The portion of Washington state south of the Olympic Mountains along the Chehalis River was investigated by Woodward-Clyde Consultants (1973) for lineaments on 1 : 250,000-scale SLAR imagery. Most of the lineaments consisted of straight stream segments and alignments of straight stream segments (Figures 4 and 5). It was recognized during the interpretation that most of these lineaments

Figure 3 SLAR imagery of the north fork of the Stanislaus River in the Sierra Nevada of California.

were not associated with geomorphic features indicative of faulting and were probably expressions of the regional jointing or fracture pattern. Published mapping and field checks of selected lineaments confirmed that these lineaments were not fault-related. The lineaments show a strong preferred orientation (Figure 6*a*), similar to what would be expected in a joint pattern. Analysis of joint and minor shear orientations (Figure 6*b*) exposed in trenches near the center of the study area shows good agreement with the lineament histogram. Although the relative importance of each set varies between the joint and lineament histograms, the preferred lineament and joint orientations coincide in most cases. The joint sets that trend N25°E and N25°W apparently control the drainage by creating planes of

Figure 4 Lineament pattern in western Washington.

weakness along joints or closely spaced series of joints that streams in the area tend to follow. The greater relative importance of other joint sets on Figure 6*b* may be caused by the local dominance of minor joint sets in the study area.

BURNING SPRINGS AREA Wing, Overby & Dellwig (1970) have studied the lineament and joint patterns in the upper plate of a regional thrust fault in the Burning Springs area of West Virginia. The SLAR imagery (Figure 7) and the aerial photography of the region show a distinctive polygonal stream pattern that suggests that the topography may be controlled by the regional joint patterns. Comparison of SLAR imagery lineament orientations with ground measurements of joint orientations (Figure 8) shows reasonably good agreement. In this example, it is possible that there may be greater scatter in the field measurements because of local varia-

Figure 5 Typical lineament in western Washington.

tions, than in the lineaments detected on SLAR imagery, where the average orientation of a linear feature is determined over a relatively long distance.

Wing, Overby & Dellwig (1970) also compared SLAR imagery lineament orientations with aerial photograph lineament orientations over part of their study area. The preferred orientations shown by this comparison are generally similar (Figure 9). It was recognized that the aerial photographs revealed the details of the polygonal features, whereas the synoptic SLAR imagery best revealed the long lineaments, alignments of short linear segments, and the overall pattern of the area.

Wing, Overby & Dellwig (1970) recognized six sets of lineaments in the upper plate of the thrust that correspond to strike, dip, and two pairs of conjugate shear fractures (Figure 9). These features are believed to be related to movement of the

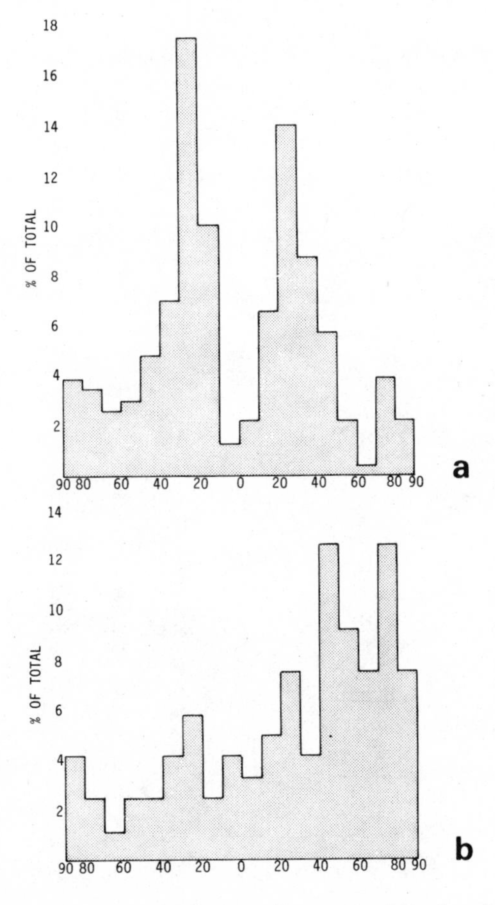

Figure 6 (*a*) Histogram of 228 SLAR imagery lineament orientations in western Washington; (*b*) histogram of 119 minor shears and joint orientations in western Washington.

thrust sheet. The rose diagram for the Burning Springs quadrangle (Figure 9) shows a 10° rotation of the preferred lineament directions. Wing, Overby & Dellwig (1970) interpret this rotation to be the result of the pileup of the leading edge of the thrust plate when the plate reached the stratigraphic limit of the Silurian salt deposits that were lubricating the sole of the thrust.

The area to the west of the thrust plate has an entirely different development of preferred orientations than the area on the thrust plate (Figure 9). Wing, Overby & Dellwig (1970) interpret the greater development of lineaments on the thrust plate to be the result of the décollement movement that allowed the maximum number of possible fracture sets to advance beyond the incipient stage of development. Because of the extensive development of fracture systems in the thrust plate, it was suggested by Wing, Overby & Dellwig (1970) that a well-developed polygonal topographic pattern may be a useful tool in defining the limits of Appalachian thrusting.

BUCHANAN COUNTY Elder, Jeran & Keck (1974) used a Ronchi grating to identify three prominent parallel sets of lineaments interpreted by them to be joints. The orientation of these lineament sets is consistent with the regional structure in Buchanan County, Virginia. A N60°E trending set was interpreted as a tensional set parallel to axes of major folds and N60°W and N25°W trending sets were inter-

Figure 7 SLAR imagery of a portion of the Burning Springs (7.5 min) quadrangle, West Virginia (from Dellwig, MacDonald & Waite 1973).

preted as shear sets parallel to major faults in the area. Complementary joint sets to the three major sets are poorly developed on the SLAR imagery. Field measurements of joint orientations showed excellent agreement with the lineaments detected on SLAR imagery.

Faults

Faults may also have pronounced linear expression on SLAR imagery. In order to be detected on SLAR imagery, a fault must significantly affect the topography or vegetation for a considerable distance. Such a fault can be seen along the west coast of Darien Province, Panama (Figure 15*a*). This fault forms a prominent lineament that extends for approximately 75 km on the imagery. The fault had not been mapped in this location prior to the acquisition of the SLAR imagery, although its extension into neighboring Columbia had been known previously (MacDonald 1969).

When using SLAR imagery to detect possible faults, it is important to note that the small scale and low resolution can combine to align several parallel linear segments, producing a single long lineament. On larger-scale imagery, these lineaments are often found to be somewhat discontinuous and the individual segments imperfectly aligned. In many instances, faults may parallel one or more major lineament trends and it may be impossible to distinguish them from non-fault-related

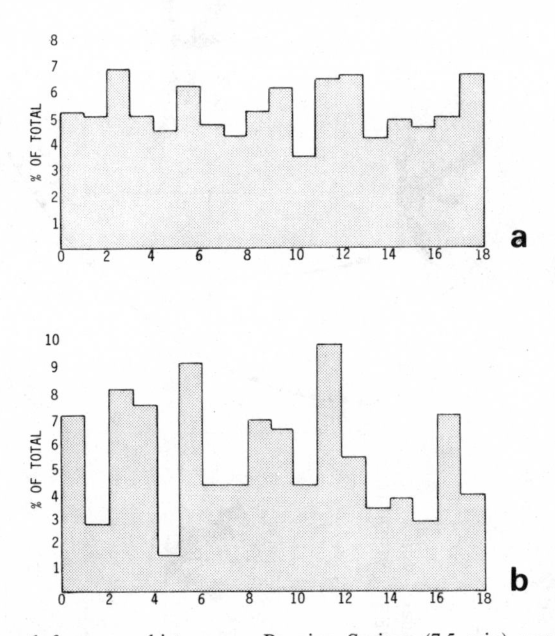

Figure 8 Azimuth-frequency histograms, Burning Springs (7.5 min) quadrangle, West Virginia (from Wing, Overby & Dellwig 1970). (*a*) Ground measurement of 796 joints; (*b*) SLAR imagery lineaments.

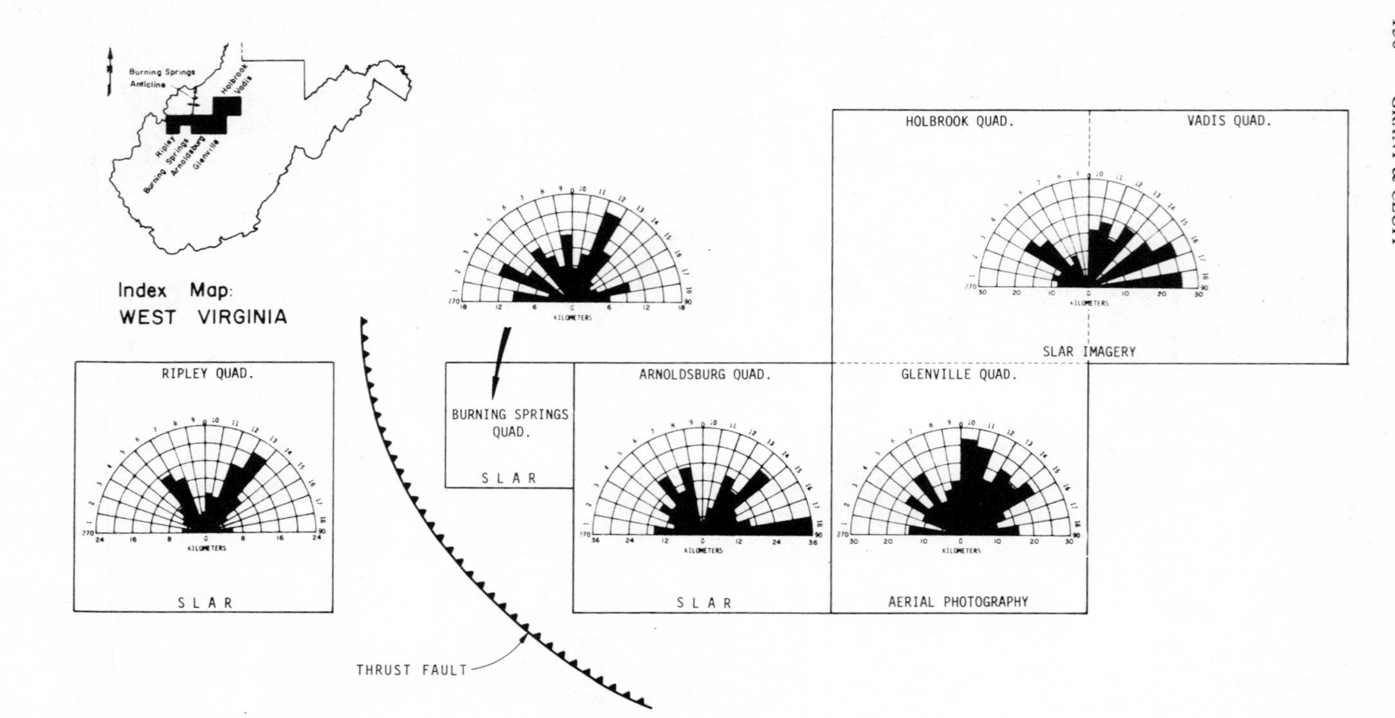

Figure 9 Rose diagrams showing cumulative lengths of lineaments on SLAR imagery in the Burning Springs area, West Virginia (from Wing, Overby & Dellwig 1970).

lineaments. For these reasons, fault interpretations must be made very carefully and must be followed by detailed studies on the ground.

Faults are often expressed as alignments of drainages. Figure 10 shows such an alignment along three faults that displace Miocene plateau basalts in southeastern Washington (Huntting et al 1961). Other features shown on SLAR imagery that may be indicative of faulting are the displacement of linear features or patterns that reflect bedding, foliation, jointing, changes in topographic texture or stream pattern along a linear contact, or changes in tone resulting from changes in vegetation type.

The small-scale, synoptic view, and terrain enhancement characteristics of SLAR may reveal fault-controlled topography not previously recognized on standard large-scale photography. Gillerman (1970) has reported on such a feature, the Roselle lineament, in southeast Missouri (Figure 11). This strong lineament on the SLAR imagery results from an alignment of stream drainages. The lineament was not detected previously on aerial photographs because each photograph had only covered a small portion of the lineament and the drainage pattern is difficult to distinguish. Geologic evidence and field checking by Gillerman (1970) indicate that the lineament is a fault and has been a major tectonic feature in this area since Precambrian time.

In many cases, faults (particularly inactive faults) will have no expression on the

Figure 10 Expression of faults on SLAR imagery of southeastern Washington.

SLAR imagery or will be no more strongly expressed than other non-fault-related lineaments because they do not significantly affect the topography or vegetation. Figure 12 shows a SLAR image of the area near Grass Valley, California, and a fault map (Burnett & Jennings 1962) of the same area. The faults in the area are part of the Foothills fault system of Mesozoic age. The fault segment between Auburn and Grass Valley is a well-developed linear feature; a highway that follows the fault trace helps to emphasize it. However, north of Grass Valley, the fault is less evident on the imagery, being expressed as a few discontinuous stream alignments. This is typical of areas where the rocks on each side of the fault do not have distinctive topographic expression. In an area where the geology and geologic structure are not strongly expressed topographically, the interpretation of SLAR imagery is difficult or impossible.

Dikes

MacDonald (1969) illustrates the strong expression on SLAR imagery of dikes in the Spanish Peaks area of Colorado. In this area, the dikes appear as linear, positive features that cut across the topography. Detectability of these features was highly dependent on look-direction. Wing & Dellwig (1970) discuss the SLAR imagery expression of the Virginia Dale ring dike complex on the Colorado–Wyoming border. The domed central core and surrounding annular dike of the structure are well expressed on the imagery. This feature was discovered independently by different workers using 1957 SLAR imagery and a 1962 high-altitude photograph mosaic. The ring dike complex probably was not discovered earlier because of its

Figure 11 Topographic expression on SLAR imagery of the Roselle lineament in southeast Missouri (from Dellwig, MacDonald & Waite 1973).

large size (9-mile diameter) (Eggler 1968), and its disclosure on small-scale synoptic imagery demonstrates the usefulness of this imagery in detecting regional structural features.

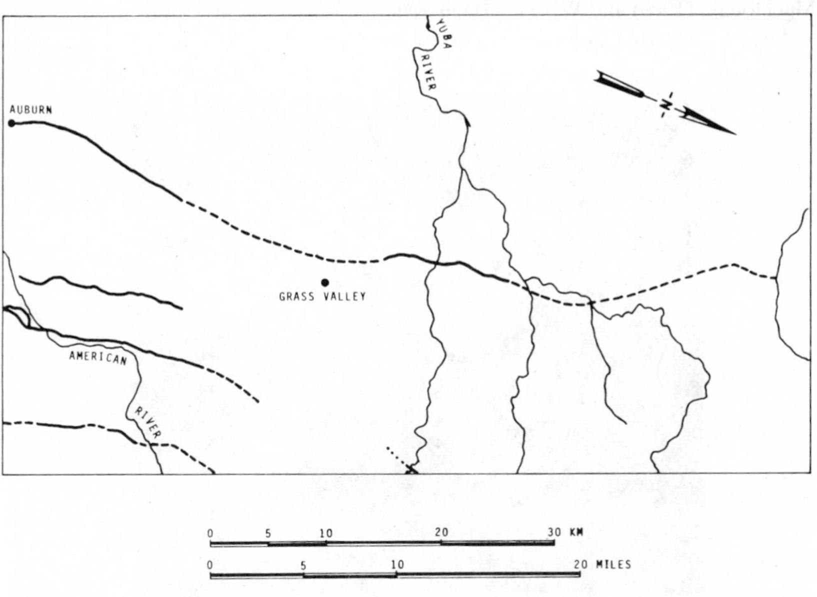

Figure 12 SLAR image and fault map (from Burnett & Jennings 1962) of region near Grass Valley, California.

Folds

Folds can be difficult to detect on SLAR imagery because folds often do not have striking topographic features associated with them. Folds are usually detected on SLAR imagery by tracing the distinctive topographic texture and tone of different lithologic units around the flanks of the fold. The Santa Fe anticline in Panama (Figure 13) is a good example of a fold defined by the distinctive topographic expression of different lithologic units. Resistant units may also form cuestas that are visible on the imagery and from which dip direction can be determined (MacDonald 1969). Cuestas formed by resistant sills and Eocene sedimentary formations (Huntting et al 1961) define an anticline and syncline on the east side of Willapa Bay, Washington (Figure 14). Stream patterns, too, can be important in the detection of folds. Streams may flow parallel to one another down the flanks of the fold (syncline in Figure 14) and flow parallel to strike around the nose of a plunging fold (Elder, Jeran & Keck 1974). Streams may also flow down the axis of synclines (Figure 14). In addition, lineament patterns sometimes may curve or "flow" around folds (Wing 1970). Although they may be difficult to recognize, the features characteristic of folds can usually be interpreted by an experienced observer.

Regional Tectonic Mapping

Some of the best examples of regional mapping using SLAR imagery are by MacDonald (1969) and Wing (1971) in eastern Panama. Because of its dense jungle and nearly perpetual cloud cover, this area had never been satisfactorily mapped at

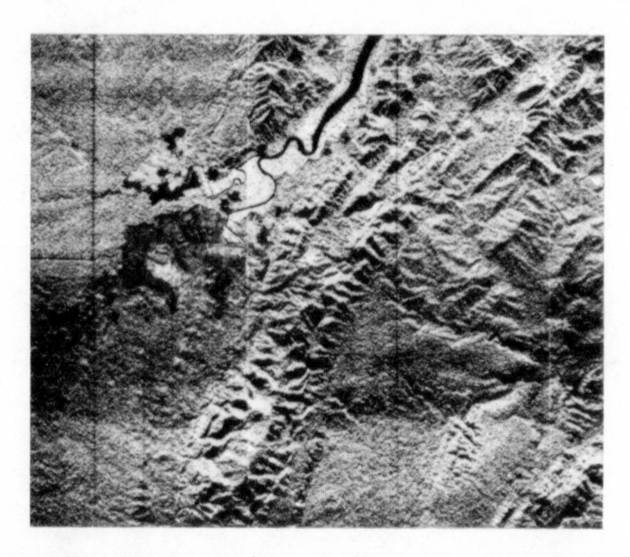

Figure 13 SLAR imagery from the Santa Fe area, Panama. The Santa Fe anticline is well expressed in the center of the image (from Dellwig, MacDonald & Waite 1973).

a reconnaissance scale, and conventional aerial photography coverage was spotty. The cloud penetration capabilities of SLAR allowed the imagery of Darien Province (Figure 15*a*) to be collected in six days. The major tectonic features of the province (Figure 15*b*) are clearly expressed on the radar mosaic, even though it is shown here at a much reduced scale. Using the SLAR imagery, MacDonald (1969) was able to prepare a geologic map of Darien Province. Differentiation between lithologic units was made on the basis of surface expressions recorded as topographic texture and tone on the SLAR imagery. Interpretation of structure was made using methods similar to those discussed in this paper. Assignments of age and rock types were made from field checks and previously published mapping in the area. The regional mapping of this area was successful because lithologic units could be differentiated from one another on the basis of topography, and the large-scale structures were topographically well expressed and not too complicated.

The northern Coast Ranges of Oregon (Figure 16) provide a good example of the ability of SLAR imagery to define large regional structures. The general structure of the range consists of a northward-plunging anticline (Wells & Peck 1961). The Eocene volcanic rock at the core of the range is clearly expressed by the rugged mountainous topography seen on Figure 16 south of Nehalem Bay. Oligocene marine rocks form a band of more subdued relief that extends northeastward from Nehalem Bay and curves around the nose of the fold. Several streams follow the strike of the beds near the Eocene-Oligocene contact and form a gently curved pattern around the nose of the fold. Miocene volcanic rocks form prominent cuestas just south of the Columbia River, indicating a gentle northward dip down

Figure 14 SLAR image of folded tertiary sediments near Willapa Bay, Washington.

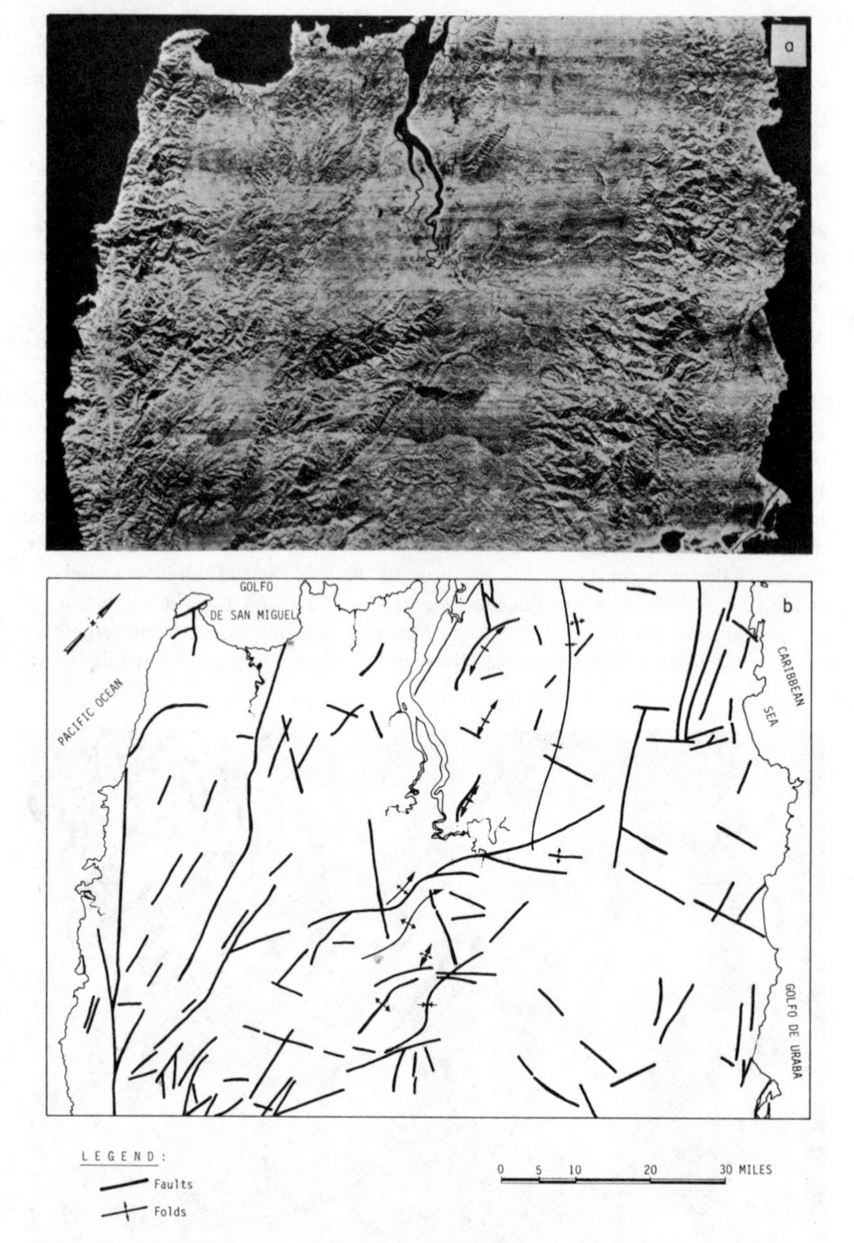

Figure 15 (*a*) SLAR mosaic of Darien Province, Panama (from MacDonald 1969); (*b*) tectonic map made using SLAR imagery of Darien Province, Panama (from MacDonald 1969).

the plunge of the fold. The Miocene rocks extend from the cuestas to Cape Falcon. The Miocene marine sediment forms areas of relatively low relief, and patches of Miocene volcanic rock form areas of high relief.

CONCLUSION

SLAR imagery has several properties that make it useful in studies of regional tectonics. The synoptic view offered by a SLAR imagery mosaic and the small-to-moderate scale allow the interpreter to detect regional patterns and structures that may not be obvious at a larger scale, wherein only small portions of a region may be studied one time, as is the case when standard aerial photographs are used. The shadowing and low resolution inherent in SLAR imagery tend to emphasize topo-

Figure 16 SLAR image of northern Coast Ranges, Oregon.

graphy and drainage and suppress vegetative detail. This highlighting can bring out geologic controls on topography that may not be evident on other forms of imagery. The fact that SLAR is an active system allows the direction of shadowing to be controlled during flight. The direction of shadowing can be important in detecting linear geologic features of specific orientations. The cloud penetration capabilities of SLAR make it possible to collect the imagery in short periods of time without regard to most local weather and climatic conditions; this is especially valuable in areas where little work could be done previously because of dense vegetation and nearly perpetual heavy cloud cover. Our experience indicates that SLAR imagery can be especially useful in fault studies because, although it is often not possible to make positive identifications of linear terrain features detected on SLAR imagery, it does serve to point out those areas that deserve further, more detailed investigations on the ground.

Interpretation of SLAR imagery is primarily based on the topographic expression of structural features and secondarily by drainage and tonal anomalies. If structures are not well expressed topographically and lithologic units cannot be differentiated, or if the structure of an area is extremely complicated, it is unlikely that SLAR interpretation will be successful. In most areas, the topographic expression of a structural feature is not conclusive enough to make an interpretation with complete confidence; this is particularly true of linear terrain features. Because of this, the final interpretation of any geologic feature detected on SLAR should only be made after a thorough ground check.

One reason SLAR imagery is not used more frequently is its limited availability; at present, the imagery is readily available for only small portions of the United States and other parts of the world. Another reason is cost. Commercial systems are available to acquire imagery on a contract basis, but the costs are high when compared with other photographic systems. The cost for SLAR imagery may range between \$3.50 and \$14.50 per square mile, depending on the size and location of the area. SLAR imagery costs begin to become comparable with those of photography for areas of more than 5000 square miles.

Literature Cited

Burnett, J. L., Jennings, C. W. 1962. *Geologic Map of California, Olaf P. Jenkins Edition, Chico Sheet,* Scale 1:250,000. Calif. Div. Mines Geol.

Clark, L. D. 1964. *Stratigraphy and Structure of Part of the Western Sierra Nevada Metamorphic Belt, California. US Geol. Surv. Prof. Paper* 410. 70 pp.

Clark, M. M. 1971. Comparison of SLAR images and small-scale low-sun aerial photographs. *Geol. Soc. Am. Bull.* 82: 1735–42

Clark, M. M. 1973. Comparison of SLAR images and small-scale low-sun aerial photographs: reply. *Geol. Soc. Am. Bull.* 84:359–62

Dellwig, L. F., MacDonald, H. C., Waite, W. P. 1973. *Radar Remote Sensing for Geoscientists, Short Course Notes.* Lawrence: Kansas Univ. Ctr. Res. 579 pp.

Dellwig, L. F., McCauley, J. 1973. Comparison of SLAR images and small-scale, low-sun aerial photographs: discussion. *Geol. Soc. Am. Bull.* 84:357–58

Eggler, D. H. 1968. Virginia Dale Precambrian ring dike complex, Colorado-Wyoming. *Geol. Soc. Am. Bull.* 79:1545–64

Elder, C. H., Jeran, P. W., Keck, D. A. 1974. *Geologic Structure Analysis Using Radar Imagery of the Coal Mining Area of Buchanan County, Va. US Bur. Mines Rep. Invest.* 7869. 29 pp.

Fisk, H. N. 1944. *Geological Investigations*

of the Alluvial Valley of the Lower Mississippi River. Vicksburg, Miss: Mississippi River Comm. 78 pp.

Gillerman, E. 1970. Roselle lineament of southeast Missouri. *Geol. Soc. Am. Bull.* 81:975–82

Huntting, M. I., Bennett, W., Livingston, V. Jr., Moen, W. 1961. *Geologic Map of Washington,* Scale 1:500,000. Wash. Div. Mines Geol.

Lattman, L. H. 1958. Techniques of mapping geologic fracture traces and lineaments on aerial photographs. *Photogramm. Eng.* 24: 568–76

Lyon, R. J. P., Mercado, J., Campbell, R. Jr. 1970. Pseudoradar. *Photogramm. Eng.* 36:1257–61

MacDonald, H. C. 1969. Geologic evaluation of radar imagery from Darien Province, Panama. *Mod. Geol.* 1:1–63

Reeves, R. G. 1969. Structural geologic interpretations from radar imagery. *Geol. Soc. Am. Bull.* 80:2159–64

Strand, R. S., Koenig, J. B. 1965. *Geologic Map of California, Olaf P. Jenkins Edition, Sacramento Sheet,* Scale 1:250,000. Calif. Div. Mines Geol.

Tator, B. A., ed. 1960. Photo interpretation in geology. In *Manual of Photographic Interpretation.* 169–342. Washington DC: Am. Soc. Photogramm.

Wells, F. G., Peck, D. L. 1961. *Geologic Map of Washington,* Scale 1:500,000. US Geol. Surv. Misc. Geol. Invest. Map I-325

Wing, R. S. 1970. Cholame area-San Andreas fault zone-California, a study in SLAR. *Mod. Geol.* 1:173–86

Wing, R. S. 1971. Structural analysis from radar imagery, eastern Panamanian Isthmus. *Mod. Geol.* 2:1–21, 75–127

Wing, R. S., Dellwig, L. F. 1970. Radar expression of Virginia Dale Precambrian ring dike complex, Wyoming/Colorado. *Geol. Soc. Am. Bull.* 81:293–98

Wing, R. S., Overby, W. K. Jr., Dellwig, L. F. 1970. Radar lineament analysis, Burning Springs area, West Virginia—an aid in the definition of Appalachian Plateau thrusts. *Geol. Soc. Am. Bull.* 81: 3437–44

Woodward-Clyde Consultants 1973. SLAR imagery interpretations based on unpublished field work. Woodward-Clyde Consultants, Oakland, California

PALEOMAGNETISM OF METEORITES

※10054

Frank D. Stacey
Physics Department, University of Queensland, Brisbane 4067, Australia

INTRODUCTION

Meteorites are the iron and stony bodies which fall to earth from time to time, apparently randomly, but from elliptical orbits, indicating that they are properly parts of the solar system and may be associated with the asteroids which are concentrated between Mars and Jupiter. The study of meteorites is central to our understanding of the chemistry and origin of the solar system in general and the terrestrial planets in particular. Not only are their compositions more representative of the earth as a whole than any terrestrial samples accessible to our surface scratching, but most meteorites have remained unchanged (except perhaps for brief shock events and cosmic ray exposure) since they were formed, with the rest of the solar system, about 4.5×10^9 years ago, whereas terrestrial evidence of the first 10^9 years of earth history has been obliterated by subsequent geological events. The properties of meteorites have been comprehensively reviewed by Mason (1962), Anders (1964), Wood (1968) and, most recently, by Wasson (1974).

Within the solar system at the present time there are magnetic fields associated with the sun (Babcock & Babcock 1955), Jupiter (Smith et al 1974, Acuna & Ness 1975) and probably Mercury (Ness et al 1975) as well as the earth. All of these fields are affected by solar wind (particle radiation from the sun) which draws out the solar field into an interplanetary field of intensity about 5×10^{-4} G (50 gammas) at the distance of the earth. Measurements on lunar rocks (Fuller 1974) indicate that the moon once had a significant field, although there are widely disparate estimates of its intensity (Banerjee & Mellema 1974, Stephenson & Collinson 1974), but the present field appears to be caused solely by the magnetizations of rocks near to the surface. Mars appears to have a weak field but Venus has no measurable field (Bridge et al 1967, Ness et al 1974), and we have no information about possible fields of the planets beyond Jupiter. This gives the impression that magnetic fields are no more than incidental to the solar system and its workings. On the other hand, large-scale and in some cases very intense magnetic fields are associated with astrophysical processes from the scale of galaxies to neutron stars. Especially relevant to the solar system problem are the magnetic fields of young stars at the stage believed to be represented by T-Tauri. The supposition that the sun had a much stronger and more extensive magnetic field at its T-Tauri stage, which probably coincided more or less with planetary formation, is important to our understanding of the solar system because it would have caused the outward transfer of angular momentum from the sun to the planetary disc (Alfvén 1954, Sonett et al 1970). Magnetic effects were evidently more significant at the time of planetary formation than at present and this is the principal reason for interest in the magnetism of meteorites.

The history of the earth's magnetic field over geological time is now well documented by studies of magnetism in rocks (paleomagnetism). Comprehensive treatments of the subject include those by Irving (1964) and McElhinny (1973) and the physics of magnetism in rocks is treated by Stacey & Banerjee (1974). Most of the techniques of rock magnetism are directly applicable to meteorites, with the proviso that destructive tests on samples of appreciable sizes should be undertaken only reluctantly. Although there are still very few reports of magnetic measurements on meteorites, familiarity with magnetism in rocks supports the essential conclusion that meteorites were magnetized by a field or fields acting on them at the time of formation or soon afterward. However, the nature of these fields remains conjectural.

METEORITE TYPES AND THEIR MAGNETIC CONSTITUENTS

Several chemically and structurally different types of meteorite are recognized. Best-known in popular literature, although not the most abundant, are the irons. These are largely metallic iron with nickel in solution averaging about 8% and usually containing grains of troilite (FeS), carbide, or phosphide, and sometimes inclusions of stony material. Although strongly magnetic, the irons have not received much attention in magnetic studies. Although the crystals are usually broken up by exsolution lamellas of nickel-rich γ phase (taenite) and nickel-poor α phase (kamacite), the iron meteorites are magnetically softer (i.e. more easily magnetized and demagnetized) than are rocks recognized as suitable for paleomagnetic work (Gus'kova 1965). Presumably, their magnetic instability would preclude significant inferences from any measured natural remanences, although Gus'kova (1965) suggested that iron meteorites do give evidence of preterrestrial magnetization.

About 85% of meteorites are classified as *chondrites*, which are predominantly stony but commonly contain grains or filaments of metal of compositions similar to the iron meteorites. Their magnetic properties are dominated by the relatively small proportion of highly magnetic metal phase. The characteristic feature of chondrites is the occurrence in them of chondrules, very finely crystalline, almost glassy spherules averaging about 1 mm in diameter, which are embedded in silicate matrix of similar composition. There are no known terrestrial samples containing chondrules and their presence in chondrites indicates a process of formation quite unlike the production of igneous rocks on the earth. An explanation by Wood (1968), at least as plausible as any, is that the chondrules were produced by transient heating of dust surrounding the sun during its T-Tauri stage, the heating being caused by the propagation of violent shock waves through the outer solar atmosphere. On this basis the chondrules are presumed to predate the accretion of planets and asteroids.

The carbonaceous chondrites are so called because they contain several percent carbon and carbon compounds. They also contain iron in the form of magnetite, which is responsible for their magnetic properties; there is little or no elemental metal present. The presence of carbon and magnetite in substantial quantities gives the carbonaceous chondrites a dark, amorphous appearance. In terrestrial collections they are very rare, but the optical reflectivities of many asteroids suggest carbonaceous chondritic compositions at least of their surface layers (Johnson & Fanale 1973, McCord & Gaffey 1974). The abundance of carbonaceous chondritic material in the asteroidal belt is probably quite high relative to the other types, which were previously assumed to be more common because they are less friable and therefore better able to survive flight through the atmosphere. The carbonaceous chondrites appear to be close in composition to the primeval dust from which the terrestrial planets accreted; this has given them a special scientific significance. In particular, members of the most primitive group, subclassified as Type I, have never been appreciably heated and contain quite a high proportion of volatiles, unlike the ordinary chondrites. The compositions of the terrestrial planets can be explained in terms of heating and reduction of Type I carbonaceous material (Ringwood 1966).

Stony meteorites, devoid of chondrules and not obviously different from terrestrial igneous rocks, form the other main group of meteorites, termed *achondrites*. Although the first, tentative paleomagnetic measurement on a meteorite (Lovering

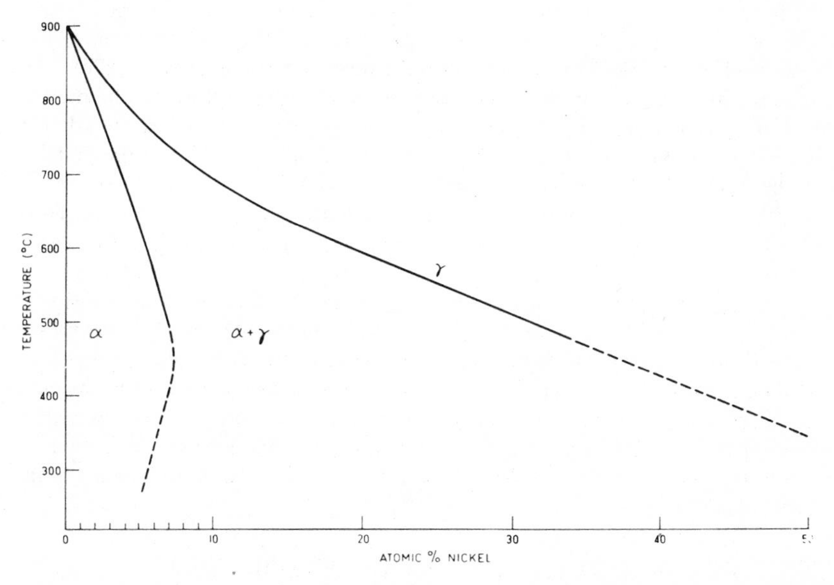

Figure 1 Low-nickel end of the nickel-iron phase diagram according to Goldstein & Ogilvie (1965).

1959) used a sample of an achondrite, the Moore County eucrite, achondrites have hardly been represented in subsequent studies, which have concentrated on the chondrites. Undoubtedly, this is at least partly because chondrites are more widely available for measurement, but partly also because the history of the chondrites is supposed to be simpler, making interpretation of chondrite measurements more meaningful. This supposition is probably true with respect to carbonaceous chondrites, but it can hardly be sustained in the case of ordinary chondrites and future work could usefully include attention to the achondrites.

The magnetic properties of ordinary chondrites are attributed to the iron-nickel phases in them and these reproduce the features seen in iron meteorites, although on a smaller scale. At temperatures above 900°C, nickel and iron are mutually soluble at all compositions in a γ (face-centered cubic) phase, which in meteorite work is referred to as taenite. At lower temperatures, progressive exsolution of a low-nickel α (body-centered cubic) phase (kamacite) occurs, and the taenite becomes enriched in Ni. By 500°C, kamacite is the dominant phase, as is seen by considering an overall composition of 7–10% Ni in the Ni-Fe phase diagram (Figure 1). It is also the magnetically important phase.

The thermomagnetic properties of kamacite are irreversible, in the sense indicated in Figure 2, which shows plots of saturation magnetization vs temperature for samples of three chondrites. Kamacite with 5–7% Ni does not have a normal Curie point or transition to a nonmagnetic state, but undergoes transformation to nonmagnetic γ phase at about 750°C. The reverse transformation occurs only at a temperature 100–200°C lower. Because the (virtual) Curie points are higher than the temperatures of formation of the metal phases, the ordinary chondrites cannot acquire conventional thermoremanent magnetization (TRM) by cooling in a magnetic field which instead produces chemical remanent magnetization (CRM, i.e. magnetization resulting from a chemical or phase change in a field). At the rates of cooling of chondrites [of order degrees per million years as indicated by diffusion boundaries of the exsolution structures (Wood 1967)], exsolution continues down to about 400°C, which emphasizes even more strongly the CRM nature of remanence in chondrites. However, the stability of CRM is similar to that of TRM (Kobayashi 1959), so that, although it is generally less intense, it is just as useful for paleomagnetism.

Magnetic susceptibilities of carbonaceous chondrites are also high (relative to rocks), generally it is because of the presence of magnetite (Fe_3O_4) rather than metal (Larson et al 1974). Being very finely divided, the magnetite is reasonably hard magnetically and is capable of supporting a stable remanence. Again, chemical remanence must be invoked because the material has not been appreciably heated. The apparently simple histories of at least some carbonaceous chondrites means that interpretation of their magnetizations is important to our understanding of the early solar system.

NATURAL MAGNETIZATIONS OF METEORITES

Natural magnetizations evidently of preterrestrial origin have been found in all classes of meteorite. The first reported measurements (Lovering 1959) were on four

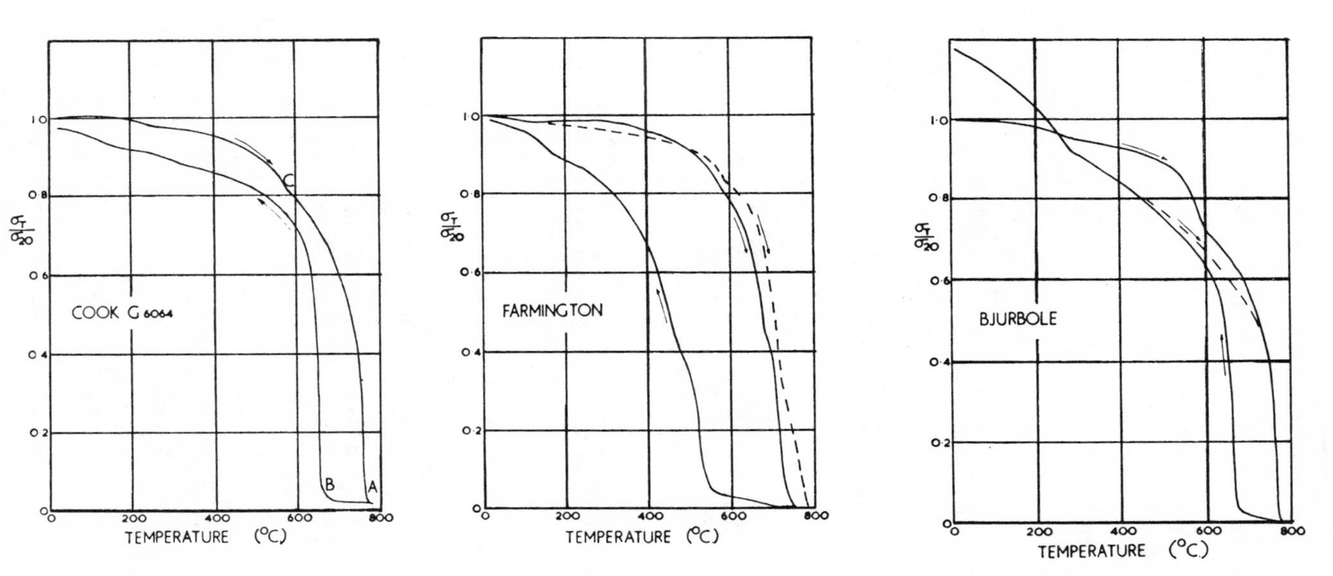

Figure 2 Thermomagnetic curves of three chondrites from Stacey et al (1961). These are plots of saturation magnetization as functions of temperature, showing thermal hysteresis of the $\alpha \rightarrow \gamma$ and $\gamma \rightarrow \alpha$ transitions of the principal phases at about 750°C and 550–650°C, respectively. Lower temperature transitions correspond to minor components and usually some permanent change. Broken lines indicate second heating of the same sample.

irons, one stony-iron and one achondrite, but the magnetic softness of the irons and stony-iron implied doubt about the significance of their magnetizations, and consequently the achondrite measurement was emphasized. Subsequent work has been concentrated on the chondrites, and early data from Stacey & Lovering (1959), Stacey et al (1961), Weaving (1962b), and Gus'kova (1963) have agreed that all ordinary chondrites examined had magnetizations of extraterrestrial origins. More

Table 1 Estimates of field intensities responsible for primary magnetizations of meteorites. Values are given in gauss ($\equiv 10^{-4}$ tesla).

Meteorite	Field intensity (gauss)	Reference
Ordinary chondrites:		
Mt. Browne	0.25	Stacey et al (1961)
Farmington	0.18	Stacey et al (1961)
Homestead	(0.9)	Stacey et al (1961)
Brewster	0.1	Weaving (1962b)
Rakovka	0.4	Gus'kova (1963)
Mordvinovka	0.4	Gus'kova (1963)
Okhansk	0.3, 0.1	Gus'kova (1963)
Pultusk	0.2, 0.25	Gus'kova (1963)
Zhovtenevy Khutor	0.15, 0.2, 0.15	Gus'kova (1963)
Abee	0.33	Brecher & Ranganayaki (1975)
Rose City	0.01	Brecher & Ranganayaki (1975)
Bald Mountain	0.01	Brecher & Ranganayaki (1975)
Andover	0.01	Brecher & Ranganayaki (1975)
Aumale	0.02	Brecher & Ranganayaki (1975)
Jelica	0.01	Brecher & Ranganayaki (1975)
Vavilovka	0.08	Brecher & Ranganayaki (1975)
Carbonaceous chondrites:		
Allende	1.1	Butler (1972). Banerjee & Hargraves (1972)
	0.25 to 1.0	Brecher (1972)
Murchison	0.18	Banerjee & Hargraves (1972)
	0.5 to 3	Brecher (1972)
Orgeuil	0.67	Banerjee & Hargraves (1972)
Renazzo	2	Brecher (1972)
Murray	0.7	Brecher (1972)
Stony-irons:		
Nechayevo	0.22 to 0.9	Gus'kova (1965)
Hainholz	0.22 to 0.9	Gus'kova (1965)
Estherville	0.22 to 0.9	Gus'kova (1965)
Steinbach	0.22 to 0.9	Gus'kova (1965)
Vaca Muerta	0.22 to 0.9	Gus'kova (1965)
Irons:		
Suggested average	~0.6	Gus'kova & Pochtarev (1967)

recently, Brecher & Ranganayaki (1975) have argued that remanences of several ordinary chondrites are less stable and the inferred primary fields weaker than the early work suggested. However, most recent emphasis has been on carbonaceous chondrites (Butler 1972, Banerjee & Hargraves 1972, Brecher 1972, Brecher & Arrhenius 1974), which are of greatest fundamental interest. A list of reported estimates of primary fields responsible for meteorite magnetizations is given in Table 1. Brecher (1972) has listed magnetic moments of a larger number of carbonaceous chondrites with some corresponding field estimates, and Herndon et al (1972) have listed magnetic moments reported in Russian literature, mainly by Gus'kova & Pochtarev (1967), for an impressive number of meteorites of all classes except carbonaceous chondrites.

The basic problem in the magnetism of meteorites—identification of the origin of natural remanence—has been tackled by various authors using different paleomagnetic techniques. Stacey & Lovering (1959) and Stacey et al (1961) favored thermal demagnetization in an attempt to retrace the acquisition of remanence and found that components of natural magnetism in ordinary chondrites survived heating to 500°C or more and behaved more like thermoremanence than isothermally induced remanence. This is clear evidence that the remanence is carried primarily by the kamacite grains and that it was acquired either directly by cooling or, more plausibly, by the formation of kamacite at a temperature of several hundred degrees (thermochemical remanent magnetization). The possibility of magnetization by the earth's field was discounted because the selected samples were all taken from observed falls and more particularly from the interior portions of meteorites, which were not heated in flight through the atmosphere (Lovering et al 1960).

The most comprehensive examination of the pattern of magnetism within a single meteorite was by Weaving (1962b). Using a line of samples taken from a cross section of the Brewster meteorite, he found that the interior part (which was not heated by flight through the atmosphere) was stably magnetized in a field of about 0.1 G, but that the reheated surface skin carried a stronger, but highly variable and less stable magnetization. Stacey (1967) pointed out that this superficial magnetization was too strong to have been induced in the earth's field and implied the circulation of substantial surface currents in the ablating meteorite, possibly generated thermoelectrically by the strong temperature gradient and perhaps extending into the surrounding ionized vapor. The essential difference between the magnetizations of the interior and surface regions is demonstrated by Weaving's (1962b) thermal demagnetization curves, reproduced in Figure 3. Whatever the precise mechanism of surface magnetization, it is clearly not extraterrestrial in origin and therefore is of little interest in the present context. The consistent magnetization of the interior portion, apparently thermal in origin, confirmed the conclusion of Stacey et al (1961) that chondrites carried magnetic evidence of their early histories.

In the case of the carbonaceous chondrite Mokoia, thermal demagnetization proved to be misleading. Stacey et al (1961) found that heating to 200°C completely destroyed its natural remanence and inferred that this was simply an unstable magnetization, which could have been terrestrially induced, and that Mokoia had no stable, primary remanence. However, like other carbonaceous chondrites,

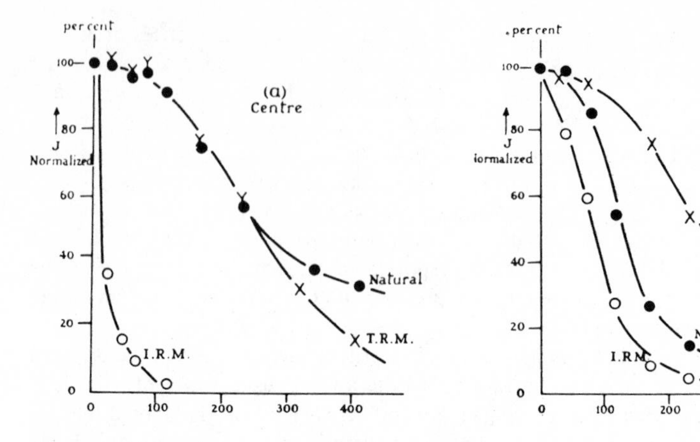

Figure 3 Thermal demagnetization of the natural remanence of two samples of Brewster meteorite, compared with laboratory-induced thermoremanence and isothermal remanence. Reproduced by permission from Weaving (1962b).

Mokoia is chemically unstable with heating, so that no inference can be drawn from thermal demagnetization. Alternating field demagnetization (Butler 1972, Banerjee & Hargraves 1972), especially combined with low temperature magnetic cleaning (Brecher 1972), demonstrates the presence of stable remanences in all carbonaceous chondrites, including Mokoia.

The low temperature experiments of Brecher (1972) are of particular interest because they demonstrate the presence of very fine, single-domained or pseudo-single-domained magnetite grains. Larger multidomained grains irreversibly lose their remanence when cooled through the magnetic isotropic point (118 K) but single-domained or pseudo-single-domained grains exhibit a memory effect (see Stacey & Banerjee 1974, pp. 26–28) whereby the initial remanence is largely recovered on rewarming to room temperature. The substantial magnetic memories found by Brecher in carbonaceous chondrites demonstrate that at least a large part of the remanence is carried by very stable, fine grains.

HYPOTHESES OF METEORITE MAGNETIZATION

It is now clear that stable remanent magnetization is a general property of all classes of meteorite with sufficiently hard magnetic constituents, including carbonaceous chondrites, and that the inducing fields were in the range 0.1–1 G (10^{-5}–10^{-4} T). As all authors reporting estimates of field intensity have been aware, this range includes the earth's field, so that it is vital to ensure that meteorites were not magnetized by the earth's field. But the conclusion that meteorites or their parent bodies were either formed or cooled in a magnetic field or fields now appears unavoidable. The nature of this field remains obscure and there is no evidence that the process of magnetization was essentially the same for all classes of meteorite. At this point conjecture takes over from observation.

An obvious inference was that the chondrites came from parent bodies with terrestrial-type fields, especially when carbonaceous chondrites were thought not to be magnetized. However, closer attention to the implications of the steady parent body field produces insuperable objections. The presence in a parent body of a substantial fluid core (capable of dynamo action) may not be entirely impossible, because the iron meteorites could be representative of the core material, but a molten core is not consistent with a mantle cool enough to acquire remanence, except for a thin surface layer. When the core solidified, the field would be switched off and the bulk of the parent body would cool without any significant field.

On the other hand, if we appeal to a process of magnetization of meteoritic material by accretion in a steady solar field, we must somehow disallow rotation of the accreting material with respect to the field, whereas the asteroids are observed to be rotating now (see e.g. Hartman & Larson 1967). We may be forced to conclude that the meteorites were magnetized by a transient process. Stacey (1967) suggested that magnetization might result from a transient field associated with the breakup of a parent body, but this does not provide a plausible explanation for magnetization of the carbonaceous chondrites, especially Type I (the least metamorphosed). Another possibility is to appeal to the T-Tauri shock waves which Wood (1968) offered as a mechanism for producing chondrules. This would use the solar field with transient shock heating to produce a thermoremanent type magnetization. Magnetization by this process does not appeal to the formation of parent bodies of any particular size and is consistent with the universality of meteorite magnetization.

Rowe (1975) raised a particular difficulty in the case of the Farmington chondrite, which appears to have suffered reheating sufficient to lose its argon only 500 million years ago. Its magnetization could hardly have survived such reheating. The event appears to have resulted from a violent shock, presumably an impact by the parent body, and suggests a return to the idea of a transient field associated with the breakup, although the intrinsic implausibility of this has not been overcome by any proposed mechanism for production of the field. The principles of symmetry disallow the possibility of magnetization by shock without a field.

Magnetic anisotropies of ordinary chondrites have been interpreted as a metamorphic effect (Stacey et al 1961). The filamentary nature of the metal grains and small sample sizes were such that a purely statistical effect could have been significant, but Weaving's (1962a) demonstration that magnetic anisotropy axes were consistent across several meteorites confirms their textural significance. Brecher (1972) reported strong anisotropies in several carbonaceous chondrites, which conflicts with the report by Stacey et al (1961) that Mokoia did not show significant anisotropy. Brecher interpreted the magnetic texture in terms of accretion of magnetic material in a field, but a quantitative examination of such a process makes it doubtful. In a field of 1 G (at the high end of the range inferred from meteorite magnetizations), a spherical single-domained magnetite grain of maximum size has a magnetic moment of about 5×10^{-14} emu and thus a field alignment energy of about 5×10^{-14} erg, which is close to the thermal excitation energy of its moment, kT, at 400 K, the accretion temperature of carbonaceous chondrites (Anders 1971).

But the rotational excitation of the whole grain would correspond more closely to kT for the more highly energetic plasma particles in the medium than to the material temperature of solid particles and significant alignment would not occur. Growth of the grain to larger size causes its subdivision into magnetic domains, so reducing its average magnetization and consequently the field energy. Further, when very fine grains stick together or to a larger grain, their magnetic interactions are much stronger than the effect of an external field of 1 G and they cluster so that their moments are opposed or form loops of flux closure and do not align parallel. This effect opposes the development of a texture. Strong anisotropy cannot be attributed to the weak field which is responsible for the remanence. The possibility that the magnetic texture of carbonaceous chondrites is a consequence of layering during accretion, as in sedimentary rocks, appears more plausible. Similarly we can discount the hypothesis (Harris & Tozer 1967) that simple magnetostatic interactions between iron grains contributed to chemical fractionation in the solar system.

We have no explanation encompassing all observations on the magnetism of meteorites and, even if we are allowed a variety of explanations, it is hard to avoid implausible assumptions. This is indicative of inadequacy in our understanding of the early solar system and suggests that the clues which magnetism provides merit much more attention than they have so far received. Meteorite magnetism provides the only real clue to the possibility of a magnetic coupling of the primitive sun to its associated planetary disc, as was hypothesized by Alfvén (1954) to explain the distribution of angular momentum between the Sun and the planets. The implication is that the planets accreted from a magnetically controlled plasma, in which case the distribution of elements would have been influenced by their ionization potentials and thus their ionization probabilities, the motion of neutral atoms being unaffected by a magnetic field. The possible role of a magnetic field in the chemistry of the solar system has so far received little attention.

Literature Cited

Acuna, M. H., Ness, N. F. 1975. Jupiter's main magnetic field measured by Pioneer 11. *Nature* 253:327–28

Alfvén, H. 1954. *On the Origin of the Solar System*. Oxford: Clarendon Press. 194 pp.

Anders, E. 1964. Origin, age and composition of meteorites. *Space Sci. Rev.* 3:583–714

Anders, E. 1971. Meteorites and the early solar system. *Ann. Rev. Astron. Astrophys.* 9:1–34

Babcock, H. W., Babcock, H. D. 1955. The sun's magnetic field, 1952–1954. *Astrophys. J.* 121:349–66

Banerjee, S. K., Hargraves, R. B. 1972. Natural remanent magnetizations of carbonaceous chondrites and the magnetic field in the early solar system. *Earth Planet. Sci. Lett.* 17:110–19

Banerjee, S. K., Mellema, J. P. 1974. Lunar paleointensity from three Apollo 15 crystalline rocks using an ARM method. *Earth Planet. Sci. Lett.* 23:185–88

Brecher, A. 1972. Memory of early magnetic fields in carbonaceous chondrites. In *Symposium on the Origin of the Solar System*, ed. H. Reeves, pp. 260–72. Paris: CNRS

Brecher, A., Arrhenius, G. 1974. The paleomagnetic record in carbonaceous chondrites: natural remanence and magnetic properties. *J. Geophys. Res.* 79:2081–2106

Brecher, A., Ranganayaki, R. P. 1975. Paleomagnetic systematics of ordinary chondrites. *Earth Planet. Sci. Lett.* 25:57–67

Bridge, H. S., Lazarus, A. J., Snyder, C. W., Smith, E. J., Davis, L., Coleman, P. J., Jones, D. E. 1967. Mariner V: plasma and magnetic fields observed near Venus. *Science* 158:1669–73

Butler, R. F. 1972. Natural remanent magnetization and thermomagnetic properties of the Allende meteorite. *Earth Planet. Sci. Lett.* 17:120–28

Fuller, M. 1974. Lunar magnetism. *Rev. Geophys. Space Phys.* 12:23–70

Goldstein, J. A., Ogilvie, R. E. 1965. A re-evaluation of the iron-rich portion of the Fe-Ni system. *Trans. Metall. Soc. AIME* 233:2083–87

Gus'kova, Y. G. 1963. Investigation of the natural remanent magnetization of stony meteorites. *Geomag. Aeron.* 3:308–12

Gus'kova, Y. G. 1965. Study of the remanent magnetism of iron and stony iron meteorites. *Geomag. Aeron.* 5:91–96

Gus'kova, Y. G., Pochtarev, V. I. 1967. Magnetic fields in space according to a study of the magnetic properties of meteorites. *Geomag. Aeron.* 7:245–50

Harris, P. G., Tozer, D. C. 1967. Fractionation of iron in the solar system. *Nature* 215:1449–51

Hartman, W. K., Larson, S. M. 1967. Angular momenta of planetary bodies. *Icarus* 7:257–60

Herndon, J. M., Rowe, M. W., Larson, E. E., Watson, D. E. 1972. Magnetism of meteorites: a review of Russian studies. *Meteoritics* 7:263–84

Irving, E. 1964. *Paleomagnetism.* New York: Wiley. 399 pp.

Johnson, T. V., Fanale, F. P. 1973. Optical properties of carbonaceous chondrites and their relationship to asteroids. *J. Geophys. Res.* 78:8507–18

Kobayashi, K. 1959. Chemical remanent magnetization of ferromagnetic minerals and its application to rock magnetism. *J. Geomag. Geoelect.* 10:99–117

Larson, E. E., Watson, D. E., Herndon, J. M., Rowe, M. W. 1974. Thermomagnetic analysis of meteorites, I.Cl chondrites. *Earth Planet. Sci. Lett.* 21:345–50

Lovering, J. F. 1959. The magnetic field in a primary meteorite body. *Am. J. Sci.* 257: 271–75

Lovering, J. F., Parry, L. G., Jaeger, J. C. 1960. Temperature and mass losses in iron meteorites during ablation in the Earth's atmosphere. *Geochim. Cosmochim. Acta* 19:156–67

Mason, B. 1962. *Meteorites.* New York: Wiley. 274 pp.

McCord, T. B., Gaffy, M. J. 1974. Asteroids: surface composition from reflection spectroscopy. *Science* 186:352–55

McElhinny, M. W. 1973. *Paleomagnetism and Plate Tectonics.* Cambridge, Engl: Cambridge Univ. Press. 357 pp.

Ness, N. F., Behannon, K. W., Lepping, R. P., Whang, Y. C., Schatten, K. H. 1974. Magnetic field observations near Venus: preliminary results from Mariner 10. *Science* 183:1301–06

Ness, N. F., Behannon, K. W., Lepping, R. P., Whang, Y. C. 1975. Magnetic field of Mercury confirmed. *Nature* 255:204–5

Ringwood, A. E. 1966. Chemical evolution of the terrestrial planets. *Geochim. Cosmochim. Acta* 30:41–104

Rowe, M. W. 1975. Constraints on magnetic field which magnetized the Farmington meteorite parent body. *Meteoritics* 10: 23–30

Smith, E. J., Davis, L., Jones, D. E., Coleman, P. J., Colburn, D. S., Dyal, P., Sonett, C. P., Frandsen, A.M.A. 1974. The planetary magnetic field and magnetosphere of Jupiter: Pioneer 10. *J. Geophys. Res.* 79:3501–13

Sonett, C. P., Colburn, D. S., Schwartz, F., Keil, K. 1970. The melting of asteroidal-sized bodies by unipolar induction from a primordial T-Tauri sun. *Astrophys. Space Sci.* 7:446–88

Stacey, F. D. 1967. Paleomagnetism of meteorites. In *International Dictionary of Geophysics,* ed. S. K. Runcorn, pp. 1141–43. Oxford: Pergamon

Stacey, F. D., Banerjee, S. K. 1974. *The Physical Principles of Rock Magnetism.* Amsterdam: Elsevier. 195 pp.

Stacey, F. D., Lovering, J. F. 1959. Natural magnetic moments of two chondritic meteorites. *Nature* 183:529–30

Stacey, F. D., Lovering, J. F., Parry, L. G. 1961. Thermomagnetic properties, natural magnetic moments and magnetic anisotropies of some chondritic meteorites. *J. Geophys. Res.* 66:1523–34

Stephenson, A., Collinson, D. W. 1974. Lunar paleointensities determined by an anhysteretic remanent magnetization method. *Earth Planet. Sci. Lett.* 23:220–28

Wasson, J. T. 1974. *Meteorites: Classification and Properties.* Berlin & New York: Springer

Weaving, B. 1962a. Magnetic anisotropy in chondritic meteorites. *Geochim. Cosmochim. Acta* 26:451–55

Weaving, B. 1962b. The magnetic properties of the Brewster meteorite. *Geophys. J.* 7: 203–11

Wood, J. A. 1967. Chondrites: their metallic minerals, thermal histories and parent planets. *Icarus* 6:1–49

Wood, J. A. 1968. *Meteorites and the Origin of Planets.* New York: McGraw-Hill. 117 pp.

GENERATION OF PLANETARY MAGNETIC FIELDS

E. H. Levy[1]
Bartol Research Foundation of The Franklin Institute, Swarthmore, Pennsylvania 19081

INTRODUCTION

The development of sophisticated astronomical techniques and the expansion of solar system exploration have greatly increased our knowledge of planetary and other astrophysical magnetic fields. Internal planetary magnetic fields are generally thought to arise through regenerative, hydromagnetic dynamo activity produced by the motion of electrically conducting fluids in the interiors of the planets. The purpose of this article is to briefly review our present understanding of the generation of these magnetic fields and to indicate where the limits and uncertainties in our understanding lie. It is worth noting that the general principles of magnetic field production discussed here are apparently applicable to a broad range of astrophysical objects (Parker 1970a, 1971a, Vainshtein & Ruzmaikin 1972, Levy & Rose 1974, Levy 1975).

One interesting feature of magnetic fields is that they leave a quasi-permanent record of their existence in the magnetic remanence of cold, solid matter. Thus the magnetization of lunar surface material and of meteorites records the presence of magnetic fields during the early years of the solar system. Both planetary fields and large-scale solar nebula fields (Sonett et al 1970) may be recorded in this paleomagnetic record. An understanding of the origin and behavior of the early magnetic fields is fundamental to a unified and accurate picture of the origin and evolution of the solar system.

It seems fair to assert that the magnetic fields of planets are explicable on the basis of the well-established laws of physics. There is, at present, no need to postulate new physical principles, beyond Newton's laws and Maxwell's equations, to account for the fields in any known astrophysical objects.

Large-scale weak magnetic fields, having intensities very much less than 10^{-6} or 10^{-7} G, appear to be one result of the cosmological phenomena that produced the universe and the processes that occurred during galaxy formation. The long

[1] Present address: Department of Planetary Science, Lunar and Planetary Laboratory, University of Arizona, Tucson, Arizona 85721.

persistence of these universal magnetic fields is a consequence of their large scales, the abundance of free electric charge, and the evident absence of free magnetic charge. The primordial magnetic fields are not discussed further in this article. We mention them because the production of magnetic fields in hydromagnetic dynamos is a process of regeneration and amplification. In each instance, the initial presence of a magnetic field, which may be arbitrarily small in magnitude, is presumed. Also in this connection, there are electrochemical and thermoelectric effects that may produce weak magnetic fields in particular objects. Some of these effects have been listed elsewhere (Elsasser 1939, Inglis 1955, Stevenson 1974).

The Planetary Evidence

It has been known since the sixteenth century that the Earth possesses an internal, dipolar magnetic field. The Earth's field is the best-studied of all astrophysical magnetic fields. It is now known to have an average strength of about one-half gauss at the surface and to deviate from a perfect dipole by 10–20% on a continental scale. The surface irregularities change with time. Individual features grow and decay, apparently at random, on a time scale of 10^3 years; the entire pattern drifts westward at a rate of about 0.2° longitude per year. The axis of the dipole is presently tilted somewhat more than 10° with respect to Earth's rotation axis.

The discovery of nonthermal radio emission from Jupiter (Burke & Franklin 1955, Sloanaker 1959) suggested that it possesses a relatively strong magnetic field. Direct exploration of the neighborhood of Jupiter with the Pioneer 10 and 11 space probes has confirmed the existence of the Jovian magnetic field (Smith et al 1974, 1975). The measurements suggest an average surface field of the order of 10 G. Jupiter's dipole moment is apparently tilted about 10° with respect to the rotation axis and offset from the center of the planet by about one tenth of the planet's radius. The deviation of Jupiter's surface magnetic field from that of a dipole appears to be larger than the deviation of Earth's magnetic field.

Recently Brown (1975) has reported hectometric radio emission from Saturn with a spectrum similar to that of Jupiter's nonthermal hectometric radiation, thus hinting that Saturn may also possess a magnetic field. From the frequency of the spectral maximum, Brown estimates that the Saturnian field is about one-eighth as intense as the Jovian field. This estimate presumes that the radiation from both planets is due to cyclotron emission.

Measurements made during the Mariner 10 encounters with Mercury revealed the presence of a well-developed magnetosphere about that planet (Ness et al 1974b, 1975a,b). The stress of the magnetospheric field deflects the solar wind and produces a magnetopause standing about 1.6 planetary radii in front of Mercury. The observations suggest that Mercury possesses an internal magnetic field, roughly dipolar in character, having an intensity of several hundred gamma at the surface ($1\gamma = 10^{-5}$ G). It has been suggested that the dipole axis is aligned within a few degrees of Mercury's spin axis and perhaps offset somewhat from the planet's center.

At present, no internal magnetic field can be detected for Venus (Bridge & Lazarus 1969, Bridge et al 1974, Dolginov et al 1969, Gringauz et al 1968, Ness et al 1974a, Van Allen et al 1969), and evidently this planet lacks a real magnetosphere. The plasma and magnetic field measurements of its bow shock are thought to be consistent with deflection of the solar wind by the planet's substantial ionosphere.

Magnetic field measurements have been made near Mars with instruments on the Mars 2 and Mars 3 spacecraft. Dolginov et al (1973) suggest that these measurements indicate a weak internal magnetic field. Possibly the observations can be accounted for on the basis of a solar wind interaction with the Martian ionosphere.

The Earth's moon is known to possess little, if any, large-scale magnetic moment (Russell et al 1974). On the other hand, lunar rock and soil samples display natural remanent magnetization (Doell et al 1970, Helsley 1970, Larochelle & Schwarz 1970, Nagata et al 1970, Runcorn et al 1970, Strangway et al 1970, Fuller 1974). The magnetizing fields are thought to have been of the order of 2000 γ, but uncertainty remains about the magnetic properties of the lunar material and different samples appear to have been magnetized by fields ranging up to 1 G (Fuller 1974, Stephenson et al 1974). In addition, the Moon is known to possess localized surface magnetic fields extending over many tens of kilometers. These fields have been observed with magnetometers both at the lunar surface (Dyal et al 1970) and carried on lunar-orbiting satellites (Coleman et al 1972), as well as through the perturbations induced in the solar wind as it blows past patches of strong field at the lunar limb (Barnes et al 1971, Mihalov et al 1971, Sonett & Mihalov 1972). The local fields have intensities ranging up to several hundred gamma and are apparently the result of permanently magnetized material. The surface magnetic fields are generally thought to arise from disruptions of the magnetized lunar crest. However, at this writing, the evidence is also consistent with fields produced by local regions of coherently magnetized material. In principle, the question can be resolved by looking at the correlation of the local fields at separated sites. If the fields originate in the disrupted crust, then the overall structure of the crustal magnetization should be discernible in the local fields.

Finally, meteoritic material often possesses substantial natural remanent magnetization (Stacey & Lovering 1959; Stacey et al 1960; Banerjee & Hargraves 1971, 1972; Butler 1972; Anders 1964; Brecher & Ranganyaki 1975). In instances when this magnetization can be interpreted as thermoremanence, the magnetizing fields appear to have been in the range of several tenths of a gauss to 1 G.

Planetary Magnetism

The collection of observational facts about planetary magnetism gives rise to a number of interesting questions. Consider even such a well-studied planet as the Earth, for which there are exceedingly good reasons to believe that the field is generated by fluid motions within the core (Elsasser 1950). The behavior of the surface field itself provides direct evidence of the existence of such fluid flow. However, despite numerous proposals, there is no firm understanding of the origin of the core motions, and it has even been suggested that much of the core is stably stratified (Higgins & Kennedy 1971), thereby inhibiting convective overturning. The situation is generally worse for the other planets, where observational data are sparse.

The Moon and Mercury present challenging and similar puzzles representative of the general problems facing the theory of magnetic field generation. As will become apparent in the ensuing discussion, magnetic fields are most readily generated in rotating bodies having large electrically conducting fluid regions. Both the Moon and Mercury rotate relatively slowly at the present time. Mercury apparently has a large iron core taking up most of the planet's interior, but the Moon has a small core, if any. The widespread magnetization of the lunar surface suggests that the Moon had an internal field some three or four billion years ago. The large and well-developed magnetosphere and bowshock associated with Mercury suggest that the present field has an internal origin and is not produced by unipolar induction with the passing solar wind field (Ness et al 1974b; C. P. Sonett, personal communication). Runcorn (1975) has pointed out that permanent magnetization induced in the crust of a planet by a centered dipole does not produce an external field after the disappearance of the magnetizing field if the magnetic susceptibility of the bulk planetary material is ignored. This reduces the possibility that a strong, external dipole field on Mercury is produced by remanent magnetization of the planet's crust. On the face of it, the simplest working hypothesis would be that the Moon had a dynamo magnetic field in the past and that Mercury has one today. It remains to be determined whether this hypothesis is sound (Levy 1972d) or whether some of the observational facts are hints of exotic physical conditions in the early solar system. A similar problem is presented by the strong magnetization of meteorites. If meteorite magnetization is caused by internal magnetic fields in the parent bodies, then a strong constraint is placed either on the parent bodies or on the theories of magnetic field production.

Research on the theory of hydromagnetic dynamos has brought us to the point where we understand the kinds of fluid motions that efficiently generate magnetic fields. The general character of the fields produced by the wide variety of fluid motions and boundary conditions that may occur in nature has not yet been fully explored. We have yet to establish, on the basis of first principles, the actual range of conditions that produce such motions with enough vigor to generate a field. We have also to fully establish the dynamical behavior of the dynamo fields, taking into account the forces that produce the fluid flow and the forces generated by the field itself.

GENERATION OF MAGNETIC FIELDS

Magnetic Fields in Conducting Fluids

We are concerned with magnetic fields embedded in fluids having high electrical conductivity and large physical dimensions. Under these conditions the behavior of a magnetic field is described by the well-known[2] hydromagnetic equation,

$$\frac{\partial \mathbf{B}}{\partial t} = \nabla \times (\mathbf{v} \times \mathbf{B}) + \eta \nabla^2 \mathbf{B}, \tag{1}$$

[2] In this article we refer to some basic results of hydromagnetic theory. The reader who has not met these ideas before will find useful introductions in most books on magnetohydrodynamics or plasma physics (e.g. see Alfvén & Fälthammer 1963 or Cowling 1957).

where $\mathbf{B}$ is the magnetic field; $\mathbf{v}$ is the velocity of the fluid and η is the magnetic diffusivity ($\eta = c^2/4\pi\sigma$, where σ is the electrical conductivity). According to Equation 1, the lines of force of the magnetic field tend to move with the fluid in which they are embedded while simultaneously diffusing through it. The degree to which the field lines are carried with the fluid is measured by the relative importance of the first and second terms on the right hand side of Equation 1. The relative magnitude of these two terms is expressed through their dimensionless ratio, which is often called the magnetic Reynolds number $R_m \equiv vl\eta^{-1}$, where l is the scale length of the fluid flow and magnetic field. In instances of very high electrical conductivity or rapid fluid motions having large physical scale, R_m is large, and the magnetic lines of force are carried efficiently. In astrophysical objects, R_m is commonly much larger than unity. The magnetic field is a vector quantity; when magnetic field lines from different parts of the fluid overlap as a result of diffusion, they must be added vectorially to obtain the net local field.

Just as the evolution of a magnetic field is influenced by the motion of the electrically conducting fluid in which it is embedded, so also is the motion of the fluid influenced by the stress of the magnetic field. The equation of fluid motion, including the magnetic stress, for an incompressible fluid is

$$\rho\frac{\partial \mathbf{v}}{\partial t} + \rho\mathbf{v}\cdot\nabla\mathbf{v} + 2\rho\mathbf{\Omega}\times\mathbf{v} = -\nabla p + \frac{1}{4\pi}(\nabla\times\mathbf{B})\times\mathbf{B} - \zeta\nabla\times\nabla\times\mathbf{v} + \mathbf{F}; \tag{2}$$

ρ is the density of the fluid, ζ is the viscosity. We have anticipated dealing with fluid motion in rotating bodies; $\mathbf{\Omega}$ is the rotation, and the centrifugal acceleration has been included in the effective pressure, p. The imposed force, $\mathbf{F}$, is responsible for the fluid motion. The magnetic stress is simply the familiar Lorentz force, $c^{-1}\mathbf{j}\times\mathbf{B}$ where the electrical current is $\mathbf{j} = (c/4\pi)\nabla\times\mathbf{B}$. The Lorentz stress can be decomposed, through a common vector identity, into

$$\frac{1}{4\pi}(\nabla\times\mathbf{B})\times\mathbf{B} = -\frac{1}{8\pi}\nabla B^2 + \frac{1}{4\pi}\mathbf{B}\cdot\nabla\mathbf{B}. \tag{3}$$

This decomposition provides a convenient conceptual picture of the forces induced in a fluid by an embedded magnetic field. The magnetic stress consists of two parts, the first of which is an isotropic magnetic pressure equal to $B^2/8\pi$. The remainder of the stress can be thought of as a tension, $B^2/4\pi$, exerted along the field lines, similar to the tension in elastic strings. The tension resists stretching and deformation of the magnetic lines of force.

The logically complete description of magnetic field generation in a natural body involves the simultaneous solution of the set of Equations 1 and 2, along with appropriate boundary conditions on the magnetic field and the fluid flow, supplemented by whatever equations are necessary to describe the forcing function, $\mathbf{F}$. This is a formidable undertaking for such a nonlinear system. The bulk of the progress so far has been through simpler and more modest approaches. The most substantial advances in our understanding of field generation in hydromagnetic dynamos have been made through a kinematical approach, in which no attempt is made to include the self-consistent dynamics of the fluid flow in the calculations. Instead, one simply postulates a particular fluid motion, thus eliminating the dynamical equations from formal consideration and leaving only the much simpler induction equation (Equation 1).

The Kinematical Approach to Magnetic Field Generation

As there are no explicit dynamical constraints to contend with, the kinematical approach permits considerable freedom in the choice of fluid flows. But because the ultimate goal is to understand generation of real magnetic fields in physical objects, special attention should be given to fluid motions reflecting the essential character of flows that one might expect to encounter in nature.

Consider convection in the core of a rotating planet. Bullard (1949) suggested that the tendency for outward and inward moving elements of material to conserve their angular momentum would produce differential rotation of the core fluid, the inner part spinning somewhat more rapidly than the outer. Indeed, the slow westward drift of identifiable features in Earth's surface magnetic field (about 0.2° longitude per year) may be a result of such nonuniform rotation and suggests a differential rotation speed of the order of 10^{-2} cm $\sec^{-1}$ between the inner and outer parts of Earth's core. The effect of such an azimuthal, differential motion is to wind the lines of force around the core, producing strong bands of toroidal magnetic field, as is shown in Figure 1. If, for example, we represent the differential rotation by

$$\mathbf{v} = r\Omega(r) \sin \theta \mathbf{e}_\phi, \tag{4}$$

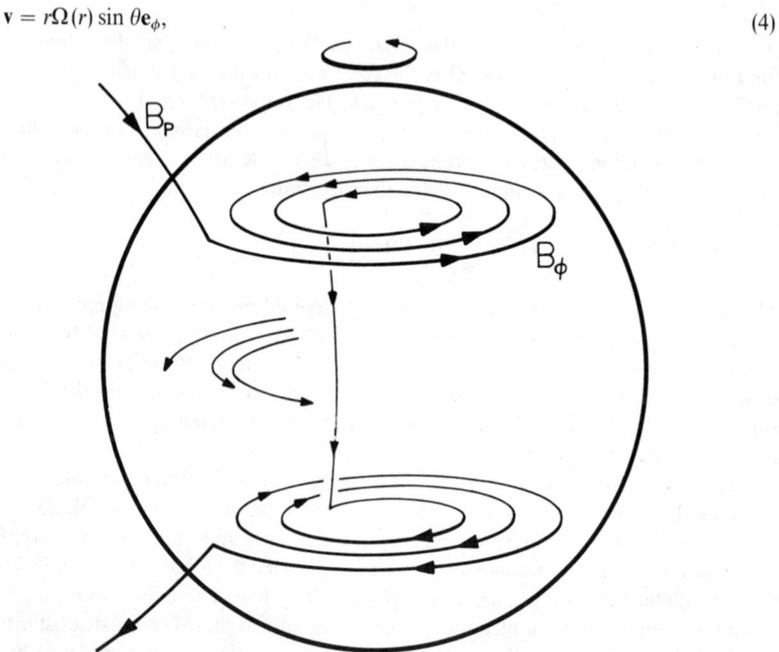

Figure 1 Nonuniform rotation of an electrically conducting fluid stretches the poloidal lines of force and winds them into strong bands of toroidal magnetic field, B_ϕ.

where $\mathbf{e}_\phi$ is the unit vector in the ϕ-direction of a standard spherical coordinate system, then

$$\nabla \times (\mathbf{v} \times \mathbf{B}) = B_r\left(\frac{\partial v_\phi}{\partial r} - \frac{v_\phi}{r}\right)\mathbf{e}_\phi, \tag{5}$$

in the case of axially symmetric magnetic fields. The ϕ-component of Equation 1 is

$$\frac{\partial B_\phi}{\partial t} - \eta\left(\nabla^2 - \frac{1}{r^2 \sin^2\theta}\right)B_\phi = B_r\left(\frac{\partial v_\phi}{\partial r} - \frac{v_\phi}{r}\right) \equiv \gamma B_r. \tag{6}$$

The other components of Equation 1 can be written compactly by defining a vector potential $\mathbf{A} = \mathbf{A}_\phi \mathbf{e}_\phi$ such that the poloidal part of the magnetic field is

$$\mathbf{B}_p \equiv (B_r, B_\theta, 0) = \nabla \times \mathbf{A}, \tag{7}$$

whence

$$\frac{\partial A_\phi}{\partial t} - \eta\left(\nabla^2 - \frac{1}{r^2 \sin^2\theta}\right)A_\phi = 0. \tag{8}$$

Equations 6–8 illustrate several points. Nonuniform rotation is very efficient in the production of strong, large-scale magnetic fields. Using Earth's core as an example, with $\eta \simeq 10^5$ cm^2 sec^{-1} and $B_r \simeq 5$ G in the core (extrapolated as a dipole from the Earth's surface), solutions of Equation 6 (Elsasser 1946a,b, 1947; Bullard & Gellman 1954) suggest that the toroidal field, $\mathbf{B}_\phi$ is as large as a few hundred gauss in the core.[3] On the other hand, note that nonuniform rotation alone is incapable of maintaining a magnetic field over long periods of time. The poloidal fields are not regenerated by $\mathbf{v}_\phi$; Equation 8 is without a source, so that $\mathbf{B}_p$ falls exponentially to zero with a time scale $\tau \sim R^2/\eta$, where R is the radius of the conducting core. In the Earth, which is a typical terrestrial planet, τ is a few tens of thousands of years. Regardless of the nonuniform rotation rate, the observable poloidal field thus falls to zero in a relatively short time. Further, since B_r is the source of the toroidal field in Equation 6, as B_r falls to zero so does B_ϕ. This illustration is a particularly simple example of the more general theorem (Elsasser 1946a, Bullard & Gellman 1954) that fluid motions with no radial component cannot maintain a magnetic field.

This theorem is one of a small class of similar theorems describing limitations on the kinds of fluid flows that produce regenerative dynamos and the structures of the magnetic fields that can result. For example, two-dimensional, as in purely axisymmetric, fields cannot be sustained in a hydromagnetic dynamo (Cowling 1934; Backus & Chandrasekhar 1956; Zel'dovich 1957; Braginskii 1965a,b; Lortz 1968). By way of interesting contrast, fluid motions that are functions of only two spatial coordinates can apparently sustain three-dimensional magnetic fields (Tverskoy 1966, Gailitis 1970, G. O. Roberts 1972). The important points made by these theorems are that radial motions are necessary to the regeneration process

[3] The toroidal field is confined to the electrically conducting regions of the planet and thus is not directly observable at the surface.

and that departures of the field from pure axisymmetry are also a central feature. Although the average magnetic field may be two-dimensional, or axisymmetric, the field in the production region must contain asymmetries essential to the generation process. Also, there is a general lower limit on the magnetic Reynolds numbers of regenerative flows. This lower limit simply states that the work done on the magnetic field by the fluid motion must be at least as great as the rate of energy dissipation in the longest-lived free decay mode of the field. In a sphere, this requirement becomes (Backus 1958) $R_m > \pi$, where R_m is calculated from the average deformation velocity of the flow. The condition is necessary but is not in itself sufficient for regeneration.

In the presence of radial convection, Equation 8 is no longer without a source. Convection in a rotating body provides a source of $\mathbf{A}_\phi$ which is proportional to $\mathbf{B}_\phi$. Then the closed system of Equations 6, 7, and 8 has the poloidal part of the field, $\mathbf{B}_p$, generated from the toroidal part $\mathbf{B}_\phi$ through the convective motions, whereas $\mathbf{B}_\phi$ is generated from $\mathbf{B}_p$ through the nonuniform rotation. The two processes together regenerate and maintain the total field.

To proceed with this illustration of magnetic field production in natural objects, separate the fluid velocity, $\mathbf{v}$, into the large-scale nonuniform rotation $\mathbf{U}$ and the small-scale convection, which we denote by $\mathbf{u}$. For simplicity, consider the case when differential rotation is strong, so that $\mathbf{B}_\phi$ is the dominant field present in the

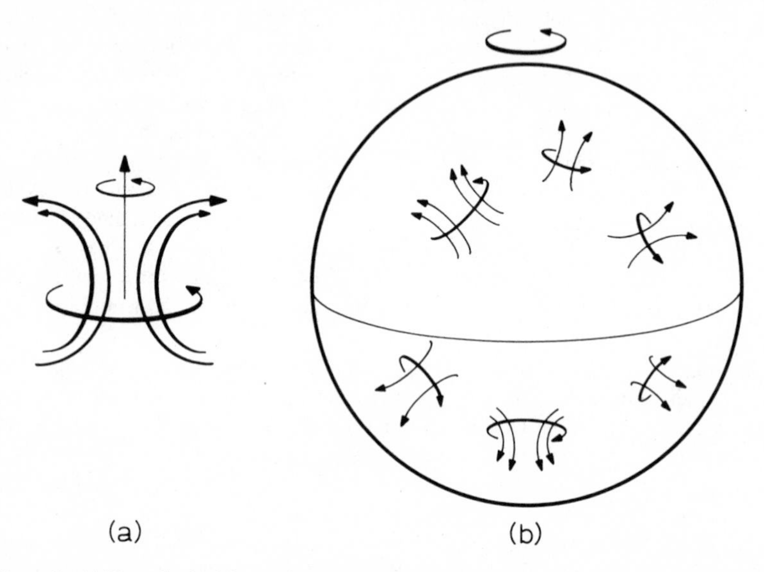

Figure 2 (*a*) Hypothetical form of a convective eddy in a rotating body. The Coriolis force acts on the converging and diverging components of the flow to impart a cyclonic component to the motion. (*b*) Convective eddies in the northern and southern hemispheres have cyclonic motions with opposite senses because the Coriolis force has a different sign in each hemisphere.

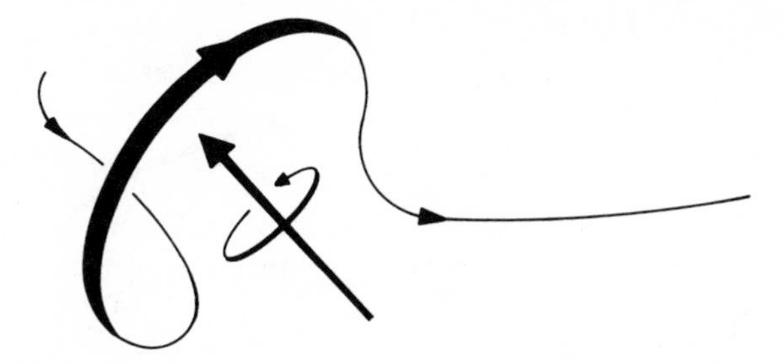

Figure 3 The effect of a cyclonic convective cell is to produce loops of toroidal field and rotate them into meridional planes.

generation region, and when the average fields are axisymmetric. Then the equation for the vector potential, A_ϕ, can be written (Parker 1955b, 1970b),

$$\frac{\partial A_\phi}{\partial t} - \eta\left(\nabla^2 - \frac{1}{r^2 \sin^2 \theta}\right) A_\phi = \langle \mathbf{u} \times \mathbf{B} \rangle_\phi. \tag{9}$$

The right-hand side of Equation 9 represents the average rate of production of vector potential in the ϕ-direction through the interaction of the convective motions with the toroidal magnetic field.

Consider the relatively small-scale convective motions in a rotating body. Parker (1955b; also Elsasser 1950) suggested that the Coriolis force acting on the converging and diverging components of the velocity causes the flow in a rising cell of fluid to be cyclonic, as shown in Figure 2*a*. He pointed out that in an incompressible fluid the average helicity $(\mathbf{u} \cdot \nabla \times \mathbf{u})$ should be positive in the northern hemisphere of a rotating body and negative in the southern (Figure 2*b*). Recall that the fluid, having high electrical conductivity, carries the magnetic lines of force with it.[4] The effect is to distort the toroidal field into twisted loops (Figure 3), which have a non-vanishing average meridional projection, as pictured in Figure 4. These loops have the same sense as the large-scale poloidal magnetic field; they diffuse outward, coalesce, and regenerate it.

We can obtain a crude estimate of the source in Equation 9 through the following simple, physical picture. Denote by n the volume rate of occurrence of cyclonic, convective eddies in the field generation region, and let $\delta\mathbf{A}$ be the magnetic vector potential associated with each loop. Suppose that the convective eddies have diameter λ; then the loops which the eddies produce from the toroidal field have a scale λ and an intensity $|\mathbf{B}_\phi|$. In order to express the source term in Equation 9 as an average over the loops, write

$$\langle \mathbf{u} \times \mathbf{B} \rangle_\phi = \left\langle \frac{(\delta \mathbf{A})_\phi}{\delta t} \right\rangle. \tag{10}$$

[4] Because of the finite conductivity of the fluid, the field lines slip back and only tend to move with it. In more precise language the velocities referred to in this part of the discussion are the velocities of the field lines. This is automatically taken into account in many of the formal calculations cited later.

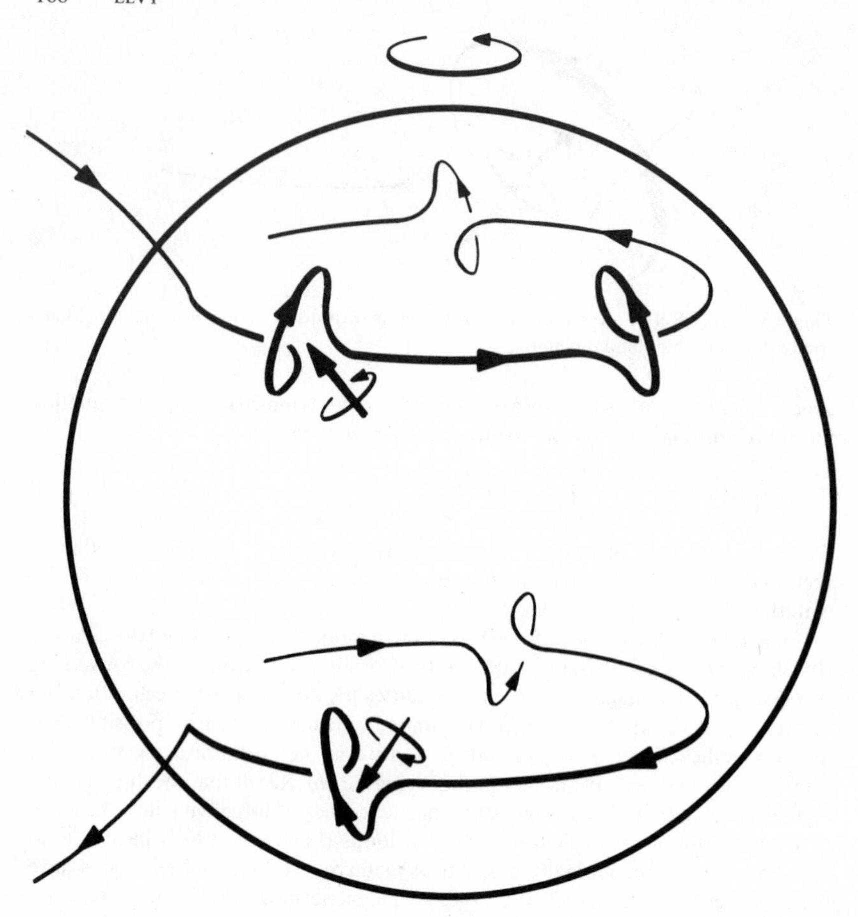

Figure 4 The rotated loops of magnetic field have the same sense as the original poloidal field—in this case a dipole. They diffuse outward and reinforce the original field.

In order of magnitude, $|\boldsymbol{\delta}\mathbf{A}| = \lambda|\mathbf{B}_\phi|$; the average direction of $\boldsymbol{\delta}\mathbf{A}$ is perpendicular to the plane of the loop and has the same sense as $\nabla\times\mathbf{B}$. If v_{cyc} is the cyclonic component of the convection velocity, each loop is twisted through an angle $\beta = v_{\text{cyc}}\,\delta t\lambda^{-1}$ radians, where δt is the lifetime of a convective eddy, then the ϕ-component of the increment of vector potential associated with a loop is $-\lambda\mathbf{B}_\phi \sin\beta$. For small β, this becomes $(\delta\mathbf{A})_\phi = -\mathbf{B}_\phi\, v_{\text{cyc}}\,\delta t$. Then, averaging over the volume of fluid,

$$\langle\mathbf{u}\times\mathbf{B}\rangle_\phi = -[n\lambda^3\,\delta t]\langle v_{\text{cyc}}\rangle B_\phi \equiv \Gamma\mathbf{B}_\phi; \tag{11}$$

$\langle v_{\text{cyc}}\rangle$ is positive or negative according to whether the previously defined helicity of the flow is positive or negative. The quantity $[n\lambda^3\,\delta t]$ is simply an efficiency factor: that fraction of the fluid undergoing cyclonic convection at any instant of time. This expression for the production of poloidal field by cyclonic convection is essentially identical to that obtained by Parker (1970b) through an integration of the hydromagnetic equations. The results of the calculations contain additional terms that represent production of toroidal field through the action of the cyclonic convection on the poloidal field, as was first emphasized by Steenbeck et al (1966) (see the following discussion). In the presence of strong nonuniform rotation, the additional terms are relatively unimportant.

Taken together, Equations 6, 7, 9, and 11 describe the generation of magnetic fields by nonuniform rotation[5] and cyclonic convection in rotating bodies. For definiteness and conceptual clarity, we have discussed the derivation of these equations in terms of the regeneration of a particular dipolar magnetic field mode. However, it should be kept clearly in mind that the equations and physical ideas are more general. The differential equations are in a broad sense microscopic equations. They express the local generation of magnetic field in terms of the local character of the fluid motion. When accompanied by appropriate boundary conditions, these dynamo equations describe a large class of both steady and time-dependent magnetic field modes. Before discussing them we briefly survey some additional work on the kinematical approach to magnetic field generation.

DYNAMO EQUATIONS A set of dynamo equations similar to that above was derived by Braginskii (1965a,b) through an expansion of the induction equation (Equation 1) in inverse powers of a large magnetic Reynolds number. Braginskii assumed a fluid velocity consisting predominantly of azimuthal nonuniform rotation, along with a smaller component of nonaxisymmetric radial convection that changes slowly with time. Although the actual mathematical and physical formulation is rather different, the regeneration process proceeds along lines similar to the previous discussion.

Several authors have explored the properties of the magnetic field generation term, $\langle \mathbf{u} \times \mathbf{B} \rangle$, in the presence of turbulent fluid motions (Steenbeck et al 1966; Rädler 1968a,b; Moffatt 1970a,b). They found that the symmetry characteristics of a turbulent flow determine whether or not it is capable of maintaining a magnetic field. As may be anticipated from our earlier discussion, an essential characteristic of the turbulence is its helicity. In the main it is found that helical, or non-mirror-symmetric, turbulence produces an electrical current proportional to the average magnetic field. The above authors calculated this induced current, in the limit of small magnetic Reynolds number, by expanding $\langle \mathbf{u} \times \mathbf{B} \rangle$ in a series of correlation products of the fluid velocity and magnetic field (see also Lerche 1971).

Consider a specified turbulent fluid velocity, $\mathbf{u}$, whose average[6] is zero, $\langle \mathbf{u} \rangle = 0$, and divide the total magnetic field into its average and fluctuating parts; $\mathbf{B} = \mathbf{B}^0 + \mathbf{b}$, where $\langle \mathbf{B} \rangle = \mathbf{B}^0$ and $\langle \mathbf{b} \rangle = 0$. With these definitions, it follows that the source in Equation 9 is $\langle \mathbf{u} \times \mathbf{B} \rangle = \langle \mathbf{u} \times \mathbf{b} \rangle$. Now putting $\mathbf{B}$ and $\mathbf{u}$ into Equation 1

[5] To be complete, Equation 6 should have the additional term $B_\theta r^{-1}(\partial v_\phi/\partial\theta - v_\phi \cot\theta)$ on the right-hand side to account for meridional gradients in the differential rotation.

[6] For economy of space we will use an informal definition of the averaging procedure. More formal definitions will be found in the literature cited.

and averaging, we obtain

$$\frac{\partial \mathbf{B}^0}{\partial t} - \eta \nabla^2 \mathbf{B}^0 = \mathbf{\nabla} \times \langle \mathbf{u} \times \mathbf{b} \rangle. \tag{12}$$

Suppose that the fluctuating quantities **u** and **b** are small; subtracting Equation 12 from Equation 1 and dropping second-order terms in small quantities, we obtain

$$\frac{\partial \mathbf{b}}{\partial t} - \eta \nabla^2 \mathbf{b} = \nabla \times (\mathbf{u} \times \mathbf{B}^0). \tag{13}$$

Equation 13 gives the fluctuating part of the magnetic field, **b**, in terms of the large-scale field and the fluid velocity. It can be solved formally for **b** in terms of tensor Green's functions. Then, crossing **b** with **u** and averaging, an expression for $\langle \mathbf{u} \times \mathbf{b} \rangle$ is obtained. The mathematics is too lengthy to reproduce here, but it is shown in the references that

$$\langle \mathbf{u} \times \mathbf{b} \rangle_i = \alpha_{ij} B_j^0, \tag{14}$$

where the quantity α_{ij} depends on the two-point correlation function of the random fluid velocity $\langle u_i(x,t) u_j(x+\delta x, t+\delta t) \rangle$. α_{ij} is nonzero in turbulence having a local net helicity. Equation 14 is the generalization of Equation 11; it arises through the same kind of distortion and twisting of the magnetic lines of force.

It is important to realize that in order for the efficient regeneration (given by Equation 11 or Equation 14) to occur, it is not necessary that the helicity of the convective flow be nonzero when averaged over the entire dynamo region. In a rotating sphere the helicity has opposite sign in the two hemispheres so that the global average vanishes, as is illustrated pictorially in Figure 2*b*. The important point is that the negative and positive helicity flows must be physically separated from one another.

Rädler (1969a) has pointed out that even when the average helicity of the fluid motion vanishes locally in a rotating fluid, there are higher order terms that contribute to the regenerative action of $\langle \mathbf{u} \times \mathbf{B} \rangle$. The higher order terms are proportional to the gradients of the large-scale field components rather than to the field itself. Solutions of the dynamo equations under these conditions have been explored by Rädler (1969b, 1970) and P. H. Roberts (1972). Gubbins (1974b) calculated the correlation characteristics of incompressible flow driven by a random, homogeneous, isotropic body force in a rotating fluid. He found that the local average of the helicity vanished, and he recovered, in addition to higher order effects, the mean regenerative term, proportional to $(\mathbf{\Omega} \cdot \mathbf{\nabla})\mathbf{B}^0$, derived earlier by Rädler. Gubbins' calculation points up again the central importance of rotation for magnetic field generation in natural objects.

DIRECT SOLUTIONS The studies discussed so far have been primarily concerned with the development of a system of equations to describe the production of magnetic fields in astrophysical objects in terms of the general character of the fluid motions occurring in them. Several investigations of magnetic field generation by flows of a somewhat more artificial nature have also been carried out. Although these fluid motions are not likely to be found in real objects, studying them helps to delimit both the class of fluid velocities capable of regenerative dynamo activity and the class of magnetic fields produced. In addition, the calculations can often be carried through with a high degree of mathematical rigor, thus putting our overall ideas about magnetic field generation on a more certain foundation.

For example, Herzenberg (1958) demonstrated that two small conducting spheres, embedded in a large conducting sphere and spinning rapidly about nonparallel axes, regenerate a magnetic field, thus establishing the principle that steady fluid motions can regenerate magnetic fields.

G. O. Roberts (1972) has shown that steady fluid motions that are functions of only two spatial coordinates regenerate fields that are functions of the three coordinates. Tverskoy (1966) and Gailitis (1970) have investigated field regeneration by annular vortices, ultimately establishing that steady axisymmetric fluid motions can regenerate nonaxisymmetric magnetic fields.

ISOTROPIC TURBULENCE The effect of homogeneous, statistically steady, isotropic, mirror-symmetric, turbulent motion of a conducting fluid on the evolution of a magnetic field has been discussed for nearly 30 years (Batchelor 1950, Biermann & Schlüter 1951; see Parker 1970a for a survey) and remains unresolved (Kraichnan & Nagarajan 1967, Krause & Roberts 1973, Lerche 1973). The problem is beset with mathematical difficulties. Kraichnan & Nagarajan have shown that intuitive statistical arguments, based for example on the equipartition of magnetic and fluid kinetic energy densities, are not capable of resolving the issue. Approximate mathematical solutions generally fail to include the higher order correlations of the field and fluid fluctuations (e.g. the terms neglected in Equation 13) and thus introduce unpredictable errors into the results of the calculations (Lerche & Parker 1973). In the case of sufficiently ordered fluid motions, such as the helical convection discussed earlier, the mathematical difficulties have been overcome, largely through physical arguments that isolated the important regenerative components of the velocity (Parker 1955b).

One important effect of turbulence on magnetic fields that does appear to be relatively well established is the increased rate of field diffusion and reconnection as a result of turbulent mixing (Rädler 1968a,b; Parker 1971b). In a static conducting fluid, a magnetic field evolves by diffusive motion of the field lines through the fluid. As mentioned in connection with Equation 1, the resistive diffusion coefficient associated with this process is $\eta_r \equiv c^2/(4\pi\sigma)$. Suppose, however, that the fluid is in a state of turbulent motion with $R_m \gg 1$. Then the field lines are carried about randomly with the fluid. This enhanced transport of the field lines as they random walk with the fluid is expressible in the usual way by a turbulent diffusion coefficient. If λ is the size of the dominant turbulent eddies and v_t is the turbulent velocity in those eddies, the turbulent component of the diffusivity can be written $\eta_t \sim \beta\lambda v_t$, where β is a dimensionless number of the order of one tenth (Rädler 1968b, Parker 1971b). The total magnetic diffusivity of a turbulent fluid can be very crudely written as $\eta = \eta_r + \eta_t$.

The significance of turbulent dissipation can be seen by comparing η_t with η_r.

In Earth's liquid iron core the electrical conductivity is thought to be of the order of 10^{15} sec^{-1}; then $\eta_r \sim 10^5$ cm^2 sec^{-1}. The fluid velocity is evidently about 10^{-2} cm sec^{-1}, so that if the core is turbulent, with the dominant eddies having a scale $\lambda \sim 10^8$ cm, then $\eta_t \sim 10^5$ cm^2 sec^{-1} also. Thus, for Earth, $\eta_t \simeq \eta_r$ and turbulent mixing does not substantially change the dissipation rate of the geomagnetic field. Similarly, in any of the terrestrial planets, turbulence is not likely to make a significant contribution to the dissipation rate of a magnetic field unless the core conductivity is very much higher than that of Earth or the fluid velocity larger. In a body as large as a Jovian planet, with fluid velocities of perhaps a few centimeters per second (Hubbard & Smoluchowski 1973) in large turbulent eddies, and an electrical conductivity as high as 10^{17} esu (Stevenson & Ashcroft 1974), turbulent mixing of the field could be more important than pure resistive diffusion by several orders of magnitude.

Dynamo Magnetic Fields

In order to investigate magnetic fields produced by the dynamo Equations 6, 7, 9, and 11, it is convenient to solve for the normal modes whose time dependence is given by $A(\mathbf{r}, t) = A(\mathbf{r})e^{\omega t}$, where ω is in general a complex number. For the study of planetary magnetic fields, the simplest appropriate geometry is that of a homogeneous fluid sphere, say, of radius R, of uniform conductivity, surrounded by vacuum. In this case, the proper boundary conditions are that the fields be regular on the axis of rotation, be continuous across $r = R$, and fall to zero at least as fast as a dipole field as $r \to \infty$. The exterior fields in $r > R$ also satisfy $\nabla \times \mathbf{B}_p = 0$ and $B_\phi = 0$, since no electrical currents flow in this region. Although this simple computational model is adequate to reveal most of the interesting behavior of planetary dynamo fields, it is not sufficient for some problems of a fundamental nature. For example, hydromagnetic coupling between the core and mantle of Earth depends on the finite electrical conductivity of the mantle, which results in penetration of B_ϕ into the lower mantle. We do not take up such topics in this article.

The fluid motions are accounted for through expressions for the rate of nonuniform rotation, $\gamma(\mathbf{r})$, in Equation 6 and through the distribution of helical convection, $\Gamma(\mathbf{r})$, in Equation 11. Since the nonuniform rotation is thought to arise from the tendency of inward and outward moving elements of fluid to conserve their angular momentum, $\gamma(\mathbf{r})$ is negative and has the same form in the northern and southern hemispheres of a rotating body. However, the Coriolis force acting on the diverging and converging components of the convection has opposite sign in the two hemispheres so that $\Gamma(\mathbf{r})$ is an odd function of distance from the equatorial plane. As mentioned earlier, in incompressible flow Γ is thought to be negative in the northern hemisphere and positive in the south (Parker 1955b). On the other hand, Steenbeck et al (1966) point out that convection in which compressible fluid elements move over distances comparable to a density scale height in a stratified medium has Γ positive in the north and negative in the south. We should remark here that the properties of hydromagnetic convection in rotating objects have not yet been fully established. Most statements on the subject should be regarded as tentative.

First consider the stationary dynamo fields, for which $\partial/\partial t = 0$. The lowest order mode has a spatial scale about equal to the radius of the sphere, R, so that Equations 6, 7, 9, and 11 can be written dimensionally, in terms of the characteristic values of their variables,

$$\eta \frac{B_\phi}{R^2} \sim \gamma B_p, \quad B_p \sim \frac{A}{R}, \quad \eta \frac{A}{R^2} \sim \Gamma B_\phi.$$

In order for this set of equations to have a solution, it is necessary that

$$N \equiv \gamma \Gamma R^3 \eta^{-2} \sim 1.$$

The dimensionless number, N, is commonly called the dynamo number. It is the product of two magnetic Reynolds numbers, that associated with the nonuniform rotation and that associated with the convection, and is a function only of the physical properties of the dynamo—its size, the fluid motions and the electrical conductivity. The sign and magnitude of N determine the character of the magnetic fields produced by a hydromagnetic dynamo having a particular geometric distribution of fluid velocity. The magnitude of the dynamo number is a measure of the vigor of the fluid motions; its sign gives the relative sense of the nonuniform rotation and the helicity of the convection. In evaluating N, it is customary to take the sign of Γ to be that in the northern hemisphere. Notice that while the criterion for field generation depends only on the product of γ and Γ, the relative intensities of the poloidal and toroidal fields depend on the ratio of Γ to γ. The behavior of the time dependent modes is also a function of the dynamo number.

Stationary states fall into a sequence of increasing complexity, each state corresponding to a discrete value of N. Generally speaking, fields corresponding to large values of $|N|$ have larger contributions from high order multipole components (Levy 1972c, Parker 1971c). Beyond this there appear to be few assertions of general applicability that one can make. For a time, it was thought that steady dipole fields were only produced when $N > 0$, which is believed to occur in incompressible convection (Parker 1955b), and that only oscillatory fields were produced when $N < 0$ (see the summary in Parker 1970a). Now this is known to be untrue. The character and behavior of fields produced in hydromagnetic dynamos depend sensitively on the distribution of fluid motions. In addition, P. H. Roberts (1972) has shown that the distortion of magnetic field lines by large-scale meridional circulation has a substantial effect on the most easily generated dynamo modes (also Braginskii 1964, 1965a). Meridional circulation is a possible component of the fluid velocity in a planetary interior as a result, for example, of Ekman pumping in the differentially rotating fluid. It might also arise as a result of the dynamical effects of the magnetic field itself. On the basis of a number of models which he computed, P. H. Roberts (see also Steenbeck & Krause 1969) suggests that oscillatory fields are the most easily produced in the absence of meridional circulation. Both dipolar and quadrupolar oscillating fields are generated with either sign of N. In the presence of sufficient meridional circulation, P. H. Roberts found that steady fields are generated—dipoles when $N > 0$ and quadrupoles when $N < 0$—by smaller fluid velocities than are oscillatory fields.

The generation of steady fields in the absence of meridional circulations is well established (Deinzer & Stix 1971; Parker 1971c; Levy 1972a,b; Roberts & Stix 1972; Stix 1973; Deinzer et al 1974). Through asymptotic solutions of a rectangular dynamo model, Parker found that steady fields were produced more easily than oscillators, with dipolelike fields occurring when $N > 0$ and quadrupolelike fields when $N < 0$. In another example, when nonuniform rotation is confined to the low latitudes of a sphere and cyclonic convection is confined to high latitudes, steady dipole fields are generated with $N < 0$ (as implied in Parker 1969). If the fluid motions are distributed more broadly, as may be representative of natural objects, steady dipoles have only been found with $N > 0$ whereas quadrupoles are produced with either sign of N (Levy 1972a,b; Stix 1973). With nonuniform rotation and cyclonic convection confined to concentric spherical shells, Deinzer et al (1974) found that steady or monotonically growing fields are excited more easily than oscillators if the distance between the shells is large. The steady fields produced by that particular computational model are dipolelike if $N > 0$ and quadrupolelike if $N < 0$. When the separation between the concentric spheres is of the order of or less than their radii, oscillating modes are the more easily excited.

As mentioned earlier, axially symmetric fluid motions can generate asymmetric magnetic fields. Similarly, axially symmetric distributions of differential rotation and cyclonic convection can give rise to asymmetric magnetic fields. The numerical calculations (Stix 1971, Roberts & Stix 1972) suggest that in some cases the low-order nonaxisymmetric modes are produced as easily as the symmetric modes.

In the absence of strong differential rotation, generation of the toroidal field occurs through the action of the helical convection on the poloidal field, as given by Equation 14. Since the regenerative cycle of hydromagnetic dynamos based on this effect involves Equation 14 twice, they are often called α^2-dynamos. The fields most easily generated by α^2-dynamos are apparently steady and of either dipole- or quadrupolelike symmetry (Steenbeck & Krause 1969, P. H. Roberts 1972).

SUMMARY OF KINEMATICAL DYNAMO FIELDS Magnetic fields produced by kinematical dynamos may either behave monotonically (i.e. grow, decay, or be steady) or oscillate (with amplitude constant, growing, or decaying). As to the magnitude of the fluid velocity necessary to generate the fields, there appears to be no overall preference for either steady or oscillatory fields; neither is it clearly easier to generate fields having the symmetry properties of dipoles rather than quadrupoles. The actual character of the most easily excited modes depends on the geometrical distribution of the fluid motions—the shape of the nonuniform rotation, the distribution of the cyclonic convection, the occurrence of meridional circulation, and the relative senses of the various components of the fluid motion.

The growth rate, as a function of dynamo number, of the monotonically evolving modes is illustrated generically in Figure 5. As $|N|$ increases from zero, a stationary state is first encountered. A slight increase in $|N|$ produces a growing magnetic field. Further increase eventually produces a maximum in the growth rate, a subsequent decline, and finally another stationary state, beyond which, at larger $|N|$, the field decays. The pattern repeats with continuing increase in the dynamo

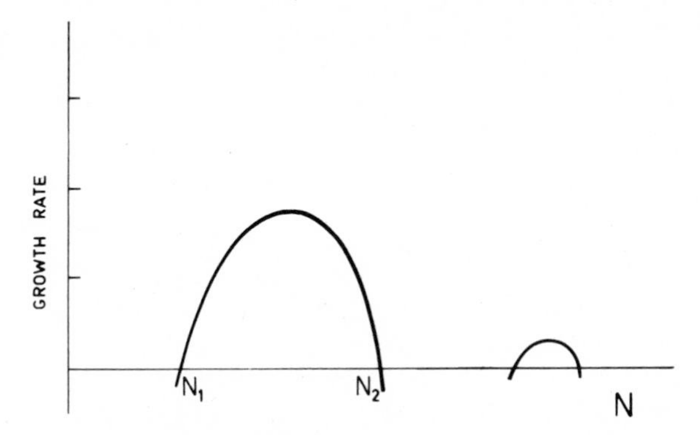

Figure 5 Generic form of the growth rate of nonoscillatory dynamo fields as a function of the dynamo number.

number. The maximum of the growth rate is typically a few times the free decay rate of the field (ηl^{-2}). Interspersed with these stationary and growing modes are the oscillatory modes (see Deinzer et al 1974). The oscillation rates of the low order oscillatory modes are typically several to as much as ten times the free decay rate (P. H. Roberts 1972).

OSCILLATING PLANETARY FIELDS Results of the calculations so far based on kinematical generation models suggest that oscillatory magnetic fields are at least as likely to be generated as steady fields. At present, it is unclear how this conclusion may be affected by a more complete understanding of the dynamical effects, but in the meantime it is instructive to consider the oscillation periods that one might expect. If one uses $\sigma \sim 10^{15}$ sec^{-4} as a crude estimate of the electrical conductivity of the iron cores in terrestrial planets, then the magnetic diffusivity, η, is about 10^5 cm^2 sec^{-1}. Cores of terrestrial planets have radii equal to a few times 10^8 cm. The magnetic diffusion time, $\tau = l^2/\eta$, is of the order of 10,000 years. A low order oscillatory mode has a period, then, of several thousand years in a terrestrial planet.

Now consider Jupiter. With the high conductivity of Jupiter's interior (Stevenson & Ashcroft 1974) and the rapid convective motions (Hubbard & Smoluchowski 1973), turbulence probably dominates the transport of the field. Taking the turbulence velocity as large as 10 cm sec^{-1} and mixing length scale of about 10^7 cm gives $\eta_t \sim 10^7$ cm^2 sec^{-1}. Then the field diffusion time is several tens of thousands of years. Thus if Jupiter's field is oscillatory, its period should be 5,000–10,000 years.[7] These long oscillation periods demonstrate the difficulty of determining by direct observation whether a particular planetary field is oscillatory or not. The paleomagnetic evidence indicates that the Earth's field is statistically steady in the

[7] By way of comparison, the short oscillation period of the sun appears to result from the high velocity (~ 1 km sec^{-1}) of the turbulence in the convective zone, which gives rise to very rapid transport and mixing of the field.

long intervals between the isolated reversals. It has been suggested that the reversals are due to interludes of oscillation (Braginskii 1964, Parker 1971c); however, this interesting phenomenon has several other possible causes (Rikitake 1958; Parker 1969; Levy 1972b,c).

DYNAMICAL PROBLEMS

In the preceding section we discussed magnetic field generation through the simplifying artifice of specifying the fluid velocity, thus passing over the dynamics of the generation process; we now turn to the dynamical questions. It is conceptually convenient to divide the dynamical considerations into two parts. The first is concerned with the purely fluid dynamical question: What physical conditions produce motions capable of generating a field? Throughout much of the previous discussion we have already alluded to the fact that convection in rotating bodies gives rise to efficient production of magnetic fields. The second part deals with the effects on the fluid of the magnetic stress (Equation 3). One of the most important of these effects is the inhibition of the fluid motion and the consequent equilibration of the field intensity.

We should warn the reader that, although this division seems natural, it does not encompass the full range of dynamical behavior of dynamo magnetic fields. In order to emphasize this point it is instructive to consider the extreme example of the solar field. The oscillating solar field is also believed to be generated through a hydromagnetic dynamo. It has been suggested (Babcock 1961, Leighton 1969, Parker 1971d) that the generation of the poloidal field is driven primarily by the buoyancy of the submerged toroidal field (Parker 1955a). If this is true, then the motions near the solar surface, in particular those most intimately responsible for generating the field, bear very little resemblance to the motions that would occur in the absence of the field. The regenerative convective motions are not inhibited by the field stress in this case; they are driven by it. This example points up especially clearly, the profoundly important effects that the nonlinear behavior of hydromagnetic dynamo fields may produce.

The instabilities induced by magnetic buoyancy are important mainly in highly compressible fluids. It is worth noting that the magnetic field may have similarly important effects on the flow of nearly incompressible fluids. The Coriolis force in a rotating fluid layer deflects the flow, and the general effect is to inhibit convective motions except insofar as the Coriolis force is balanced by viscosity (Chandrasekhar 1961). However, in the case when both rotation and a magnetic field are present, as is true in a hydromagnetic dynamo, part of the Coriolis force can be balanced by the Lorentz force, $(4\pi)^{-1}(\mathbf{\nabla} \times \mathbf{B}) \times \mathbf{B}$. The overall result is to facilitate convective motions (Chandrasekhar 1961, Eltayeb 1972) and perhaps produce a final dynamo state that is much different from what would be expected on the basis of motions occurring in the absence of the field. At this writing, knowledge of the nonlinear development of such effects has not progressed to the point where substantive conclusions can be drawn about their contribution to the equilibria of dynamo magnetic fields.

Inhibition of the Fluid Motions

Regenerative dynamo processes always involve stretching and twisting deformations of the magnetic field lines similar to those in the earlier figures. Recall from our discussion of the equation of fluid motion (Equation 2) that the stress of the magnetic field exerts a tension, equal to $B^2/4\pi$, along the lines of force. As the strength of the magnetic field increases, the magnetic tension resists the deformations of the field lines with greater efficacy until the fluid motions are slowed to the point at which the field amplitude is steady. Thus, if the magnitude of the dynamo number is slightly larger than N_1 in Figure 5, the field strength grows until the fluid velocity falls to where the dynamo number is equal to N_1.

Equilibration of the magnetic field intensity can occur through reduction of the helical or cyclonic convection velocity of the nonuniform rotation, or through both. Moffatt (1972) and Vainshtein (1972) have discussed the inhibition of the helical motions. Consider the linearizations of Equations 1 and 2, where $\mathbf{B} = \mathbf{B}^0 + \mathbf{b}$; $\mathbf{b}$ and $\mathbf{u}$ are small quantities and $\mathbf{B}^0$ is constant over the small scales characteristic of the turbulent, convective motions.

$$\frac{\partial \mathbf{b}}{\partial t} = \mathbf{B}^0 \cdot \nabla \mathbf{u} + \eta \nabla^2 \mathbf{b}, \tag{15}$$

$$\rho \frac{\partial \mathbf{u}}{\partial t} + 2\rho \mathbf{\Omega} \times \mathbf{u} = -\nabla p + (4\pi^{-1})\mathbf{B}^0 \cdot \nabla \mathbf{b} + \zeta \nabla^2 u + \mathbf{F}, \tag{16}$$

where p is the total pressure, fluid plus magnetic. Also suppose that the fluid is incompressible, and that $\mathbf{F}$ is chosen so as to produce a regenerative fluid motion and that $\nabla \cdot \mathbf{F} = 0$ (Moffatt 1972). Then we have

$$\nabla \cdot \mathbf{b} = \nabla \cdot \mathbf{u} = \nabla \cdot \mathbf{F} = 0. \tag{17}$$

Denoting by $\tilde{\omega}$ the Fourier transform of ω, Equations 15, 16, and 17 become

$$-i\omega \tilde{\mathbf{b}} = i\mathbf{B}^0 \cdot \mathbf{k}\tilde{\mathbf{u}} - \eta k^2 \tilde{\mathbf{b}}, \tag{18}$$

$$-i\rho\omega \tilde{\mathbf{u}} + 2\rho \mathbf{\Omega} \times \tilde{\mathbf{u}} = -i\mathbf{k}\tilde{p} + i(4\pi^{-1})\mathbf{B}^0 \cdot \mathbf{k}\tilde{\mathbf{b}} - \zeta k^2 \tilde{\mathbf{u}} + \tilde{\mathbf{F}}, \tag{19}$$

$$\mathbf{k} \cdot \tilde{\mathbf{u}} = \mathbf{k} \cdot \tilde{\mathbf{b}} = \mathbf{k} \cdot \tilde{\mathbf{F}} = 0, \tag{20}$$

where $\mathbf{k}$ and ω are the Fourier conjugates of the space and time variables. Now the set of algebraic Equations 18, 19, and 20 can be solved for $\tilde{\mathbf{u}}$ in terms of the rotation, the body force, and the large-scale magnetic field. Fourier inversion of $\tilde{\mathbf{u}}$ recovers $\mathbf{u}$, the turbulent convective velocity, including the restraining influence of the magnetic stress. The velocity $\mathbf{u}$ is then used to calculate the regenerative source term, Equation 14. The dynamo equations containing the field-dependent regeneration term are no longer linear and homogeneous. As a result, the equilibrium field intensity is determined by the equations. Moffatt (1972) carried this analysis through with a number of simplifying assumptions to facilitate the mathematics. An interesting specific result of the calculations is his expression for the equilibrium ratio of the magnetic field energy density to the fluid kinetic energy density,

$$\frac{\langle B^2 \rangle/4\pi}{\rho\langle u^2 \rangle/2} \sim \left(\frac{\Omega}{\omega_0}\right)^{1/2} \left(\frac{\zeta}{\eta}\right)^{1/2} \frac{L}{l}. \tag{21}$$

Among the assumptions used in deriving this expression are $\Omega \gg \omega_0$, where ω_0 is the characteristic frequency of the random motions, and $\zeta \ll \eta$. Equation 21 shows that, when the scale, L, of the average magnetic field is large in comparison with the scale, l, of the fluid forcing function, the magnetic energy density exceeds the kinetic energy density in Moffatt's formal example.

Moffatt's calculation is an example of the failure of the arguments, which we mentioned earlier, for statistical equipartition between the fluid kinetic energy density and the magnetic field energy density. The reasons these arguments may be irrelevant to the large-scale fields of astrophysical bodies can be understood on the basis of the dynamo process that we have discussed. Magnetic field generation in natural objects depends on the organized character of the fluid flow as determined by the imposed body forces and the ordering effect of the Coriolis force. Then the equilibrium intensity of the field is that at which the Lorentz force causes deflection of the fluid velocities sufficient to halt the growth of the field. This suggests that the maximum field strength produced by a hydromagnetic dynamo corresponds to an approximate equality between the Lorentz force and the Coriolis force or the forces driving the fluid motion. Of course, in any particular instance, the saturation value of the field may be smaller (Vainshtein 1972). Note that if the westward drift of surface features of Earth's field represents nonuniform rotation of the core, then the dynamical balance in Earth's core is probably between the Lorentz and Coriolis forces (Bullard & Gellman 1954, Levy 1974).

Tension in the magnetic lines of force also resists stretching by nonuniform rotation. The toroidal field is like a wound-up spring (Figure 1) held in check against its tendency to come unwound by the forces that produce the nonuniform rotation. Malkus & Proctor (1975) have pointed out that even magnetic fields generated, in the absence of nonuniform rotation, by the so-called α^2-effect discussed earlier, have their intensities limited by the tendency of the toroidal field to unwind. In this case, the dynamical balance results from continual production of poloidal and toroidal field in the form of a wound-up magnetic spring countered by steady unwinding of the spring.

Dynamical Models

The earlier-defined conceptual division of the dynamical considerations can be translated directly into a computational algorithm. A number of authors (e.g. Gilman 1969a,b; Busse 1973) have followed this approach. Busse considers a particularly simple example of two dimensional convection rolls produced by a temperature difference, ΔT, between two horizontal boundaries. Superimposed on the convection is a large-scale velocity shear along the axes of the rolls. The initial fluid motion can be expressed as

$$\mathbf{u} = \nabla \times (\mathbf{e}_z v_0 \sin \alpha y \sin \pi x) + \mathbf{e}_z \omega_0 \cos 2\pi x. \tag{22}$$

The amplitude, v_0, of the convective motions is a function of the Rayleigh number, $\gamma g d^3 \rho \Delta T/\kappa\zeta$; here γ is the thermal expansion coefficient of the fluid, d the separation between the boundaries, g the acceleration due to gravity and κ the thermal conductivity. Busse uses Equation 22 in the hydromagnetic induction equation (Equation 1) and solves for the fastest growing magnetic field mode through a series expansion. He then iterates the equation of fluid motion (Equation 2) by including the linearized magnetic field stress. The main effect of the magnetic stress in this illustrative model is to inhibit the convective motions and the field comes to equilibrium at an intensity determined by the Rayleigh number.

Busse (1975) has extended this analysis, to make it more directly applicable to planetary field generation, by fitting the two-dimensional array of convection rolls into a rotating cylindrical annulus. The axes of the rolls are now parallel to the rotation axis and the large-scale shear motion is taken to be meridional circulation induced by Ekman suction. Although the mathematics is more complicated, the iterative logic is the same as in the plane, two-dimensional model. Lack of space precludes a detailed discussion of the foundations of the calculations so we describe only the main conclusions. In the initial linearized solution of the fluid mechanical equations, the flow is closely geostrophic, that is, the Coriolis force is nearly balanced by the pressure

$$2\rho\mathbf{\Omega}\times\mathbf{u} = -\nabla p. \tag{23}$$

The fluid motions produce a field that is aligned with the rotation axis and is generally dipolar in character. The main action of the Lorentz stress is to reduce the amplitude of the convection rolls and this effect stabilizes the intensity of the field.

Several interesting results emerge from the model. The linearized convection does not produce strong differential rotation; consequently a strong toroidal field is not generated. In the stationary equilibrium, the flow remains largely geostrophic and the Lorentz force remains small compared to the Coriolis force. The Earth is the best available paradigm with which to compare our ideas about magnetic field production. Busse's model suggests that the total field in the core is not much larger than the poloidal field extrapolated from the surface, i.e. 5–10 G. This lowers the total energy required to maintain the field—by one or two orders of magnitude when compared with the energy requirements of models incorporating strong differential rotation. On the other hand, it requires fluid velocities of about 0.4 cm sec^{-1}, considerably faster than the velocities usually thought characteristic of the Earth's core.

It remains to be established whether such a linearized model is directly representative of magnetic field generation in any real objects. But in any case it serves as a valuable point of departure and suggests questions that need to be included in future studies. First consider the boundary conditions on the flow. In the model, fluid motions are driven by a fixed temperature difference between the inner and outer boundaries. The magnetic stress slows the rate of convective overturning and diminishes the heat flow. Suppose now that convection is driven by the continuous generation of heat in the interior of a body. Then the proper boundary condition on the flow is that the heat flux remain constant. In a convecting body, most of the total heat transport is by convection. Under the boundary conditions then, the convective motions cannot be easily slowed. One can speculate that the equilibrium state of motion, including the effect of the magnetic field, may consists of a somewhat reduced convection rate coupled with an increased thermal gradient to keep the heat flux constant. This suggests that the equilibrium value of the magnetic field will be higher than that estimated from fluid motions driven by a fixed temperature differential. Formal calculations will of course be necessary to determine if and under what circumstances the field strength becomes great enough to balance the Coriolis force. Finally, there remains the question raised at the start of this section: whether the field will grow strong enough to change the stability character of the convection and give rise to a final flow that does not resemble the onset of linearized convection.

Dynamical Stability and Oscillations of Dynamo Fields

Restraint of the regenerative fluid motions by magnetic stresses has several additional interesting and important consequences. Stix (1972) has investigated an example of the effect on oscillatory dynamo modes. He used a simple expression for the source term, Γ, in Equation 11, setting it equal to zero when the magnitude of the field exceeded some arbitrary value, B_c, and setting it equal to a constant large enough to be regenerative when the field fell below B_c. Stix then followed the evolution of oscillatory modes numerically. His main conclusion was that the periods of strongly driven oscillatory modes may be longer than the periods obtained from the purely kinematical analysis (Γ = constant), by a factor of about two. Braginskii (1970) used the same kinematical device to simulate large-amplitude dynamical oscillations of stationary dynamo fields about their equilibrium states.

Levy (1974) has explored the dynamical stability and oscillations of stationary dynamo fields through perturbations of the equilibrium states, taking into account the established results of the kinematical analyses. In the neighborhood of stationary states (e.g. N_1 and N_2 in Figure 5), the behavior of the field magnitude, $\mathscr{B}$, can be expressed as (Levy 1974),

$$\frac{\mathrm{d}\mathscr{B}}{\mathrm{d}t} = \nu\mathscr{B}, \tag{24}$$

where ν is the magnetic field growth rate. The equation of fluid motion (Equation 2) can be reduced to

$$\rho Q \frac{\mathrm{d}\mathscr{V}}{\mathrm{d}t} = R - \mathscr{B}^2 S, \tag{25}$$

where $\mathscr{V}$ is the amplitude of the fluid motion, R is the work per unit of velocity done on the fluid by the imposed forces, and $\mathscr{B}^2 S$ is the work per unit of velocity done on the magnetic field by the fluid; Q is roughly equal to volume of the dynamo region. Setting $\mathscr{B} = \mathscr{B}_0 + \delta\mathscr{B}$ and $\mathscr{V} = \mathscr{V}_0 + \delta\mathscr{V}$ in Equations 24 and 25 and linearizing, we obtain finally

$$\delta\mathscr{B} \propto e^{\pm i\omega t} \tag{26}$$

where

$$\omega = \left\{ \mathscr{B}_0^2 \frac{2S}{\rho Q} \left(\frac{\partial \nu}{\partial \mathscr{V}} \right)_{\mathscr{V}_0} \right\}^{1/2} \tag{27}$$

Thus the restraining influence of the magnetic stresses renders stationary dynamo equilibria dynamically stable or unstable according to whether $\partial v/\partial \mathcal{V}$ (or equivalently $\partial v/\partial N$) is positive or negative in the neighborhood of the equilibrium point. In Figure 5, perturbations about the equilibrium at N_1 result in stable oscillations; perturbations about N_2 grow exponentially. For dynamical reasons, then, N_2 could not be a stationary state of a magnetic field in a real object. A crude estimate of the time scale in either case suggests that the dynamical oscillation time or instability growth time, for infinitesimal perturbations, is the geometric mean of the magnetic field resistive decay time and the time for an Alfven wave to cross the dynamo region.

CONCLUDING REMARKS

The aim of this article is to review the general status of our ideas about magnetic field generation by hydromagnetic dynamos in planets. As a consequence of the limited purview, a number of important and relevant topics have been either omitted or passed over briefly. We have not dealt in any detail with attempts to solve the hydromagnetic induction equation numerically, nor with numerical experiments on the dynamical problem. Neither have we emphasized particular mathematical formalisms, such as expansion of the hydromagnetic equations in multiple scales. A recent survey of such topics, among others, has been given by Gubbins (1974a). In confining our attention to the central questions of magnetic field generation, we have neglected a number of issues that are more directly applicable to specific objects. We have not, for example, detailed the many ideas concerning the origin of geomagnetic reversals or of the westward drift of the geomagnetic field. In our opinion, there are a number of tenable theories, and the task of deciding between them is intimately connected with the task of determining the precise magnetohydrodynamical state of the Earth's core.

So far as we are presently aware, the ultimate source of interior motions has not been firmly established for any planet. Quantitative application of the ideas reviewed here to the elucidation of planetary magnetic fields will require an improved understanding of the physical, chemical, and thermal states of planetary interiors.

In principle, the observational facts we surveyed in the introduction provide the best available empirical foundation for our theories of magnetic field generation. Unfortunately, our present notions about the evolution and interior states of planets are not so firmly fixed that the observational facts can yet be unambiguously interpreted. Similarly, these facts provide important clues about the evolution of the solar system and planets. But the theories of magnetic field generation are not yet sufficiently developed to interpret the magnetic clues unambiguously. For a time, then, we expect that our ideas about magnetic field generation and our picture of solar system cosmogenesis will progress together, each contributing to the other.

In summary, the kinematical approach to the study of magnetic field generation has provided a working knowledge of the kinds of fluid flows that efficiently generate fields and a heuristic picture of the conditions producing such flows. Generally speaking, sufficiently intense convection or similar overturning motion of electrically conducting fluid in a rotating body produce a magnetic field. Although there is room for quantitative improvement in calculations of the regenerative action of these fluid motions, the dynamical problems are now the major barriers to our understanding. The outstanding dynamical mysteries include the origin and character of planetary fluid motions, the effect of the magnetic forces on the fluid motions, and the nature of the dynamical balance in the equilibrium states of dynamo magnetic fields.

NOTE ADDED IN PROOF Preliminary reports have recently appeared, reporting surface magnetic determinations for Mars—64γ (S. S. Doginov)—and for Venus—30γ (C. T. Russell).

Literature Cited

Alfvén, H., Fälthammer, C.-G. 1963. *Cosmical Electrodynamics.* Clarendon: Oxford Univ. Press. 228 pp.

Anders, E. 1964. Origin, age, and composition of meteorites. *Space Sci. Rev.* 3:583

Babcock, H. W. 1961. The topology of the sun's magnetic field and the 22-year cycle. *Ap. J.* 133:572

Backus, G. E. 1958. A class of self-sustaining dissipative spherical dynamos. *Ann. Phys.* 4:372

Backus, G. E., Chandrasekhar, S. 1956. On Cowling's theorem on the impossibility of self-maintained axisymmetric dynamos. *Proc. Nat. Acad. Sci.* 42:105

Banerjee, S. K., Hargraves, R. B. 1971. Natural remanent magnetization of carbonaceous chondrites. *Earth Planet. Sci. Lett.* 10:392

Banerjee, S. K., Hargraves, R. B. 1972. Natural remanent magnetizations of carbonaceous chondrites and the magnetic field in the early solar system. *Earth Planet. Sci. Lett.* 17:110

Barnes, A., Cassen, P. M., Mihalov, J. D., Eviatar, A. 1971. Permanent lunar surface magnetism and its deflection of the solar wind. *Science* 172:716

Batchelor, G. K. 1950. On the spontaneous magnetic field in a conducting liquid in turbulent motion. *Proc. R. Soc. London, Ser. A* 201:405

Biermann, L., Schlüter, A. 1951. Cosmic radiation and cosmic magnetic fields. II. Origin of cosmic magnetic fields. *Phys. Rev.* 82:863

Braginskii, S. I. 1964. Kinematic models of Earth's hydromagnetic dynamo. *Geomagn. Aeron.* 4:732

Braginskii, S. I. 1965a. Self-excitation of a magnetic field during the motion of a highly conducting fluid. *Sov. Phys. JETP* 20:726

Braginskii, S. I. 1965b. Theory of the hydromagnetic dynamo. *Sov. Phys. JETP* 20:1462

Braginskii, S. I. 1970. Oscillation spectrum of the hydromagnetic dynamo of the Earth. *Geomagn. Aeron.* 10:172

Brecher, A., Ranganyaki, R. P. 1975. Paleomagnetic systematics of ordinary chondrites. *Earth Planet. Sci. Lett.* 25:57

Bridge, H. S., Lazarus, A. J. 1969. Mariner V: Plasma and magnetic fields observed near Venus. *Science* 158:1669

Bridge, H. S., Lazarus, A. J., Scudder, J. D., Ogilvie, K. W., Hartle, R. E., Asbridge, J. R., Bame, S. J., Feldman, W. C., Siscoe, G. L. 1974. Observations at Venus encounter by the plasma science experiment on Mariner 10. *Science* 183:1293

Brown, L. W. 1975. Saturn radio emission near 1 MHz. *Ap. J.* 198:L89

Bullard, E. C. 1949. The magnetic field within the Earth. *Proc. R. Soc. London, Ser. A* 197:433

Bullard, E. C., Gellman, H. 1954. Homogeneous dynamos and terrestrial magnetism. *Phil. Trans. R. Soc. London, Ser. A* 247:213

Burke, B. F., Franklin, K. L. 1955. Observations of a variable radio source associated with the planet Jupiter. *J. Geophys. Res.* 60:213

Busse, F. H. 1973. Generation of magnetic fields by convection. *J. Fluid Mech.* 57(3):529

Busse, F. H. 1975. A model of the geodynamo. *Geophys. J. R. Astron. Soc.* 42:437

Butler, R. F. 1972. Natural remanent magnetization and thermomagnetic properties of the Allende meteorite. *Earth Planet. Sci. Lett.* 17:120

Chandrasekhar, S. 1961. *Hydrodynamic and Hydromagnetic Stability.* Clarendon: Oxford Univ. Press. 652 pp.

Coleman, P. J. Jr., Schubert, G., Russell, C. T., Sharp, L. R. 1972. Satellite measurement of moon's magnetic field: A preliminary report. *Moon* 4:419

Cowling, T. G. 1934. The magnetic field of sunspots. *Mon. Not. R. Astron. Soc.* 94:39

Cowling, T. G. 1957. *Magnetohydrodynamics.* New York: Wiley-Interscience. 115 pp.

Deinzer, W., Kusserow, H.-U., Stix, M. 1974. Steady and oscillatory $\alpha\omega$-dynamos. *Astron. Astrophys.* 36:69

Deinzer, W., Stix, M. 1971. On the eigenvalues of Krause-Steenbeck's solar dynamo. *Astron. Astrophys.* 12:111

Doell, R. R., Grommé, C. S., Thorpe, A. N., Senftle, F. E. 1970. Magnetic studies of Apollo 11 lunar samples. *Geochim. Cosmochim. Acta* 34: Suppl. 1, p. 2097

Dolginov, S. S., Eroshenko, E. G., Lewis, L. 1969. Nature of the magnetic field in the neighborhood of Venus. *Kosmicheskie Issledovaniya.* 7:747. Transl. in *Cosmic Res.* 7:675

Dolginov, S. S., Yeroshenko, Y. G., Zhuzgov, L. N. 1973. Magnetic field in the very close neighborhood of Mars according to data from the Mars 2 and Mars 3 spacecraft. *J. Geophys. Res.* 78:4779

Dyal, P., Parkin, C. W., Sonett, C. P. 1970. Apollo 12 magnetometer: Measurement of a steady magnetic field on the surface of the Moon. *Science* 169:762

Elsasser, W. M. 1939. On the origin of the Earth's magnetic field. *Phys. Rev.* 55:489

Elsasser, W. M. 1946a. Induction effects in terrestrial magnetism. Part I: Theory. *Phys. Rev.* 69:106

Elsasser, W. M. 1946b. Induction effects in terrestrial magnetism. Part II. The secular variation. *Phys. Rev.* 70:202

Elsasser, W. M. 1947. Induction effects in terrestrial magnetism. Part III. Electric modes. *Phys. Rev.* 72:821

Elsasser, W. M. 1950. The Earth's interior and geomagnetism. *Rev. Mod. Phys.* 22:1

Eltayeb, I. A. 1972. Hydromagnetic convection in a rapidly rotating fluid layer. *Proc. R. Soc. London, Ser. A* 236:229

Fuller, M. 1974. Lunar magnetism. *Rev. Geophys. Space Phys.* 12:23

Gailitis, A. 1970. Self-excitation of a magnetic field by a pair of annular vortices. *Magn. Gidrodin.* 6:19

Gilman, P. A. 1969a. A Rossby wave dynamo for the Sun, I. *Solar Phys.* 8:316

Gilman, P. A. 1969b. A Rossby wave dynamo for the Sun, II. *Sol. Phys.* 9:3

Gringauz, K. I., Bezrukikh, V. V., Muscator, L. S., Breus, T. K. 1968. Plasma measurements conducted in the vicinity of Venus by the cosmic apparatus "Venus 4." *Komicheskie Issledovaniya.* 6:411

Gubbins, D. 1974a. Theories of the geomagnetic and solar dynamos. *Rev. Geophys. Space Phys.* 12:137

Gubbins, D. 1974b. Dynamo action of isotropically driven motions of a rotating fluid. *Stud. Appl. Math.* 53:157

Helsley, C. E. 1970. Magnetic properties of lunar 10022, 10069, 10084 and 10085 samples. *Geochim. Cosmochim. Acta* 34: Suppl. 1, p. 2213

Herzenberg, A. 1958. Geomagnetic dynamos. *Phil. Trans. R. Soc. London, Ser. A* 250:543

Higgins, G., Kennedy, G. C. 1971. The adiabatic gradient and the melting point gradient in the core of the Earth. *J. Geophys. Res.* 76:1870

Hubbard, W. B., Smoluchowski, R. 1973. Structure of Jupiter and Saturn. *Space Sci. Rev.* 14:599

Inglis, D. R. 1955. Theories of the Earth's magnetism. *Rev. Mod. Phys.* 27:212

Kraichnan, R. H., Nagarajan, S. 1967. Growth of turbulent magnetic fields. *Phys. Fluids.* 10:859

Krause, F., Roberts, P. H. 1973. Some problems of mean field electrodynamics. *Ap. J.* 181:977

Larochelle, A., Schwarz, E. J. 1970. Magnetic properties of lunar samples 10048. *Geochim. Cosmochim. Acta* 34: Suppl. 1, p. 2305

Leighton, R. B. 1969. A magneto-kinematic model of the solar cycle. *Ap. J.* 156:1

Lerche, I. 1971. Kinematic-dynamo theory. *Ap. J.* 166:627

Lerche, I. 1973. Kinematic-dynamo theory. V. Comments on diverse matters including historical development, isotropic turbulence and expansion techniques. *Ap. J.* 181:993

Lerche, I., Parker, E. N. 1973. On the improper neglect of certain terms in random function theory. *J. Math. Phys.* 14:1949

Levy, E. H. 1972a. Effectiveness of cyclonic convection for producing the geomagnetic field. *Ap. J.* 171:621

Levy, E. H. 1972b. Kinematic reversal schemes for the geomagnetic dipole. *Ap. J.* 171:635

Levy, E. H. 1972c. On the state of the geomagnetic field and its reversals. *Ap. J.* 175:573

Levy, E. H. 1972d. Magnetic dynamo in the Moon: A comparison with the Earth. *Science* 178:52

Levy, E. H. 1974. Dynamical stability of stationary dynamo magnetic fields. *Ap. J.* 187:361

Levy, E. H. 1975. Generation of magnetic fields in the cosmos. *Proc. Nagata Conf. Mag. Fields,* ed. R. M. Fisher, M. Fuller,

J. A. Schmidt, P. J. Wasilewski, p. 30. Pittsburgh, Penn: Univ. Pittsburgh Press

Levy, E. H., Rose, W. K. 1974. Production of magnetic fields in the interiors of stars and several effects on stellar evolution. *Ap. J.* 193:419

Lortz, D. 1968. Impossibility of steady dynamos with certain symmetries. *Phys. Fluids.* 11:913

Malkus, W. V. R., Proctor, M. R. E. 1975. The macrodynamics of α-effect dynamos in rotating fluids. *J. Fluid Mech.* 67(3):417

Mihalov, J. D., Sonett, C. P., Binsack, J. H., Moutsoulas, M. D. 1971. Possible lunar fossil magnetism inferred from satellite data. *Science* 171:892

Moffatt, H. K. 1970a. Turbulent dynamo action at low magnetic Reynolds number. *J. Fluid Mech.* 41:435

Moffatt, H. K 1970b. Dynamo action associated with random inertial waves in a rotating conducting fluid. *J. Fluid Mech.* 44:705

Moffatt, H. K. 1972. An approach to a dynamic theory of dynamo action in a rotating fluid. *J. Fluid Mech.* 53(2):385

Nagata, T., Ishikawa, Y., Kinoshita, H., Kono, M., Syono, Y., Fisher, R. M. 1970. Magnetic properties and natural remanent magnetization of lunar materials. *Geochim. Cosmochim. Acta* 34: Suppl. 1, p. 2325

Ness, N. F., Behannon, K. W., Lepping, R. P., Whang, Y. C., Schatten, K. H. 1974a. Magnetic field observations near Venus: Preliminary results from Mariner 10. *Science* 183:1301

Ness, N. F., Behannon, K. W., Lepping, R. P., Whang, Y. C., Schatten, K. H. 1974b. Magnetic field observations near Mercury. *Science* 185:151

Ness, N. F., Behannon, K. W., Lepping, R. P., Whang, Y. C. 1975a. The magnetic field of Mercury, Part 1. *J. Geophys. Res.* 80:2708

Ness, N. F., Behannon, K. W., Lepping, R. P., Whang, Y. C. 1975b. Magnetic field of Mercury confirmed. *Nature* 185:151

Parker, E. N. 1955a. The formation of sunspots from the Sun's toroidal field. *Ap. J.* 121:491

Parker, E. N. 1955b. Hydromagnetic dynamo models. *Ap. J.* 122:293

Parker, E. N. 1969. The occasional reversal of the geomagnetic field. *Ap. J.* 158:815

Parker, E. N. 1970a. The origin of magnetic fields. *Ap. J.* 160:383

Parker, E. N. 1970b. The generation of magnetic fields in astrophysical bodies. I. The dynamo equations. *Ap. J.* 162:665

Parker, E. N. 1971a. The generation of magnetic fields in astrophysical bodies. II. The galactic field. *Ap. J.* 163:255

Parker, E. N. 1971b. The generation of magnetic fields in astrophysical bodies. III. Turbulent diffusion of fields and efficient dynamos. *Ap. J.* 163:279

Parker, E. N. 1971c. The generation of magnetic fields in astrophysical bodies. IV. The solar and terrestrial dynamos. *Ap. J.* 164:491

Parker, E. N. 1971d. The generation of magnetic fields in astrophysical bodies. VIII. Dynamical considerations. *Ap. J.* 168:239

Rädler, K.-H. 1968a. On the electrodynamics of conducting fluids in turbulent motion. I. The principles of mean field electrodynamics. *Z. Naturforsch.* 23a: 1841

Rädler, K.-H. 1968b. On the electrodynamics of conducting fluids in turbulent motion. II. Turbulent conductivity and turbulent permeability. *Z. Naturforsch.* 23a:1851

Rädler, K.-H. 1969a. On the electrodynamics of turbulent fluids under the influence of Coriolis forces. *Monatsber. Deut. Akad. Wiss. Berlin* 11:194

Rädler, K.-H. 1969b. A new turbulent dynamo, I. *Monatsber. Deut. Akad. Wiss. Berlin* 11:272

Rädler, K.-H. 1970. A new turbulent dynamo, II. *Monatsber. Deut. Akad. Wiss. Berlin* 12:468

Rikitake, T. 1958. Oscillations of a system of disk dynamos. *Proc. Cambridge Phil. Soc.* 54:89

Roberts, G. O. 1972. Dynamo action of fluid motions with two-dimensional periodicity. *Phil. Trans. R. Soc. London, Ser. A* 271:411

Roberts, P. H. 1972. Kinematic dynamo models. *Phil. Trans. R. Soc. London, Ser. A* 272:663

Roberts, P. H., Stix, M. 1972. α-effect dynamos, by the Bullard-Gellman formalism. *Astron. Astrophys.* 18:453

Runcorn, S. K. 1975. An ancient lunar magnetic dipole field. *Nature* 253:701

Runcorn, S. K., Collinson, D. W., O'Reilly, W., Battey, M. H., Stephenson, A., Jones, J. M., Manson, A. J., Readman, P. W. 1970. Magnetic properties of Apollo 11 lunar samples. *Geochim. Cosmochim. Acta* 34: Suppl. 1, p. 2369

Russell, C. T., Coleman, P. J. Jr., Lichtenstein, B. R., Schubert, G. 1974. The permanent and induced magnetic dipole moment of the Moon. *Geochim. Cosmochim. Acta* 38: Suppl. 5, p. 2747

Sloanaker, R. M. 1959. Apparent temperature of Jupiter at a wavelength of 10 cm. *Astron. J.* 64:346

Smith, E. J., Davis, L. Jr., Jones, D. E., Colburn, D. S., Coleman, P. J. Jr., Dyal, P., Sonett, C. P. 1974. Magnetic field of Jupiter and its interaction with the solar wind. *Science* 183:305

Smith, E. J., Davis, L. Jr., Jones, D. E., Coleman, P. J., Colburn, D. S., Dyal, P., Sonett, C. P. 1975. Jupiter's magnetic field, magnetosphere, and interaction with the solar wind: Pioneer 11. *Science* 188:451

Sonett, C. P., Colburn, D., Schwartz, K., Keil, K. 1970. The melting of asteroidal-sized bodies by unipolar dynamo induction from a primordial T-Tauri Sun. *Astrophys. Space Sci.* 7:446

Sonett, C. P., Mihalov, J. D. 1972. Lunar fossil magnetism and perturbation of the solar wind. *J. Geophys. Res.* 77:588

Stacey, F. D., Lovering, J. F. 1959. Natural magnetic moments of two chondritic meteorites. *Nature* 183:529

Stacey, F. D., Lovering, J. F., Parry, L. G. 1960. Thermomagnetic properties, natural magnetic moments, and magnetic anisotropies of some chondritic meteorites. *J. Geophys. Res.* 66:1523

Steenbeck, M., Krause, F., Rädler, K.-H. 1966. A calculation of the mean electromotive force in an electrically conducting fluid in turbulent motion, under the influence of Coriolis forces. *Z. Naturforsch.* 21a:369

Steenbeck, M., Krause, F. 1969. On the dynamo theory of stellar and planetary magnetic fields, II. D.C. dynamos of the planetary type. *Astron. Nachr.* 291:271

Stephenson, A., Collinson, D. W., Runcorn, S. K. 1974. Lunar magnetic field paleointensity determinations on Apollo 11, 16, and 17 rocks. *Geochim. Cosmochim. Acta* 3: Suppl. 5, p. 2859

Stevenson, D. 1974. Planetary magnetism. *Icarus* 22:403

Stevenson, D. J., Ashcroft, N. W. 1974. Conduction in fully ionized liquid metals. *Phys. Rev. A.* 9:782

Stix, M. 1971. A non-axisymmetric α-effect dynamo. *Astron. Astrophys.* 13:203

Stix, M. 1972. Non-linear dynamo waves. *Astron. Astrophys.* 20:9

Stix, M. 1973. Spherical $\alpha\omega$-dynamos by a variational method. *Astron. Astrophys.* 24:275

Strangway, D. W., Larson, E. E., Pearce, G. W. 1970. Magnetic studies of lunar samples—Breccia and Fines. *Geochim. Cosmochim. Acta* 34: Suppl. 1, p. 2435

Tverskoy, B. A. 1966. Theory of hydrodynamic self-excitation of regular magnetic fields. *Geomagn. Aeron.* 6:7

Vainshtein, S. I. 1972. Nonlinear problem of the turbulent dynamo. *Sov. Phys. JETP* 34:327

Vainshtein, S. I., Ruzmaikin, A. A. 1972. Generation of the large scale galactic magnetic field. *Sov. Astr. A. J.* 15:714

Van Allen, J. A., Krimigis, S. M., Frank, L. A., Armstrong, T. P. 1969. Venus: An upper limit on intrinsic magnetic dipole moment based on absence of a radiation belt. *Science* 158:1673

Zel'dovich, Y. B. 1957. The magnetic field in the two-dimensional motion of a conducting turbulent liquid. *Sov. Phys. JETP* 4:460

CHEMICAL FRACTIONATION AT THE AIR/SEA INTERFACE

Robert A. Duce and Eva J. Hoffman
Graduate School of Oceanography, University of Rhode Island, Kingston, Rhode Island 02881

INTRODUCTION

The ocean is one of the major natural sources of atmospheric particulate matter, or aerosols. However, there is considerable evidence that the chemical composition of atmospheric particles over the ocean is often significantly different from that of bulk seawater. It has often been suggested that the relative difference in chemical composition of the atmospheric particles and seawater may be due to chemical and physical processes occurring during the production of these particles at the air/sea interface. The occurrence, or nonoccurrence, of this selective chemical fractionation or enrichment has been the subject of considerable research and controversy in the field of atmospheric chemistry during the past several decades. Unfortunately, inadequate analytical techniques and poorly designed field and laboratory studies have in many cases led to the postulation of novel but unlikely theories to explain inaccurate data. The primary intent of this review is to evaluate the data for common trace substances present in marine aerosols in an attempt to ascertain whether their source may be the ocean or some other natural or anthropogenic process.

Atmospheric Sea Salt Particles and the Marine Aerosol

It has been estimated that the ocean produces between 10^{15} and 10^{16} g/yr of atmospheric sea salt particles with radii less than approximately 20 μm (Eriksson 1959, 1960, Blanchard 1963). Estimates of the total annual production of atmospheric particles from all natural sources, incorporating the lower value of 10^{15} g/yr from the sea, suggest that the ocean may contribute 30–75% of the total production (Hidy & Brock 1970, Robinson & Robbins 1971, Peterson & Junge 1971). Eriksson (1959, 1960) estimated that approximately 90% of the atmospheric sea salt particles are removed over the ocean, with 10% being deposited on land as the cyclic salts subsequently carried by streams and groundwater. This would result in from 10^{14} to 10^{15} g/yr of continental cyclic salts, according to these two production estimates. Livingstone (1963) calculated that the annual global input of cyclic salt Na from river runoff into the ocean is approximately 10^{14} g/yr, which is equivalent to a total salt content of approximately 3×10^{14} g/yr. Atmospheric sea salt apparently plays a significant role in atmospheric and terrestrial geochemical cycles. Significant enrichment of trace substances on these particles, compared to the source seawater, may be an important factor in the overall geochemical cycles of these substances.

Most of the sea salt particles with atmospheric residence times longer than a few minutes are believed to be produced by bubbles breaking at the sea surface. Blanchard & Woodcock (1957) investigated various mechanisms for bubble production in the ocean, for example, breaking waves or raindrops and snowflakes striking the water surface. They suggested that, except under local conditions, breaking waves, or whitecaps, are by far the most important source of bubbles. Boyce (1951) showed that relatively few salt particles were produced by the mechanical disintegration of the water in a breaking wave, but that a considerably greater number of particles were produced a few seconds later when the air bubbles resulting from wave action burst at the sea surface. Using high speed photography, Kientzler et al (1954) showed that a bubble at a seawater surface forms a jet, which ejects two to five droplets into the air. Blanchard (1963) found that the diameter of these jet droplets was approximately 10% of the diameter of the bubbles from which they were formed. Mason (1957) and Blanchard (1963) also found that a significant but variable number of smaller droplets, called film droplets, were also produced by the shattering bubble film cap.

Relatively little is known of bubble size and number distribution on the open sea. Blanchard & Woodcock (1957) measured the distribution of bubbles between about 75 μm and 750 μm diameter in waves breaking on a beach. They pointed out that

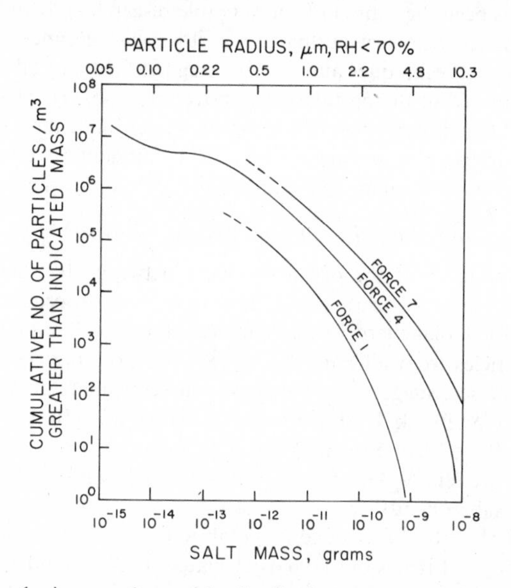

Figure 1 Atmospheric sea salt particle distribution several hundred meters above the ocean surface near Hawaii at various wind forces. After Woodcock (1953, 1972).

their methods were biased against the millimeter-sized bubbles, which they indicated were present, but in low abundance. Medwin (1970) investigated the bubble distribution in the ocean under relatively light wind conditions, but there has unfortunately been no work on the complete bubble distribution at various wind speeds in open ocean areas. Blanchard (1963) estimated that on a global basis, 3–4% of the ocean surface is covered by whitecaps at any time, resulting in an overall oceanic production rate of approximately 0.1 jet droplets cm^{-2} sec^{-1} and 0.07 film droplets cm^{-2} sec^{-1}. These figures are based on individual bubbles bursting at the sea surface, using the bubble distribution of Blanchard & Woodcock (1957). Little information is available on the production of atmospheric particles by the bursting of bubble clusters or foams, although this may be an important source for these particles.

Woodcock (1953) made numerous measurements of the size and number distribution of atmospheric sea salt particles with salt masses between 10^{-8} g and 10^{-12} g. Figure 1 illustrates a typical particle distribution several hundred meters above the ocean surface near Hawaii for force 1, 4, and 7 winds. Woodcock (1972) extended his earlier work down to particles with a salt mass of 10^{-15} g, and these results are also shown in Figure 1 for force 4 winds. Woodcock found that the number of particles and their mass median diameter increased with increasing wind speed and with decreasing altitude above the sea. He attributes the discontinuity in the distribution shown in Figure 1 for force 4 winds to a zone of transition from a bubble jet to a bubble film source for the droplets in this size range.

Junge (1972) summarized our understanding of the distribution of atmospheric particles in marine air and pointed out that, in addition to sea salt particles, a number of other components can be distinguished in the undisturbed marine aerosol. These include large, apparently organic particles with radii greater than 20 μm; mineral dust particles; particles with radii less than 0.03 μm, possibly the result of trace gas reactions or the reactions of small ion clusters (Vohra et al 1970, Mohnen 1970); and the so-called tropospheric background aerosol. The latter is a rather homogeneous aerosol uniformly distributed within the troposphere over the ocean (and continents) at about 200–400 particles per cm^3 of air, with a maximum in the range of $\sim$0.1 μm radius. Sulfate is a likely major component of the background aerosol. Perhaps the most important point made by Junge (1972) is that the marine aerosol is composed of many substances that have sources other than the sea. This has often been ignored in field and laboratory studies of chemical fractionation of atmospheric sea salt particles. Distinguishing between chemical substances present on sea salt particles injected into the marine atmosphere during the bubble-breaking process and those present on other components of the marine aerosol is very difficult, but is necessary in order to properly assess the presence and importance of chemical fractionation processes. We shall return to this question a number of times during our discussion.

Standardized Chemical Fractionation Notation

The Working Symposium on Sea-to-Air Chemistry, held at Fort Lauderdale, Florida, January 30–February 4, 1972, recommended a standardized nomenclature to quantitatively describe apparent chemical fractionation at the air/sea interface (Duce et al 1972a). This nomenclature, used throughout this paper, is as follows:

$$\text{Fractionation: } F_{Na}(X) = \frac{(X/Na)_{atm}}{(X/Na)_{sea}} \tag{1}$$

$$\text{Enrichment: } E_{Na}(X) = \frac{(X/Na)_{atm}}{(X/Na)_{sea}} - 1, \tag{2}$$

where $(X/Na)_{atm}$ is the mass ratio of substance X to sodium in any atmospheric sample, e.g. rain, particles, etc, and $(X/Na)_{sea}$ is the ratio of substance X to sodium in bulk (subsurface) seawater.

Occasionally Na is not measured in field and laboratory studies and such elements as Cl or Mg are used as seawater reference elements. They then replace Na in the notation above. Sodium is preferred as a reference element over chlorine, primarily because a vapor phase of Cl is present in marine air and because the exchange of Cl between the vapor and particle phases may confuse the interpretation of enrichments based on Cl. It might be noted that fractionation, F, is the same as the term enrichment factor, EF, used a number of papers in the literature. A value for $E_{Na}(X)$ equal to 0 indicates no enrichment, positive or negative; whereas values of $F_{Na}(X)$ equal to one indicate no enrichment for substance X.

The calculation of a positive enrichment for any substance according to Equation 2 does not indicate that this increased concentration is caused by some process occurring during sea salt particle production at the air/sea interface. Enrich-

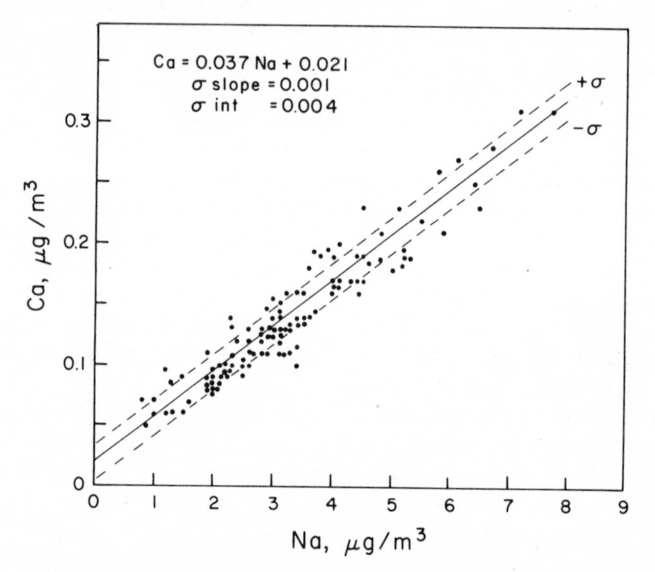

Figure 2 Atmospheric Na vs Ca concentrations observed from a 20 m high tower on the coast of windward Oahu, Hawaii. After G. Hoffman & Duce (1972).

ment as defined above gives no information as to the source of the substance of interest. This relationship simply points out whether or not the substance is present in a relative abundance similar to that in bulk seawater. Several investigators have attempted to distinguish between the apparent enrichment of some substance due to the addition of that substance to the sample from a non–sea salt source and a true enrichment due to either physical or chemical processes occurring during particle production at the sea surface. One approach (Lazrus et al 1970, G. L. Hoffman & Duce 1972) is to calculate simple linear regression equations to describe the relationship between Na and the substance of interest. For any substance X, the slope of the regression line should represent the X/Na ratio associated with the sea salt component of X, while the intercept of the regression line represents the quantity of the non–sea salt component. For example, consider Figure 2, which is a plot of 110 samples collected in Hawaii and analyzed for Ca and Na by G. L. Hoffman & Duce (1972). The mean total Ca/Na ratio for these samples was 0.045 ± 0.011. With a seawater Ca/Na ratio of 0.0381, this results in an apparent enrichment—i.e. a value for E_{Na}(Ca)—of 0.20 ± 0.20. However, from the slope of the regression equation and the standard deviation of this slope (0.037 ± 0.001), the calculated value for E_{Na}(Ca) becomes -0.03 ± 0.03. The intercept, 0.021, suggests that 0.021 $\mu g/m^3$ of Ca has some source other than that associated with the Na. It must be pointed out that this approach assumes that the nonmarine intercept is constant in time and space, which is quite unlikely. In addition, it assumes that the sea salt Na overwhelms any Na contribution from other sources. Therefore, this approach must be used with caution, but it is at least a first step in differentiating between a sea salt and non–sea salt source for various substances.

Sampling location is a critical factor in any field studies of chemical fractionation at the air/sea interface. Atmospheric samples collected near the continental margins obviously have a high probability of containing significant quantities of particulate matter of nonmarine origin. Midocean sampling sites are less disturbed. Atmospheric sampling from ships is easily subject to contamination from the ship itself unless great care is taken. Aircraft can often provide suitable sampling platforms, as can midocean islands. Even here, however, care must be taken to avoid collection of local surf-produced particles, which can be chemically quite different from the bubble-produced particles formed on the open sea (Duce & Woodcock 1971). Finally, samples collected even a few km inland from the coast, on either islands or continents, will almost certainly be contaminated with significant quantities of land-produced aerosols—natural, anthropogenic, or both. These precautions appear obvious to most people working in the field today, but it is surprising and discouraging to realize how many past investigations (some very recent) have produced data that were rather useless in evaluating the significance of chemical fractionation because one or more of these factors was ignored.

BUBBLES AND THE SEA SURFACE MICROLAYER

When bubbles break at the surface of the ocean they skim off a very thin layer of the air/sea interface to produce the atmospheric film and jet droplets. MacIntyre (1968,

1972) investigated this microtome effect for jet droplets using a combination of experimental and theoretical approaches. Figure 3*a* presents a simulated time sequence of what occurs after the collapse of a dyed bubble (MacIntyre 1972). This sequence illustrates the flow of surface material down the interior of the bubble cavity, transported by an irrotational single capillary ripple, and concentration of this material into the top of the jet before the first jet droplet breaks off. Subsequent jet droplets are composed of material that was present in consecutively deeper concentric shells in the bubble cavity, as shown in Figure 3*b* (MacIntyre 1972). MacIntyre calculated that the material present in the top jet droplet was originally spread over the interior of the bubble surface at a thickness equal to approximately 0.05% of the bubble diameter, and he assumes this relationship is valid for all bubble sizes. Thus the top jet droplet from a 100 μm diameter bubble is composed of material from a surface layer only 0.05 μm thick. The second jet droplet is produced from the next 0.05 μm layer, etc. With a bubble size distribution in the sea ranging from ~50 μm to perhaps 1500 μm diameter, the top jet droplets produced from these bubbles strip off approximately the top 0.025–0.75 μm of the air/water interface. Iribarne & Mason (1967) reached a similar conclusion in their studies of charge transfer during bubble bursting. The chemical composition of this extremely thin stripped surface layer is determined both by the composition of the ocean surface layer before the bubble arrives at the surface (the surface microlayer) and the composition of the bubble skin itself.

It is interesting to note that when the bubble cap or film shatters to produce film droplets, it is probably no thinner than 2 μm, since these bubble caps show no visible interference patterns which become apparent for thinner films (MacIntyre 1972). As MacIntyre (1974a) points out, this suggests that the jet droplets, which are generally larger than the film droplets, may be sampling, or may be composed of, a much thinner layer of the water surface than the smaller film droplets. The smaller jet droplets are apparently composed of material from a thinner layer of the surface

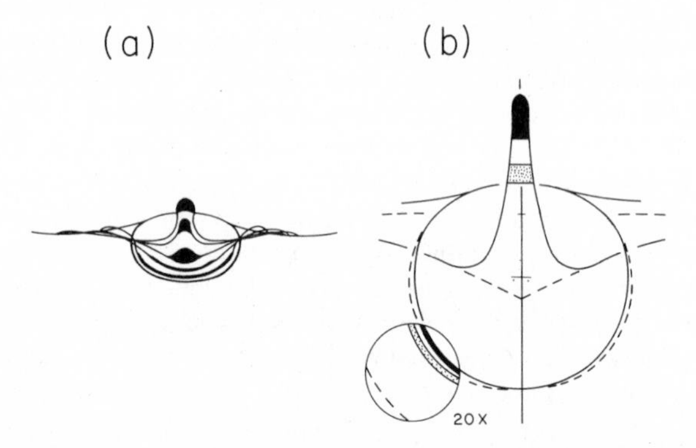

Figure 3a Simulated time sequence of jet droplet formation after the collapse of a dyed bubble. From MacIntyre (1972), with permission of the author. *b* Source of material present in the first three jet droplets produced from a bursting bubble. From MacIntyre (1972), with permission of the author.

than the larger jet droplets. Unfortunately, although we now have a rather good qualitative picture of these processes occurring in the upper few hundred micrometers of the ocean surface, the detailed hydrodynamics remain largely unknown. MacIntyre (1974a) presents an excellent review of our present understanding of these processes.

Since the breaking bubbles produce atmospheric sea salt particles from such a thin layer of the ocean surface, our attention should turn briefly to the chemical composition of the thinnest surface layer of the oceans, the surface microlayer. The surface microlayer thickness is operationally defined by the type of microlayer collector used. The term microlayer itself implies no particular thickness, chemical composition, or concentration, but simply refers physically to the thin surface layer being sampled. A variety of systems have been used to sample the marine surface microlayer. These systems include, among others, the use of plastic or metal screens (Garrett 1965), to collect the top 100–300 μm of the water surface; glass plates (Harvey & Burzell 1972) for the top 30–60 μm; rotating drum collectors mounted on rafts (Harvey 1966) for the top 50–60 μm; germanium prisms (Baier 1972) to sample organic layers as thin as monomolecular for subsequent analysis by internal reflection infrared spectroscopy; and the bursting bubble itself (Fasching et al 1974) to generate atmospheric sea salt particles from bursting bubbles produced in a closed system at sea. Hatcher & Parker (1974) have compared the collection characteristics of several of these collectors.

Because the various collectors sample different thicknesses of the ocean surface, the principles by which they collect surface material often differ, and the sea state conditions under which they can be used vary widely, comparison of studies using different types of collectors has been difficult. However, the chemical composition of the surface microlayer is evidently significantly different from that of water 10–20 cm or more below the surface. Among the substances significantly concentrated in the surface microlayer are a wide variety of organic substances, including fatty acids and alcohols (Garrett 1967, Duce et al 1972b, Wade & Quinn 1975, Miget et al 1974, Marty & Saliot 1974); hydrocarbons (Garrett 1967, Baier 1972, Baier et al 1974, Ledet & Laseter 1974, Morris 1974); glycoproteins and proteoglycans (Baier et al 1974); chlorinated hydrocarbons such as PCB (Seba & Corcoran 1969, Duce et al 1972b, Bidleman & Olney 1974); particulate and dissolved organic carbon, nitrogen, and phosphate (Williams 1967, Nishizawa & Nakajima 1971); and trace metals (Duce et al 1972b, Piotrowicz et al 1972, Szekielda et al 1972, Barker & Zeitlin 1972, G. Hoffman et al 1974). Numerous marine organisms have also been found concentrated in the surface microlayer (Sieburth 1971, Harvey 1966, Parker & Barsom 1970, Tsiban 1971, Marumo et al 1971, Bezdek & Carlucci 1972). Thus the potential certainly exists for atmospheric sea salt particles produced from this thin layer of the ocean to have a chemical composition significantly different from that of subsurface water.

The air/sea interface and air/sea exchange processes in general can also be significantly modified by human activities, particularly in nearshore areas. Oil spills and man-made sea slicks can affect certain physical characteristics of the air/sea interface. These slicks cause damping of capillary waves and retardation of evaporation, as well as changes in gas exchange rates and the number and size distribution of the atmospheric sea salt particles produced from these waters (Blanchard 1963, Garrett 1972). High concentrations of such potentially harmful substances as pesticides, petroleum hydrocarbons, and heavy metals have been observed in the microlayer. Bacteria can be transported to the water surface by rising bubbles and then ejected into the atmosphere (Blanchard & Syzdek 1972). The potential health hazard from this type of exchange was pointed out by Adams & Spendlove (1970), who measured coliform bacteria on aerosols downwind from a sewage treatment plant. Other pollutants concentrated in the surface layers of the water may be similarly injected very efficiently into the atmosphere.

The surface of the ocean is generally quite turbulent, and it is difficult to imagine the surface microlayer as anything but dynamic. MacIntyre (1974a) points out that swells and ripples on the ocean surface do not destroy the enriched surface layer but only change its thickness. However, as winds increase and waves break there will be a constant mixing of surface material into subsurface water and transport of surface material back upward. G. Hoffman et al (1974) have calculated that the average lifetime of particulate iron in the top 100–300 μm of the ocean surface under 5–8 m/sec winds is only a few seconds before it is mixed back into subsurface waters. Even with this short surface residence time, however, particulate iron was always significantly enriched in the top 100–300 μm compared to water 20 cm below the surface.

What causes substances to be enriched at the air/sea interface? Some solid substances are present simply because of density differences and some lipoids, such as hydrocarbons, are present because they are practically insoluble in seawater, but most of the material enriched at the surface is either surface active or associated in some way with surface-active material. Surface-active substances are characterized by a molecular structure containing both hydrophobic and hydrophilic portions, hence their affinity for an air/water interface. The chemical composition of these surfactants concentrated at the ocean surface has been the subject of some controversy. Garrett (1967), investigating chloroform soluble surfactants in the surface microlayer, suggested the surfactants are largely lipid material, such as fatty esters, free fatty acids, and fatty alcohols. Baier et al (1974), on the other hand, report that the major organic components of the marine surface microlayer are glycoproteins and proteoglycans. According to the latter investigators, their technique of internal-reflection infrared spectroscopy is capable of nondestructively determining petroleum, lipids, proteins, and polysaccharides, and they conclude that lipids constitute only a small fraction of the organic matter in the microlayer.

Although many of the enriched substances in the surface microlayer are not surface-active themselves, they may be associated with surfactants and thus be transported to the surface with the surface-active agent. Examples include mineral particles and marine organisms such as bacteria or plankton covered with a surface-active coating, and trace metals complexed with functional groups in proteinaceous surfactants. Most likely, under breaking wave conditions, the primary mechanisms of transport of surfactants and their associated substances to the ocean surface is by bubbles. Considerable evidence indicates that bubbles are quite effective in transporting this material in seawater (reviewed in subsequent sections). Thus the chemical composition of the ocean surface microlayer, the surface of bubbles rising through the water, and the jet and film droplets are closely interrelated and interdependent, although the overall geochemical importance of these interrelationships is not well understood in many cases.

A number of other mechanisms that would cause higher concentrations of various ionic species at the air/sea interface compared to subsurface water have been suggested. MacIntyre (1974a) has reviewed these mechanisms in detail: (*a*) Gibbs surface adsorption, where inorganic ions that lower the surface free energy are concentrated and those that raise the surface free energy are depleted at the surface (Bloch & Luecke 1968); (*b*) the Ludwig-Soret effect, where ions migrate to different regions of a thermal gradient field (Komabayasi 1962); (*c*) alteration of the surface water structure, where ions that disrupt this structure because of their size and/or charge are rejected from the air/water interface whereas those that alter the structure least are preferred (Horne & Courant 1970); and (*d*) electrical double-layer effects, where size and charge of ions may be related to the degree of attraction to or repulsion from the surface double layer (Glass & Matteson 1973). Although these processes can probably cause increased concentration of certain ions at an air/water interface, MacIntyre (1974a) points out that these processes probably only affect ionic concentrations to a depth of a few molecular layers. Their geochemical importance in causing the large enrichments often measured experimentally in the surface microlayer and atmospheric sea salt particles, which represent layers hundreds to thousands of times thicker, may therefore be marginal.

EXPERIMENTAL DATA

Marine Organisms and Organic Carbon

If significant chemical fractionation of trace substances occurs during the production of atmospheric sea salt particles at the air/sea interface, it very likely results from the association of these substances with organic matter, living or dead, in the ocean. Our attention should focus first on what is known about the ocean/atmosphere exchange of organic material. The atmospheric transport of organisms has been investigated since the 1850s, primarily because of interest in the spread of disease. A number of studies report the presence of marine organisms in the atmosphere. Zobell & Mathews (1936) found that the type of bacteria in air over coastal locations was a function of wind direction. Marine forms dominated during onshore wind conditions. Woodcock (1948) suggested that the red tide toxin, produced by the dinoflagellate *Gymnodinium breve*, was transferred from the Gulf of Mexico to the atmosphere as a component of sea spray droplets produced during the bursting of air bubbles in the water. The presence of this material, apparently associated with the atmospheric particles, resulted in human respiratory irritation along the Florida coast and inland during red tide blooms. Stevenson & Collier (1962) identified three species of diatoms and four flagellates in the atmosphere, all in the nannoplankton size range ($< 5\ \mu m$), and Maynard (1968) found several species of dinoflagellates in shipboard atmospheric samples and several diatoms in atmospheric samples from higher altitudes. Aubert (1974) identified terrestrial bacteria in coastal locations of the Mediterranean sea and marine bacteria over open ocean areas. The metabolic and physiological characteristics of the bacteria differed depending on whether the samples were collected over the Mediterranean Sea or the Atlantic Ocean.

Carlucci & Williams (1965) investigated the transport of bacteria to the seawater surface in a laboratory foam column. They found that the presence of particulates enhanced the transport of bacteria and the efficiency of bubble transport varied from species to species. Also using a laboratory foam column, Wallace et al (1972) demonstrated that rising bubbles are capable of transporting a substantial portion of the particulate matter found in surface seawater samples to the air/sea interface. These authors suggested that phytoplankton in stationary and post-stationary phases of growth are apparently more subject to flotation than cells in the log phase of growth. In a subsequent study, Wallace & Duce (1975) found that 30–59% of the particulate organic carbon (POC) could be scavenged from samples of surface water from Narragansett Bay by rising bubbles in the foam column.

Sutcliffe et al (1963) reported that *Phaeodactylum tricornutum* were enriched in droplets produced by bursting bubbles in the laboratory by over an order of magnitude compared to bulk seawater. In subsequent studies, Blanchard & Syzdek (1970, 1972) showed that the freshwater bacterium *Serratia marcescens* could be concentrated on jet droplets by up to a factor of 10^4 relative to the water from which the droplets were produced. The magnitude of this concentration factor was a function of jet droplet (and thus bubble) size, bacteria concentration in the bulk solution, and bubble age, i.e. the length of time available for the bubble to scavenge bacteria. Increasing bubble age, up to about 20 sec, resulted in significantly higher concentrations of bacteria on the jet droplets, but the bubble surface apparently became saturated after this time, as no further concentration was observed. Blanchard & Syzdek (1974) calculated that the bubble collection efficiency (i.e. the number of bacteria scavenged by a bubble rising through the water divided by the number of bacteria present in the volume of water swept out by the rising bubble) for transporting these bacteria to the surface was $\sim$0.12%. The efficiency of the transfer of these bacteria from the water surface to the jet droplets was estimated by Blanchard & Syzdek (1974) to vary between 14% and 85%, depending on the age of the bubbles.

Bezdek & Carlucci (1972) found that the concentration factors for the marine bacterium *Serratia marinorubra* on laboratory-produced jet droplets varied as a function of droplet size, with the greatest concentration factor generally found on intermediate size droplets. Concentration factors using naturally occurring bacteria in seawater ranged from 33 to 250 and appeared to depend on the type of bacteria. Carlucci & Bezdek (1972) suggest that the scavenging of bacteria by bubbles is not the critical step in jet droplet enrichment of these organisms but that the bacteria are enriched in the sea surface microlayer before the bubbles arrive. Thus they believe the bacteria found on the jet droplets are largely from the microlayer.

There have been relatively few studies of organic carbon (OC) in the marine atmosphere considering the potential importance of organic carbon to chemical fractionation in general. In addition, different fractions or classes of organic material were measured in these investigations, making intercomparison of the results very difficult. Lodge et al (1960) reported a mean concentration of 1.6 $\mu g/m^3$ of benzene-soluble organic material in samples collected over the Pacific Ocean. The benzene-extractable organic material/salt mass ratio of 0.29 was equivalent to an enrichment of over 1000. Blanchard (1964) reported the presence of surface-active organic material on sea salt particles collected on thin platinum wires on the coast of Hawaii Island. In a later study, Blanchard (1968) estimated that the surface-active organic material/salt mass ratio varied from 0.3 to 0.7 on these particles. In a study from a coastal tower on Oahu, Hawaii, Barger & Garrett (1970) found the chloroform extractable organic (lipoid) material in atmospheric particles collected on glass fiber filters, as measured by weighing the material remaining after the chloroform was evaporated, ranged from 0.7 $\mu g/m^3$ to 6.3 $\mu g/m^3$, with a lipoid/salt mass ratio of 0.07 to 0.14. They found that the lipoid content increased with increasing salt content. E. Hoffman & Duce (1974) determined the total organic carbon content of sea salt particles collected on glass fiber filters from a coastal tower in Bermuda, using the wet oxidation analytical technique of Menzel & Vaccaro (1964). The total organic carbon content varied from 0.15 $\mu g/m^3$ to 0.47 $\mu g/m^3$ and the organic carbon/salt mass ratio varied from 0.01 to 0.19, equivalent to $E_{Na}(OC)$ values of $\sim$300 to 6000. In contrast to the study of Barger & Garrett (1970), E. Hoffman & Duce found that the organic carbon concentration was relatively constant, resulting in an inverse relationship between the organic carbon/salt ratio and the salt content. This inverse relationship has also been confirmed by R. Chesselet (personal communication) for samples collected from ships over the north Atlantic Ocean.

The inverse relationship between the organic carbon/salt ratio and sea salt concentration in the atmosphere may be explained if the organic carbon is present primarily on the smaller particles with longer atmospheric residence times and thus more uniform spacial and temporal concentration patterns. Indeed, Barker & Zeitlin (1972) showed qualitatively that the relative concentration of organic carbon increases with decreasing particle size. The source for the proposed small particle carbon may be the ocean or it may be gas phase reactions of natural or anthropogenic hydrocarbons or adsorption of these substances on other particles present in the atmosphere (Goetz & Klejnot 1972, Hoppel & Dinger 1973, Lovelock & Penkett 1974). About 50% of the background aerosol is volatile at 320°C (Twomey 1971). Although Twomey (1971) and Rosen (1971) attributed the volatility to loss of $(NH_4)_2SO_4$ or H_2SO_4, several authors (Blanchard 1971, Pueschel et al 1973) have suggested the volatile substances could be organic compounds. Clearly the determination of the significance of non–sea salt components of organic carbon in the marine atmosphere is critical before we can assess the importance of chemical fractionation of organic carbon at the air/sea interface. We can, however, estimate a maximum input of organic carbon from the ocean to the atmosphere on sea salt particles. If we assume the mean organic carbon/salt ratio of 0.05 observed in Bermuda by E. Hoffman & Duce (1974) is a reasonable estimate for the marine atmosphere as a whole and represents the organic carbon/salt ratio on sea salt particles and if we use the atmospheric sea salt production estimates of 10^{15} g/yr (Eriksson 1959, 1960) to 10^{16} g/yr (Blanchard 1963), the organic carbon transport from ocean to atmosphere ranges from 5×10^{13} g/yr to 5×10^{14} g/yr. This is from 0.25% to 2.5% of the estimated total organic carbon productivity of the oceans, which is 2×10^{16} g/yr (Ryther 1969). In the same range is Rasmussen & Went's (1965) estimate of the natural atmospheric production of hydrocarbons from terrestrial vegetation: 4.4×10^{14} g/yr.

There have been several laboratory investigations of the transport of organic material from the ocean to the atmosphere. Bezdek & Carlucci (1974) studied the transport of insoluble monomolecular films to the atmosphere by bursting bubbles. They found that ^{14}C stearic acid was incorporated into the jet droplets and that the smaller droplets were more enriched in this material than larger droplets.

E. Hoffman & Duce (1976) investigated several factors in the laboratory that influence the quantity and enrichment of organic material transported into the atmosphere by bursting bubbles. E_{Na} (OC) was found to be a function of bubble path length and the particular seawater used as a source. Water from Narragansett Bay, which is relatively shallow and characterized by a high dissolved and particulate organic carbon content, a high suspended load, and a large phytoplankton population, produced sea salt particles with OC/Na ratios averaging 0.098 ± 0.053. Water from the Sargasso Sea, which is characterized by lower levels of organic carbon, a lower suspended load, and a low phytoplankton population, produced sea salt particles with OC/Na ratios of 0.008 ± 0.005. The difference between the organic carbon concentration of these two bodies of water is one cause of the marked differences in the aerosol OC/Na ratios. However, when the OC/Na ratios of the aerosols are divided by the OC/Na ratio in the different seawaters to calculate E_{Na} (OC) values, these values are 250 ± 145 for Narragansett Bay aerosols and 73 ± 27 for Sargasso Sea aerosols. The significantly different enrichment factors for aerosols from these two different water masses suggest that not only is the quantity of organic material in the seawater important, but also the chemical form of the organic material is perhaps even more critical in determining the amount of organic carbon on the sea salt particles.

The effect of bubble age noted by Blanchard & Syzdek (1972, 1974) for bacteria was also confirmed for total organic carbon in the experiments of E. Hoffman & Duce (1976). An increase of bubble path length by a factor of three before bubble bursting increased E_{Na} (OC) by a factor of approximately two. Generally there was no difference in E_{Na} (OC) using either filtered or unfiltered seawater, suggesting that the dissolved or colloidal fraction of the organic material was primarily responsible for the bulk of the organic carbon in the sea salt particles produced in this investigation. Since the possible enrichment of virtually all substances during bubble breaking at the ocean surface depends on association with organic carbon, further studies of the character and source of this material are urgently needed.

Alkali and Alkaline Earth Metals

One of the most confusing and controversial questions in the air/sea chemical exchange literature over the past few decades has been the presence or absence of significant chemical fractionation of the alkali/alkaline earth elements Mg, Ca, K, and Sr. A number of early papers indicated that many or all of these elements were significantly fractionated in marine aerosols compared to seawater (e.g. see Sugawara et al 1949, Koyama & Sugawara 1953, Oddie 1960, Komabayasi 1962). Enrichments of greater than ten were not uncommon. Most of these earlier field studies were conducted in areas with potentially significant contributions from nonmarine sources to the marine aerosol. Indeed, the data in several of the Japanese papers above show that the enrichment of these elements increased as the sampling sites were moved progressively further inland. However, relatively little consideration was given at the time to the possibility that the high enrichments observed for these elements might be due to non–sea salt particles in the samples collected.

In more recent field studies involving these elements in marine aerosols, awareness of the importance of the atmospheric sampling site has grown. Atmospheric samples have been collected at several midocean locations by a number of groups during the past few years and a different picture is beginning to emerge relative to chemical fractionation of these elements. Enrichment values obtained for Mg, Ca, K, and Sr in some of these recent studies are presented in Table 1. There is no significant enrichment of Mg or Sr reported in any of these recent studies, nor, in most cases, any enrichment of Ca or K. Where apparent enrichment has been observed it has generally been explained by the presence of crustal material in the samples. Several different approaches have been used in these papers to identify and correct for a non–sea salt component for these elements. In studies in Hawaii and over the North Atlantic, G. Hoffman & Duce (1972) and E. Hoffman et al (1974) calculated enrichments from considerations of the slope of the linear least squares regression line calculated from plots of Mg, K, Ca, and Sr vs Na, as described previously. As seen in Table 1, enrichments calculated in this manner are very close to zero, indicating that the sea salt–associated components of these elements are present on the particles in seawater composition, within a very few percent. This strongly suggests little or no chemical fractionation in the areas sampled.

In a different approach to the same problem, several workers (Tsunogai et al 1972; Wilkniss & Bressan 1972, Wada & Kokubu 1973, Buat-Menard et al 1974) have either analyzed only the water-soluble component of the samples they collected or the water-soluble and water-insoluble fraction separately. The basic assumption is that the K, Ca, Mg, and Na in the sea salt particles are water soluble but the K, Ca, etc, in other components of the marine aerosol, particularly mineral dust, are not. However, Tsunogai et al (1972) found significant enrichment of K and Ca in the water-soluble fraction, which was attributed to a water-soluble component of land-derived Ca and K. Wada & Kokubu (1973) also believed that the contribution of water-soluble Ca and K from land sources had to be evaluated before drawing conclusions about the chemical composition of sea salt particles. These conclusions are not surprising. Far from land, it is unlikely that atmospheric sea salt and mineral dust particles exist as discrete particle populations. Coagulation processes, such as cloud and raindrop formation and evaporation, etc, will destroy these discrete distributions relatively quickly. The extent and rate of leaching of the various alkali/alkaline earth metals from the dust particles in the saline droplets has not been measured, but at the expected pH of 5–6 in these particles, it may be rather significant.

For samples collected over the North Atlantic, Buat-Menard et al (1974) found that $E_{Na}(Mg)$ in the water-soluble fraction of the aerosol was generally near zero and that $E_{Na}(Ca)$, which ranged from zero to ten, was related to the quantity of the water-insoluble fraction (see the section on sulfate). Although some of the high K enrich-

Table 1 Some representative values for alkali and alkaline earth enrichments in marine aerosols

Location	No. of samples	$E_{Na}(Mg)$	$E_{Na}(Ca)$	$E_{Na}(K)$	$E_{Na}(Sr)$	Remarks	Reference
Norwegian Sea cruise	6	—	—	$3.4+2.1$	—	total aerosol analyzed	Chesselet et al (1972b)
Antarctica	5	0.06 ± 0.06	0.33 ± 0.08	0.35 ± 0.11	—	total aerosol analyzed	Chesselet et al (1972b)
Windward coast of Oahu	119[a,b]	-0.02 ± 0.01	-0.03 ± 0.03	$0.05+0.08$	-0.11 ± 0.04	total aerosol analyzed, samples contain <50 ng/m^3 Fe	G. Hoffman & Duce (1972)
North and South Pacific cruises	25	0.13 ± 0.15	1.7 ± 0.9	1.0 ± 0.7	—	water-soluble aerosol only	Tsunogai et al (1972)
Atlantic, Caribbean, and Pacific cruises	9	—	—	0.26 ± 0.18	—	water-soluble aerosol only	Wilkniss & Bressan (1972)
Mid-Pacific cruise	18	-0.02 ± 0.05	0.46 ± 0.83	$0.17+0.46$	—	water-soluble aerosol only	Wada & Kokubu (1973)
Caribbean to West Africa cruise	19	-0.34 ± 0.18	1.3 ± 1.7	0.15 ± 0.20	—	water-soluble aerosol only	Buat-Menard et al (1974)
West Africa, Gulf of Guinea coastal cruise	17	0.00 ± 0.07	2.3 ± 3.3	1.2 ± 1.6	—	water-soluble aerosol only	Buat-Menard et al (1974)
North Atlantic cruises	5–8[a]	0.02 ± 0.05	-0.03 ± 0.09	-0.03 ± 0.04	-0.15 ± 0.21	total aerosol analyzed, samples contain <10 ng/m^3 Fe	E. Hoffman et al (1974)
North Atlantic cruises	36–38[c]	0.07 ± 0.02	0.22 ± 0.04	-0.03 ± 0.03	-0.16 ± 0.08	total aerosol analyzed, samples contain <250 ng/m^3 Fe	E. Hoffman et al (1974)

[a] Enrichments based on regression line slope and standard deviation of the slope
[b] 56 samples for K
[c] 10 samples for K

ments also appeared related to the amount of water-insoluble material in their samples, the largest values for $E_{Na}(K)$ were found in samples collected over the more biologically productive areas off the West African coast, especially over the upwelling region in the Gulf of Guinea. Buat-Menard et al (1974) speculated that the K enrichment in this area was due to true chemical fractionation at the air/sea interface resulting from the association of K with high concentrations of surface-active material in these biologically rich waters.

The effect of dust on $F_{Na}(K)$ for particles collected during onshore wind conditions in Bermuda is shown in Figure 4 (Hoffman 1975). Note that $F_{Na}(K)$ is used in this figure rather $E_{Na}(K)$, because it is impossible to plot zero and negative values on log-log plots. Using atmospheric Fe as an indicator of dust, this figure clearly shows the dependence of $F_{Na}(K)$ on the Fe/Na ratio. When the dust contribution is relatively insignificant (i.e. Fe/Na approaches zero), $F_{Na}(K)$ approaches one. If this data is replotted as K/Na vs Fe/Na, the slope of the resulting regression equation (excess K/Fe) is 0.326 ± 0.005. This suggests a crustal source for the excess K, since the mean K/Fe ratio in soil is ~ 0.36 (Vinogradov 1959). Air mass trajectories calculated for these Bermuda samples indicate that the dust collected can originate from the Sahara desert and the east coast of North America. Notice the magnitude of the effect of dust on $F_{Na}(K)$ in Figure 4. During onshore wind conditions, apparent fractionation up to ten is found, even at this coastal location over 1000 km from the nearest major land mass. This appears to be entirely the result of the additional K component present in continental dust that has been transported from 1000 to 5000 km across the ocean.

It has been suggested that the enrichment of alkali and alkaline earth metals in

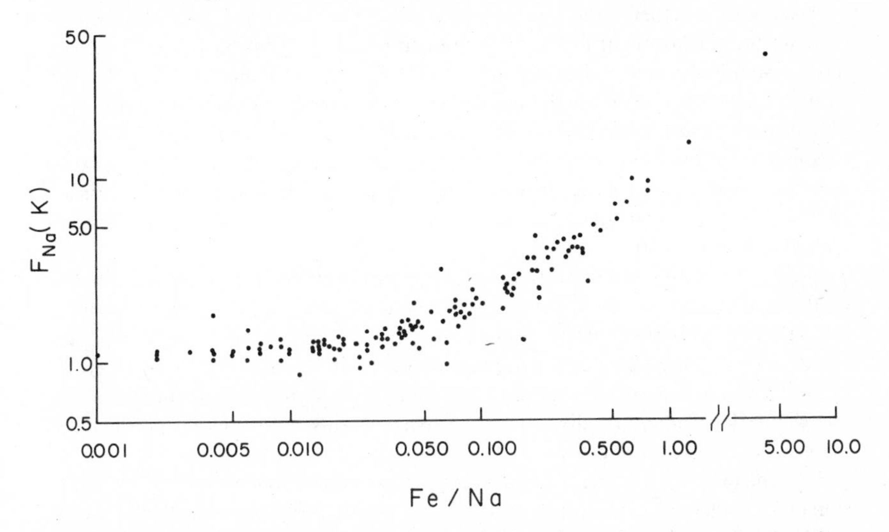

Figure 4 The effect of dust, as indicated by the Fe/Na ratio, on the apparent fractionation of potassium in marine aerosols collected from a 20 m high tower on the coast of Bermuda (E. Hoffman 1975).

sea salt particles is a function of particle size, with only the smallest particles being significantly enriched. Indeed, if the smaller particles are more enriched than the larger particles, this would explain apparent enrichments of these elements in particles at high altitudes (Chesselet et al 1971) and in particles and snow collected at the South Pole (Chesselet et al 1972a). The cascade impactor data of Barker & Zeitlin (1972) from a coastal tower on Oahu, Hawaii, indicate that enrichment of K, Ca, and Mg on the small particles cannot be ruled out. However, their reported analytical uncertainties are so large that their data cannot confirm this either. Chesselet et al (1972b) also investigated the particle size distribution of K and Ca over the Norwegian Sea, but again the reported analytical uncertainties in the data make a definitive interpretation impossible. Andren & Harriss (1971) found enrichment of Ca and depletion of Mg on all the particle size ranges investigated 10 km inland in Puerto Rico and on board a ship in the Carribbean. They found the smaller particles were less enriched in Ca than the larger particles. Mineral dust may again be involved since several of their samples were collected inland, and the general area is often under the influence of the Sahara dust plume. Winkler (1975) found that the mass ratio of the elements Cl:K:Ca is significantly different from seawater on particles $\leqq 0.2$ μm radius over the Atlantic Ocean. Winkler is uncertain as to the cause of the relatively high K and Ca concentrations on these small particles and notes that chemical fractionation is a possibility. He points out that mineral dust and anthropogenic sources may also be significant.

A vast literature on the composition of rain in coastal locations routinely shows enrichment of the alkali/alkaline earth elements relative to seawater, but it has been shown that particulate matter above two km in height is fairly uniform in composition, whether collected over continental or marine areas (Blifford & Gillette 1972, Delany et al 1973). As mentioned previously, this background aerosol has a maximum number distribution at ~ 0.1 μm radius (Junge 1963, 1972) and a rather long residence time in the atmosphere. Thus its origin relative to a continental or marine source is obscured. Since the contribution of this background aerosol becomes more and more important with increasing altitude, the use of high altitude aerosols, oceanic cloud water, or oceanic rain water (e.g. Bloch et al 1966) to indicate chemical fractionation at the air/sea interface is often misleading and certainly complicates the question.

In conclusion, the often-observed apparent enrichment of the alkali and alkaline earth metals relative to Na probably results in most cases from the inclusion of non-sea salt particles, particularly continental dust, in the environmental samples. Fractionation of these elements in the environment is likely restricted to a few percent in most areas of the world ocean. The possibility that K may be fractionated at the air/sea interface in biologically productive areas remains, but the global geochemical importance of this is unknown.

There have been a number of laboratory studies of the possible chemical fractionation of these elements. Several have shown significant enrichment, particularly of K and Ca, whereas others have shown little or no enrichment (e.g. see Komabayasi 1964; Bloch & Luecke 1968, 1972; Bruyevich & Korzh 1970; Glass & Matteson 1973; Morelli et al 1974; E. Hoffman 1975). Many earlier studies have been reviewed by

MacIntyre (1974a). These laboratory model studies differ from each other and from the natural environment in so many ways that it is extremely difficult to evaluate them in terms of natural processes occurring at the ocean surface, and we make no attempt to do so. Suffice it to say that there is no real reason to doubt the results in any of these studies, but on the basis of the most recent field data above, whatever mechanism or mechanisms caused the observed laboratory enrichments of these elements appears to be of limited importance in the environment.

Phosphate

In the laboratory study perhaps most responsible for the recent interest in bubble scavenging of surfactants and associated substances and their injection into the atmosphere, Baylor et al (1962) found that the inorganic dissolved phosphate in filtered seawater could be efficiently stripped from the seawater by bubbling. In their investigation, the concentration of phosphate in seawater at any time, C_t was given by

$$C_t = C_o e^{-Kt}, \tag{3}$$

where C_o = initial $PO_4^{\equiv}$ concentration, t = time after initiation of experiment, and K = velocity or removal constant. K was a function of bubbling rate and the total surface area of the bubbles. Baylor et al (1962) believed that the phosphate removed from the water column was largely transferred into the atmosphere. The phosphate remaining in the seawater after bubbling was primarily associated with organic material and was filterable, that is, it no longer passed through a 0.45 μm pore size filter. In a subsequent study, Sutcliffe et al (1963) found that the atmospheric particles produced by laboratory bubbling of filtered seawater were indeed highly enriched in phosphate and that most of the phosphate in the aerosol was particulate rather than dissolved and was organically associated. There was considerable interest in whether collapsing bubbles could generate organic aggregates in the sea, as the implications to the marine food chain were obvious. It is now realized, however, that very small particles are necessary for the nucleation of the larger aggregates in seawater; these small particles are most likely colloidal in nature and therefore were not trapped by the 0.45 μm pore size filters used (Batoosingh et al 1969). Nevertheless, these experiments clearly showed that bubble scavenging of certain dissolved and/or colloidal organic materials and other associated substances, such as $PO_4^{\equiv}$, significantly enrich these other substances in atmospheric particles.

Bruyevich & Kulik (1967a) also found significant enrichment of $PO_4^{\equiv}$ in aerosols produced by bubbling seawater in the laboratory. MacIntyre & Winchester (1969) and MacIntyre (1970) followed up these initial studies with a detailed laboratory investigation of phosphate enrichment on the atmospheric sea salt particles. Using radioactive $^{32}PO_4^{\equiv}$ and $^{22}Na^+$ tracers and a variety of aqueous bubbling solutions, MacIntyre (1970) always found excess phosphate in the aerosol. $E_{Na}(PO_4)$ ranged from 0.01 to 10 for jet droplets and as high as 1000 for what he believed were droplets from a well-drained film cap. Figure 5 shows the results of three runs using filtered seawater, distilled water, and distilled water with a negatively charged surfactant (sodium lauryl sulfate) added. The highest enrichment was observed for the intermediate size atmospheric particles, approximately 1–6 μm radius, which MacIntyre

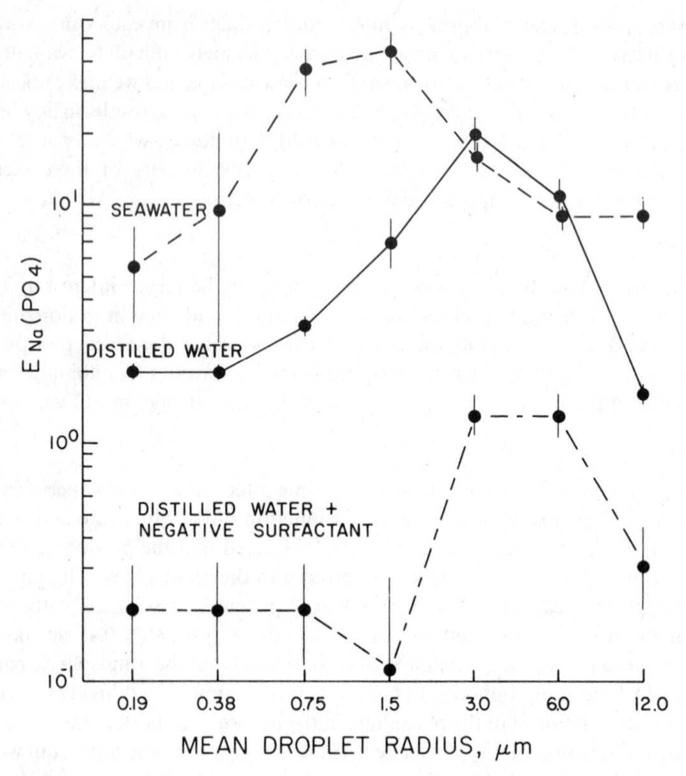

Figure 5 Enrichment of phosphate as a function of particle size for atmospheric particles produced in the laboratory by bubbles bursting in various "ocean" solutions. After MacIntyre (1970).

believed were the film droplets because they were produced from very large bubbles with a large film cap area. Similar studies with numerous smaller bubbles, which produced primarily jet droplets, showed little particle size dependence of $E_{Na}(PO_4)$. The use of seawater resulted in higher values for $E_{Na}(PO_4)$ than distilled water, probably because organic surfactants have a higher concentration in seawater. However, when the surfactant sodium lauryl sulfate was added to the distilled water, the enrichment dropped by an order of magnitude. MacIntyre (1970) pointed out that for this surfactant the surface active moiety is the lauryl sulfate anion, which should attract positive ions, rather than the negative $PO_4^{\equiv}$ ion. Thus $PO_4^{\equiv}$ enrichment is significantly decreased. When a positive surfactant, such as cetyltrimethylammonium bromide, was used in distilled water, values for $E_{Na}(PO_4)$ were ten or greater. MacIntyre (1970) suggested that ions with high ionic potential, i.e. large charge and small size, are preferred as counterions, or ions with a charge opposite to that of the surfactant. Thus, for example, Ca^{2+} and Mg^{2+} are preferred over Na^+. MacIntyre (1970) suggested that this counterion mechanism plays a major role in the association of surface-active substances with inorganic ions, which in turn could result in the enrichment of the inorganic ions in atmospheric sea salt particles. Koske (1974) suggests this mechanism may be responsible for the barium enrichment that he observed on sea salt particles produced by distillation of seawater in the laboratory.

Surprisingly, we have been able to find no data on phosphate enrichment in the ambient marine aerosol. Although there is considerable data on sodium and such nutrients as phosphate, nitrate, etc, in rainfall, most of these studies were undertaken as part of general ecological and biological field programs and are not in suitable locations for collection of true marine rainfall. Perhaps most representative of marine air is the rain data of Allen et al (1967) from the small island of Signy, off the coast of Antarctica in the South Orkney Islands. As part of a nutrient balance study, these authors concluded that the primary source of phosphate on the island was precipitation and that the phosphate and other nutrients in the precipitation came from the sea. The average value for $E_{Na}(PO_4)$ in the Signy Island precipitation was 170. Allen et al (1968) undertook a similar study in England. At two stations ranging from 2 to 8 km from the sea, their data suggest $PO_4^{\equiv}$ enrichments of over 1000. However, the air over the highly populated, agricultural, and industrialized island of Great Britain is certainly not representative of uncontaminated marine air. The general geochemical importance of the atmospheric transport of phosphate to and from marine areas remains to be determined.

Nitrogen Compounds

With the demonstration that phosphate can be efficiently transported from the sea to the atmosphere on particles from bursting bubbles, it is rather surprizing that so little research in this context has been carried out on nitrogen containing nutrients. We have found no data on organic nitrogen in marine aerosols. Two investigations in New Zealand, however, suggest the potential importance of the chemical fractionation of organic nitrogen compounds. Wilson (1959) investigated the organic nitrogen content of snows collected 1600–2600 m above sea level on remote mountains located 20–160 km from the sea. Because the samples were freshly collected from regions where no plants or animals exist, Wilson expected no contamination from these sources. He could see no insoluble inorganic matter in the samples with the naked eye and believed that no (or very little) crustal weathering products were present. He felt strongly that the source of the organic nitrogen in these samples was the ocean.

The organic nitrogen (ON) concentration in the snow ranged from 20 mg/liter to 200 mg/liter. A single coastal rain sample contained 200 mg/liter organic nitrogen, whereas typical seawater off New Zealand contained 8 μg/liter. These concentrations, combined with Na measurements made on the same samples, result in $E_{Na}(ON)$ values ranging from 4×10^4 to 1×10^6. These are extremely high enrichments and, if correct, suggest that the transport of organic nitrogen from the sea to coastal areas could be significant in the nutrient balance of some regions. The high organic nitrogen concentration in coastal marine rain was confirmed by Dean (1963), who found 170 mg/liter in rain at Taita in coastal New Zealand. Organic nitrogen concentrations of ~25 mg/liter were found by Williams (1967) in rain over the ocean off California and near Samoa. High enrichments relative to seawater were also found for NH_4^+ in the rain and snow samples of Wilson (1959), and an enrichment for total nitrogen of 2×10^4 was found in a foam sample collected on a rocky coast north of Wellington, New Zealand.

Are such high enrichments in the atmosphere possible from a marine source? As a crude attempt to answer this question, we can make the following calculation. Assuming the organic nitrogen in the atmosphere was present on sea salt particles when they were produced at the sea surface, assuming the average value of E_{Na}(ON) for particles in the atmosphere near New Zealand is 2×10^5, and assuming the organic nitrogen concentration in the mixed layer of the ocean near New Zealand is ~8 μg/liter, the concentration of organic nitrogen in the surface layer stripped by the bursting bubbles, i.e. the top 0.05–0.5 μm, must be ~1.6 g/liter. (This concentration could be attained in this surface layer by the transport of a quantity of organic nitrogen equivalent to that present in approximately 1–10 cm of subsurface water. This is perhaps not completely unreasonable.) Williams (1967) found the concentration of both dissolved and particulate organic nitrogen ranged from 100 μg/liter to 500 μg/liter in the top 150 μm of the ocean surface off Peru and off California. We assume an average concentration of 300 μg/liter in the 150 μm microlayer. The microlayer concentrations observed were 1.5–50 times concentrations present at depths of 10–30 meters. However, if this organic nitrogen in the 150 μm microlayer is actually concentrated in a much thinner layer from 0.05 μm to 0.5 μm thick, which the bursting bubble strips into jet droplets, the concentration in this thinner layer would be ~0.1–1 g/liter—close enough to the required concentration of 1.6 g/liter calculated above that we cannot dismiss the possibility that this mechanism is potentially significant in the environment. However, these calculations and any interpretation of this field data in terms of chemical fractionation must be viewed cautiously because our understanding of the interactions of such gaseous species as NO, NO_2 and NH_3 with aerosols and rain and cloud droplets is limited and because chemical reactions of these inorganic nitrogen species with organic material in these particles and droplets are possible.

Williams (1967) discussed the general implications of enriched nitrogen transport into the atmosphere on sea salt particles to the nitrogen cycle in the sea and suggested that, if highly enriched nitrogen is a general phenomenon, this "closed system of nitrogen recycling would greatly reduce the net input of nitrogen into the sea" from continental sources. The implications to the overall nitrogen cycle are potentially significant, but this subject awaits considerable additional research.

In a different aspect of the nitrogen problem, Bloch & Luecke (1970) showed in the laboratory that inorganic NH_4^+ is enriched in the particles produced by boiling artificial salt solutions of NH_4Cl and collecting the condensate and spray produced. They explained that their NH_4^+ enrichments, which ranged up to ~1000, were due to hydrolysis of NH_4^+ in the seawater, with the subsequent enrichment of NH_4OH at the water surface due to Gibbs adsorption. As mentioned above, the question of NH_4^+ in atmospheric aerosols is complicated by the interaction of gas phase NH_3 with the saline droplets. While NH_4^+ is apparently enriched in marine aerosols and rain relative to seawater (Junge 1957, Wilson 1959, Menzel & Spaeth 1962, Williams 1967, Tsunogai 1971), there is no strong evidence to suggest that this NH_4^+ originates on the particles produced at the sea surface.

The Halogens

CHLORINE Chlorine is the major mass component of atmospheric sea salt and is relatively easy to analyze. One might expect this element to be the most thoroughly understood relative to chemical fractionation. However, there is still considerable uncertainty regarding the marine atmospheric chemistry of chlorine. A Cl gas phase exists in marine air, and considerations of the Cl/Na ratio and Cl enrichment during particle production are complicated by the possibility of Cl gain or loss from the aerosol after its formation. Most of the data on Cl/Na ratios are from rain, with relatively little information for atmospheric particles in open ocean areas. Wilkniss & Bressan (1972) measured Cl/Na ratios on atmospheric particles in midocean and near-continent marine air. The mean value for $E_{Na}(Cl)$ from their midocean samples was -0.07 ± 0.21, although for particles >2 μm radius $E_{Na}(Cl)$ was $+0.04 \pm 0.19$. Along the east coast of the United States, they found $E_{Na}(Cl)$ values ranging from -0.99 to -0.31. They note that these negative enrichments are undoubtedly due to the simultaneous collection of continental aerosols with relatively high Na concentrations. Chesselet et al (1972a) found the mean value of $E_{Na}(Cl)$ to be -0.04 ± 0.02 on particles from several midocean locations, whereas Peirson et al (1974) found an average value of -0.09 during 1972 and 1973 at Lerwick in the Shetland Islands. On a cruise in the mid-Pacific, Wada & Kokubu (1973) found a mean value of -0.03 ± 0.03 for $E_{Na}(Cl)$, and the data of Buat-Menard et al (1974) over the North Atlantic indicate a mean of -0.12 ± 0.11.

Perhaps the most valuable data to date is that of Martens et al (1973), who investigated the variation of the Cl/Na ratio with particle size in Puerto Rico. On the

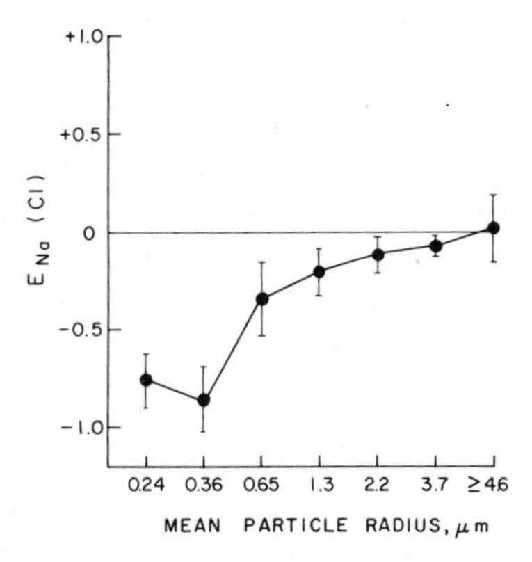

Figure 6 Enrichment of Cl as a function of particle size for atmospheric samples collected at coastal and inland sites in Puerto Rico. After Martens et al (1973).

basis of the Cl/Na ratio in the particles, corrected for any Na contribution from weathered crustal material, they calculated a 7–25% depletion of Cl relative to the expected seawater composition. The average $E_{Na}(Cl)$ for their bulk marine aerosol was -0.12 ± 0.09. Most importantly, they found this Cl depletion was very particle-size dependent, with increasing depletion of Cl with decreasing particle size. Figure 6 presents a composite of $E_{Na}(Cl)$ vs particle size for their four samples at coastal and inland Puerto Rico sites. Martens et al (1973) ascribe this marked decrease of $E_{Na}(Cl)$ on the small particles to a loss of Cl to the gas phase. Based on similar studies in the San Francisco Bay area, where NO_2 was simultaneously measured, these authors conclude that HCl is released from the particles after uptake of gaseous HNO_3 (and perhaps SO_2). Assuming that the HNO_3 and SO_2 uptake is controlled by diffusion to the particle surfaces, under nonequilibrium conditions the smaller particles, with their larger surface area to volume ratio, will be most affected by the addition of these gases and subsequent loss of HCl. If chemical equilibrium is ultimately reached between the gases and the particles, there should be no difference in the Cl/Na ratios on particles of different sizes. The existence of such a difference suggests that, if this mechanism is valid, equilibrium is not attained.

The mean $E_{Na}(Cl)$ for the six midocean studies above is -0.08 ± 0.04, equivalent to an apparent 8% loss of Cl to the gas phase, assuming no chemical fractionation of Cl relative to Na during particle production. A gaseous Cl component does exist in the marine atmosphere. Although its chemical form is still uncertain, the concentrations are generally 1–5 $\mu g/m^3$ near sea level, with the gaseous/particulate Cl ratio averaging about 0.5 (Junge 1957, Duce et al 1965, Chesselet et al 1972a, Rahn et al 1976). Thus, if this ganeous Cl is generated from the atmospheric sea salt particles, it must have an atmospheric residence time approximately six times that of particulate Cl, or ~2–3 weeks. Delany et al (1973, 1974) measured the vertical profile of Cl/Na in aerosols over continental and marine areas up to 9 km. Over the sea the ratio fluctuated around the seawater value, but over continental areas the ratio generally increased with height, which Delany and co-workers believed was caused by interaction between the continental aerosol and gaseous Cl.

There have been a number of suggestions for the reactions that may release gaseous Cl from sea salt particles. These include the uptake of SO_2 (Eriksson 1959, 1960) or HNO_3 (Robbins et al 1959) to release HCl, described previously, the reactions of Cl^- with O_3 to produce Cl_2 (Cauer 1938), the uptake and reaction of NO_2 with NaCl to produce NOCl (Schroeder & Urone 1974), and photochemical reactions in either the sea or atmospheric particles to produce the pernitrite ion, which subsequently reacts with Cl to release Cl_2 and/or NOCl (Petriconi & Papee 1972). Valach (1967) and Lazrus et al (1970) suggest that volcanoes are the major source for marine atmospheric gaseous Cl. Zafiriou (1975) has suggested that significant quantities of CH_3Cl may be generated in seawater by the reaction of CH_3I with Cl^-. Zafiriou points out that CH_3Cl should be quite stable in seawater and should exchange with the atmosphere, where it would be photostable and probably have a relatively long residence time. CH_3Cl (and CCl_4) is a major component of the light chlorocarbons present in the atmosphere in the state of Washington (Grimsrud & Rasmussen 1975) and it does not appear to have an anthropogenic source. In addition, at a coastal

Bermuda site, Rahn et al (1976) have found that only about 25% of the gaseous Cl is inorganic (as measured by capture on LiOH impregnated filters), and 75% is apparently organic (as measured by capture on activated charcoal, which is quite efficient for CH_3Cl). Determination of the inorganic and organic speciation of gaseous Cl in marine air will be critical in understanding the interaction between atmospheric sea salt particles and these gaseous components.

There have been a few laboratory studies of Cl fractionation. Chesselet et al (1972b) found no deviation of the Cl/Na ratio on bubble-produced sea salt particles in the laboratory as compared to seawater. Wilkniss & Bressan (1972), in a similar study using a cascade impactor, found no deviation from the seawater ratio on the largest particles, but a slight negative enrichment on the submicron particles produced. There appears to be no strong evidence to suggest that there is any significant fractionation, either positive or negative, for Cl during particle production.

There is extensive data on the Cl/Na ratio in rain, from both coastal and inland areas. The early data was reviewed by Junge (1963). Most results show a negative enrichment for Cl, with an average $E_{Na}(Cl)$ of approximately -0.05 to -0.10. Recent studies of rain in Hawaii at various altitudes, reported by Seto et al (1969), show no significant Cl enrichment. In studies of the Cl/Na ratio in cloud water 1000 m above sea level in Puerto Rico, Lazrus et al (1970) likewise found no statistical difference from seawater. Many past samples were collected in areas with potential for significant contributions of continental material to the rain. Owing to the presence of a gas phase, possible incorporation of continental and background aerosols from higher altitudes, and the fact that rainout and washout efficiency is a function of both particle size and particle composition, any general attempt to understand the chemistry of atmospheric particles through rain analysis is fraught with problems; this is especially true for particulate chlorine.

BROMINE There is little evidence to suggest that bromine is fractionated during sea salt particle production. However, gaseous bromine may be released from sea salt particles to a greater extent than Cl after the particles enter the atmosphere, resulting in a Br/Cl ratio on the particles that is slightly depleted relative to seawater. Duce et al (1967) and Moyers & Duce (1972a) investigated the Br/Cl ratio as a function of particle size from 20 m high towers directly on the coast of the islands of Hawaii and Oahu. Figure 7 presents the mean values for $F_{Cl}(Br)$ for 12 cascade impactor samples from Oahu. To facilitate graphical display of this data, F rather than E values are presented. Values for $F_{Cl}(Br)$ were generally less than unity for all particle sizes. The mean value for $F_{Cl}(Br)$ for the total particle population suggests that, relative to Cl, 40% more of the Br has apparently been lost to the gas phase. Moyers & Duce (1972a) measured the gaseous Br concentration in Hawaii, and it averaged approximately 50 ng/m^3, or 4–10 times the particulate bromine concentration. This suggests that, if the gaseous bromine does result from sea salt particle release, the residence time of the gaseous species is several times that of particulate Br. Martens (1973) has investigated $F_{Na}(Br)$ as a function of particle size at both coastal and inland stations in Puerto Rico. His results agree with the Hawaii data presented in Figure 7 for particles with radii >1 μm. For particles between 0.5 μm and 1 μm, however, $F_{Na}(Br)$ ranged from 2 to 11 for four samples. Martens (1973) notes that the high fractionation values for Br on the smallest particles may be due to local contamination by the presence of Br associated with Pb from the combustion of tetraethyllead and ethylene bromide in gasoline. Measured Br/Pb ratios on these small particles in Puerto Rico support this suggestion.

Duce et al (1973) have measured atmospheric bromine at the geographic South Pole at 2800 m above sea level, and $F_{Cl}(Br)$ is greater than 12 for the total aerosol population sampled. Gaseous bromine was found to be approximately 20 times the particulate Br concentrations. Similar high value for $F_{Cl}(Br)$ were found by Cadle (1972) in the midlatitude, northern hemisphere stratosphere, at an altitude of approximately 10 km, and by Delany et al (1973) at altitudes from 1 km to 9 km over marine and continental areas. The latter group suggested that the high $F_{Cl}(Br)$ values at higher altitudes may result from gas phase reactions yielding NH_4Br aerosol. Cadle suggested that the high stratospheric Br/Cl ratio might result from increased Br from the combustion of ethyl fluid in gasoline. Another possibility is the uptake of natural gaseous Br on the background aerosol. In summary, it appears that there is no strong evidence for bromine fractionation at the air/sea interface and that deviations of the Br/Na or Br/Cl ratio from seawater values are probably caused by the exchange of a natural gaseous bromine phase with the sea salt particles and

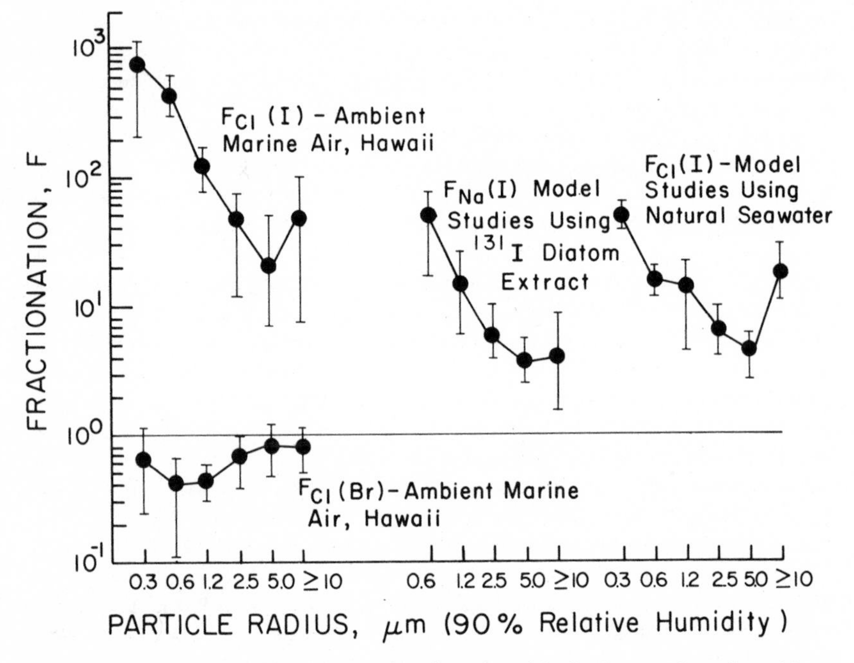

Figure 7 Fractionation of I and Br as a function of particle size for samples collected from a 20 m high tower on windward Oahu, Hawaii and from a laboratory model ocean-atmosphere system. From Moyers & Duce (1972a,b) and Seto & Duce (1972).

Table 2 Some representative values for iodine enrichment in marine aerosols

Location	Reference	E(I)	Reference
Sugashima Island, Japan, 200 m	Cl	10–100	Komabayasi (1962)
Tokyo, Japan, near sea level	Cl	2500	Miyake & Tsunogai (1963)
New Zealand coast, sea level to 35 m	Cl	35–500	Dean (1963)
Hawaii Island, sea level to 2000 m	Cl	10–200	Duce et al (1965)
Barrow, Alaska coast, 10 m (winter)	Cl	55–10,000	Duce et al (1966)
Oahu, Hawaii coast, 20 m	Cl	80–300	Moyers & Duce (1972b)
McMurdo, Antarctica coast, 50 m	Cl	750–28,000	Duce et al (1973)
South Pole, Antarctica, 2800 m	Cl	$\geqq$ 12,000	Duce et al (1973)
Puerto Rico, 20–450 m	Na	220	Martens (1973)

background aerosol or by the addition of Br-rich particles from the combustion of ethyl fluid.

IODINE Iodine is one element for which there is considerable evidence for chemical fractionation at the air/sea interface. A number of studies have shown that iodine enrichment in marine aerosols generally ranges from 100 to 1000. Some representative values for iodine enrichment are presented in Table 2. The sea is the accepted source for atmospheric iodine present in marine aerosols, but the mechanism for enrichment remains uncertain. It is likely that either (*a*) a gaseous form of iodine, probably I_2 or CH_3I, is released from the ocean surface and exchanges with the particles, or (*b*) surface-active organic material enriched in iodine in the sea is scavenged by rising bubbles and injected preferentially into the atmosphere. Evidence supports both mechanisms and indeed both are probably important in the environment.

In marine air, iodine is generally found on smaller particles than is Br or Cl. In studies from 20-meter-high towers on the coasts of Hawaii and Oahu, Duce et al (1967) and Moyers & Duce (1972b) found that the major mass of iodine is generally on particles with radii of 0.5–1.0 μm whereas the major mass of Cl is present on particles with radii of 1–5 μm at 90% relative humidity. This results in a general increase in the I/Cl ratio, or $F_{Cl}(I)$, with decreasing particle size, as illustrated in Figure 7. Duce et al (1967) point out that, if the high iodine enrichment is the result of gaseous uptake by the particles, the overall effect of the addition of gaseous iodine to the particle chemical composition would be inversely proportional to the particle radius. Using their experimental data, Robbins (1970) developed a mathematical diffusion model for the reaction of gaseous iodine with particle surfaces that could explain quantitatively the particle size dependence of the I/Cl ratio. His model, however, required that all the particles have the same, rather short, residence time of approximately 15 min. Moyers & Duce (1972b) suggested that the enrichment variation with particle size may be due to a rather slow interaction of the gas phase with the particles, so that the residence time of the particles becomes the controlling factor in the uptake of gaseous iodine. The smaller particles, with their longer atmospheric residence times (Esmen & Corn 1971, Junge 1972) would thus have higher iodine enrichment. If the gas-particle interactions are inorganic in nature and involve such species as I_2, I^-, and IO_3^-, Moyers & Duce (1972b) showed that thermodynamic equilibrium would probably never be attained and the particles would continue to act as a sink for gaseous iodine as long as they remain in the atmosphere.

An obviously critical factor in this discussion is the presence of gaseous iodine over the ocean. Duce et al (1965) and Moyers & Duce (1972b) found that the general level of gaseous iodine in surface air near Hawaii ranged from 5–20 ng/m^3, with the gaseous species generally two to four times particulate iodine. In Antarctica, both at coastal and inland sites, the gaseous concentration was 2–3 ng/m^3, again two to four times particulate iodine. The chemical form of the gaseous iodine is still quite uncertain. Miyake & Tsunogai (1963) added $^{131}I^-$ to a model ocean system and found I_2 released when the system was irradiated with light at 300–500 nm. They attributed the I_2 release to:

$$2I^- + 1/2\,O_2 + H_2O \xrightarrow{h\nu} I_2 + 2OH^- \tag{4}$$

Martens & Harriss (1970) and Seto & Duce (1972) found similar results using $^{131}I^-$ tracer in the laboratory.

In their model ocean-atmosphere studies, Seto & Duce (1972) also cultured the diatom species *Phaeodactylum tricornutum* in the presence of $^{131}I^-$. Organic material incorporating the $^{131}I^-$ was extracted from the diatom culture and added to the model ocean system; the seawater was bubbled and the particles produced were collected using a cascade impactor. $F_{Cl}(I)$ values for these size-separated laboratory aerosol samples are presented in Figure 7. Additional model studies were performed on fresh seawater with no radioactive iodine added, with the natural iodine present analyzed directly in the particles produced. These results are also presented in Figure 7. It is seen that the pattern of iodine enrichment on the particles from both model studies is very similar to that in the marine environment, that is, general increase in iodine enrichment, or fractionation, with decreasing particle size, except for the very largest particles. However, the magnitude of the enrichment is about 20 times greater in the natural environment. These results suggest strongly that at least part of the iodine enrichment in atmospheric sea salt particles results directly from chemical fractionation at the air/sea interface during particle production, probably caused by association of some of the iodine with surface-active organic material. These conclusions agree with those of Dean (1963), who found considerable organic material (primarily algae and plankton debris) in sea spray in New Zealand and believed the high iodine enrichment he measured was due to iodine associated with the organic material.

In their model studies using natural seawater with no added radioactive iodine, Seto & Duce (1972) found that, during bubbling, gaseous iodine comprised 90% of the total iodine released (the rest was particulate) when the surface of the model ocean was irradiated with ultraviolet light and decreased to 75% when the ultraviolet light was absent. When the same experiments were run without bubbling, the absolute concentration of gaseous iodine released decreased significantly, but was not zero and was approximately the same with and without ultraviolet light. Thus a gaseous form of iodine was released from the model ocean surface, but its release was apparently not entirely dependent upon photochemical reactions and was enhanced by bubbling. Further laboratory studies in which the form of this gaseous iodine is determined would be most valuable.

Moyers & Duce (1972b) assumed the gaseous species was I_2 for their thermodynamic equilibria calculations, but Zafiriou (1974) has pointed out that the mean atmospheric lifetime of I_2 would be only a few seconds because of photodissociation during daylight hours. He felt that molecule-aerosol reactions might be important at night but that photodissociation followed by atom reactions (e.g. $X + O_3 \rightarrow XO \times O_2$) would be prevalent during daylight. Lovelock et al (1973) found methyl iodide over the North Atlantic at concentrations of ~ 7 ng I/m^3. They suggested that CH_3I produced biologically in the sea escapes into the atmosphere and is destroyed photolytically after a residence time of a few days. Zafiriou (1975) has shown by reaction kinetic calculations and laboratory studies that CH_3I in seawater reacts with Cl^- to produce CH_3Cl at about the same rate that CH_3I escapes into the atmosphere. He also showed that reaction of CH_3I with Cl^- in the salt particles after they reach the atmosphere could not account for the high I enrichments observed in the marine aerosol. If CH_3I is photolytically destroyed in the atmosphere, however, as suggested by Lovelock et al (1973) and Zafiriou (1975), the products of this photolysis are probably scavenged rapidly by the marine aerosol and may be the cause of the high iodine enrichments observed. Clearly, the chemical species of gaseous iodine present in the marine atmosphere need to be determined and gas/particle iodine exchange must be studied to fully elucidate the marine atmospheric chemistry of this element.

FLUORINE There have been very few studies of fluorine in the uncontaminated marine atmosphere. Carpenter (1969) summarized the results of rain analyses for F and pointed out that $E_{Cl}(F)$ in these rain samples, virtually all of which were collected in continental areas, ranged from 10 to 1000. Carpenter (1969) and Sugawara (1965) concluded that F is preferentially injected into the atmosphere from the ocean surface. Wilkniss & Bressan (1971, 1972) have investigated the chemistry of F and Cl in rain as well as atmospheric particles in the field and in the laboratory. At mid-ocean sites, far from the influence of land, they found a mean value for $E_{Na}(F)$ of -0.62 ± 0.07, indicating no positive chemical fractionation of fluorine during sea salt particle production. This was supported by model studies in the laboratory. $E_{Na}(F)$ values in atmospheric particulate samples collected over the ocean but near the east coast of the United States by these authors ranged from 2 to 25. Wilkniss & Bressan note that these apparent high enrichments are probably caused by incorporation of continental material into the marine aerosol they were sampling. Wada & Kokubu (1973) reached a similar conclusion to explain their mean $E_{Na}(F)$ value of 12 ± 9 in the western Pacific. Analyses of aerosol samples collected at various altitudes above sea level in Hawaii support this conclusion (Wilkniss & Bressan 1972). $E_{Na}(F)$ in these samples was near zero up to about 500 m, but increased considerably at 3000 m. The increase is probably related to a relatively greater mass of continental or background aerosol at the higher level. $E_{Na}(F)$ values in rain in Hawaii were often as high as ten and were inversely proportional to the Na concentration in the rain. Wilkniss & Bressan suggest that the high fluorine enrichment values in rain result from the incorporation of significant quantities of the background aerosol in the precipitation. However, Bewers & Haysom (1974) present mass budget calculations suggesting that terrigenous dust can account for only a few percent of the fluorine in the atmosphere. They postulate that the excess F in the rain may be entering the atmosphere from the ocean as a gas, possibly HF, or on particles with radii less than 0.1 μm. Wilkniss & Bressan (1972) presented laboratory evidence suggesting that gaseous F is released from the atmospheric sea salt particles. The relation, if any, between gaseous F and the high $E_{Na}(F)$ values in rain is not known. Apparently no attempt has been made to measure gaseous F, except for the freons, in the marine atmosphere, and the geochemical importance of the sea or of sea salt particles as a source for gaseous F has not been investigated. There is no strong evidence for positive enrichment of fluorine during atmospheric sea salt particle production, but additional work on the atmospheric fluorine cycle is clearly needed.

Sulfate

Although several workers have reported occasional samples where the $SO_4^=/Na$ ratio in marine aerosols and rain is similar to seawater (e.g. Koyama & Sugawara 1953, Buat-Menard et al 1974), most studies have reported $E_{Na}(SO_4)$ values greater than zero for marine aerosols (Sugawara et al 1949, Koyama & Sugawara 1953, Junge & Werby 1958, Junge 1963, Tsunogai et al 1972, Buat-Menard et al 1974, Cuong et al 1974). However, in most cases the apparent enrichment has been attributed to $SO_4^=$ sources other than sea salt. As with some other substances, the question of possible $SO_4^=$ enrichment on atmospheric sea salt particles is complicated by the conversion of sulfur gases of both marine and nonmarine origin, e.g. SO_2, H_2S, and perhaps organic sulfides, to particulate $SO_4^=$.

A variety of sulfate-containing particles not originating from the ocean have been detected in the marine atmosphere. For example, the major mass of the background aerosol is probably sulfate, presumably in the form of $(NH_4)_2SO_4$ or H_2SO_4 (Dinger et al 1970). The source of the sulfate in the background aerosol is believed to be the oxidation of SO_2. Anthropogenic $SO_4^=$ (also resulting from the oxidation of SO_2) released by burning of fossil fuels has also been shown to be transported some distance over the ocean. Koide & Goldberg (1971) and Weiss et al (1975) found that non–sea salt $SO_4^=$ in a Greenland glacier began to increase in the mid-twentieth century. The $SO_4^=$ in the snows of 1964–1965 also correlated with the Pb in these samples, which suggests a pollution source for a portion of the excess sulfate (Murozumi et al 1969, Koide & Goldberg 1971). The magnitude of this anthropogenic effect in glacier ice is disputed, however. Some workers (Hamilton & Langway 1967, Junge 1960) found no increase with age in the $SO_4^=/Na$ ratios of their ice samples from Greenland. However, marine samples taken downwind of large industrial and urban sites will undoubtedly contain significant quantities of excess anthropogenic $SO_4^=$ (Brosset & Åkerström 1972).

Another source of $SO_4^=$ in particles over the ocean is continental weathering material or mineral dust. Koyama & Sugawara (1953) found that $E_{Cl}(SO_4)$ in aerosols increased as samples were collected further inland. In addition, Buat-Menard et al (1974) found that $SO_4^=/Na$ ratios in most of their samples from off the coast of northwest Africa correlated with Ca/Na ratios. They attribute enrichment of both substances in their samples to gypsum ($CaSO_4 \cdot 2H_2O$) from the desert areas of Africa. Particles resulting from volcanic activity may also contribute significant amounts of $SO_4^=$ to the atmosphere, particularly in marine areas near volcanoes (Pueschel et al 1973) and in the stratosphere up to several years after major eruptions (Cadle et al 1971).

In addition to these nonmarine sources, the sea may release gaseous sulfur compounds into the atmosphere. Release of hydrogen sulfide from the ocean appeared necessary to balance calculated atmospheric sulfur budgets (e.g. Eriksson 1963, Robinson & Robbins 1968). Kellogg et al (1972), however, pointed out that the rapid oxidation of H_2S in surface seawater meant that release of H_2S from the sea would only be significant from biological processes in coastal areas, particularly tidal flats. Hitchcock & Wechsler (1972) showed that H_2S can be liberated from water through the reduction of $SO_4^=$ by anaerobic bacteria from muddy lake bottoms, wet soils and swamps, coastal wet lands and estuaries, and some anoxic fiords.

Because of the anaerobic conditions necessary for H_2S evolution, Lovelock et al (1972) recently suggested that organic sulfides, which can be produced by bacteria under normal aerobic conditions, may be a more significant oceanic source of sulfur to the atmosphere. They reported the presence of dimethylsulfide in seawater and in varieties of the marine alga *Laminaria*. Rasmussen (1974) found that dimethyldisulfide and methyl mercaptan, as well as dimethylsulfide, are produced by aquatic bacteria and Pacific tidal zone seaweeds. Even under anaerobic conditions, when H_2S was released, the organic sulfides were produced in concentrations 100 times greater than H_2S. Rasmussen (1974) concludes that, since organic sulfides are produced during both aerobic and anaerobic conditions, the organic sulfide production by the oceans may be a major source of sulfur in the atmosphere. Dimethylsulfide has yet to be detected in the marine atmosphere, however.

The possibility of a biogenic source for the excess $SO_4^=$ may be indirectly confirmed by the data of Buat-Menard et al (1974), who found significant $SO_4^=$ enrichment over the biologically productive upwelling waters off the Gulf of Guinea. Because this area is south of the intertropical convergence zone and thus not dominated by Sahara dust, they attribute these enrichments to either of two possible mechanisms: (*a*) biological production of sulfur gases which are subsequently converted to SO_3 and H_2SO_4, or (*b*) sea surface enrichment of sulfate due to the association of sulfur containing compounds with surface active organic material.

Although the $SO_2/SO_4^=$ ratio over "unpolluted" land areas varies from 1.3 to 10 (Georgii 1970, Rõdhe 1972), recent work by Cuong et al (1974) shows that the $SO_2/SO_4^=$ ratio over oceanic areas is approximately 0.1, suggesting a rapid conversion of SO_2 to $SO_4^=$ or rapid removal of SO_2 from the atmosphere. They state that in some areas, such as the Mediterranean Sea, the ocean is a sink for SO_2. Laboratory studies by Spedding (1972) and Beilke & Lamb (1974) also suggest the ocean is a major sink for SO_2.

A few laboratory studies have investigated $SO_4^=$ fractionation at the air/sea interface. With this problem in mind, Russian investigators have undertaken several studies that found positive enrichments for sulfate (Bruyevich & Kulik 1967b, Bruyevich & Korzh 1970). These authors believe that the sulfate enrichment observed in the marine aerosol and rain result from fractionation processes occurring at the air/sea interface. Their experimental design did not approximate environmental conditions, however, and evaluation of the results relative to processes occurring at the natural air/sea interface, is rather difficult. Further laboratory studies of $SO_4^{=}$ fractionation are needed.

In conclusion, although most workers concede that there is a sulfate excess in particles collected over the oceans, the excess has generally been attributed to $SO_4^{=}$ input from the background aerosol, anthropogenic sources, continental dust, volcanic emissions, and particles derived from oxidation of sulfur gases released by the ocean. The significance of sulfate fractionation at the air/sea interface has not yet been adequately evaluated because of the complicated atmospheric sulfur cycle, which hampers simple environmental studies, and because the laboratory data is meager. Isotopic studies of sulfur in atmospheric sulfate particles can distinguish sources in urban areas (Nielsen 1974). Perhaps, with further work on isotopic ratios, the significance of these various sources to the total composition of sulfur in the marine aerosol will be evaluated.

Heavy Metals

In the past few years interest in the heavy metal content of marine aerosols has increased, largely through concern that man may be affecting the heavy metal composition of atmospheric aerosols on a global basis, as well as through a general interest in the possible importance of eolian transport of trace metals to marine sedimentation. A complete discussion of heavy metals in the marine atmosphere is beyond the scope of this review, but a brief overview, considering the possible importance of the ocean relative to atmospheric trace metals, is worthwhile.

In contrast to studies of alkali and alkaline earth metals, virtually all studies of heavy metals in marine air have assumed that the metals have a continental source. Thus enrichment values such as those calculated for other chemical substances (e.g. $E_{Na}(X)$, where X = heavy metal) have not generally been determined. Comparison of the atmospheric concentration of the heavy metals with that expected on the basis of metal/Na ratios in bulk seawater results in very large enrichments (10^2 to 10^4). With crustal weathering as the main expected source for these elements, enrichments relative to average crustal material have usually been calculated, where enrichment of any element X, $E_{crust}(X)$, is defined as follows

$$E_{crust}(X) = \frac{(X/\mathrm{Al})_{atm}}{(X/\mathrm{Al})_{crust}} - 1, \tag{5}$$

where X/Al is the mass ratio of X to aluminum in the atmosphere or the Earth's crust. A representative sampling of the crustal enrichment factors, as defined in Equation 5, observed for some selected trace metals in marine aerosols collected on filters is presented in Table 3. To facilitate comparisons, all the reported enrichments in the studies listed in Table 3 have been recalculated using Al as the reference element and utilizing the average crustal abundances of Taylor (1964). Since Al was not determined in the Gulf of Guinea study (Crozat et al 1973), Sc was used as a reference element and enrichments were calculated on the basis of Al, assuming a value of $E_{crust}(Sc)$ of -0.2. (Additional valuable information on the distribution of trace metals in soil-sized particles (radius greater than $\sim 4\ \mu m$) over the world ocean is given by Chester & Stoner (1974) and Chester et al (1974) but is not included in Table 3.)

Judging from the data in Table 3, a number of the heavy metals have weathered crustal material as their source (e.g. Sc, Al, Fe, Th, Mn, Ce, Co, and perhaps Cr). Of course, mean crustal ratios can only be used to crudely approximate the relative composition of crustal material aerosols, owing to different types of crustal material and soils in various source areas and uncertainties concerning chemical fractionation processes during rock weathering. Thus variations of $E_{crust}(X)$ up to an order of magnitude may still indicate a crustal source for these elements. These and other problems in ascertaining the source of trace metals in aerosols utilizing crustal ratios have been discussed in some detail by Rahn (1975) and Duce et al (1975). However, a number of elements are highly enriched relative to both the crust and seawater: Zn, Cu, Cd, Hg, Sb, As, Pb, and Se, among others. Zoller et al (1974) pointed out that these highly enriched elements were rather volatile and suggested that high temperature processes, either natural or anthropogenic, might be responsible for the high concentrations of the elements observed. Indeed, Bertine & Goldberg (1971) and Natusch et al (1974) noted many volatile metals, including As, Sb, Cd, Pb, and Se, that are probably volatilized during coal burning, subsequently adsorbing or condensing primarily on the smallest particles. Duce et al (1975), however, suggested that the similar enrichment factors found for samples collected using identical techniques over the North Atlantic and at the South Pole, coupled with the short residence times of atmospheric particles (on the order of one week or so), long tropospheric mixing time between northern and southern hemisphere (6–12 months), and predominant source of anthropogenic particles in the northern hemisphere ($\sim 90\%$), indicates that the source of the high enrichments for these elements may be natural. Possibilities include volcanism and biological mobilization. Some of these elements may be biologically methylated in either the terrestrial or marine biosphere. Wood (1974) notes that methylated forms of As, Hg, S, and Se have been found in the marine environment. He predicted methylated forms of Hg and As in the atmosphere, and these have been detected by Johnson & Braman (1974; also unpublished data). Clearly, the importance of the biosphere, and in particular the marine biosphere, as a source for these elements in the atmosphere must be investigated in more detail.

Of particular interest to this review is the possibility that some of these elements may actually undergo chemical fractionation during atmospheric sea salt particle production at the sea surface. We have already seen that many heavy metals are concentrated in the surface microlayer, and Piotrowicz et al (1972) suggested that this should result in enrichment of these trace metals on the sea salt particles produced by bursting bubbles. These authors also point out, however, that the effects of any fractionation of trace metals during the bubble-bursting process may be overwhelmed by the presence of these metals in atmospheric particles having a nonmarine source. Relatively little information on the nature of the chemical associations of these

Table 3 Atmospheric heavy metal enrichments $[E_{crust}(X)]$ in marine areas

	Location					
Element	South Pole	Atlantic N of 30°N	Lerwick, Shetland Is. (1972)	North coast of Norway	Gulf of Guinea	Oahu, Hawaii
Sc	−0.2	−0.2	−0.1	−0.4	−0.2	—
Al	0.0	0.0	0.0	0.0	0.0	0.0
Fe	1.1	0.4	0.7	0.9	0.7	1.6
Th	−0.1	—	2.8	1.8	—	—
Mn	0.4	1.6	3.6	4.0	—	1.6
Ce	3.4	—	0.7	1.3	1.6	—
Co	3.7	1.4	2.8	5.3	3.0	—
V	0.4	16	23	23	—	14
Cr	5.9	10	14	11	5.8	—
Zn	68	110	790	240	—	—
Cu	92	120	<77	78	—	450
Cd	—	730	—	1200	—	—
Hg	—	—	<700	—	1800	—
Sb	1300	2300	3600	3600	2400	—
As	—	—	1500	1900	—	—
Pb	2500	2200	3300	800	—	2900
Se	18,000	10,000	12,000	10,000	—	—
Reference	Zoller et al (1974)	Duce et al (1975)	Peirson et al (1974)	Rahn (1975)	Crozat et al (1973)	G. Hoffman et al (1972)

heavy metals in the sea surface microlayer is available. The possibilities are numerous, ranging from physical adsorption or chemisorption of trace metals on biological and inorganic particulate matter to dissolved metallo-organic coordination complexes with such atoms as N, O, etc. There have been numerous papers published on trace metals in seawater, but, except for some recent results on particulate trace metals, much of the data are rather unreliable; it is difficult to collect uncontaminated samples and to prevent subsequent contamination during storage and analysis. Since most literature on dissolved trace metals is somewhat questionable, data on the speciation of these elements in seawater, particularly with regard to organic complexes, is even less reliable. One possible exception is several papers indicating that a significant fraction of the copper in nearshore, coastal waters may be associated with organic material (Slowey et al 1967, Alexander & Corcoran 1967, Williams 1969, Foster & Morris 1971). Suffice it to say that this will be an interesting and important, as well as difficult, area for future research, both relative to chemical fraction considerations and marine chemistry in general.

A few laboratory studies were designed to investigate certain aspects of chemical fractionation of heavy metals. Van Grieken et al (1974) added carrier-free quantities of inorganic ^{65}Zn, ^{75}Se, and ^{22}Na to unfiltered coastal seawater and collected the particles produced by bursting bubbles of air generated below the water surface. Both ^{75}Se and ^{65}Zn showed enrichment in the bulk aerosol relative to the seawater, with ^{65}Zn generally having the greater enrichment. Higher enrichments were often observed when the bubble path length before bursting was increased from 1 cm to 10 cm. ^{65}Zn enrichments were generally less than ten and ^{75}Se less than three.

Wallace & Duce (1975) investigated the transport of particulate organic carbon (POC) and particulate trace metals (PTM) to the sea surface by rising bubbles in samples of surface water from Narragansett Bay using an adsorptive bubble separation technique. Nitrogen bubbles of $\sim$1 mm diameter were allowed to rise through the sample contained in an all-glass column 70 mm in outside diameter and 2.4 m high. The froth or foam that accumulated at the surface was analyzed for particulate materials, as was the original sample and the bubble-stripped residue. Recoveries of POC in the foam ranged from 30% to 59%, whereas those of particulate Al, Mn, Fe, V, Cu, Zn, Ni, Pb, Cr, and Cd were generally greater than 50%. Wallace & Duce (1975) extrapolated their laboratory results to obtain a crude order of magnitude extimate of the bubble transport of POC and PTM to the ocean surface under open ocean conditions and suggested that this may be the primary mechanism by which the sea surface microlayer is enriched in particulate trace metals. Their study and that of Van Grieken et al (1974) illustrate the potential importance of bubble transport of certain forms of some heavy metals to the air/sea interface and suggest that further work is necessary before ruling out chemical fractionation at the air/sea interface as a cause for the apparently anomalous atmospheric concentrations of a number of these metals.

Boron

There has been relatively little interest in the atmospheric chemistry of boron, although several Japanese workers have reported $E_{Cl}(B)$ values in rain that vary from

50 to several hundred in marine areas (Sugawara 1948, Muto 1952, 1953, 1956). Attempting to explain these results, Gast & Thompson (1959) distilled seawater containing boric acid in the laboratory and found considerable boric acid–boron but no detectable chloride in their distillate. They proposed that boric acid vaporizes from the sea surface into the atmosphere and is subsequently scavenged by rain. To support this suggestion, they passed air over the surface of dilute solutions of boric acid in distilled water at 25–32°C and collected the condensate from the subsequent evaporation. The condensate contained considerable quantities of boric acid, equivalent to ~ 60 μg B/liter. Gast & Thompson point out that the evaporation of boric acid from a seawater surface would be a function of both the boric acid concentration in seawater and the water temperature and suggest that this temperature effect might have caused the environmental differences observed by the Japanese investigators. Creac'h & Point (1966) measured boric acid in the gas phase in surface air on the coast of France and found a mean concentration of 56 ± 22 μg B/m^3, which is extremely high. They believed the source of the H_3BO_3 was the ocean.

Nishimura & Tanaka (1972) repeated the evaporation experiments of Gast & Thompson (1959) using seawater containing 20–60 times as much boric acid–boron as normal seawater. Assuming that the vapor pressure of boric acid varied linearly with concentration, they extrapolated their results to seawater containing normal boron concentrations, concluding that at equilibrium the evaporation condensate should contain $\leqq 1.1$ μg B/liter at 25°C, considerably less than that found by Gast & Thompson (1959). The condensate concentration of $\leqq 1.1$ μg B/liter is less than the excess B (i.e. the quantity of boron above what is expected on the basis of the B/Na ratio in seawater) found in precipitation over the ocean, 4.7 ± 1.1 μg B/liter. Nishimura & Tanaka (1972) therefore concluded that the ocean is not a source of boron but a sink.

Both investigations considered only evaporation of boron to the atmosphere and not chemical fractionation of boron on particles produced during bubble bursting. No studies have evaluated this possible mechanism for boron enrichment. However, Gast & Thompson (1959) note that atmospheric boron has been attributed to sources other than the sea, including dust, volcanic activity, pollution, and evaporation of boron compounds from plants. Our understanding of boron in the marine atmosphere is minimal at best.

CONCLUSIONS

Evaluation of the occurrence and extent of chemical fractionation of various substances during the production of atmospheric sea salt particles by bursting bubbles is difficult indeed. Two factors emerging from our discussion of the available data are perhaps most responsible for this difficulty:

1. Atmospheric particles that are not produced by the ocean but may contribute significantly to the concentration of a number of trace substances in the marine aerosol are present throughout the atmosphere over the world ocean. These particles include the so-called background aerosol and mineral dust, among others. Care in

the selection of sampling sites and the collection of samples may greatly minimize the effects of non–sea salt aerosols, but these particles are consistently present throughout the marine atmosphere.

2. Gaseous forms of many of the trace substances of interest in the particles coexist in the marine atmosphere, and extensive gas-particle exchange is likely. This may be most important for the halogens, sulfate, nitrogen compounds, boron, and even organic carbon and some trace metals such as Se, Hg, As, etc.

Certainly, the well-documented enrichment in the ocean surface microlayer of many of the substances we have discussed suggests that they should be enriched on the particles produced when bubbles burst through this layer. However, in many cases enrichments are likely to be swamped by the high concentrations of these substances already present in the atmosphere from other sources. Based on the available data, the only substances having positive enrichments of more than a few percent in the atmosphere that can be clearly ascribed to chemical fractionation occurring during sea salt particle production are I, $PO_4^{\equiv}$, probably organic nitrogen and organic carbon, and possibly some heavy metals and K in biologically productive waters. The evidence suggests that the other substances discussed either show no enrichment in the environment, or if an enrichment is observed, it is apparently due to the presence of particles with a nonmarine source or to gas/particle interactions.

Further research in a number of areas should significantly enhance understanding of chemical fractionation processes. For example, we know virtually nothing about the chemical composition of particles less than about 0.5 μm radius over the ocean. Although many of these particles are known to have a nonmarine origin, the size of the smallest particles that can be produced by the ocean remains unknown. Investigation of the chemical composition of these very small particles is also critical in evaluating the importance of gas/particle interactions as a source for the apparent enrichment observed for some substances. Obviously simultaneous measurements of the gas and particulate phase of these substances are required in the field.

Surface-active organic material in the surface layers of the ocean plays a major role in the transport of certain trace substances from the ocean to the atmosphere. However, our knowledge of the chemical form and quantity of this surface-active material in ocean surface water is fragmentary. We know even less about the concentration and composition of organic material in the atmosphere and the chemical and physical nature of its associations with other trace substances. Further work is urgently needed in this area.

Finally, continued laboratory studies of air/sea exchange of matter in both directions are certainly necessary, and chemical fractionation experiments should be continued as well. A word of warning is crucial here, however. MacIntyre (1974b) has recently pointed out the many dangers and pitfalls that can arise in laboratory studies, particularly with regard to contamination of the model water system with surface active organic material—especially organic material with properties unlike those of surfactants actually found in the sea. MacIntyre (1974b) points out that only once in a decade has he been able to produce "a solution surface which was not detectably dirty by bubble-experiment criteria." His description of how this was accomplished only once should impress the problems of laboratory simulation of air/sea exchange processes indelibly on the minds of all investigators working in this field.

As in all areas of natural science, studies of chemical fractionation at the air/sea interface require carefully planned field and laboratory programs. Many of the problems that result from improper sampling sites, poor collection techniques, inadequate analytical procedures, and contamination in all phases of field and laboratory work are now well documented. New problems will certainly arise, but the marked increase in overall quality of studies in this field over the past five years suggests that the next five to ten years should result in some definitive answers to many of the most important questions concerning chemical fractionation at the air/sea interface.

ACKNOWLEDGMENTS

The authors have been extremely fortunate in being able to work with A. H. Woodcock, J. W. Winchester, D. C. Blanchard, F. MacIntyre, and R. Chesselet at various times over the past decade. Their insight and encouragement relative to the problems and potentials of chemical fractionation at the air/sea interface has been most stimulating. We would also like to acknowledge the Office of Naval Research (contract NONR 3748(04)) and the National Science Foundation (Grants GA 20000, GA 20010, GX 28340, GA 31918, GV 33335, GX 33777, and GA 36513) for their generous support of research in this area in our laboratory.

Literature Cited

Adams, A. P., Spendlove, J. C. 1970. Coliform aerosols emitted by sewage treatment plants. *Science* 169:1218–20

Alexander, J. E., Corcoran, E. F. 1967. The distribution of copper in tropical seawater. *Limnol. Oceanogr.* 12:236–42

Allen, S. E., Carlisle, A., White, E. J., Evans, C. C. 1968. The plant nutrient content of rainwater. *J. Ecol.* 56:497–507

Allen, S. E., Grimshaw, H. M., Holdgate, M. W. 1967. Factors affecting the availability of plant nutrients on an Antarctic island. *J. Ecol.* 55:381–94

Andren, A. W., Harriss, R. C. 1971. Anomalies in the size distribution of magnesium, calcium, and sodium in marine aerosols. *J. Appl. Meteorol.* 10:1349–50

Aubert, J. 1974. Les aerosols marins, vecteurs de microorganismes. *J. Rech. Atmos.* 8:541–54

Baier, R. E. 1972. Organic films on neutral waters: their retrieval, identification, and modes of elimination. *J. Geophys. Res.* 77:5062–75

Baier, R. E., Goupil, D. W., Perlmutter, S., King, R. 1974. Dominant chemical composition of sea surface films, natural slicks, and foams. *J. Rech. Atmos.* 8:571–600

Barger, W. R., Garrett, W. D. 1970. Surface active organic material in the marine atmosphere. *J. Geophys. Res.* 75:4561–66

Barker, D. R., Zeitlin, H. 1972. Metal-ion concentrations in sea-surface microlayer and size-separated atmospheric aerosol samples in Hawaii. *J. Geophys. Res.* 77:5076–86

Batoosingh, E., Riley, G. A., Keshwar, B. 1969. An analysis of experimental methods for producing particulate organic matter in seawater by bubbling. *Deep-Sea Res.* 16:213–19

Baylor, E. R., Sutcliffe, W. H., Hirschfeld, D. C. 1962. Adsorption of phosphates onto bubbles. *Deep-Sea Res.* 9:120–24

Beilke, S., Lamb, D. 1974. On the absorption of SO_2 in ocean water. *Tellus* 26:268–71

Bertine, K. K., Goldberg, E. D. 1971. Fossil fuel combustion and the major sedimentary cycle. *Science* 173:233–35

Bewers, J. M., Haysom, H. H. 1974. The terrigenous dust contribution to fluoride and iodide in atmospheric precipitation. *J. Rech. Atmos.* 8:689–97

Bezdek, H. F., Carlucci, A. F. 1972. Surface concentration of marine bacteria. *Limnol. Oceanogr.* 17:566–69

Bezdek, H. F., Carlucci, A. F. 1974. Concentration and removal of liquid microlayers from a seawater surface by bursting bubbles. *Limnol. Oceanogr.* 19:126–34

Bidleman, T. F., Olney, C. E. 1974. Chlorinated hydrocarbons in the Sargasso Sea atmosphere and surface water. *Science* 183:516–18

Blanchard, D. C. 1963. Electrification of the atmosphere by particles from bubbles in the sea. *Progress in Oceanography*, 1: 71–202. Oxford: Pergamon

Blanchard, D. C. 1964. Sea to air transport of surface active material. *Science* 146: 396–97

Blanchard, D. C. 1968. Surface active organic material on airborne salt particles. *Proc. Int. Conf. Cloud Physics, Toronto, Canada*, pp. 25–29

Blanchard, D. C. 1971. The oceanic production of volatile cloud nuclei. *J. Atmos. Sci.* 28:811–12

Blanchard, D. C., Syzdek, L. D. 1970. Mechanism for the water-to-air transfer and concentration of bacteria. *Science* 170:626–28

Blanchard, D. C., Syzdek, L. D. 1972. Concentration of bacteria in jet drops from bursting bubbles. *J. Geophys. Res.* 77:5087–99

Blanchard, D. C., Syzdek, L. D. 1974. Bubble tube: apparatus for determining rate of collection of bacteria by an air bubble rising in water. *Limnol. Oceanogr.* 19: 133–38

Blanchard, D. C., Woodcock, A. H. 1957. Bubble formation and modification in the sea and its meteorological significance. *Tellus* 9:145–58

Blifford, I. H., Gillette, D. A. 1972. The influence of air origin on the chemical composition and size distribution of tropospheric aerosols. *Atmos. Environ.* 6: 463–80

Bloch, M. R., Kaplan, D., Kertes, V., Schnerb, J. 1966. Ion separation in bursting air bubbles: an explanation for the irregular ion ratios in atmospheric precipitations. *Nature* 209:802–3

Bloch, M. R., Luecke, W. 1968. Uneinheitliche Verschiebungen der Ionenverhaltnisse zwischen Meereswasser und Niederschlagen durch Gischtbildung. *Naturwissenschaften* 9:441–43

Bloch, M. R., Luecke, W. 1970. The origin of fixed nitrogen in the atmosphere. *Is. J. Earth-Sci.* 19:41–49

Bloch, M. R., Luecke, W. 1972. Geochemistry of ocean water bubble spray. *J. Geophys. Res.* 77:5100–5

Boyce, S. G. 1951. Source of atmospheric salts. *Science* 113:620–21

Brosset, C., Åkerström, Å. 1972. Long distance transport of air pollutants—measurements of black air-borne particulate matter (soot) and particle-borne sulphur in Sweden during the period of Sept.–Dec. 1969. *Atmos. Environ.* 6:661–73

Bruyevich, S. V., Korzh, V. C. 1970. Main patterns of salt exchange between the ocean and the air. *Dokl. Akad. Nauk. SSSR* 190:208–12. (English version)

Bruyevich, S. V., Kulik, Y. Z. 1967a. Chemical interaction between the ocean and the atmosphere (salt exchange). *Oceanology* 7:279–93

Bruyevich, S. V., Kulik, Y. Z. 1967b. Changes of principal salt constituents as it passes into the atmosphere. *Dokl. Akad. Nauk. SSSR* 175:190–92. (English version)

Buat-Menard, P., Morelli, J., Chesselet, R. 1974. Water soluble elements in atmospheric particulate matter over tropical and equatorial Atlantic. *J. Rech. Atmos.* 8:661–73

Cadle, R. D. 1972. Composition of the stratospheric "sulfate" layer. *EOS, Trans. Am. Geophys. Union* 53:812–20

Cadle, R. D., Wartburg, A. F., Grahek, F. E. 1971. The proportion of sulfate to sulfur dioxide in Kilauea Volcano fume. *Geochim. Cosmochim. Acta* 35:503–7

Carlucci, A. F., Bezdek, H. F. 1972. On the effectiveness of a bubble for scavenging bacteria from seawater. *J. Geophys. Res.* 77:6608–10

Carlucci, A. F., Williams, P. M. 1965. Concentration of bacteria from seawater by bubble scavenging. *J. Cons. Perm. Int. Explor. Mer* 30:28–33

Carpenter, R. 1969. Factors controlling the marine geochemistry of fluorine. *Geochim. Cosmochim. Acta* 33:1153–67

Cauer, H. 1938. Einiges uber den Einfluss des Meeres auf den Chemismus der Luft. *Balneologe* 5:409–15

Chesselet, R., Morelli, J., Buat-Menard, P. 1971. Sur la distribution d'aerosols d'origine marine dans la basse troposphere. *C. R. Acad. Sci. Paris, Ser. B* 272: 1221–24

Chesselet, R., Morelli, J., Buat-Menard, P. 1972a. Some aspects of the geochemistry of marine aerosols. *Nobel Symposium 20: The Changing Chemistry of the Oceans*, 93–114. New York: Wiley

Chesselet, R., Morelli, J., Buat-Menard, P. 1972b. Variations in ionic ratios between reference sea water and marine aerosols. *J. Geophys. Res.* 77:5116–31

Chester, R., Aston, S. R., Stoner, J. H., Bruty, D. 1974. Trace metals in soil sized particles from the lower troposphere over the world ocean. *J. Rech. Atmos.* 8:777–89

Chester, R., Stoner. J. H. 1974. The contribution of Mn, Fe, Cu, Ni, Co, Ga, Cr, V, Ba, Sr, Sn, Zn, and Pb in some soil sized particulates from the lower troposphere over the world ocean. *Mar. Chem.* 2:157–88

Creac'h, P. V., Point, G. 1966. Mise en evidence, dans l'atmosphere d'acide borique gazeux provenant de l'evaporation de l'eau de mer. *C. R. Acad. Sci. Paris, Ser. B* 263:89–91

Crozat, G., Domergue, J. L., Bogui, V. 1973. Etude de l'aerosol atmospherique en Cote d'Ivoire et dans le Golfe de Guinee. *Atmos. Environ.* 7:1103–16

Cuong, N. B., Bonsang, B., Lambert, G. 1974. The atmospheric concentration of sulfur dioxide and sulfate aerosols over antarctic, subantarctic areas and oceans. *Tellus* 26:241–49

Dean, G. A. 1963. The iodine content of some New Zealand drinking waters with a note on the contribution from sea spray to the iodine in rain. *N. Z. J. Sci.* 6:208–14

Delany, A. C., Pollock, W. H., Shedlovsky, J. P. 1973. Tropospheric aerosol: the relative contribution of marine and continental components. *J. Geophys. Res.* 78: 6249–65

Delany, A. C., Shedlovsky, J. P., Pollock, W. H. 1974. Statospheric aerosol: the contribution from the troposphere. *J. Geophys. Res.* 79:5646–50

Dinger, J. E., Howell, H. B., Wojciechowski, T. A. 1970. On the source and composition of cloud nuclei in a subsident air mass over the North Atlantic. *J. Atmos. Sci.* 27:791–97

Duce, R. A., Hoffman, G. L., Zoller, W. H. 1975. Atmospheric trace metals at remote northern and southern hemisphere sites: pollution or natural. *Science* 187:59–61

Duce, R. A., Stumm, W., Prospero, J. M. 1972a. Working symposium on sea-air chemistry: summary and recommendations. *J. Geophys. Res.* 77:5059–61

Duce, R. A., Quinn, J. G., Olney, C. E., Piotrowicz, S. R., Ray, B. J., Wade, T. L. 1972b. Enrichment of heavy metals and organic compounds in the surface microlayer of Narragansett Bay, Rhode Island. *Science* 176:161–63

Duce, R. A., Winchester, J. W., Van Nahl, T. W. 1965. Iodine, bromine, and chlorine in the Hawaiian marine atmosphere. *J. Geophys. Res.* 70:1775–99

Duce, R. A., Woodcock, A. H., Moyers, J. L. 1967. Variation of ion ratios with size among particles in tropical oceanic air. *Tellus* 19:369–79

Duce, R. A., Woodcock, A. H. 1971. Difference in chemical composition of atmospheric sea salt particles produced in the surf zone and on the open sea in Hawaii. *Tellus* 23:427–35

Duce, R. A., Zoller, W. H., Moyers, J. L. 1973. Particulate and gaseous halogens in the Antarctic atmosphere. *J. Geophys. Res.* 78:7802–11

Eriksson, E. 1959. The yearly circulation of chloride and sulfur in nature: meteorological, geochemical, and pedological implications. Part I. *Tellus* 11:375–403

Eriksson, E. 1960. The yearly circulation of chloride and sulfur in nature; meteorological, geochemical, and pedological implications. Part II. *Tellus* 12:63–109

Eriksson, E. 1963. The yearly cycle of sulfur in nature. *J. Geophys. Res.* 68:4001–8

Esmen, N. A., Corn, M. 1971. Residence time of particles in urban air. *Atmos. Environ.* 5:571–78

Fasching, J. L., Courant, R. A., Duce, R. A., Piotrowicz, S. R. 1974. A new surface microlayer sampler utilizing the bubble microtome. *J. Rech. Atmos.* 8:649–52

Foster, P., Morris, A. W. 1971. The seasonal variation of dissolved ionic and organically associated copper in the Menai Straits. *Deep-Sea Res.* 18:231–36

Garrett, W. D. 1965. Collection of slick-forming materials from the sea surface. *Limnol. Oceanogr.* 10:602–5

Garrett, W. D. 1967. The organic chemical composition of the ocean surface. *Deep-Sea Res.* 14:221–27

Garrett, W. D. 1972. Impact of natural and man-made surface films on the properties of the air/sea interface. In *The Changing Chemistry of the Oceans. Nobel Symp.* 20:75–91. New York: Wiley

Gast, J. A., Thompson, T. G. 1959. Evaporation of boric acid from sea water. *Tellus* 11:344–47

Georgii, H. W. 1970. Contribution to the atmospheric sulfur budget. *J. Geophys. Res.* 75:2365–71

Glass, S. J., Matteson, M. J. 1973. Ion enrichment in aerosols dispersed from bursting bubbles in aqueous salt solutions. *Tellus* 25:272–80

Goetz, A., Klejnot, O. J. 1972. Formation and degradation of aerocolloids by ultraviolet radiation. *Environ. Sci. Technol.* 6:143–51

Grimsrud, E. P., Rasmussen, R. A. 1975. Survey and analysis of halocarbons in the atmosphere by gas chromatography—mass spectrometry. *Atmos. Environ.* 9: 1014–17

Hamilton, W. L., Langway, C. C. Jr. 1967. A correlation of microparticle concentrations with oxygen isotope ratios in 700 year old Greenland ice. *Earth Planet. Sci.*

Lett. 3:363–66
Harvey, G. W. 1966. Microlayer collection from the sea surface: a new method and initial results. *Limnol. Oceanogr.* 11:608–13
Harvey, G. W., Burzell, L. A. 1972. A simple microlayer method for small samples. *Limnol. Oceanogr.* 17:156–57
Hatcher, R. F., Parker, B. C. 1974. Laboratory comparisons of four surface microlayer samples. *Limnol. Oceanogr.* 19:162–65
Hidy, G. M., Brock, J. R. 1970. An assessment of the global sources of tropospheric aerosols. *Clean Air Congr., 2nd, Washington DC.* 36 pp.
Hitchcock, D. R., Wechsler, A. E. 1972. *Biological Cycling of Atmospheric Trace Gases. Final Report.* Washington DC: NASA
Hoffman, E. J., 1975. *Chemical fractionation at the air/sea interface: alkaline earth metals and total organic carbon.* PhD thesis. Univ. Rhode Island, Kingston.
Hoffman, E. J., Duce, R. A. 1974. The organic carbon content of marine aerosols collected on Bermuda. *J. Geophys. Res.* 79:4474–77
Hoffman, E. J., Duce, R. A. 1976. *Factors Influencing the Organic Carbon Content of Atmospheric Sea Salt Particles: A Laboratory Study.* Presented at 55th Nat. Meet. Am. Meteorol. Soc. January 20–23, Denver, Colorado.
Hoffman, E. J., Hoffman, G. L., Duce, R. A. 1974. Chemical fractionation of alkali and alkaline earth metals in atmospheric particulate matter over the North Atlantic. *J. Rech. Atmos.* 8:675–88
Hoffman, G. L., Duce, R. A. 1972. Consideration of the chemical fractionation of alkali and alkaline earth metals in the Hawaiian marine atmosphere. *J. Geophys. Res.* 77:5161–69
Hoffman, G. L., Duce, R. A., Hoffman, E. J. 1972. Trace metals in the Hawaiian marine atmosphere. *J. Geophys. Res.* 77:5322–29
Hoffman, G. L., Duce, R. A., Walsh, P. R., Hoffman, E. J., Ray, B. J., Fasching, J. L. 1974. Residence time of some particulate trace metals in the oceanic surface microlayer: significance of atmospheric deposition. *J. Rech. Atmos.* 8:745–59
Hoppel, W. A., Dinger, J. E. 1973. Production of cloud nuclei by ultraviolet radiation. *J. Atmos. Sci.* 30:331–34
Horne, R. A., Courant, R. A. 1970. The structure of seawater at the air/sea interface. Presented at Int. Symp. Hydrogeochem. Biogeochem., Tokyo
Iribarne, J. V., Mason, B. J. 1967. Electrification accompanying the bursting of bubbles in water and dilute aqueous solutions. *J. Chem. Soc. Faraday Trans.* 63:2234–45
Johnson, D. L., Braman, R. S. 1974. Distribution of atmospheric mercury species near ground. *Environ. Sci. Technol.* 8:1003–9
Junge, C. E. 1957. Chemical analysis of aerosol particles and of gas traces on the island of Hawaii. *Tellus* 9:528–37
Junge, C. E. 1960. Sulfur in the atmosphere. *J. Geophys. Res.* 65:227–37
Junge, C. E. 1963. *Air Chemistry and Radioactivity.* New York: Academic. 382 pp.
Junge, C. E. 1972. Our knowledge of the physico-chemistry of aerosols in the undisturbed marine environment. *J. Geophys. Res.* 77:5183–5200
Junge, C. E., Werby, R. T. 1958. The concentration of Cl^-, Na^+, K^+, Ca^{++}, and $SO_4^=$ in rain water over the U.S. *J. Meteorol.* 15:417–25
Kellogg, W. W., Cadle, R. D., Allen, E. R., Lazrus, A. L., Martell, E. A. 1972. The sulfur cycle. *Science* 175:587–96
Kientzler, C. F., Arons, A. B., Blanchard, D. C., Woodcock, A. H. 1954. Photographic investigation of the projection of droplets by bubbles bursting at a water surface. *Tellus* 6:1–7
Koide, M., Goldberg, E. D. 1971. Atmospheric sulfur and fossil fuel combustion. *J. Geophys. Res.* 76:6589–96
Kombayasi, M., 1962. Enrichment of inorganic ions with increasing atomic weight in aerosol, rainwater, and snow in comparison with sea water. *J. Meteorol. Soc. Jpn.* 40:25–38
Kombayasi, M. 1964. Primary fractionation of chemical components in the formation of submicron spray drops from sea salt solution. *J. Meteorol. Soc. Jpn.* 42:309–16
Koske, P. H. 1974. Surface structure of aqueous salt solutions and ion fractionation. *J. Rech. Atmos.* 8:623–37
Koyama, T., Sugawara, K. 1953. Separation of the components of atmospheric salt and their distribution. *Bull. Chem. Soc. Jpn.* 26:123–26
Lazrus, A. L., Baynton, H. W., Lodge, J. P. 1970. Trace constituents in oceanic cloud water and their origin. *Tellus* 22:106–13
Ledet, E. J., Laseter, J. L. 1974. Alkanes at the air-sea interface. *Science* 186:261–63
Livingstone, D. A. 1963. Chemical composition of rivers and lakes. *US Geol. Surv. Prof. Paper 440-G,* pp. 1–64
Lodge, J. P. Jr., MacDonald, A. J. Jr., Vihman, E. 1960. A study of the composition of marine atmospheres. *Tellus* 11:184–87
Lovelock, J. E., Maggs, R. J., Rasmussen,

R. A. 1972. Atmospheric dimethyl sulfide and the natural sulfur cycle. *Nature* 237:452–53

Lovelock, J. E., Maggs, R. J., Wade, R. J. 1973. Halogenated hydrocarbons in and over the Atlantic. *Nature* 241:194–96

Lovelock, J. E., Penkett, S. A. 1974. PAN over the Atlantic and the smell of clean linen. *Nature* 249:434

MacIntyre, F. 1968. Bubbles: a boundary layer "microtome" for micron thick samples of a liquid surface. *J. Phys. Chem.* 72:589–92

MacIntyre, F. 1970. Geochemical fractionation during mass transfer from sea to air by breaking bubbles. *Tellus* 22:451–62

MacIntyre, F. 1972. Flow patterns in breaking bubbles. *J. Geophys. Res.* 77:5211–28

MacIntyre, F. 1974a. Chemical fractionation and sea-surface microlayer processes. *The Sea, Volume 5, Marine Chemistry,* 245–99. New York: Wiley

MacIntyre, F. 1974b. Non-lipid-related possibilities for chemical fractionation in bubble film caps. *J. Rech. Atmos.* 8:515–27

MacIntyre, F., Winchester, J. W. 1969. Phosphate ion enrichment in drops from breaking bubbles. *J. Phys. Chem.* 73:2163–69

Martens, C. S. 1973. Ion ratio variations with particle size in Puerto Rican aerosols. *J. Geophys. Res.* 78:8867–71

Martens, C. S. Harriss, R. C. 1970. Mechanisms of iodine injection from the sea surface. *Precipitation Scavenging (1970). AEC Symp. Ser.* 22:319–24

Martens, C. S., Harriss, R. C. 1973. Chemistry of aerosols, cloud droplets, and rain in the Puerto Rican marine atmosphere. *J. Geophys. Res.* 78:949–57

Martens, C. S., Wesolowski, J. J., Harriss, R. C., Kaifer, R. 1973. Chlorine loss from Puerto Rican and San Francisco Bay area marine aerosols. *J. Geophys. Res.* 78:8778–92

Marty, J. C., Saliot, A. 1974. Etude chimique comparée du film de surface et de l'eau de mer sous-jacent: acide gras. *J. Rech. Atmos.* 8:563–70

Marumo, R., Taga, N., Nakai, T. 1971. Neustonic bacteria and phytoplankton in surface microlayers of the equatorial waters. *Bull. Plankton Soc. Jpn.* 18:36–41

Mason, B. J. 1957. The oceans as a source of cloud-forming nuclei. *Geofis. Pura Appl.* 36:148–55

Maynard, N. G. 1968. The significance of airborne algae. *Z. Allg. Mikrobiol.* 8:225–26

Medwin, H. 1970. In situ acoustic measurements of bubble population in coastal ocean waters. *J. Geophys. Res.* 75:599–611

Menzel, D. W., Spaeth, J. P. 1962. Occurrence of ammonia in Sargasso Sea waters and in rain water at Bermuda. *Limnol. Oceanogr.* 7:159–62

Menzel, D. W., Vaccaro, R. F. 1964. The measurement of dissolved organic and particulate organic carbon in seawater. *Limnol. Oceanogr.* 9:138–42

Miget, R., Kator, H., Oppenheimer, C., Laseter, J. L., Ledet, E. J. 1974. New sampling device for the recovery of petroleum hydrocarbons and fatty acids from aqueous surface films. *Anal. Chem.* 46:1154–57

Miyake, Y., Tsunogai, S. 1963. Evaporation of iodine from the ocean. *J. Geophys. Res.* 68:3989–93

Mohnen, V. A. 1970. Preliminary results on the formation on negative small ions in the troposphere. *J. Geophys. Res.* 75:1717–21

Morelli, J., Buat-Menard, P., Chesselet, R. 1974. Production experimentale d'aerosols a la surface de la mer. *J. Rech. Atmos.* 8:961–86

Morris, R. J. 1974. Lipid composition of surface films and zooplankton from the eastern Mediterranean. *Mar. Pollut. Bull.* 5:105–9

Moyers, J. L., Duce, R. A. 1972a. Gaseous and particulate bromine in the marine atmosphere. *J. Geophys. Res.* 77:5330–38

Moyers, J. L., Duce, R. A. 1972b. Gaseous and particulate iodine in the marine atmosphere. *J. Geophys. Res.* 77:5229–38

Murozumi, M., Chow, T. J., Patterson, C. C. 1969. Chemical concentrations of pollutant lead aerosols, terrestrial dusts, and sea salts in Greenland and Antarctic snow strata. *Geochim. Cosmochim. Acta* 33:1247–94

Muto, S. 1952. Geochemical studies of boron (III). Boron content of rain waters. *J. Chem. Soc. Jpn., Pure Chem. Sect.* 73:446–48

Muto, S. 1953. Geochemical studies of boron (VI). Comparison of chemical composition of rain and snow at Kiriu. *J. Chem. Soc. Jpn., Pure Chem. Sect.* 74:420–23

Muto, S. 1956. Geochemical studies of boron (XI). The source of boron in rain water. *J. Chem. Soc. Jpn., Pure Chem. Sect.* 77:1770–73

Natusch, D.F.S., Wallace, J. R., Evans, C. A. Jr. 1974. Toxic trace elements: preferential concentration in respirable particles. *Science* 183:202–4

Nielsen, H. 1974. Isotopic composition of the major contributors to atmospheric sulfur. *Tellus* 26:213–21

Nishimura, M., Tanaka, K. 1972. Sea water may not be a source of boron in the atmosphere. *J. Geophys. Res.* 77:5239–42

Nishizawa, S., Nakajima, K. 1971. Concentration of particulate organic material in the sea surface layer. *Bull. Plankton Soc. Jpn.* 18:12–19

Oddie, B.C.V. 1960. The variation in composition of sea salt nuclei with mode of formation. *Quart. J. Roy. Meteorol. Soc.* 86:549–51

Parker, B. C., Barsom, G. 1970. Biological and chemical significance of surface microlayers in aquatic ecosystems. *Bioscience* 20:87–93

Peirson, D. H., Cawse, P. A., Cambray, R. S. 1974. Chemical uniformity of airborne particulate material, and a maritime effect. *Nature* 251:675–79

Peterson, J. T., Junge, C. E. 1971. Sources of particulate matter in the atmosphere. *Man's Impact on the Climate* 310–20. Cambridge, Mass: MIT Press.

Petriconi, G. L., Papee, H. M. 1972. On the photolytic separation of halogens from sea water concentrates. *Water, Air Soil Pollut.* 1:117–31

Piotrowicz, S. R., Ray, B. J., Hoffman, G. L., Duce, R. A. 1972. Trace metal enrichment in the sea-surface microlayer. *J. Geophys. Res.* 77:5243–54

Pueschel, R. F., Bodhaine, B. A., Mendonca, B. G. 1973. The proportion of volatile aerosols on the island of Hawaii. *J. Appl. Meteorol.* 12:308–15

Rahn, K. A. 1975. *The Chemical Composition of the Atmospheric Aerosol. Tech. Rep.* Univ. Rhode Island, School of Oceanography Kingston, R. I. 203 pp.

Rahn, K. A., Borys, R. D., Duce, R. A. 1976. Tropospheric halogen gases: inorganic and organic components. *Science.* In press

Rasmussen, R. A. 1974. Emission of biogenic hydrogen sulfide. *Tellus* 26:254–60

Rasmussen, R. A., Went, F. W. 1965. Volatile organic material of plant origin in the atmosphere. *Proc. US Nat. Acad. Sci.* 53:216–20

Robbins, J. A. 1970. Model for variations with particle size of halogen ion ratios in marine aerosols. *Precipitation Scavenging (1970). AEC Symp. Ser.* 22:325–31

Robbins, R. C., Cadle, R. D., Eckhardt, D. L. 1959. The conversion of sodium chloride to hydrogen chloride in the atmosphere. *J. Meteorol.* 16:53–56

Robinson, E., Robbins, R. C. 1968. *Sources, Abundance, and Fate of Gaseous Atmospheric Pollutants,* pp. 11–48. New York: Am. Pet. Inst.

Robinson, E., Robbins, R. C. 1971. *Emissions, Concentration, and Fate of Particulate Atmospheric Pollutants. Am. Pet. Inst. Publ. No. 4076.* 108 pp.

Rodhe, H. 1972. A study of the sulfur budget for the atmosphere over Northern Europe. *Tellus.* 24:128–38

Rosen, J. M. 1971. The boiling point of stratospheric aerosols. *J. Appl. Meterol.* 10:1044–46

Ryther, J. H. 1969. Photosynthesis and fish production in the sea. *Science* 166:72–76

Schroeder, W. H., Urone, P. 1974. Formation of nitrosyl chloride from salt particles in air. *Environ. Sci. Technol.* 8:756–58

Seba, D. B., Corcoran, E. F. 1969. Surface slicks as concentrators of pesticides in the marine environment. *Pestic. Monit. J.* 3:190–93

Seto, Y. B., Duce, R. A., Woodcock, A. H. 1969. Sodium-to-chlorine ratio in Hawaiian rains as a function of distance inland and of elevation. *J. Geophys. Res.* 74:1101–3

Seto, Y. B., Duce, R. A. 1972. A laboratory study of iodine enrichment on atmospheric sea salt particles. *J. Geophys. Res.* 77:5339–49

Sieburth, J. M. 1971. Distribution and activity of oceanic bacteria. *Deep-Sea Res.* 18:1111–21

Slowey, J. F., Jeffrey, L. M., Hood, D. W. 1967. Evidence for organic complexed copper in seawater. *Nature* 214:377–78

Spedding, D. J. 1972. Sulfur dioxide absorption by seawater. *Atmos. Environ.* 6:583–86

Stevenson, R. E., Collier, A. 1962. Preliminary observations on the occurrence of airborne marine phytoplankton. *Lloydia* 25:89–93

Sugawara, K. 1948. Chemistry of precipitation. *Kogaku* 18:485–92

Sugawara, K. 1965. Exchange of chemical substances between air and sea. *Oceanogr. Mar. Biol. Ann. Rev.* 3:59–77

Sugawara, K., Oana, S., Koyama, T. 1949. Separation of the components of atmospheric salt and their distribution. *Bull. Chem. Soc. Jpn.* 22:47–52

Sutcliffe, W. H. Jr., Baylor, E. R., Menzel, D. W. 1963. Sea surface chemistry and Langmuir circulation. *Deep-Sea Res.* 10:233–43

Szekielda, K.-H., Kupferman, S. L., Klemas, V., Polis, D. F. 1972. Element enrichment in organic films and foam associated with aquatic frontal systems. *J. Geophys. Res.* 77:5278–82

Taylor, S. R. 1964. Abundance of chemical elements in the continental crust: a new table. *Geochim. Cosmochim. Acta* 28:1273–86

Tsiban, A. V. 1971. Marine bacterioneuston. *J. Oceanogr. Soc. Jpn.* 27:56–66

Tsunogai, S. 1971. Ammonia in the oceanic atmosphere and the cycle of nitrogen compounds through the atmosphere and the hydrosphere. *Geochem. J.* 5:57–67

Tsunogai, S., Saito, O., Yamada, K., Nakaya, S. 1972. Chemical composition of oceanic aerosol. *J. Geophys. Res.* 77:5283–92

Twomey, S. 1971. The composition of cloud nuclei. *J. Atmos. Sci.* 28:377–81

Valach, R. 1967. The origin of the gaseous form of natural atmospheric chlorine. *Tellus* 19:509–16

Van Grieken, R. E., Johansson, T. B., Winchester, J. W. 1974. Trace metal fractionation effects between sea water and aerosols from bubble bursting. *J. Rech. Atmos.* 8:611–21

Vinogradov, A. P. 1959. *The Geochemistry of Rare and Dispersed Chemical Elements in Soils.* New York: Consultants Bureau. Inc. 209 pp. 2nd ed.

Vohra, K. G., Vasudevan, K. N., Nair, P. V. N. 1970. Mechanisms of nucleus-forming reactions in the atmosphere. *J. Geophys. Res.* 75:2951–60

Wada, S., Kokubu, N. 1973. Chemical composition of maritime aerosols. *Geochem. J.* 6:131–39

Wade, T. L., Quinn, J. G. 1975. Hydrocarbons in the Sargasso Sea surface microlayer. *Mar. Pollut. Bull.* 6:54–57

Wallace, G. T. Jr., Duce, R. A. 1975. Concentration of particulate trace metals and particulate organic carbon in marine surface waters by a bubble flotation mechanism. *Mar. Chem.* 3:157–81

Wallace, G. T. Jr., Loeb, G. I., Wilson, D. F. 1972. On the flotation of particulates in sea water by rising bubbles. *J. Geophys. Res.* 77:5293–5301

Weiss, H. V., Bertine, K., Koide, M., Goldberg, E. D. 1975. The chemical composition of a Greenland glacier. *Geochim. Cosmochim. Acta* 39:1–10

Wilkniss, P., Bressan, D. J. 1971. Chemical processes at the air-sea interface: the behavior of fluorine. *J. Geophys. Res.* 76:736–41

Wilkniss, P., Bressan, D. J. 1972. Fractionation of the elements F, Cl, Na, and K at the sea-air interface. *J. Geophys. Res.* 77:5307–15

Williams, P. M. 1967. Sea surface chemistry: organic carbon and organic and inorganic nitrogen and phosphorus in surface film and subsurface waters. *Deep-Sea Res.* 14:791–800

Williams, P. M. 1969. The association of copper with dissolved organic matter in seawater. *Limnol. Oceanogr.* 14:156–58

Wilson, A. T. 1959. Surface of the ocean as a source of air-borne nitrogenous material and other plant nutrients. *Nature* 184:99–101

Winkler, P. 1975. Chemical analysis of Aitken particles (<0.2 μm radius) over the Atlantic Ocean. *Geophys. Res. Lett.* 2:45–48

Wood, J. M. 1974. Biological cycles for toxic elements in the environment. *Science* 183:1049–52

Woodcock, A. H. 1948. Note concerning human respiratory irritation associated with high concentrations of plankton and mass mortality of marine organisms. *J. Mar. Res.* 7:56–61

Woodcock, A. H. 1953. Salt nuclei in marine air as a function of altitude and wind force. *J. Meteorol.* 10:363–71

Woodcock, A. H. 1972. Smaller salt particles in oceanic air and bubble behavior in the sea. *J. Geophys. Res.* 77:5316–21

Zafiriou, O. C. 1974. Photochemistry of halogens in the marine atmosphere. *J. Geophys. Res.* 79:2730–32

Zafiriou, O. C. 1975. Reaction of methyl halides with seawater and marine aerosols. *J. Mar. Res.* 33:75–81

Zobell, C. E., Mathews, H. M. 1936. A qualitative study of the bacterial flora of sea and land breezes. *Proc. Nat. Acad. Sci.* 22:567–72

Zoller, W. H., Gladney, E. S., Duce, R. A. 1974. Atmospheric concentrations and sources of trace metals at the South Pole. *Science* 183:198–200

PACIFIC DEEP-SEA MANGANESE NODULES: THEIR DISTRIBUTION, COMPOSITION, AND ORIGIN

Stanley V. Margolis
Department of Oceanography, Hawaii Institute of Geophysics, University of Hawaii, Honolulu, Hawaii 96822

Roger G. Burns
Department of Earth and Planetary Sciences, Massachusetts Institute of Technology, Cambridge, Massachusetts 02139

INTRODUCTION

Since their first recovery from the ocean floor in the 1870s during the cruises of the H.M.S. *Challenger* (Murray & Renard 1891), manganese nodules have been found to cover vast portions of the bottoms of the Pacific, Atlantic, and Indian oceans. Despite an enormous amount of research during the ensuing 100 years, the genesis of these deposits remains an enigma. For decades deep-sea manganese nodules were only a scientific curiosity, but economic and political attention has recently been focused on them. In the northeastern equatorial Pacific (Figure 1) they have been shown to exhibit copper and nickel concentrations of around 3 wt% and seafloor surface coverages with potential for mineral resource exploitation (Hammond 1974a,b).

In response to this economic impetus, the international scientific community has launched major research programs during the last three years. Many of the technological leaders, including the United States, Germany, Japan, France, New Zealand, and the USSR, have conducted one or more major scientific cruises to the central Pacific to map the distribution of nodules in relation to seafloor geological features and to recover nodule samples for analysis. In the United States, the principal scientific effort on manganese nodules has been the work of the Manganese Nodule Project, funded by the Seabed Assessment Program of the International Decade of Oceanographic Exploration Division (IDOE) of the National Science Foundation (NSF). This project focuses on ferromanganese deposits that are or may be considered potential mineral resources, that is, that contain sufficiently high concentrations of copper, nickel, and/or cobalt to be of ore grade at present market values. Rather than addressing itself to the broad question of how manganese nodules form in all regions of the ocean, the Project is concerned with an identification of processes that could result in high concentrations of copper, nickel, and/or cobalt in nodules in specific regions. It is obvious, however, that this goal cannot be achieved without making assumptions on the mode of accretion.

Another factor in evaluating manganese nodules as a potential source of copper and nickel is the variability of the deposits, which must be known before any mining operation is conceived. To answer these questions, the Manganese Nodule Project has gathered together scientists in many relevant disciplines from ten universities. From the project's inception, industry has cooperated in defining scientific questions of economic interest, and other countries have participated in joint cruises and conferences.

As participants in this program, the authors, who specialize in marine geology and sedimentology (Margolis) and geochemistry and mineralogy (Burns), present their view of advances in the field of manganese nodule research that have resulted from the increased interest in these seafloor deposits during the last three years. Because of the spectrum of fields encompassed, this paper is necessarily restricted to those topics of which the authors have first hand knowledge.

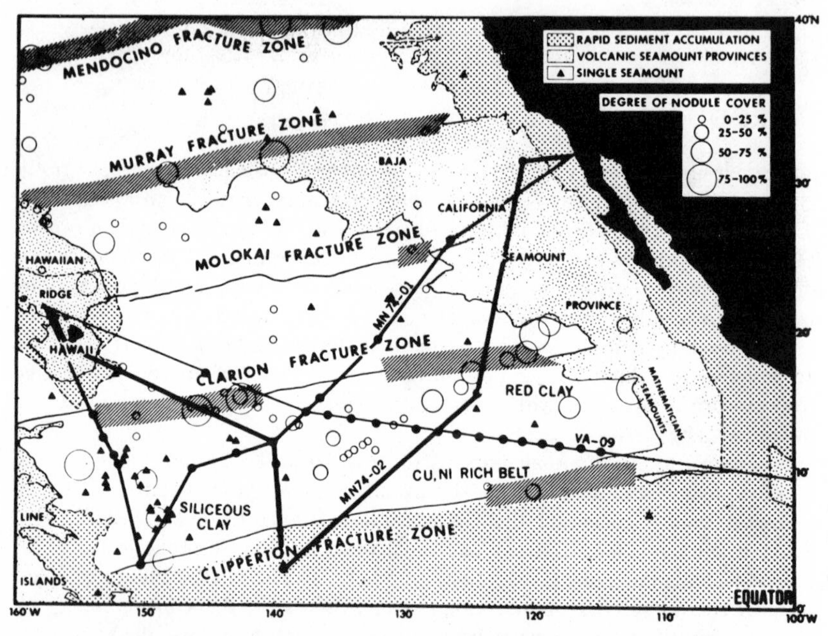

Figure 1 Map of northeastern equatorial Pacific showing regional seafloor structure, sediment type, cruise tracks, and degree of nodule cover. The copper-nickel-rich belt is bounded by the Mathematician Seamounts and the Line Islands on the east and west and by the Clarion and Clipperton fracture zones on the north and south. The water depths in this belt range between 5800 and 4000 meters. Map is modified from Horn et al (1973).

DISTRIBUTION OF MANGANESE NODULES IN THE PACIFIC OCEAN

From the early works of Murray & Renard (1891) and Agassiz (1901, 1906) it became apparent that most portions of the Pacific Ocean floor (except for trenches, continental margins, and extreme polar regions) contained manganese deposits in the form of either nodules or crusts. Exceptionally high concentrations of nodules were noted in the eastern Pacific during the *Albatross* Expedition (Agassiz 1901, 1906), where almost every dredge contained manganese nodules. During the 1950s and 1960s our knowledge of the distribution of manganese nodules in the Pacific expanded with the work of Dietz (1955) in the North Pacific, Menard & Shipek (1958) in the Central and South Pacific, and Zenkevitch & Skornyakova (1961) in this same region.

Based on then-available data, Mero (1965) divided the Pacific Ocean floor into three nodule regions, defined by concentrations of surficial manganese nodules as indicated by sampling and bottom photography. According to Mero, the eastern region averages about 0.78 g of manganese nodules per square centimeter of sea-floor, the central region about 1.45 g/cm^2, and the western region about 0.86 g/cm^2. Differences in nodule concentration seem to be related to sedimentation rates, with nodules concentrated in areas of slowly accumulating (1–3.5 mm per 10^3 yr) brown clays and radiolarian oozes, far removed from sources of terrigenous sediments. The bulk of the current interest has been focused in the Central Pacific. Figure 1 shows the tracks of cruises by the Manganese Nodule Project and a joint cruise with the German R/V *Valdivia* nodule research group, which provide the data base for our discussion of the factors influencing nodule distribution in the Central Pacific.

CHEMICAL COMPOSITION OF MANGANESE NODULES FROM THE PACIFIC

Manganese nodules are heterogeneous mixtures of very fine-grained iron and manganese oxides, detrital mineral grains, and biogenic components deposited in semiconcentric layers around a central nucleus of variable composition. These variations are believed to reflect changes in the environment of formation. It is not surprising, therefore, that there are systematic differences between the average composition of manganese nodules from the Atlantic and Pacific oceans (Table 1). Likewise, there are chemical differences in nodules from different regions within the Pacific (Table 2), but some general trends in nodule chemistry can be related to geological setting (Figure 1). Manganese appears most abundant in areas of slow pelagic sedimentation below the calcium carbonate compensation depth, like those found in the northeast Equatorial, and southeastern Pacific. Lowest manganese concentrations occur in nodules from the western and northwestern Pacific (Cronan 1972, Mero 1965). There are many exceptions to these trends, however, because of topographic, volcanic, and structural influences. Most published chemical data on manganese nodules indicate a reciprocal relationship between iron and manganese.

Manganese-iron ratios are therefore often used as a measure of chemical variability.

Regional studies of the bulk chemistry of manganese nodules (e.g. Cronan 1972, Horn et al 1973a,b, Glasby & Lawrence 1974. Skornyakova & Andrushchenko 1974, Friedrich et al 1974) reaffirm that marine nodules are enriched in many metals, including Mn, Fe, Co, Ni, Cu, Zn, Pb, Mo, and Ti. Correlation coefficients computed from recent chemical analyses (e.g. Ostwald & Frazer 1973, Friedrich et al 1974, Cronan 1975) corroborate the pronounced interelement correlations previously established between manganese and Ni, Cu, Co, Zn, Ba, Mo, and Sr, and between Fe and Ti.

Biogenic sediment grains often are incorporated in the nodule matrix during periods of low metal oxide accretion or of higher pelagic sedimentation. Variable amounts of biogenic silica, calcium carbonate, and phosphate may be found in manganese nodules and dilute the metal oxide phases. Since bulk nodule analyses frequently include the nuclei, variations in nucleus type, (e.g. shark's teeth, volcanic rock fragments, lithified sediment, nodule fragments, etc) can also be a source of chemical variability between regions, as different varieties of seeds are frequently found.

Table 1 Maximum, minimum, and average weight percentages of principal elements contained in manganese nodules from the Pacific and Atlantic Oceans (From Mero 1965)

	Weight percentages (dry weight basis)[a]					
Element	Pacific Ocean statistics on 54 samples			Atlantic Ocean statistics on 4 samples		
	maximum	minimum	average	maximum	minimum	average
Na	4.7	1.5	2.6	3.5	1.4	2.3
Mg	2.4	1.0	1.7	2.4	1.4	1.7
Al	6.9	0.8	2.9	5.8	1.4	3.1
Si	20.1	1.3	9.4	19.6	2.8	11.0
K	3.1	0.3	0.8	0.8	0.6	0.7
Ca	4.4	0.8	1.9	3.4	1.5	2.7
Ti	1.7	0.11	0.67	1.3	0.3	0.8
Mn	41.1	8.2	24.2	21.5	12.0	16.3
Fe	26.6	2.4	14.0	25.9	9.1	17.5
Co	2.3	0.014	0.35	0.68	0.06	0.31
Ni	2.0	0.16	0.99	0.54	0.31	0.42
Cu	1.6	0.028	0.53	0.41	0.05	0.20
Zn	0.08	0.04	0.047	—	—	—
Ba	0.64	0.08	0.18	0.36	0.10	0.17
L.O.I.[b]	39.0	15.5	25.8	30.0	17.5	23.8

[a]As determined by X-ray emission spectrography.
[b]L.O.I. = Loss on ignition at 1100°F for one hour. The L.O.I. figures are based on a total weight of air-dried sample basis.

Table 2 Average composition of surface nodules from different areas within the Pacific Ocean

	a	b	c	d	e	f	g	h	i
Mn	15.85	33.98	22.33	19.81	15.71	16.61	16.87	13.96	12.29
Fe	12.22	1.62	9.44	10.20	9.06	13.92	13.30	13.10	12.00
Ni	0.348	0.097	1.080	0.961	0.956	0.433	0.564	0.393	0.422
Co	0.514	0.0075	0.192	0.164	0.213	0.595	0.395	1.127	0.144
Cu	0.077	0.065	0.627	0.311	0.711	0.185	0.393	0.061	0.294
Ba	0.306	0.171	0.381	0.145	0.155	0.230	0.152	0.274	0.196
Ti	0.489	0.060	0.425	0.467	0.561	1.007	0.810	0.773	0.634
L.O.I.[j]	24.78	21.96	24.75	27.21	22.12	28.73	25.50	30.87	22.52
Depth (m)	1146	3003	4537	4324	5049	3539	5001	1757	4990

[a]Southern borderland seamount province
[b]Continental borderland off Baja California
[c]Northeast Pacific
[d]Southeast Pacific
[e]Central Pacific
[f]South Pacific
[g]West Pacific
[h]Mid-Pacific mountains
[i]North Pacific
[j]See footnote *b* of Table 1

Data from Cronan (1972) in weight percent, air dried weight. Geographic regions are shown in Figure 1.

Mero (1965) recognized chemical provinces within the Pacific based on nodule composition. The equatorial northeastern Pacific nodules, for instance, show combined copper-nickel values that total 3% and more, and are therefore economically significant. Although bathymetric, nodule distribution, and sediment maps with corresponding chemical composition for nodules have been prepared for this region (Horn et al 1973a,b, Fraser & Arrhenius 1972), most such maps are generalizations and are not sufficiently detailed to show relationships between seafloor microtopography, sediment type and age, and nodule distribution and chemistry. The following section discusses what we have learned about small-scale variability in nodule composition within the zone of copper-nickel enrichment during cruises of the IDOE Manganese Nodule Project.

GEOLOGICAL INFLUENCES ON MANGANESE NODULE DISTRIBUTION

The bathymetry of the northeastern equatorial Pacific (Figure 1) is dominated by abyssal hills that typically show a few hundred meters relief. The structural trend of the region between the Clipperton and Clarion fracture zones seems to parallel the East Pacific Rise (Andrews et al 1974). These abyssal hills were initially formed by repeated volcanic extrusions, which produced an irregular, basement structure that influenced the shape of crests and slopes (Luyendyk 1969). A rugged bottom topography is evident in areas only thinly covered by red clay. This rugged bathymetry is often overlain by a variable thickness of siliceous clay and ooze, which smooths the originally rough basement structure. Near bottom surveys in this region have reported slopes of between 6° and 30° (Mudie et al 1972). The variable sediment cover indicates movement of sediment from topographic highs to valleys by either slumping or winnowing. Luyendyk (1970) states that the valleys between abyssal hills in this region show a sediment thickness twice that of adjacent hill

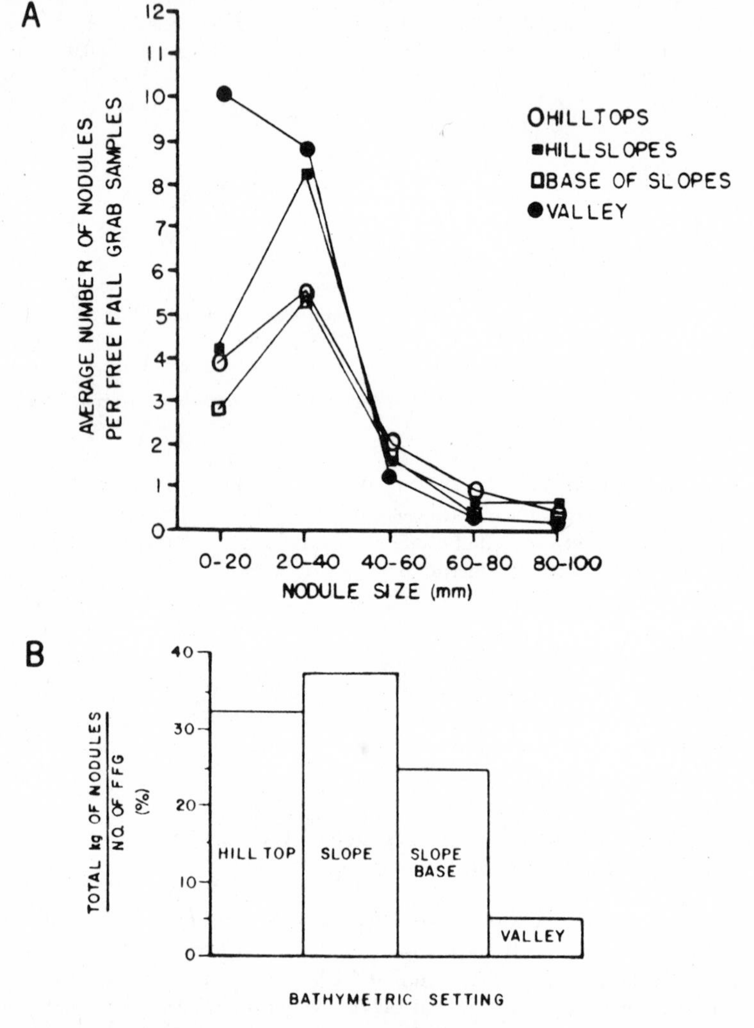

Figure 2 *A:* Nodule size vs bathymetric setting. Data is based upon observations made during the three cruises shown in Figure 1. (From J. D. Craig, in preparation). *B:* Weight percent nodules from different bathymetric settings. (From J. D. Craig, in preparation).

crests, and others have noted Tertiary outcrops on hill crests and mixed Tertiary and Quaternary faunas in valleys. The regional sedimentary patterns between the Clarion and Clipperton fracture zones were primarily produced by changes in the calcium carbonate compensation depth and by planktonic productivity associated with the westward spreading and sinking of the ocean crust and northward migration of the Pacific Plate during the last 45 million years (Van Andel & Heath 1973).

Observations during reflection profiling and sampling of sediment and nodules in this region indicate a tremendous variability in nodule size, morphology, coverage density, and sediment type over short distances. This variability has been noted by other workers (Horn et al 1973a,b, Meyer 1973), but most maps do not show sufficient detail to offer an explanation.

A discussion of our observations from detailed surveys of the distribution of nodules with respect to bathymetry follows. Abyssal hill topography can be categorized by changes in slope, i.e. hill tops, hill slopes, bases of slopes, and valleys. Figure 2 shows the relationship of nodule size and number to bathymetric setting. Note the higher number of smaller nodules occurring in all environments, and that about three times as many larger nodules occur on elevated areas. Other data indicate that hill slopes show about seven times the nodule coverage of valleys and that valleys contain a much higher percentage of nodule fragments.

Although normal pelagic sedimentation would uniformly cover these low-lying abyssal hills, the low shear strengths (Kogler 1975) and high water contents (Horn et al 1973a,b) indicate that they were redistributed by bottom currents or slow downslope movements. This would produce a higher apparent sedimentation rate in valleys than on slopes. Bottom currents have resulted in considerably steeper average slopes and generally thinner stratigraphic sections on the south- and southwest-facing sides of abyssal hills in this region. Antarctic bottom water is believed to be flowing in from these directions.

Nodule distribution is strongly influenced by small-scale variations in sediment accumulation rates, since intermittent burial by slumping and uncovering by erosion will affect the nodules position relative to the sediment-water interface, where growth conditions are optimal. Topographic highs would be the least affected by such processes, and this could explain the greater number, weight, and size of nodules in such areas.

Although variations in nodule distribution can be explained by local differences in bathymetric setting and sedimentation, this does not seem to be reflected as variations in nodule chemistry. Nodules recovered from areas of red clay show the same enrichment in Cu and Ni as those from the belt of siliceous ooze deposition, (discussed in the next section).

REGIONAL CHEMICAL VARIATIONS IN MANGANESE NODULES

Sampling of manganese nodules from the region of copper-nickel enrichment was undertaken during three cruises, shown in Figure 1. Stations were selected to cover all the different structural settings, sediment types, and water depths within this region. Free-fall grabs obtained detailed nodule samples from a seafloor area of 0.08 m^2. Figure 3 shows a typical nodule assortment obtained by a free-fall grab and a bottom photograph from the same region. During detailed mapping and surveying the sampling site is presumed to be directly beneath the drop site. Grabs were normally launched successively along a survey track at intervals of 1–2 nautical miles, and the precise bathymetric positionings (i.e. abyssal hill top, slope, etc) were noted.

Table 3 summarizes the results of bulk chemical analyses from cruises Mn-74-01, Mn-74-02, and VA-09, for Mn, Fe, Cu, and Ni. Most noteworthy are the similar regional values of Mn, Cu, and Ni for all three sets of samples, although Fe and therefore Mn/Fe values are somewhat more variable. Table 4 shows the measured standard deviations of nodules analyzed from individual free-fall grab samples from cruise Mn-74-02. Results demonstrate a relatively consistent metal composition over the wide area of this survey. Individual nodules varied little in any given free-fall grab sample, between different free-fall grab samples at the same station, or between samples from widely separated stations.

Cu + Ni content has often been observed to increase as Mn/Fe increases over the entire range of nodule composition. This general correlation holds true for nodules analyzed from Mn-74-02 (Figure 4). However, within the compositional range of nodules analyzed, as the Mn/Fe increases, the Ni/Cu ratio decreased (Figure 5). The question is whether the Cu enrichment prevails because of higher percentages of Mn or lower values of Fe. No significant increase of Ni with increasing Mn

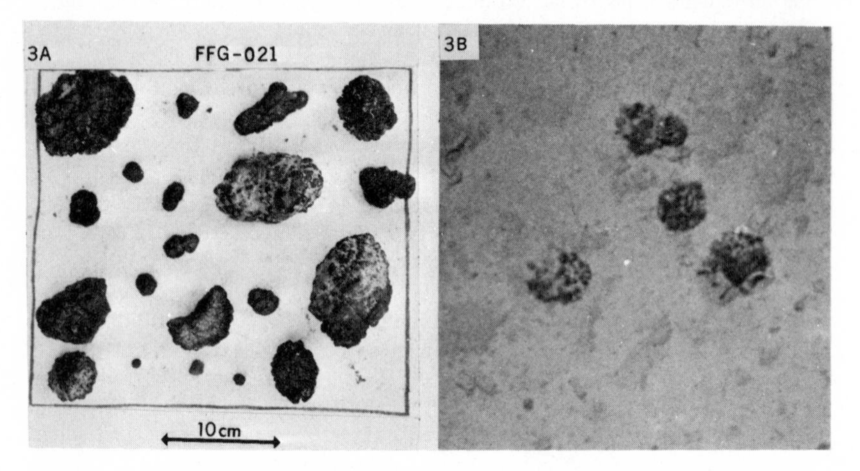

Figure 3 *A:* Nodules collected in free-fall grab, collected during cruise Mn-74-02 from long. 140°W, lat. 09°N, from a depth of 4800 meters. Note the variations in nodule size and shape. Irregular nodule in lower center is a fragment. Note also the sediment adhering to surface of nodule. *B:* Bottom photograph taken at approximately the same position as *A*, showing the in situ position of nodules on the seafloor. Note how they are partially buried in the sediment. Many of the smaller nodules seen in *A* are probably covered up by sediment in *B*. Scale is the same as *A*.

content was found (Figure 6). On the other hand, a slight general increase of Cu, especially at higher values of Mn, was apparent. More significant was the relationship of Ni and Cu with Fe (Figure 7). Ni remains relatively constant over the whole Fe range covered by the analyses. However, Cu demonstrates a noticeable increase with decreasing Fe content. So far, the apparent chemical trends suggest that the lack of Fe rather than the abundance of Mn may control the minor element enrichment, especially of Cu, in manganese nodules from this region of the Pacific.

Variability in nodule chemistry with respect to these four metals is surprisingly small considering the broad region covered. The higher standard deviation in Fe values suggests a possibly variable input of this element, related perhaps to localized

Table 3 Chemical variability of metals in manganese nodules from the zone of Cu and Ni enrichment

Regional average for entire cruise[a]	Cruise		Mn	Fe	Mn/Fe	Cu	Ni	Cu+Ni
	VA-09	X	29.85	6.09	5.02	1.24	1.45	2.69
	(20)	s	1.34	0.73	0.71	0.14	0.12	0.18
	Mn-74-01	X	29.25	5.52	5.26	1.29	1.52	2.79
	(20)	s	1.88	1.17	0.95	0.15	0.20	0.27
	Mn-74-02	X	28.69	8.08	3.62	1.04	1.71	2.74
	(58)	s	2.00	0.97	0.67	0.13	0.13	0.22
Local averages for individual stations[a]								
Sediment type	VA-09							
Siliceous ooze	Loc. 2	X	28.48	6.00	4.80	1.19	1.41	2.59
	(9)	s	1.86	0.58	0.69	0.12	0.03	0.14
Siliceous ooze	Loc. 4	X	26.81	6.64	4.09	1.04	1.43	2.47
	(7)	s	3.11	0.60	0.73	0.10	0.10	0.20
Siliceous clay	Loc. 11	X	28.78	6.48	4.47	1.22	1.41	2.59
	(9)	s	2.47	0.92	0.87	0.19	0.13	0.28
Red clay	Loc. 17	X	31.66	6.39	4.98	1.23	1.46	2.69
	(9)	s	1.19	0.44	0.43	0.04	0.06	0.09
	Mn-74-01							
Siliceous ooze	Sta. 6	X	28.70	5.33	5.71	1.22	1.62	2.84
	(10)	s	2.75	1.25	1.66	0.25	0.20	0.44
	Mn-74-02							
Siliceous ooze	Sta. 15	X	28.62	7.45	3.93	1.09	1.73	2.82
	(10)	s	2.86	0.90	0.85	0.12	0.11	0.17

[a]( ) = number of sample analyses, X = mean of the sample, and s = standard deviation.

Table 4 Bulk chemical analyses of four metals for nodules collected during Cruise Mn-74-02 showing arithmetic means and standard deviations for individual samples, for station areas, and for all samples. Station 13 A, B, C are from red clay region and Stations 15 and 16 are from siliceous ooze region

Sample[a]		Mn	Fe	Ni	Cu	Mn/Fe	Ni+Cu
FFG-002 (4)	*X*	29.23	7.78	1.73	0.95	3.79	2.68
	s	1.90	0.74	0.26	0.14	0.55	0.31
FFG-004 (5)	*X*	28.09	8.19	1.80	1.03	3.44	2.83
	s	0.68	0.43	0.03	0.04	0.24	0.07
STA. 13A (9)	*X*	28.60	8.01	1.77	0.99	3.59	2.76
	s	1.40	0.59	0.17	0.10	0.42	0.21
FFG-005 (4)	*X*	29.47	8.52	1.73	1.01	3.46	2.74
	s	0.96	0.07	0.06	0.05	0.14	0.11
FFG-008 (6)	*X*	30.06	7.63	1.66	1.13	3.96	2.78
	s	1.27	0.48	0.04	0.03	0.43	0.06
STA. 13B (11)	*X*	29.27	8.09	1.69	1.07	3.66	2.75
	s	2.13	0.65	0.06	0.08	0.53	0.09
FFG-009 (6)	*X*	28.32	8.94	1.61	0.89	3.19	2.50
	s	1.12	0.72	0.11	0.10	0.36	0.21
FFG-011 (5)	*X*	27.53	8.65	1.61	0.95	3.20	2.56
	s	2.62	0.51	0.17	0.13	0.46	0.28
FFG-012 (7)	*X*	27.98	8.77	1.78	1.03	3.24	2.80
	s	1.58	0.90	0.13	0.14	0.55	0.25
DREDGE-1	*X*	29.02	8.41	1.68	1.07	3.45	2.75
	s	1.03	0.35	0.04	0.03	0.04	0.04
ALL DREDGES	*X*	29.24	8.55	1.66	1.07	3.42	2.72
	s	0.86	0.47	0.04	0.02	0.11	0.05
STA. 13-C (23)	*X*	27.97	8.80	1.68	0.96	3.21	2.63
	s	1.72	0.71	0.15	0.13	0.44	0.27
FFG-021 (4)	*X*	30.37	7.20	1.67	1.20	4.29	2.87
	s	2.14	0.95	0.07	0.06	0.76	0.11
FFG-023 (4)	*X*	26.89	8.02	1.76	1.02	3.36	2.78
	s	1.35	0.50	0.14	0.04	0.19	0.18
STA. 15 (10)	*X*	28.62	7.45	1.73	1.09	3.93	2.82
	s	2.86	0.90	0.11	0.12	0.85	0.17
FFG-030 (3)	*X*	30.59	6.20	1.82	1.25	5.00	3.07
	s	1.01	0.79	0.14	0.10	0.77	0.24
STA. 16 (5)	*X*	29.81	6.46	1.79	1.18	4.67	2.96
	s	2.12	0.66	0.11	0.16	0.73	0.25
TOTAL (58)	*X*	28.69	8.08	1.71	1.04	3.62	2.74
	s	2.00	0.97	0.13	0.13	0.67	0.22

[a]( ) = number of nodules analyzed, X = mean of the sample, and s = standard deviation.

volcanic sources. The relative uniformity in Cu + Ni values over a broad range of Mn/Fe values indicates that perhaps this enrichment is caused by some large-scale regional oceanic process rather than any localized geological phenomenon influenced by sediment type or bathymetry, both of which can vary greatly over short distances. A relationship between nodule chemistry and morphology has been noted but the reasons for this are unclear. Meyer (1973) noted that larger discoidal

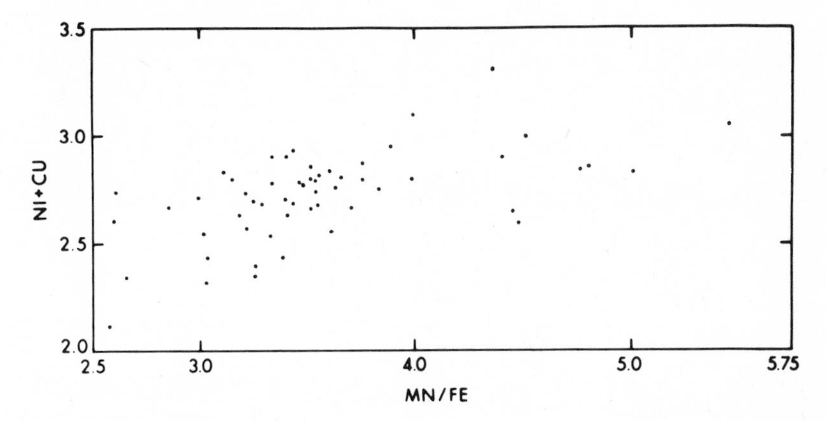

Figure 4 Plot of Mn/Fe ratio vs combined weight percent Cu and Ni, based on bulk analyses performed on nodules collected on cruises Mn-74-01 and Mn-74-02 (Figure 1). Figure from S. V. Margolis and C. Bowser (in preparation).

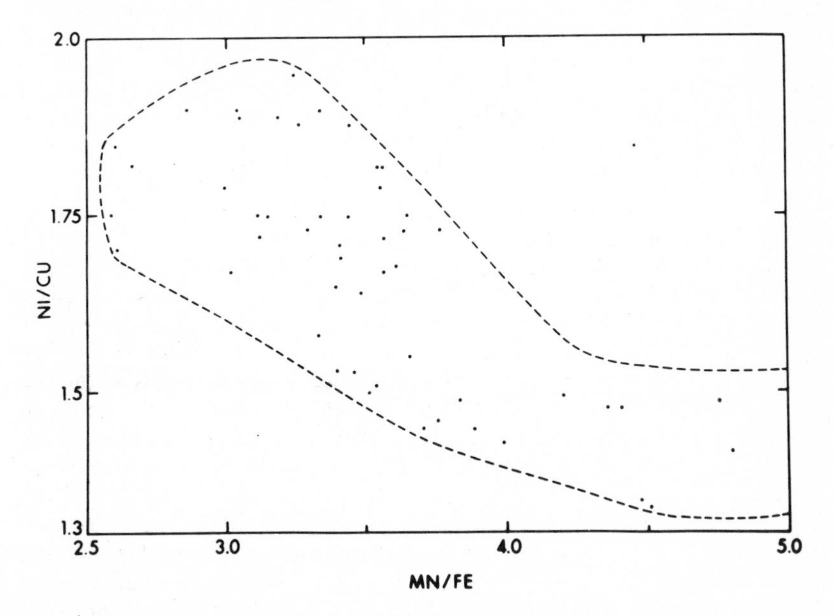

Figure 5 Plot of Mn/Fe vs Ni/Cu based on same data as Figure 4.

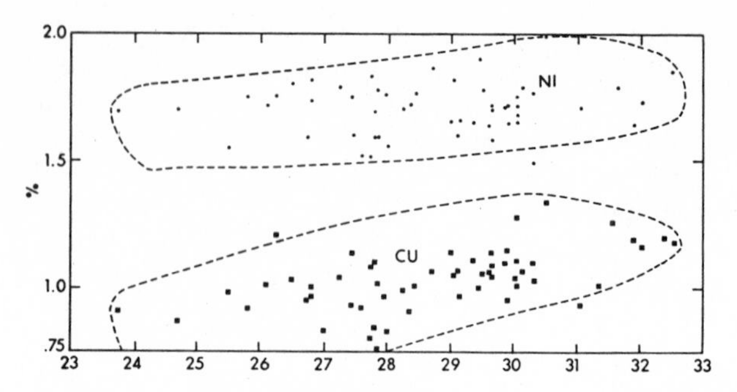

Figure 6 Plot of weight percent Ni and weight percent Cu, vs weight percent Mn.

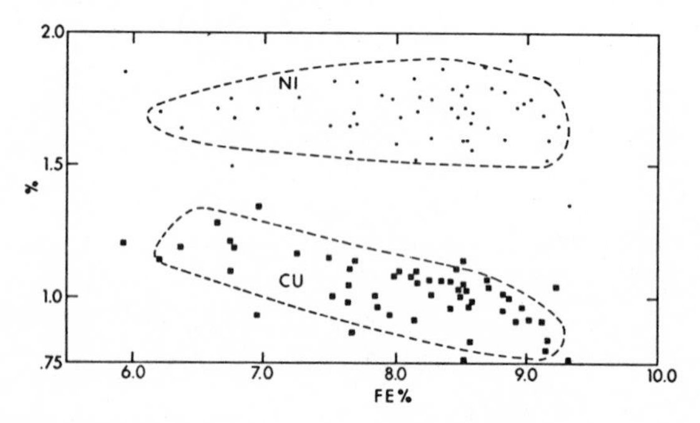

Figure 7 Weight percent Ni and Cu plotted vs Fe. Note decrease in Cu with increasing Fe.

nodules are higher in Mn, Mn/Fe, and Cu + Ni and lower in Fe content than smaller more irregular nodules both locally and regionally. This problem is discussed more fully in the section on nodule microchemistry.

MORPHOLOGY AND STRUCTURE OF MANGANESE NODULES

Since the initial description of nodules recovered by the *Challenger* expedition (Murray & Renard 1891), no one has proposed a widely accepted terminology or classification system to deal with the variety of nodules recovered from the deep sea. Recently, Meyer (1973) presented a classification scheme based on nodule recovery operations of the R/V *Valdivia* in the zone of copper-nickel enrichment. The nodule types listed by Meyer are distinguished by size, primary shape, surface texture, intergrowth frequency, and internal structure. Many nodules, however, show transitional features and do not fit into any one category.

Another classification of nodules allowed for the apparent morphological variability (Meylan 1974). The rationale was the need for a brief, unambiguous, and yet informative nomenclature for field description of manganese nodules. Nodules are classified by size, primary morphology, and surface texture, with no attempt at internal description or inferred genesis (Table 5). Frequently nodules are difficult to separate into distinct categories, in which case they may be jointly classified.

A summary table of nodules recovered by Nodule Project cruises (Mn-74-01, Mn-74-02) shows the regional occurrence of nodule types (Table 6). The frequency

Table 5 Field classification of Mn nodules[a]

Prefix	Primary Morphology	Suffix
s = small = < 3 cm nodule	[S] = Spheroidal	*s* = smooth (smooth or microgranular)
	[E] = Ellipsoidal	
m = medium = 3–6 cm size	[D] = Discoidal (or tabular-discoidal form)	*r* = rough (granular or microbotryoidal) surface texture
l = large = > 6 cm maximum diameter	[P] = "Poly" (coali-spheroidal or botryoidal form)	*b* = botryoidal
	[B] = Biological (shape determined by tooth, vertebra, or bone nucleus)	
	[T] = Tabular	
	[F] = Faceted (polygonal form due to angular nucleus or fracturing)	

[a] Examples: *l*[D]*b* = large discoidal nodule with botryoidal surface; *m-l*[E]^{s}*r-b* = medium-to-large ellipsoidal nodules with smooth tops, rough-to-botryoidal bottoms.

Table 6 Relationship between nodule morphological type and sediment type. Data based upon cruises illustrated in Figure 1 (from J. D. Craig, in preparation)

Siliceous region		Red clay region		Nodule type
Number of nodules	%	Number of nodules	%	
262	37.5	76	20.9	D
144	20.6	42	11.6	E
129	18.6	33	9.1	S
131	18.8	27	7.4	P
17	2.4	178	49.0	F
8	1.1	1	0.3	B
7	1.0	6	1.7	T
698	100.0	363	100.0	

Table 7 Average number of nodules from each sample of a particular size range, collected from bathymetric "highs" (hill tops and slopes) and "lows" (base of slopes and valleys). Based on all stations from cruise VA-09.

Nodule size (mm)	Highs	Lows
0–20	8.1	12.9
20–40	13.8	14.1
40–60	3.4	2.9
60–80	1.6	0.6
80+	1.1	0.4

of nodule types from the red clay region contrasts with that of the average nodule from the siliceous region. Relative proportions of major nodule types are fairly constant in both regions and tabular, faceted, and biological nodule types are less common.

Table 7 shows a summary of the sizes of recovered nodules. The greatest number of individual nodules occur in the smaller-sized fractions; the relative number of nodules that are 40 mm in diameter and greater decreases rapidly. This trend of decreasing number with increased size may result from winnowing mechanisms or selective burial of the larger nodules, fragmentation of larger nodules into smaller ones, or perhaps improved, conditions for nodule nucleation and accretion during the Quaternary. It is equally plausible that few areas of longterm stability (low sediment accumulation) conducive to the continuous accretion of larger nodules presently exist on the seafloor.

Several generalities regarding primary shape, size, and surface texture greatly simplify the seemingly endless variety of nodules (see Figures 3 and 8). First is the relationship of nodule type and size: apparently most larger nodules are of [D] and [E] types, although large [S] types have been reported elsewhere (Glasby 1972, 1973). These larger types are often transitional, i.e. [D-E]. Internal examination of most of these large nodules revealed that the nuclei consist of fragments of pre-existing nodules (Figure 9) and that the primary shape is often controlled by the fragment shape. Excellent examples of this phenomenon are presented by Sorem (1967). Sorem & Foster (1972), and Raab (1972). Many of the larger, irregularly shaped nodules have clearly resulted from the removal of a piece of a whole nodule and subsequent regrowth over the fractured surface. Smaller nodules occur mostly in [S] and [E] types, with uniform thicknesses of ferromanganese crust surrounding a nucleus that commonly consists of palagonite tuff. However, as in the larger nodules, the overall shape conforms to the shape of the nucleus. Type [P] nodules are probably formed by the coalescence of small type [S] or [E] nodules and owe their genesis to close contact and subsequent bonding of several nodules by ferromanganese oxides.

INTERNAL STRUCTURE AND SURFACE TEXTURE

Surface texture on nodules is primarily determined by closely spaced hemispherical protrusions called botryoids (Figure 8), which produce different magnitudes of

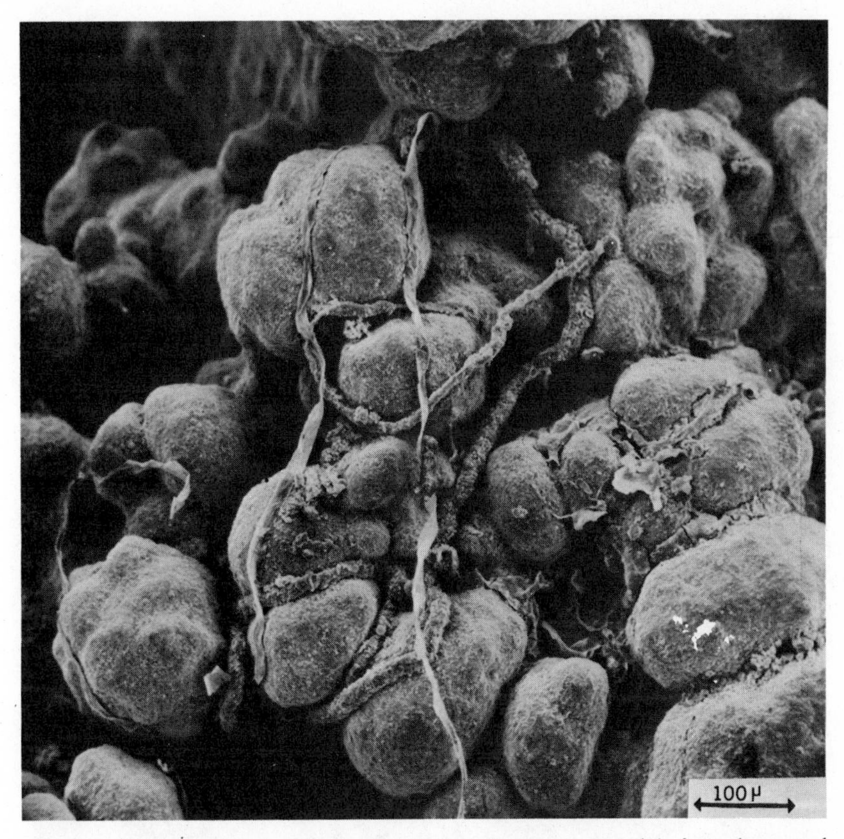

Figure 8 SEM photograph of botryoidal surface of manganese nodule from the central Pacific, showing tubes of benthic foraminifera and other organic material coating surface.

relief ranging from smooth to granular to botryoidal, depending on their size. Small nodules (<20 mm) typically exhibit a rather uniform surface texture and internal layering (Figure 10), although individuals may vary somewhat in their degree of surface roughness. Larger nodules tend to be very distinctively zoned internally and externally, often exhibiting a range of smooth to botryoidal surface texture and internal structure in a single nodule. The complex internal layering of these larger nodules (Figure 11) suggests frequent reorientation on the seafloor during growth. Large nodules (40 mm diameter and greater) commonly exhibit a prominent extended equatorial belt. Internal examination of these nodules reveals accretion rinds in this zone that typically are twice as thick as on the top and bottom (Figure 12). The botryoidal nature of the surface suggests that relatively faster accretion rates produce this type of high-relief surface. Often top and bottom portions of nodules can be recognized by the position of sediment on the nodule's outer surface (mudlines). Nodule tops are generally smoother than the bottom surface, which often displays a high-relief, botryoidal texture, but the complex

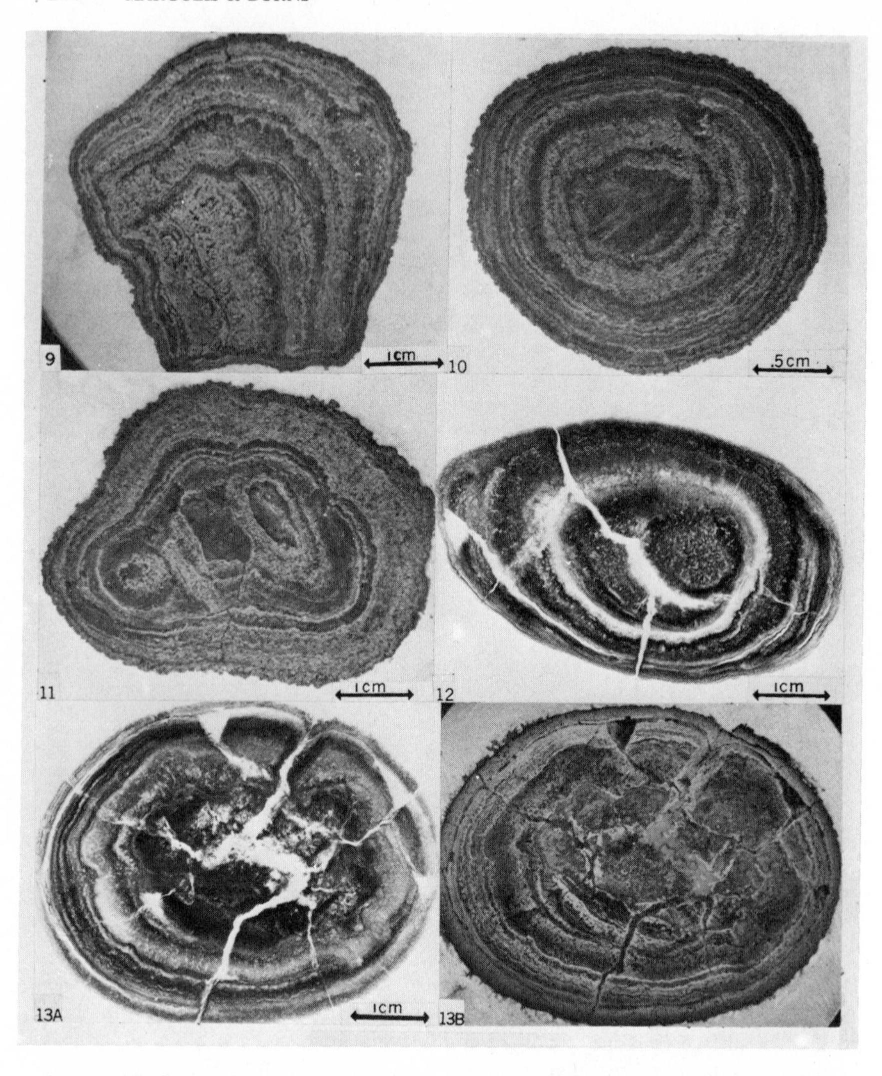

Figure 9 Reflected light micrograph of nodule from Cu, Ni-rich belt, showing that the nucleus consists of a fragment of a nodule, which is encrusted by a new layer of ferromanganese oxides. *Figure 10* Light micrograph of nodule showing concentric laminations around a palagonite nucleus. Note the alternating bands of different reflectivity. *Figure 11* Light micrograph of polished section of nodule showing how the shape of nucleus controls the shape of whole nodule. Note the complex internal layering. *Figure 12* X radiograph of nodule showing equatorial thickening of layers. Note prominent cracks (white areas) that go through the entire nodule. *Figure 13* *A:* X radiograph of nodule showing varied zonation of internal structures of different density and cracks through entire nodule. *B:* Light micrograph of *A*, showing filled-in cracks.

internal stratigraphy of such nodules indicates that orientations are by no means permanent: zones of thickened layers can be seen in many quadrants of the nodules.

Our observations of nodule morphology and internal structure suggest the following sequence of events in nodule genesis. The actual time frame of this growth is discussed in the section on nodule dating.

During the juvenile stage, nodules are small, exhibit uniform surface texture, and contain a visible nucleus that usually consists of palagonite tuff or biogenic material. Their overall shape is closely controlled by the shape of the nucleus material. As a nodule grows into a mature stage, it is less subject to reorientation on the seafloor and tends to grow fastest along the equatorial belt, slightly slower on the bottom, and slowest on the top. This differential growth masks the original nucleus shape, and the nodule becomes discoidal.

It has been suggested that fragmentation of mature nodules is initiated by internal stress due to possible recrystallization or hydration reactions or perhaps to impact during mass movement of sediment. Alternatively, fragmentation may occur during long periods of nongrowth, when the nodules may crumble. We have observed that these cracks frequently heal by secondary infilling of ferromanganese oxides during new episodes of accreton and thus prevent the nodule from fragmenting (Figure 13). Fragments provide the irregularly shaped nuclei found within the largest nodules of the old-age stage. During this last stage of development, nodules are subject to recurring cycles of partial or complete fragmentation and subsequent regrowth by ferromanganese oxides. Nodule populations may expand in this manner. Nodule samples recovered from individual free-fall grabs often contain many different types and all size ranges, suggesting the complex growth conditions of these deposits.

Mineralogy of Manganese Nodules

The minerals constituting manganese nodules continue to be a source of confusion and debate. The nodules are known to be composed of a variety of cryptocrystalline materials of biological, detrital, and authigenic origin (Mero 1965). They include a variety of hydrated oxides of manganese and iron; accessory minerals such as quartz, feldspar, calcite, apatite, clays, and zeolites: organic and colloidal matter; and igneous and metamorphic rocks in varying degrees of degradation. Characterizing the predominant manganese and iron oxide-hydroxide phases is difficult: these minerals are so intimately intergrown with one another and with the accessory and detrital materials that it is virtually impossible to extract a homogeneous phase for identification by X-ray powder diffraction measurements, let alone isolate a single crystal for structure determinations by conventional X-ray crystallography. Moreover, the phases are frequently so poorly crystalline as to be amorphous to X-rays. Recently, significant progress has been made in identifying and deducing the crystal structures of the minerals in manganese nodules by scanning electron microscopy (SEM), transmission electron microscopy (TEM), and electron diffraction measurements on corroborative synthetic phases and terrestrial mineral analogues.

The terminology for the manganese oxide minerals in manganese nodules—10Å manganite, 7Å manganite, and δ-MnO_2—was originally proposed (Buser & Grütter 1956, Buser 1959) for three different crystalline fractions giving distinctive X-ray powder diffraction patterns that could be correlated with those of certain synthetic manganese oxides. This terminology has long been considered unsatisfactory because it leads to confusion with the mineral manganite (γ-MnOOH), which has not been identified in manganese nodules. Recent studies of synthetic phases (Giovanoli et al 1970a,b 1971) produced an alternative complex chemical nomenclature for the various manganese (IV) oxides found in manganese nodules. A third nomenclature scheme, based on the close agreement of certain lines in the X-ray powder patterns of manganese nodules with those of the terrestrial minerals todorokite and birnessite, is now generally adopted (Burns & Burns 1975b). Recently it was recommended that the name δ-MnO_2 be retained for a disordered birnessite phase having minimal structural periodicity (Burns et al 1974); the validity of todorokite as a homogeneous phase is currently under debate. Electron diffraction measurements of synthetic phases have been interpreted (Giovanoli et al 1971) as indicating that todorokite is a mixture of primary buserite (10Å manganite) partly dehydrated to birnessite and partly reduced to manganite (γ-MnOOH). The γ-MnOOH crystals are so extremely small that they are undetectable by X-ray diffraction analysis (Giovanoli & Burki 1976). Synthetic buserite was accepted as a new mineral in place of the 10Å-manganite phase in manganese nodules in 1970 by the International Mineralogical Association (IMA) Commission on New Minerals (Hey & Embrey 1974). Several mineralogists, however, believe that much stronger evidence is required before the naturally occurring mineral todorokite is discredited in favor of the synthetic phase buserite (Burns et al 1974). It is confusing to find todorokite, so widely and voluminously distributed in the lithospheric and oceanic crust, giving reproducible X-ray powder patterns with no evidence of the strongest lines for manganite and birnessite. Further work is required to resolve the todorokite-buserite problem. Other manganese (IV) oxide minerals reported in manganese nodules, in addition to the dominant phases todorokite, birnessite, and δ-MnO_2, include ramsdellite, nsutite, and psilomelane (Burns & Burns 1975b). Mineralogical data for these and other pertinent manganese oxide minerals are summarized in Table 8.

The iron-bearing minerals in manganese nodules are usually amorphous to X-rays. The literature reveals confusion and misidentification of the hydrated iron oxide-hydroxide phases occurring in nodules (Burns & Burns 1975b). It is now generally accepted that the most common iron-bearing phases in nodules are goethite and material variously called amorphous $Fe(OH)_3$, iron (III) oxide hydrate gel, colloidal ferric species, and hydrated ferric oxide polymer, which may be identical to the new mineral ferrihydrite (Chukhrov et al 1973). The crystallites of this mineral rarely exceed 100Å, which accounts for its amorphicity to X-rays.

An area of active research centers around the structures of the host manganese (IV) oxides and iron (III) oxyhydroxide phases of manganese nodules because their crystal structures are fundamental to understanding the processes of nucleation, growth, and fractionation of heavy metals. Most of the structural information is

Table 8 Selected manganese (IV) oxide minerals[a].

Mineral	Formula	Crystal class (space group)	Cell parameters (Å)	Structure type (isostructural compounds)
pyrolusite	β-MnO_2	tetragonal ($P4_2/mn2$)	$a_o = 4.39$; $c_o = 2.87$	rutile
[b]ramsdellite	MnO_2	orthorhombic (Pbnm)	$a_o = 4.53$; $b_o = 9.27$; $c = 2.87$	ramsdellite (diaspore, goethite)[b]
[b]nsutite	$(Mn^{2+},Mn^{4+})(O,OH)_2$	hexagonal	$a_o = 9.65$; $c_o = 4.43$	pyrolusite + ramsdellite intergrowths
hollandite	$(Ba,K)_{1\text{-}2}Mn_8O_{16}.xH_2O$	tetragonal (I4/m)	$a_o = 9.96$; $c_o = 2.86$	hollandite (cryptomelane, manjiroite, coronadite, α-MnO_2, akaganèite)
cryptomelane	$K_{1\text{-}2}Mn_8O_{16}.xH_2O$	tetragonal (I4/m)	$a_o = 9.84$; $c_o = 2.86$	hollandite
[b]psilomelane (romanchèite)	$(Ba,K,Mn,Co)_2Mn_5O_{10}.xH_2O$	monoclinic (A2/m)	$a_o = 9.56$; $b_o = 2.88$; $c_o = 13.85$; $\beta = 92°\ 30'$	psilomelane
[b]todorokite (10Å manganite)	$(Na,Ca,K,Ba,Mn^{2+})Mn_3O_7.xH_2O$	monoclinic	$a_o = 9.75$; $b_o = 2.849$; $c_o = 9.59$; $\beta = 90°$	unknown
lithiophorite	$[Mn_5^{4+}Mn^{2+}O_{12}][Al_4Li(OH)_2]$	monoclinic (C^2/m)	$a_o = 5.06$; $b_o = 8.70$; $c_o = 9.61$; $\beta = 100°\ 7'$	lithiophorite
chalcophanite	$ZnMn_3O_7.3H_2O$	triclinic ($P\bar{1}$)	$a_o = 7.54$; $b_o = 7.54$; $c_o = 8.22$; $\alpha = 90°$; $\beta = 117°\ 12'$; $\gamma = 120°$	chalcophanite
synthetic birnessite ("7Å manganite")	$Na_4Mn_{14}O_{27}.9H_2O$	orthorhombic	$a_o = 8.54$; $b_o = 15.39$; $c_o = 14.26$	incompletely determined
[b]natural birnessite	$(Ca,Na)(Mn^{2+},Mn^{4+}).xH_2O$	unknown	—	not determined
[b]δ-MnO_2	$(Mn,Co^{3+})Mn_6O_{13}.xH_2O$	hexagonal	—	disordered birnessite ($FeOOH.xH_2O$)[b]

[a]Selected data from tabulation by Burns & Burns (1975a).
[b]Constituent of manganese nodules.

derived from crystal morphologies observed with SEM and TEM and from electron diffraction measurements of synthetic analogues. SEM study of manganese nodules reveals that todorokite consists of acicular and bladed crystals, whereas birnessite crystallizes as thin platelets and lamellae (Margolis & Glasby 1973, Margolis 1973, Fewkes 1973, Woo 1973). Synthetic birnessites also consist of platelets (Giovanoli et al 1970a,b). Moreover, electron diffraction measurements of synthetic birnessite (Giovanoli et al 1970a,b) indicate that its structure is similar to that of chalcophanite ($ZnMn_3O_7.3H_2O$). The dominant feature of chalcophanite and other manganese (IV) oxide minerals is that their crystal structures consist of edge-shared [MnO_6] octahedra linked in various degrees of complexity (Burns & Burns 1975b). Single chains of edge-shared octahedra are found in pyrolusite,

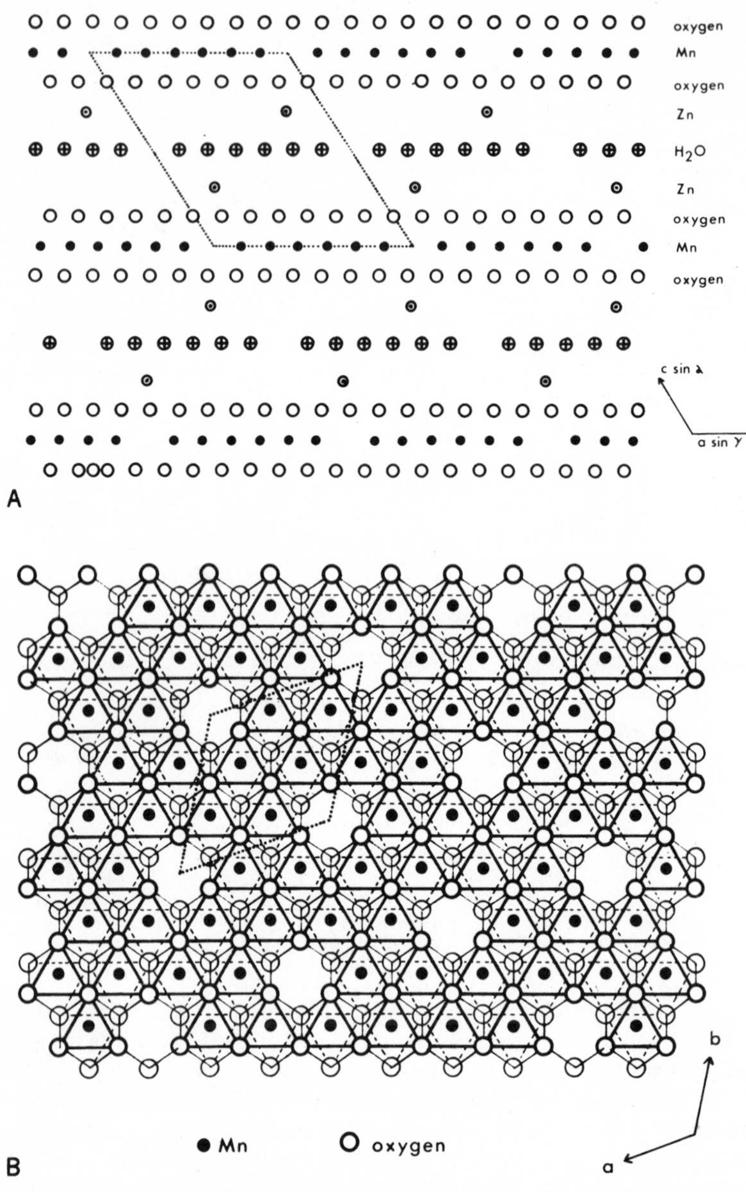

Figure 14 The chalcophanite crystal structure (after Wadsley 1955). *A:* Projection along the *b* axis. Vacancies in the Mn layer define the rhombus unit cell. Note that one out of every seven Mn atoms is a vacancy. *B:* The edge-shared [MnO_6] layer viewed normal to the basal plane. The vacant octahedral sites at the origin are the corners of a rhombus outlining the plane of the Mn atoms. Note that each Mn atom is adjacent to a vacancy.

double chains occur in ramsdellite, and cross-linked double or triple chains are present in the hollandite group or psilomelane in tunnel structures exhibiting cation exchange properties. Layers of edge-shared $[MnO_6]$ octahedra characterize the sheet structures of lithiophorite, chalcophanite, and birnessite.

The chalcophanite structure (Wadsley 1955), which serves as a structural model for birnessite (Figure 14) consists of single sheets of water molecules between layers of edge-shared $[MnO_6]$ octahedra, with Zn atoms located between the water layer and oxygens of the $[MnO_6]$ layer. The stacking sequence along the *c* axis is thus -O-Mn-O-Zn-H_2O-Zn-O-Mn-O- and the perpendicular distance between two consecutive $[MnO_6]$ layers is about 7.17Å. The water molecules are grouped in open double-hexagonal rings and vacancies exist in the layer of linked $[MnO_6]$ octahedra, so that six out of every seven octahedral sites are occupied by manganese. Each $[MnO_6]$ octahedron shares edges with five neighboring octahedra and is adjacent to a vacancy (Figure 14). The Zn atoms, located above and below the vacancies in the manganese layer, are coordinated to three oxygens of the $[MnO_6]$ layer. Each Zn atom completes its coordination with three water molecules, forming an irregular coordination polyhedron. The chemical compositions of natural chalcophanites differ significantly from the ideal formula, $Zn^{2+}Mn_3^{4+}O_7.3H_2O$. Not only is the water content variable but also Mn^{4+} ions are deficient and the number of cations usually exceeds four per formula unit. These trends indicate that some Mn^{2+} replaces Mn^{4+} in the linked octahedra, accounting for the larger average Mn-O distance of 1.95Å (compared to 1.88Å in pyrolusite or ramsdellite). Additional cations also occur in interstitial positions between the H_2O layers and oxygens of the $[MnO_6]$ layers.

By analogy with chalcophanite, birnessite contains sheets of water molecules and hydroxyl groups located between layers of edge-shared $[MnO_6]$ octahedra, which are separated by about 7.2Å along the *c* axis. One out of every six octahedral sites in the layer of linked $[MnO_6]$ octahedra is unoccupied (compared to one out of seven in chalcophanite), and Mn^{2+} or Mn^{3+} ions appear to lie above and below these vacancies. These low valence manganese ions are coordinated to oxygens in both the $[MnO_6]$ layer and the (H_2O,OH) sheet.

The phase δ-MnO_2 is generally considered a disordered variety of birnessite. X-ray patterns of δ-MnO_2 typically contain only two broad diffuse lines at about 2.40Å and 1.42Å; there is little or no suggestion of the addition of lines around 7.0–7.2Å and 3.5–3.6Å, which are the diagnostic basal plane reflections for birnessite. Thus, the structure of δ-MnO_2 probably contains fragments of the layers of edge-shared $[MnO_6]$ octahedra with small periodicities of stacking along the *c* axis found in chalcophanite (Figure 14) and birnessite. As a result, a high proportion of vacancies on exterior surfaces of δ-MnO_2 are available for sorption of cations on each side and within the sheets of edge-shared $[MnO_6]$ octahedra.

Another way of viewing the structures of δ-MnO_2, birnessite, and chalcophanite is in terms of a hexagonal close-packed oxygen framework in which Mn^{4+} ions and vacancies are distributed in an ordered array among the octahedral sites. The structure of hydrated ferric oxyhydroxide polymer ($FeOOH.xH_2O$) is also believed to contain Fe^{3+} ions in the octahedral sites of a hexagonal close-packed oxygen network (Feitknecht et al 1973); however, the Fe^{3+} ions are almost randomly distributed in these sites with only a slight degree of order. This has led to the postulate that δ-MnO_2 and $FeOOH.xH_2O$ are highly susceptible to epitaxial intergrowths, which probably initiate nucleation and lead to the intimate association of Mn and Fe oxide phases in manganese nodules (Burns & Burns 1975a). This intergrowth of δ-MnO_2 and $FeOOH.xH_2O$ is believed to inhibit the formation of the ordered layer structure of birnessite.

The structure of todorokite is not as well characterized as that of birnessite. The acicular habits of todorokite crystals in manganese nodules and in crustal rocks suggest that its structure resembles that of psilomelane and the hollandite group, which have chain structures of linked $[MnO_6]$ octahedra rather than sheet structures like birnessite, chalcophanite, or lithiophorite (Burns & Burns 1975b). This hypothesis represents a significant departure from the original suggestion (Buser & Grütter 1956) that 10Å manganite (i.e. todorokite or buserite) has a structure analogous to that of platey lithiophorite, but conforms with results from synthesis experiments (Giovanoli et al 1973), which failed to prepare lithiophorite-buserite solid-solutions.

MICROMORPHOLOGY AND MICROSTRUCTURE

The internal layers observed optically in manganese nodules reveal much about the growth history of individual nodules. Variations in reflectivity seen in light micrographs are related to differences in ferromanganese oxide mineral components in the nodules. According to Sorem & Foster (1972), the white, dense, highly reflective zones consist of crystalline Mn-Fe oxides such as todorokite and birnessite (see section on mineralogy); zones of moderate reflectivity and smooth gray appearance are usually composed of amorphous Fe-Mn oxides; and dark pitted areas with low reflectivity are composed of clay and biogenic detritus. In addition to degrees of reflectivity there are also several distinct types of growth zones. The first are massive highly reflective zones of microcrystalline todorokite and birnessite that sometimes show faint laminations. Mottled zones contain the same minerals as the massive zones but also exhibit impurities of iron oxide, clay, and biogenic material. Columnar zones are commonly found in nodules and have also been called cuspate or botryoidal growth zones. This type of growth is composed of X-ray amorphous Fe-Mn oxides, with thin crystalline seams of todorokite and birnessite (Sorem & Foster 1972). Compact and laminated zones are also recognized in nodules. These are similar to massive zones in reflectivity, but have better defined laminations and consist of X-ray amorphous oxide material rich in Fe.

X-ray radiographs (Figures 12 and 13*A*) show similar zonations of growth structure but record variations in density, which are related to chemical composition and degree of hydration or compaction. All these structural zones reflect the complex depositional history of nodules. They are probably produced by variations in growth rate, influx of sediments, source of metals, position of the nodule relative to sediment-water interface, bottom currents, and biological activity on and adjacent to the nodule (Margolis & Glasby 1973).

SEM study of manganese nodules has revealed still more detail in microstructure, both on surfaces and internally. Nodule surfaces have been shown to be covered by small crystals of manganese oxide, micro- and nannofossils, amorphous clay aggregates, and attached benthic fauna (Figures 8, 15, and 16). These assorted particles are found clumped together to form microbotryoids on the nodule surface, which are reflected internally in columnar growth zones.

Internal examination of nodule x sections using the SEM reveals microlamina-

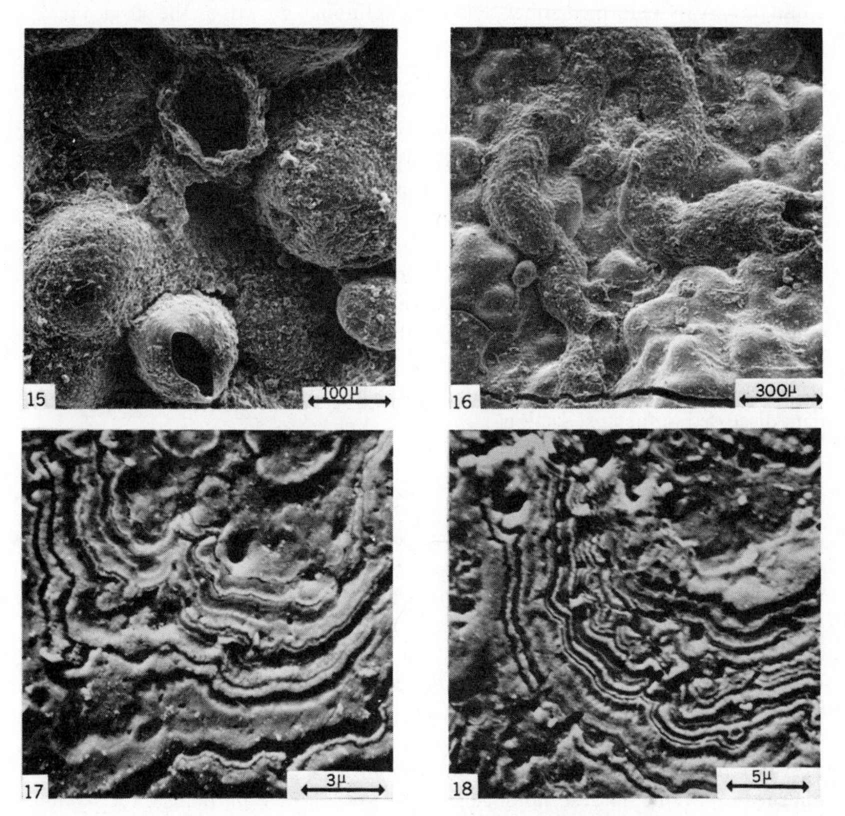

Figure 15 SEM photograph of botryoidal surface of manganese nodule from the central Pacific, showing dome-shaped blisters built by benthic foraminifera. *Figure 16* SEM photograph of nodule surface showing curved tube of benthic foraminifera in center, with aperture at right-hand margin. Nodule is from the central Pacific, and foraminifera has been identified as *Tolypammina sp.* (B. K. Dugalinsky, personal communication). *Figures 17 and 18* SEM photographs of polished sections of manganese nodules from the area of Cu and Ni enrichment in the central Pacific. Note the variations in thickness of different microlaminations and how each remains well-defined over some distance. These microlaminations are believed to be the fundamental growth units of nodules. Microprobe analysis has shown that each layer exhibits a distinct range in chemical composition.

tions ranging in size from 0.25 μm to greater than 10 μm (Figures 17 and 18). These are found in both the cuspate columnar growth zones and the zones of straight laminations and appear to be the smallest structural growth units in manganese nodules (Margolis & Glasby 1973). The layers visible in the light micrographs and radiographs are usually composed of many microlaminations, which can be traced for great distances around the nucleus and can grade from straight lamellae zones to cuspate ones without any change in thickness. Microlaminations can be formed either by (*a*) discontinuous growth at the surface of the nodule produced by short-term changes in the growth rate of the nodule or periods of nondeposition, or (*b*) diagenetic recrystallization or ordering phenomena. We believe that these layers are primary growth features because they are found in all portions of nodule interiors and in nodules from all regions of the ocean, regardless of growth history. Assuming growth rates of 1–10 mm per 10^6 yr (Ku & Broecker 1969), the time scale for the formation of individual microlaminae is 25 to 10,000 years, depending on thickness and growth rate assumed. What factors may fluctuate over this time scale to produce the microlaminations is a matter of speculation; two possibilities are bottom currents, which can control sediment accumulation (Margolis & Glasby 1973), and submarine volcanism from oceanic rises, which could periodically provide a source of metals (Fe, Mn, Cu, Ni) for nodule accretion (Bonatti et al 1972). This microstructural evidence takes on added significance relative to nodule genesis and composition when correlated with microchemical data discussed in the next section.

NODULE MICROCHEMISTRY

Both wavelength and energy-dispersive electron microprobe systems have been utilized to obtain chemical information on micron-scale elemental variations in manganese nodules (Burns & Fuerstenau 1966, Burns & Brown 1972). Using these techniques, the concentrations of as many as 16 elements can be profiled within individual microlaminations and across the assortment of growth zones found in nodules. The element associations mentioned in the section on bulk chemistry are further confirmed by recent microchemical analysis (Ostwald & Frazer 1973, Margolis et al 1976). The latter techniques are particularly important because they provide chemical data for micron and submicron areas, which approach the dimensions of the largest homogeneous growth layers in manganese nodules (Margolis & Glasby 1973). Metal concentrations fluctuate widely across the nodules, well above and below the bulk chemical analyses for each element in the nodules.

Particularly significant in the interiors of the manganese nodules is a manganese oxide phase enriched in Ni and Cu, deposited along the edges of healed cracks inside the nodules (Sorem 1967; Burns et al, unpublished results). This indicates that remobilization of elements occurred inside the nodules and that Cu and Ni were taken up after the nodule grew. Most microprobe analyses of nodules from the copper-nickel rich belt indicate that these metals are also concentrated in thin, highly reflective laminations high in Mn and low in Fe and Co. There is considerable variation, however, in the Cu-Ni content and overall chemical composition along the length and width of these thin laminations, although uniform Ni/Mn and Cu/Mn ratios are often observed. Although bulk chemical analysis of nodules from the mining area run between 1.0 wt% and 1.5 wt% each of copper and nickel, microprobe analyses show concentration of each of these elements to be as high as 5 wt% in layers high in Mn and low in Fe, and Fe-rich layers contain much lower amounts of Cu and Ni. This nonuniformity in Cu-Ni enrichment may result from either periodic outpouring of these metals from submarine volcanic sources or secondary migration of the metals during recrystallization in the nodule interior. The larger ($>100\mu$) optically recognized layers in nodules, which consist of numerous microlamellae, exhibit internally consistent interelement ratios; the periodicities of these layers may serve as marker horizons within individual nodules and may be traced from one nodule to another to identify a series of accretion events within a given nodule deposit. The microlamellae contained within the larger sizes, however, show inhomogeneities in the distribution of these elements of the same magnitude both along strike and between adjacent layers. This argues against microlamellae being precipitated in a monolayer from a uniform solution and instead favors a particle-by-particle accretion of flocs from multiple sources. Furthermore, individual Fe-Mn oxide flocs, clay mineral aggregates, and detrital and biogenic grains have been identified by the SEM on the surface of and within nodules from this and other regions. Another observable chemical trend is that the layer immediately outside a non-nodule fragment nucleus is usually higher in Fe and lower in Cu and Ni than the rest of the nodule, as is the outer surface of nodules, although upper surfaces of oriented nodules may frequently have higher Cu+Ni than the portion buried in the sediment. This, however, may be caused by adhering sediment. The chemical differences between upper halves vs lower halves of nodules, where the orientation is documented, do not show any preferential concentration of Cu or Ni. Nodules with enriched upper portions are as common as those with enriched lower portions, which further attests to the frequent reorientation of nodules during growth, so well-demonstrated by the internal structural evidence found in most nodules. It would be necessary to obtain a nodule that showed evidence of being undisturbed in one growth position for the majority of its history (a difficult but not impossible task) in order to prove that the source of Cu and Ni came from the overlying water column or from the sediment beneath the nodule.

Dark, poorly reflective bands within nodules are usually high in Fe and contain inclusions of clay and microfossils; these areas are typically low in Mn, Cu, and Ni. As we discussed in reference to nodule bulk chemistry, the absence of Fe seems to control high Cu and Ni concentrations more than the abundance of Mn. Thin, highly reflective zones within nodules from areas in the Pacific outside the northeastern equatorial region also showed similar high Cu and Ni concentrations, but such zones were sparsely distributed in the nodules, and their bulk analyses yielded low total Cu and Ni. This indicates that whatever the process producing this enrichment, it also occurs elsewhere in the Pacific, but it either occurs less frequently or is masked by some other chemical input during nodule growth. Further clues as to the mineral structure of these highly reflective layers, which contain the Cu and Ni, are presented in the next section.

TRACE METAL UPTAKE IN HOST MANGANESE OXIDE MINERALS

The interelement relationships for manganese nodules form the basis for understanding the mechanism of uptake of the metals into the constituent manganese oxide minerals (Burns & Burns 1975b). Because manganese nodules form in oxidizing environments, cations with higher valencies predominate over reduced oxidation states. Results from Mössbauer spectroscopy (e.g. Carpenter & Wakeham 1973) show that iron occurs principally as Fe^{3+} ions in manganese nodules, whereas electron spin resonance measurements (Wakeham & Carpenter 1974) suggest the presence of only small amounts of Mn^{2+} ions, indicating that most of the manganese is present in the Mn(IV) oxidation state. Spectroscopic techniques are not developed enough to identify the oxidation states of Co, Ni, or Cu in the nodules, but interelement relationships and crystal chemical data indicate the presence of Cu^{2+}, Ni^{2+}, Co^{2+}, and Co^{3+} in manganese nodules. The strong positive correlations between Mn, Ni, Cu, and often Co suggest that divalent cations substitute for Mn^{2+} ions in todorokite and birnessite. Oxidation of Co^{2+} to Co^{3+} occurs in the marine environment, and evidence (Burns 1976) supports the hypothesis that low-spin Co^{3+} ions (ionic radius 0.53Å) replace equidimensional Mn^{4+} ions (0.54Å) in manganese (IV) oxide minerals.

Knowledge of the essential features of the birnessite structure has led to an interpretation of the uptake of transition metal ions by δ-MnO_2 and birnessite (Burns & Burns 1975b, Burns 1976). Hydrated divalent cations in seawater are initially adsorbed onto the surfaces of δ-MnO_2 in the vicinity of the vacancies in the edge-shared $[MnO_6]$ octahedra (Figure 14). These metals may also be held in lieu of Mn^{2+} ions between the octahedral layers of incipient birnessite (Figure 14). The sorption capacity of δ-MnO_2 is diminished as the periodicity of the stacked edge-shared $[MnO_6]$ layers increases (i.e. as the proportion of birnessite increases). Any Co^{2+} ions adsorbed near the vacancies in the $[MnO_6]$ layer are susceptible to oxidation. For example, note the following reaction:

$$2Co^{2+}_{(aq)} + 2OH_{(aq)} + 2H_2O_{(l)} + \delta\text{-}MnO_{2(s)} \rightarrow 2Co(OH)_{3(s)} + Mn^{2+}_{(aq)};$$
$$\Delta G^{\circ}_{298} = -17.1 \text{ kcal}$$

Co^{2+} is spontaneously oxidized by δ-MnO_2, releasing manganese ions in solution in accord with experimental observations (Loganathan & Burau 1973). It is unlikely, however, that a separate Co(III) oxide phase is formed. Instead, low-spin Co^{3+} ions are expected to occupy sites in the edge-shared $[MnO_6]$ octahedral layer vacated by Mn^{4+} ions involved in the redox reactions, or to fill the vacancies. Such Co^{3+} ions acquire a very high crystal field stabilization energy in octahedral sites, which accounts for their strong enrichment and resistance to leaching (Burns 1976). Insufficient structural data exist for todorokite to explain the uptake of Ni^{2+} and Cu^{2+} into this phase in manganese nodules. If todorokite has a tunnel structure similar to hollandite or psilomelane, then the cation exchange properties and isomorphous substitution for Mn^{2+} in this mineral are readily explained.

ACCRETION RATES AND DATING OF MANGANESE NODULES

One of the most critical and controversial areas of manganese nodule research is concerned with growth rates of individual nodules. It is most difficult to accept that the sediment substrate on which nodules are found accumulates at the rate of about 3 m per 10^6 yr, whereas the nodules grow at about 3 mm per 10^6 yr. This surprisingly slow nodule growth rate has been confirmed by radiometric dating of outer portions of nodules, based on the distribution of the uranium series isotopes Th^{230}, Pa^{231}, U^{234}, U^{232}, and Th^{232}; measurements yielded rates of 2–4 mm per 10^6 yr (T. L. Ku, in preparation). Further proof for this slow growth rate, 1000 times less than that of sediment accumulation, was obtained by the K/Ar method used to date volcanic fragments found as the nucleii of some nodules. Assuming that manganese accretion began quite soon after arrival of the fragment on the seafloor, rates of 1–3 mm per 10^6 yr have been shown (Broecker 1974). A new technique using Be-10, which has a half-life of 2.5 million years, has been used to date entire manganese nodule cross-sections, yielding similar conclusions on nodule growth rates. According to Ku, the observed rapid drop-off of Th^{230} and Pa^{231} activities in surface layers has sometimes been interpreted as the result of processes other than radioactive decay. For example, nodules may accrete by a rapid succession of precipitation events, presumably because of submarine volcanic eruptions. During subsequent long exposure to seawater between eruptions, the nodule surfaces absorb high concentrations of Th^{230} and Pa^{231}. The surface gradients of these nuclides could then be generated by inward diffusion (either through the porous nodules or by the dilution effect that results from mixing surface material) to greater depth naturally through cracks or artificially during sample preparation procedures before analysis. However, the above-mentioned possibility is not supported either by (*a*) concordant rate results from Th^{230}, Pa^{231}, U^{234}, and Be^{10}, or (*b*) the distribution of Th^{230} as revealed by the means of the nuclear emulsion plates.

The rapid accumulation hypothesis has been proposed to circumvent the problem of keeping nodules at the sediment-water interface without getting buried by the adjacent sediments deposited at a much faster rate. It is shown from the amount of Th^{230} found in the nodules, however, that the above Th^{230} absorption argument still leaves the burial problem unsolved.

A further approach to the nodule age question is to extend measurements of the U-series isotopes from surface to deeper layers close to the nuclei. T. L. Ku (in preparation) has made such measurements of four nodules. Within experimental error, radioactive equilibrium between Th^{230} and U^{234} is found in most of the deeper layers examined. In several cases, disequilibrium does occur However, this can be explained as being caused by contamination of more recent sedimentary material introduced diagenetically along fissures, and in one case, possibly by limited post-depositional migration U^{234}.

Arguments against the low growth rates gained some plausibility from recent work of Russian and French scientists. In the inner portions of several nodules, Cherdyntsev et al (1971) and Lalou et al (1972) observed the presence of Th^{230}. These workers suggest young ages (as opposed to ages of millions of years as depicted by the reported low rates) for the Th^{230} and Pa^{231}. The picture may therefore be complicated by postdepositional remobilization of these nuclides.

Detailed measurements have been made recently by T. L. Ku on both the surface layers and inner core portion (deeper than about 1 cm) of a nodule. A sharp gradient of Th^{230} in the outer 3 mm was observed in a North Pacific nodule, giving an accretion rate of 4 mm per 10^6 year. However, in the nucleus, about 2.5 cm deep, excess Th^{230} over U^{234} (activity ratio $Th^{230}/U^{234} = 1.35 \pm 0.06$) appears to be present. Because this finding bears important implications on the dating method, further analyses testing the closed system assumption used in this type of nodule dating are in progress.

ORGANIC INFLUENCE ON NODULE GROWTH

It was found that nodules that had been carefully collected during Manganese Nodule Project cruises and handled so as to avoid destruction of delicate surface structures reveal an entire community of benthic life forms that use the nodule surface as a substrate. In addition, tests of planktonic foraminifera, radiolarians, and diatoms found on nodule surfaces in underlying sediments show evidence of replacement by manganese oxide as they become incorporated into the nodule matrix.

The organic role in nodule genesis has been speculated upon for some time (Correns 1941, Graham 1959, Ehrlich 1972) but without conclusive evidence. Current research, however, may provide new insight into this aspect of nodule genesis. Recent work by Greenslate (1974) and Wendt (1974) has suggested that benthic agglutinated foraminifera may play an active role in the actual construction of the nodules on which they live. These single-celled animals attach themselves to the surface of manganese nodules, which frequently present the only hard, stable substrate found on the deep-ocean floor. They construct their tubular tests out of biogenic and mineral debris found on and adjacent to the nodules; these particles are then held together by a cementing agent secreted by the animal. The shape and composition of these tests, which eventually get incorporated into the nodule matrix, are a function of individual species, but until recently only a few genera had been identified on manganese nodules, and no chemical work had been done to test compositions.

Investigations of well-preserved nodules from several ragions within the Pacific basins have identified more than 17 different species of attached benthic foraminifera and there are many unidentified forms. The greatest number and highest diversity of these animals were found in the equatorial northestern Pacific, where nodules are enriched in copper and nickel. The majority of specimens were attached to the equatorial rim of the nodule, which is actually at the sediment-water interface. The foraminifera are almost always concentrated in the rough, irregular areas between botryoids (Figures 8 and 16). Comparison of chemical analyses of foraminiferal tubes and nodule surfaces indicated that, in addition to the obviously higher silica content produced by the siliceous biogenic debris used for test construction, there were interesting relationships in the Fe, Mn, and trace metal concentrations. The foraminiferal tubes all showed higher Fe, Co, and Ti and lower Mn, Cu, and Ni than the nodule surface on which they were growing. In addition, the interior linings of the tubes show the highest concentrations of iron, and micronodules are occasionally found incorporated in the foraminiferal tubes. What then can be said about the significance of this data relative to nodule genesis and composition? We of course speculate here because we lack hard data, but several observations are apparent. First, it seems the benthic foraminifera do not actually precipitate Mn, Cu, and Ni themselves, although they deposit Fe in the form of a cementing agent (Hedley 1963, Dudley 1976); Co and Ti may be codeposited with the iron or may be in a detrital phase along with the other sediments scavenged by the foraminifera in constructing the test. Their contribution to the bulk composition of the nodules appears to be as a sediment dilutant. Cross sections of benthic foraminiferal tubes are occasionally found within the interior of nodules, and microprobe analyses of such features reveal them to be high in silica and Fe, i.e. similar to tubes formed on the surface of the nodules. Analyses have also shown pockets and layers of high Si and Fe material dispersed throughout most nodules, as mentioned in the section on nodule microchemistry. It may be that during periods of intermittant or slow manganese oxide accretion, the agglutinated benthic foraminifera flourish on the nodule surface; however, when an episode of rapid metal oxide accretion takes place, the tubes are buried in the nodule matrix and appear as a layer relatively rich in iron and silica in comparison with the metal oxide phases. Although this may explain some of the chemical variability in nodules, it does not explain the relationship between higher abundance and diversity of benthic foraminifera and high copper plus nickel values in the northeastern equatorial Pacific. Perhaps this is a coincidence or perhaps their abundance in this area discloses something about the relationship between organic productivity and trace metal concentration. In order to link these phenomena, we must begin at the top of the marine food chain with the accumulation of trace metals from seawater by phytoplankton in oceanic surface layers. These trace metals are most likely present as organometallic complexes serving metabolic functions within organisms as well as structural functions in the hard tests of microfossils.

Ingestion of phytoplankton by filter feeding zooplankton results in a concentration of the organometallic complexes in fecal pellets, which being larger particles settle faster through the water column and thus accelerate transport of trace metals from the surface to the floor of the ocean. This rain of organic, rich detritus serves as food for the benthic foraminifera. The organic complexes are metabolized with the subsequent release of trace metals (Graham 1959). These trace metals build up concentrations within the animal's protoplasm and Fe, Ti, and Co may be utilized in the production of tube cement. Toxic trace metals (Mn, Ni and Cu), which serve no metabolic or constructive purpose, are possibly excreted into the surrounding water where they might be absorbed into ferromanganese oxides in the nodule surface layer. Furthermore, Graham & Cooper (1959) considered that the protein-rich exudate of the foraminifera could provide a favorable substrate for the growth of other biota, including bacteria capable of extracting trace elements from seawater. Periods of high organic productivity in the surface waters and increased biogenic sedimentation may result in high activity on nodule surfaces and could possibly explain layers of high trace metal concentration in manganese nodules. Trace metals would be concentrated more in nodules than in surrounding sediments, because of the activities of benthic foraminifera.

Another possible role of these benthic agglutinated foraminifera is in the initiation of nodule growth. They attach to any hard substrate on the seafloor, such as fragments of volcanic rocks or glass, bones, or shark's teeth. After attachment, the iron-rich cement used in the tube would produce an iron-rich layer covering the outer surface of the fragment or tooth. This iron-rich layer would serve as a site for the accretion of manganese and metal ions as described by Burns & Brown (1972). Electron microprobe analyses have indeed shown the existence of such an iron-rich layer adjacent to the nuclei of many nodules (Burns & Brown 1972, Morgenstein 1972), although this may also be produced by leaching of iron from the nucleus material.

On a regional scale, benthic agglutinated foraminifera are found from the poles to the equator, so their mere presence on nodules would not explain the relatively high Cu-Ni content in nodules from the equatorial northeastern Pacific zone. Examination of benthic foraminifera on nodules from various localities, however, has indicated that nodules from the high Cu-Ni region showed the highest concentration of foraminifera on nodule surfaces; for example, there were as many as 22 individuals per 10 cm nodule from the area of Cu and Ni enrichment, as opposed to only 1 to 2 individuals per 10 cm nodule from sites elsewhere in the Pacific. This greater number of benthic agglutinated foraminifera could be explained by the high organic productivity in the overlying equatorial water mass and the resultant higher sedimentation rate of organic rich detritus available as a food supply for a high density of benthic animals. The scavenging activities of the benthic foraminifera may also keep the nodules from being buried by the rain of sediments and could explain the anomaly between nodule growth rates and surrounding sediment accumulation rates initially recognized by Bender et al (1970).

If high biological productivity in the surface layers of the ocean is the critical factor in trace metal concentration in manganese nodules, then why do nodules from areas of oceanic upwelling and high biological productivity not have Cu and Ni enrichment? For example, why would nodules from the diatom ooze band beneath the Antarctic convergence at about 55°S (Ewing et al 1969) not be so enriched? A possible reason for the low Cu-Ni values in Antarctic nodules (Goodell et al 1971) is dilution by terrigenous sediments transported by bottom currents and by ice-rafting from Antarctica; these factors are not applicable in the equatorial northeast Pacific, which is free of significant terrigenous sediments. Coastal upwelling regions of high biological productivity are also close to large sources of terrigenous detritus.

The following factors present in the northeastern equatorial Pacific manganese nodule belt may combine to produce Cu-Ni and Mn enrichment:

(*a*) There is high biological productivity in the overlying surface waters.

(*b*) The bottom lies below the calcium carbonate compensation depth, so carbonate microfossils dissolve to liberate any manganese and other metals present in their tests.
(*c*) Radiolarians and diatoms are abundant as a source of siliceous detritus, which may act as nuclei for Mn oxide precipitation as they dissolve.
(*d*) Abundant organic detritus in bottom sediments could support benthic in- and epifauna.
(*e*) Bottom water movement is sufficient to maintain a mild oxidizing condition and keep some fine material in suspension, but current activity is not enough to erode or transport coarse biogenic material.
(*f*) The area is far from any source of terrigenous material that would serve as a dilutant.
(*g*) There are enough particles such as rock fragments, shark's teeth, etc, to serve as hard substrate (seed) for attachment of benthic agglutinated foraminifera.

We are not proposing that concentration of trace metals by benthic agglutinated foraminifera and precipitation and replacement by manganese of planktonic foraminifera, radiolarian, and diatom tests are the sole mechanisms in operation in growing manganese nodules. We do, however, feel that the large number of these animals and their remains present on the nodules from the copper-nickel-rich zones of the northeastern equatorial Pacific, are somehow linked to the unique chemical composition of nodules from this area.

Further investigation of species distribution in the benthic fauna on manganese nodules, the microchemistry of their tests, and the organic chemistry of the nodules and associated sediments are in progress.

CONCLUSIONS

Understanding of structural and sedimentological factors that control the localized distribution of nodules within the copper-nickel-rich belt in the northeastern equatorial Pacific has greatly increased. We know more about the internal structure, bulk, and microchemistry of individual nodules and nodule deposits in this region than in any other. Study of the nodule mineralogy in this and other regions has shown that copper and nickel appear to be concentrated in the mineral todorokite, which is more abundant in nodules from the northeastern equatorial Pacific than from adjacent areas, but we do not know why this is so. The environmental parameters that combine to produce the unique nodule composition found in this region are not firmly established. The exact mechanism of nodule growth—be it related to the overlying water column, the sediments below, the periodic activity of nearby volcanic sources, or some biochemically controlled process—remains elusive. Manganese nodule growth rates are still puzzling; if they grow more slowly than the surrounding sediments accumulate, we cannot explain how they keep from getting buried. Although there are still many unanswered questions, we have made significant progress in the last few years. Current studies include the following:

1. Cruises during 1975 utilized an acoustical navigation system (ATNAV) to pinpoint the position of bathymetric survey and nodule sampling locations,

obtaining precise information on the relationship of structure to nodule distribution.

2. A near-bottom Deeptow system will be used for bottom current and sediment distribution mapping as well as close-up photography on a cruise early in 1976. In conjunction with the ATNAV, this will enhance our ability to detect small-scale variations in the ocean bottom environment.

3. The Manganese Nodule Project is using a bottom ocean monitor (BOM) during 1975 to take long time-series observations of nodule fields and bottom-ocean environmental parameters. Several photographs a day are taken of a small area of the seafloor that contains nodules. Conducted over a period of many months, this documents the effect of currents and the activity of benthic organisms as possible agents in moving or overturning nodules, or possibly in keeping them from being buried by sediment. Long-term records of current speed and direction are also kept, as are records of variations in light scattering, which indicate the amount of suspended material found near the ocean floor. This instrument is placed in an area containing nodules that were previously found to be high in Cu and Ni; it should retrieve valuable information on nodule environment.

4. After an initial period of intercalibration, water chemists in the Manganese Nodule Project are sampling the water column and pore waters from sediments. Analyses performed during 1975 should show whether Mn, Fe, Cu, and Ni are migrating up from below to be precipitated on nodules at the sediment-water interface, or whether transition metals are enriched in the particulate or dissolved phases in surface, intermediate, or bottom waters.

5. The question of organic influence is being approached by taxonomic and microchemical analysis of the benthic fauna on nodules and by organic chemical analysis of nodules, associated sediments and dissolved and suspended organic material in seawater to determine whether organic complexing concentrates transition metals.

6. Detailed radiometric and paleontological dating are being performed on nodules and sediments collected in situ from box cores. Proponents of the slow growth theory (T. L. Ku) and those who believe in faster growth (C. Lalou) will work together during the coming year to resolve this controversy.

7. Microchemical and mineralogical analyses are being performed on individual layers in nodules of the same size and morphology, aiming to correlate them from nodule to nodule within a deposit and so define growth events. Once these are established on a local scale (10 nm^2), an attempt will be made to correlate them on a regional scale. Variations in the chemistry of individual layers in the nodules may record variations in world climate, much as tree rings do, or perhaps metal-rich emanations from oceanic-rise spreading centers (Bonatti et al 1972).

Possibly the tremendous increase in global volcanism during the last 2.5 million years is related to cooling of world climates, as suggested by Kennett & Thunell (1975). It is interesting to note that during this same time period manganese nodule production was apparently higher than during previous times: nodules are more abundant on the present-day seafloor surface than in deep-sea sediment cores. Whether these factors are directly related or whether there is an indirect link, by way of climatic cooling, that would increase the velocity in Antarctic bottom water circulation (Kennett & Watkins 1975) will be resolved only by paleoenvironmental study of sediments associated with nodule deposits.

ACKNOWLEDGMENTS

The authors thank the participants of the Manganese Nodule Project for their suggestions and discussions. Special thanks are given to J. D. Craig, B. K. Dugalinsky, W. C. Dudley, and M. A. Meylan for shipboard and laboratory data that provided the scientific basis for many of the discussions in this paper. L. Olsen performed the X-ray and light microscope photography. This research was funded by the Manganese Nodule Project of the National Science Foundation (IDOE). Grant GX-34659.

Literature Cited

Agassiz, A. 1901. *Albatross* expedition preliminary report, *Mem. Mus. Comp. Zool. Harvard* 26:1–111

Agassiz, A. 1906. *Albatross* expedition reports. *Mem. Mus. Comp. Zool. Harvard* 33:1–50

Andrews, J. E. et al. 1974. *Ferromanganese Deposits of the Ocean Floor. Cruise Report Mn 74-01*. Hawaii Inst. Geophys. Rep. HIG 74-9 Honolulu. 193 pp.

Bender, M. L., Ku., T. L., Broecker, W. S. 1970. Accumulation ratios of manganese in pelagic sediments and nodules. *Earth Planet. Sci. Lett.* 8:143–48

Bonatti, E., Kraemer, T., Rydell, H. 1972. Classification and genesis of submarine iron-manganese deposits. In *Ferromanganese Deposits on the Ocean Floor* ed. D. R. Horn, 149–66. Washington DC: NSF-IDOE

Broecker, W. S. 1974. *Chemical Oceanography*. New York: Harcourt, Brace, Jovanovich. 214 pp.

Burns, R. G. 1976. The uptake of cobalt into ferromanganese nodules, soils, and synthetic manganese (IV) oxides. *Geochim. Cosmochim. Acta.* In press

Burns, R. G., Brown, B. A. 1972. Nucleation and mineralogical controls on the composition of manganese nodules. *Ferromanganese Deposits of the Ocean Floor*. ed. D. R. Horn, 51–62. Washington DC: NSF-IDOE

Burns, R. G., Burns, V. M., Sung, W., Brown, B. A. 1974. Ferromanganese nodule mineralogy: suggested terminology of the principal manganese oxide phases. *Geol. Soc. Am. Ann. Meet., Miami* 6:1069–71. (Abstr.)

Burns, R. G., Burns, V. M. 1975a. Manganese nodule authigenesis: mechanism for nucleation and growth. *Nature* 255:130–31

Burns, R. G., Burns, V. M. 1975b. Mineralogy of ferromanganese nodules. In *Marine Manganese Deposits*, ed. G. P. Glasby, Chap. 7. Amsterdam: Elsevir. In press

Burns, R. G., Fuerstenau, D. 1966. Electron microprobe determinations of inter-element relations in manganese nodules. *Am. Mineral.* 51:895–902

Buser, W. 1959. The nature of the iron and manganese compounds in manganese nodules. *Int. Oceanogr, Congr., AAAS*, pp. 962–63. (Preprints)

Buser, W., Grütter, A. 1956. Über die Natur der Manganknollen. *Schweiz. Mineral. Petrogr. Mitt.* 36:49–62

Carpenter, R., Wakeham, S. 1973. Mössbauer studies of marine and fresh water manganese nodules. *Chem. Geol.* 11:109–16

Cherdyntsev, V. V., Kadyror, N. B., Novichkova, N. 1971. Origin of manganese nodules of the Pacific Ocean from radioisotope data. Geokhimiya 3:339–54

Chukhrov, F. V., Zvyagin, B. B., Gorshkov, A. I., Yermilova, L. P., Balashova, V. V. 1973. On ferrihydrite (hydrous ferric oxide). *Akad. Nauk. SSSR, Ser. Geol. No. 4*, pp. 23–33 (in Russian)

Correns, W. 1941, Beiträge zur Geochemie des Eisens und Mangans. *Nachr. Akad. Wiss. Goettingen, Math. Phys. Kl.* 5:219–30

Cronan, D. S. 1972. Regional geochemistry of ferromanganese nodules in the world ocean. In *Ferromanganese Deposits on the Ocean Floor*, ed. D. R. Horn, 51–61. Washington DC: NSF-IDOE. 293 pp.

Cronan, D. S. 1975. Zinc in marine manganese nodules. *Inst. Min. Metall. Trans.* 84:30–32

Dietz, R. S. 1955. Manganese deposits on the northeast Pacific seafloor. *Calif. J. Mines Geol.* 51:209–20

Dudley, W. C. 1976. Concentration of iron

and cementation in Foraminifera on manganese nodules. *J. Foraminiferal Res.* 10 pp.
Ehrlich, H. L. 1972. The role of microbes in manganese nodules genesis and degradation. In *Ferromanganese Deposits on the Ocean Floor,* ed. D. R. Horn, pp. 63–70. Washington DC: NSF-IDOE
Ewing, M., Houtz, R., Ewing, J. 1969. South Pacific sediment distribution. *J. Geophys. Res.* 74:2477–93
Feitknecht, W., Giovanoli, R., Michaelis, W., Müller, M. 1973. Über die Hydrolyse von Eisen (III) Salzlosungen 1. Die Hydrolyse der Lösungen von Eisen (III) Chlorid. *Helv. Chim. Acta* 56:2847–56
Fewkes, R. H. 1973. External and internal features of marine manganese nodules as seen with the SEM and their implications in nodule origin. *The Origin and Distribution of Manganese Nodules in the Pacific and Prospects for Exploration,* ed. M. Morganstein, 21–26. Honolulu: Univ. Hawaii Press. 175 pp.
Fraser, J. Z., Arrhenius, G. 1972. Worldwide distribution of ferromanganese nodules and elemental concentrations in selected Pacific Ocean nodules. *Tech. Rep. No. 2 N.S.F. GX-34659,* Nat. Sci. Found., Washington DC. 51 pp.
Friedrich, G. H. W., Kunzendorf, H., Pluger, W. L. 1974. Ship-borne geochemical investigations of deep-sea manganese nodule deposits in the Pacific using a radioisotope energy dispersive X-ray system. *J. Geochem. Expl.* 3:303–17
Giovanoli, R., Bühler, H., Sokolowska, K. 1973. Synthetic lithiophorite: electron microscopy and X-ray diffraction. *J. Microsc., Oxford* 18:27184
Giovanoli, R., Burki, P. 1976. Comparison of X-ray evidence of marine manganese nodules and non-marine manganese ore deposits. *Chimea* In press
Giovanoli, R., Feitknecht, W., Fischer, F. 1971. Über Oxidhydroxide des vierwertigen Mangans mit Schichtengitter. 3. Reduktion von Mangan (III)-Manganat (IV) mit Zimtalkohol. *Helv. Chim. Acta* 54:1112–24
Giovanoli, R., Stahli, E., Feitknecht, W. 1970a. Über Oxidhydroxide des vierwertigen Mangans mit Schichtengitter. 1. Mitteillung: Natrium-Mangan (II, III) Manganat (V). *Helv. Chim. Acta,* 53:209–20
Giovanoli, R., Stahli, E., Feitknecht, W. 1970b. Über Oxidhydroxide des vierwertigen Mangans mit Schichtengitter. 2. Mangan (II)-Manganat (IV). *Helv. Chim. Acta* 53:453–64
Glasby, G. P. 1972. The Mineralogy of Manganese Nodules from a range of marine environments. *Mar. Geol.* 13:57–72
Glasby, G. P. 1973. Manganese deposits in the Southwest Pacific. In *Ferromanganese Deposits of the Ocean Floor,* Phase 1 report, pp. 137–69. Washington DC: NSF-IDOE
Glasby, G. P., Lawrence, P. 1974. Manganese deposits in the South Pacific Ocean. *NZ Oceanogr. Inst. Chart, Misc. Ser.,* 33.
Goodell, H. G., Meylan, M. A., Grant, B. 1971. Ferromanganese deposits of the South Pacific Oceans, Drake Passage and Scotia Sea. In *Antarctic Oceanography I. Antarct. Res. Ser.,* ed. J. L. Reid, 27–92. 343 pp.
Graham, J. W. 1959. Metabolically induced precipitation of trace elements from seawater. *Science* 129:1428–29
Graham, J. W., Cooper. S. C. 1959. Biological origin of manganese-rich deposits of the seafloor. *Nature* 183:1050–51
Greenslate, J. L. 1974. Manganese and biotic debris associations in some deep-sea sediments. *Science* 186:529–31
Hammond, A. L. 1974a. Manganese nodules (I): Mineral resources on the seabed. *Science* 183:502–3
Hammond. A. L. 1974b. Manganese nodules (II): Prospects for deep-sea mining. *Science* 183:644–46
Hedley, R. H. 1963. Cemented iron in the arenaceous Foraminifera. *Micropaleontology* 9:433–41
Hey, M. H., Embrey, P. G. 1974. Twenty-eighth list of new mineral names. *Mineral. Mag.* 39:903–32
Horn, D. R., Horn, B. M., Delach, M. N. 1973a. *Factors Which Control the Distribution of Ferromanganese Nodules and Proposed Research Vessels track, North Pacific. Tech. Rep. No. 8, N.S.F. GX-33616,* Nat. Sci. Found., Washington DC. 20 pp.
Horn, D. R., Horn, B. M., Delach, M. N. 1973b. *Ocean Manganese Nodules: Metal Values and Mining Sites. Tech. Rep. No. 4, GX-33616,* Nat. Sci. Found.–IDOE. Washington DC. 57 pp.
Kennett, J. P., Thunell, R. C. 1975. Global increase in Quaternary explosive volcanism. *Science* 187:497–502
Kennett, J. P., Watkins, N. D. 1975. Deep sea erosion and manganese nodule pavement development in the South and Indian Ocean. *Science* 188:1011–13
Kogler, F. C. 1975. Sediment-physical properties of three deep-sea cores from the Central Pacific Ocean. *Mar. Technol.*

5:199–201

Ku, T. L., Broecker, W. S. 1969. Radiochemical studies on manganese nodules of deep-sea origin. *Deep Sea Res.* 16: 625–37

Lalou, C., Brichet, E., Wyart, M. J. 1972. Significant des mesures radiochimiques dans l'evaluation de la vitesse de croissance des nodules de manganese. *CR Acad. Sci. Paris, Ser. D,* pp. 815–18

Loganathan, P., Burau, R. G. 1973. Sorption of heavy metal ions by a hydrous manganese oxide. *Geochim. Cosmochim. Acta* 37:1277–93

Luyendyk, B. P. 1969. Origin and history of abyssal hills in the Northeastern Pacific Ocean. *Geol. Soc. Am. Bull.* 81: 2237–60

Luyendyk, B. P. 1970. *Geological and geophysical observations in an abyssal hill area using a deep tow package.* PhD thesis. Univ. Calif., San Diego. 225 pp.

Margolis, S. V. 1973. Manganese deposits encountered during deep-sea drilling project leg 29 in sub-Antarctic water. In *The Origin and Distribution of Manganese Nodules in the Pacific and Prospects for Exploration,* ed. M. Morgenstein, 109–13. Honolulu: Univ. Hawaii Publ. 175 pp.

Margolis, S. V., Andrews, J. E., Dudley, W. C. 1976. Microstructure and microchemistry of Pacific manganese nodules. *Proc. Circum-Pacific Energy Resources Symp. Am. Assoc. Petrol. Geol. Spec. Publ.* 15 pp. In press

Margolis, S. V., Glasby, G. P. 1973. Microlaminations in marine manganese nodules as revealed by scanning electron microscopy. *Geol. Soc. Am. Bull.* 83: 3601–10

Menard, H. W., Shipek, C. J. 1958. Surface concentrations of manganese nodules. *Nature* 182:1156–58

Mero, J. L., 1965. *The Mineralogical Resources of the Sea.* Amsterdam: Elsevier. 312 pp.

Meyer, K. 1973. Surface sediment and manganese nodule facies encountered on R/V *Valdivia* cruises 1972/1973. In *The Origin and Distribution of Manganese Nodules in the Pacific and Prospects for Exploration,* ed. M. Morgenstein, 125–30. Honolulu: Univ. Hawaii Publ. 175 pp.

Meylan, M. A. 1974. Field description and classification of manganese nodules. See Andrews et al. 1974, pp 158–68

Morgenstein, M. 1972. Manganese accretion at the sediment-water interface at 400 to 2400 m. depth, Hawaiian Archipelago. *Ferromanganese deposits on the ocean floor,* ed. D. R. Horn, 131–38. Washington DC: NSF-IDOE

Murray, J., Renard, A. 1891. Manganese nodules. In *Report on the Scientific Results of the Voyage of the H.M.S. Challenger, Vol 5 Deep-Sea Deposits,* ed. C. W. Thompson, 341–78. London: Eyre & Spottiswoode. 525 pp.

Ostwald, J., Frazer, F. W. 1973. Chemical and mineralogical investigations on deep-sea manganese nodules from the Southern Ocean. *Miner. Deposita* 8:303–11

Raab, W. 1972. Physical and chemical features of deep-sea manganese nodules and their implications to the genesis of nodules. In *Ferromanganese Deposits on the Ocean Floor,* ed. D. R. Horn, 31–49. Washington DC: NSF

Skornyakova, N. S., Andrushchenko, P. F. 1974. Iron-manganese concretions in the Pacific Ocean. *Int. Geol. Rev.* 16:863–919

Sorem, R. K. 1967. Manganese nodules: nature and significance of internal structure. *Econ. Geol.* 62:141–47

Sorem, R. K., Foster, A. R. 1972. Internal structure of manganese nodules and implications in beneficiation. In *Ferromanganese Deposits of the Ocean Floor,* ed. D. R. Horn, 167–79. Washington DC: NSF-IDOE

Van Andel, T. J., Heath, G. R. 1973. The central-equatorial Pacific west of the East Pacific Rise. In *Geological Results of Leg 16. Initial Reports of the Deep Sea Drilling Project,* Vol. 16, pp. 19–312. Washington DC

Wadsley, A. C. 1955. The crystal structure of chalcophanite, $AnMn_3O_7.3H_2O$. *Acta Crystallogr.* 8:165–72

Wakeham, S., Carpenter, R. 1974. Electron spin resonance spectra of marine and fresh-water manganese nodules. *Chem. Geol.* 13:39–47

Wendt, J. 1974. Encrusting organisms in deep-sea manganese nodules. *Spec. Publ. Int. Assoc. Sedimentol.* 1:437–47

Woo, C. C. 1973. Scanning electron micrographs of marine manganese micronodules, marine pebble-sized nodules, and fresh-water manganese nodules. In *Papers on the Origin and Distribution of Manganese Nodules in the Pacific and Prospects for Exploration,* ed. M. Morgenstein, 165–71. Honolulu: Univ. Hawaii Publ. 175 pp.

Zenkevitch, N., Skornyakova, N. S. 1961. Iron and manganese on the ocean bottom. *Natura USSR* 3:47–50

HYDROGEN LOSS FROM THE TERRESTRIAL PLANETS

Donald M. Hunten
Kitt Peak National Observatory, Tucson, Arizona 85726

Thomas M. Donahue
Department of Atmospheric and Oceanic Science, The University of Michigan, Ann Arbor, Michigan 48104

INTRODUCTION

The idea of escape of light atmospheric gas is older than the kinetic theory, having been introduced by J. L. Waterston in 1846 (Chamberlain 1963). The thermal escape concept was revived by G. J. Stoney around the turn of the century and given a definitive form by Jeans (1925), whose name is usually associated with it. Spitzer (1952) was the first to realistically incorporate atmospheric structure into the problem. Observation in 1955 of Lyman-α scattering from the geocorona, the atomic-hydrogen cloud around the Earth, and many more recent satellite and space-probe measurements have led to a number of interpretive and theoretical papers, which Chamberlain (1963) and Tinsley (1974) have reviewed. Composition and structure of other atmospheres are discussed by Hunten (1971b) and Ingersoll & Leovy (1971).

Escape from a planet is intimately linked to the nature of the exosphere, the region in which the mean free path is so long that collisions can be neglected for many purposes. Most atoms in the exosphere are in free ballistic orbits; a few of the fastest ones may be on escape orbits. The bottom of the exosphere is the critical level, or exobase, defined as where the mean free path (in the horizontal direction) is equal to the scale height. The conventional approximation of the critical level is that the atmosphere is fully collisional below it and collisionless above; Chamberlain (1963) discusses the validity of this approximation. The planetary corona can be defined as the hydrogen component of the exosphere.

Global averages are used throughout most of the present review. The exosphere is actually asymmetric to a considerable degree, as are the escape processes. Nevertheless, the long mean free path permits a great deal of lateral flow, which helps to justify the approximation of spherical symmetry (McAfee 1967; Vidal-Madjar & Bertaux 1972; Tinsley 1974; Tinsley, Hodges & Strobel 1975).

Atmospheric escape is an interesting phenomenon, particularly for its effect on the evolution of atmospheres. Loss of hydrogen creates free oxygen at significant rates, for example, twice the amount now in the Earth's atmosphere if the current rate has always been valid. On Mars, it is probable that the oxygen also escapes nonthermally, in such a way that the hydrogen loss is balanced. The recent appreciation of nonthermal processes has totally changed our picture of atmospheric escape. This has been accompanied by a simple means of describing the important diffusion effects that link the upper and lower atmospheres (Hunten 1973b). Space limitations have kept the descriptions brief; many sections must be regarded as little more than guides to the literature.

PROCESSES

In this section we discuss the various individual processes that enter into the picture of hydrogen loss: the escape mechanisms themselves, the production of H and H_2 from other molecules, and the ionization of H atoms.

Thermal or Jeans Escape

The escape flux ϕ_J is readily derived with the postulate of a critical level, a surface from which all the escaping atoms are regarded as originating. A convenient variable is the ratio of gravitational to thermal energy:

$$\lambda = \frac{GMm}{KTr} = \frac{V_{\text{esc}}^2}{U^2} = \frac{r}{H}. \tag{1}$$

G is the Newtonian gravitational constant, M the mass of the planet, m the mass of the atom, k Boltzmann's constant, T the temperature, r the distance from the center of the planet, v_{esc} the escape velocity at r, $U = (2kT/m)^{\frac{1}{2}}$, a thermal velocity, and $H = kT/mg$ the scale height (g is the acceleration of gravity at r). The Maxwell-Boltzmann velocity distribution is weighted by the cosine of the zenith angle and integrated over the upper hemisphere, and from v_{esc} to infinity, to give the flux:

$$\phi_J = n_c w_J = n_c BU(1+\lambda_c)\exp(-\lambda_c)/2\pi^{\frac{1}{2}}. \tag{2}$$

Subscript c refers to the critical level; n_c is the number density of H atoms at the critical level, and w_J a velocity defined by the second equality, usually called the effusion velocity. The factor B, usually between 0.5 and 1, is discussed below.

Taking account of the variation of gravity with height, the barometric equation for an isothermal upper atmosphere reads

$$n = n_c \exp(\lambda - \lambda_c). \tag{3}$$

According to Equation 3, the density at infinity has the finite value $n_c \exp(-\lambda_c)$, another indication that escape must be important, unless λ_c is very large. Despite this defect in Equation 3, it is a useful approximation at the heights of primary interest. Corrections are discussed by Herring & Kyle (1961), Chamberlain (1963), and Hunten (1973b). Substitution of Equation 3 into Equation 2 gives

$$\phi_J = nBU(1+\lambda_c)\exp(-\lambda)/2\pi^{\frac{1}{2}}. \tag{4}$$

Except for the slowly varying term $(1+\lambda_c)$, Equation 4 does not contain any reference to the critical level; it may be evaluated at any convenient height with a temperature equaling that of the critical level that is connected to it by Equation 3. This result illustrates why the exact definition of the critical level is unimportant. The simplest possible definition may therefore be adopted: at the critical level, the mean free path is equal to the scale height, and the probability is $1/e$ that a particle travelling straight up can either escape totally from the planet or can go into a free elliptical orbit. In terms of the collision cross-section $Q \simeq 3.3 \times 10^{-15}$ cm^2, the total density at the critical level is given by

$$n_{ct} H_{ct} Q = 1. \tag{5}$$

For the Earth, the corresponding height is 450–500 km.

A detailed discussion of the validity of Equations 2 and 5 has been given by Chamberlain (1963). He confirms their accuracy and shows that they are intimately linked to the existence of the Maxwell-Boltzmann velocity distribution. Because hydrogen atoms in the high-velocity tail are escaping, this tail must be at least slightly depressed. Graphic illustrations of this effect are given by Öpik & Singer (1961).

Subsequently, several authors attempted to calculate the corresponding reduction of the escape flux. Brinkmann (1971) and Chamberlain & Smith (1971) finally obtained convergence. The effect depends on the ratio of masses of the escaping and background gases; with a large ratio, energy exchange during a collision is inefficient and the deviation larger. The deviation also increases as escape becomes more rapid, or λ_c smaller. Typical values for the Jeans ratio B (i.e., actual flux divided by Jeans flux) are 0.50 for Mars (H in CO_2), 0.72 for Earth (H in O), and close to unity for equal masses.

Equation 2, corrected in this way, gives a relation between the escape flux and the density n_c. Until recently, it was assumed that n_c was determined by the atmospheric structure and that the flux can thus be obtained. It is often the reverse, however; Hunten first realized this for Titan (Hunten 1973a), and it was later applied to the Earth (Hunten 1973b, Hunten & Strobel 1974, Liu & Donahue 1974a). The hydrogen flux can be determined by processes acting hundreds of kilometers below the critical level, and the Jeans equation then serves to adjust the density appropriately. This was recognized much earlier for hydrogen atoms above 100 km, but the full implications were not realized at the time.

In fact, both Hunten & Strobel and Liu & Donahue found the thermal escape flux is only one-third to one-half what was espected. Liu & Donahue (1974b,c) therefore proposed that escape of nonthermal atoms produced by charge exchange with fast protons, along with a smaller contribution by direct proton escape at polar latitudes compensated for the difference. These processes are discussed in the next two sections.

Charge Exchange

The Earth's plasmasphere contains a component of hot protons trapped in the magnetic field. According to Serbu & Maier (1970), the temperatures range from

5000° to 20,000°K and the density can be as large as 10^4 cm^{-3} at $r = 1.5$ Earth radii on the day side. These protons cannot escape, but they can exchange charge with exospheric hydrogen atoms:

$$H^+ + H \rightarrow H + H^+. \tag{6}$$

The result is a thermal proton and a fast atom that is almost certain to escape unless directed at the Earth. The cross-section is large: 4×10^{-15} cm^2. The significance of escape by charge exchange was pointed out by Cole (1966), and a crude quantitative estimate has been made by Tinsley (1973). He stresses that the Serbu & Maier ion densities may be somewhat too large. The rate of reaction, Equation 6, per unit volume is $kn(r)n^+(r)$, where k is the rate constant calculated by Dalgarno (1960) and n^+ is the proton density. Tinsley multiplied this rate by the geometrical factor, $(r/r_c)^2$, to normalize the flux to the critical level, and another factor, g, to account for the fraction of particles travelling in directions that will permit them to escape. Thus he obtained

$$\phi_E = \int g(r/r_c)^2 knn^+ \, dr \tag{7}$$

as the contribution to the charge exchange flux from $1.5R_E$ to infinity. Using a hydrogen density at the critical level of about 3×10^4 cm^{-3}, which is appropriate to a thermal escape flux of 10^8 cm^{-2} sec^{-1} at 1250°K (the day side of this model), Tinsley found the corresponding charge exchange escape flux to be 4×10^8 cm^{-2} sec^{-1}, averaged over the globe, or four times the thermal flux. Thus, even if Tinsley's estimate is too high, it seems likely that the charge-exchange process is is somewhat more important than the Jeans process.

Proton Escape

Because of the magnetic field, protons cannot normally escape from the Earth. There is one exception: at high magnetic latitudes, the field lines are "open," being drawn out into a long tail by the solar wind. Proton escape along these lines is so rapid as to constitute a "polar wind" (Banks & Holzer 1969). The flow is greatly aided by an upward electric field caused by the more mobile electrons escaping, combined with O^+ ions remaining in the ionosphere. Velocities typically approach 10 km/sec at $r = 2$ earth radii. For a particular choice of neutral-hydrogen density, Banks & Holzer obtain ion fluxes in the neighborhood of 5×10^8 cm^{-2} sec^{-1}. Liu & Donahue (1974b) estimate that the fraction of the global area is 6%, and the mean flux is therefore 3×10^7 cm^{-2} sec^{-1}. Modified estimates, adapted to specific conditions, are given below in the discussion of the Earth.

Solar Wind Sweeping and Accretion

The solar wind, for the present purpose, may be considered a fully ionized hydrogen plasma that carries an embedded magnetic field and sweeps past the planets at a typical velocity of 320 km/sec. At the Earth's orbit, the proton and electron densities are 8 cm^{-3} and the fluxes therefore $F = 2.5 \times 10^8$ cm^{-2} sec^{-1} (Hundhausen 1972). At Venus and Mars, the densities and fluxes are 1.9 and 0.37 as great.

Because of the magnetic field, any ions created in the solar wind are swept away. The details of the process involve the motional electric field, which causes the new ions to move on a cycloidal trajectory. Such details are of interest near the limb of a planet, where the transverse motion may carry the ions into the atmosphere instead of downstream, but the total mass loss is affected very little (Wallis 1972; Cloutier, Daniell & Butler 1974). We therefore adopt the simple description of Michel (1971): new ions are caught up in the solar wind and swept away.

The geometry of the flow past the various planets is discussed below, along with the shielding of the Earth by the geomagnetic field. First, however, we discuss the creation of ions in the outer atmosphere.

ION PRODUCTION One limit on the loss rate from a planet is the rate at which ions can be created. The planetary hydrogen corona is optically thin to solar ionizing radiation, and the ionization coefficient J can be calculated by integrating the product of solar photon flux and the ionization cross-section over all frequencies. The result is $J = 5 \times 10^{-8}$ sec^{-1} for the Earth; the mean lifetime of an atom is $1/J = 2 \times 10^{7}$ sec, or about eight months. The column abundance of H above the exobase is about 5×10^{12} atoms/cm^2 (Meier & Mange 1973), and the corresponding ion production rate is 2.5×10^{5} ions cm^{-2} sec^{-1}. Allowance for shadowing on the night side reduces this number to around 1.5×10^{5} cm^{-2} sec^{-1} as a global mean. With the total escape rate over 10^{8} cm^{-2} sec^{-1}, the photoionization rate is trivial by comparison.

For the Earth, a much more potent process exists: ionization of O atoms followed by the charge exchange

$$O^{+} + H \leftrightarrows O + H^{+}. \tag{8}$$

Reaction 8 happens to be resonant, and can proceed in either direction at a rapid rate (Fehsenfeld & Ferguson 1972). Because the total production of O^{+} is enormous compared to the photoionization of H, the net direction is very much to the right. The large polar wind flow discussed above essentially originates through Reaction 8. On Mars and Venus, any such processes are much less important, and the potential of ion escape is correspondingly much more limited.

SWEEPING The capacity of the solar wind to pick up atmospheric ions is limited (Biermann, Brosowski & Schmidt 1967; Cloutier, McElroy & Michel 1969). If the added mass flux exceeds the incident mass flux by a factor $\sigma \simeq 9/16$, a shock forms, and the solar wind is diverted around the planet. Michel (1971) has estimated the maximum possible mass addition to this diverted flow, using Figure 1 as a model. The lowest flow line to return to space is determined by the new atmospheric ions as just outlined; therefore the mass loss is essentially independent of atmospheric density. If all ions above the dividing level are lost, the height integral gives a factor H and the integral around the limb gives $2\pi r \sin\theta$. If F is the proton flux in the solar wind, the maximum swept-flux is σF, and the total number of protons carried away per second is $(\sigma F)(2\pi r H \sin\theta)$. This may be divided by $4\pi r^2$ to get the global-mean flux. Michel estimates that $\sigma \sin\theta$ is approximately 0.34; the result is therefore

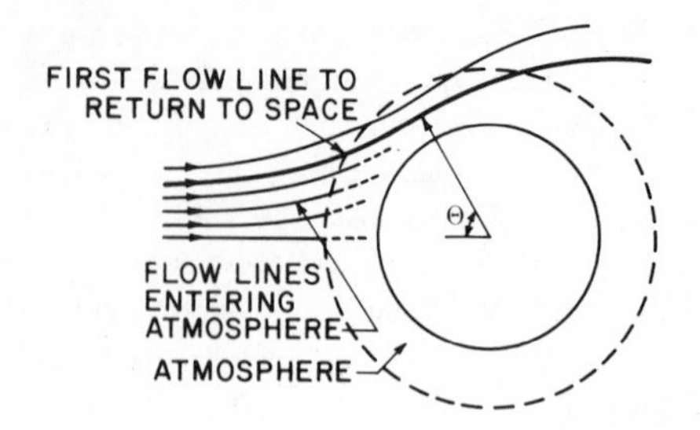

Figure 1 Michel's model for flow of solar wind past an unmagnetized planet. Probably the angle θ of closest approach is considerably smaller than the sketch.

$$\phi_s \simeq 0.34FH/2r = 0.17F/\lambda, \tag{9}$$

where $\lambda = r/H$ as before. For Venus, $\lambda \simeq 15$ at a temperature of 400°K, and ϕ_s therefore 5×10^6 cm^{-2} sec^{-1}. It is not much less for Mars, where F and λ are both reduced by a similar factor.

For Earth, the solar wind is deflected by the magnetosphere at a great distance, and the picture outlined here does not apply. Instead, we have the more potent processes discussed in earlier sections. The magnetosphere protects the atmosphere from a direct interaction, but evidently the indirect effects become much larger; in a sense, the magnetosphere gathers energy from a large cross section of solar wind and makes some of it available for escape.

ACCRETION In Figure 1, the solar wind flow below the dividing line runs into the planet; the separation of the flow lines is exaggerated for clarity. Without this distortion, the corresponding area upsteam in the solar wind is probably much less than πr^2. If the factor is ~ 0.1, the number of protons accreted per second is $\sim 0.1\pi r^2 F$; the ratio of gain to loss is $\sim \lambda/7 \sim 1$. Most likely, then, for an unmagnetized planet with an atmosphere the gain of hydrogen from the solar wind is of the same order as the corresponding loss.

On the Earth, the loss seems to be dominant. Johnson & Axford (1969), in a discussion of the He^3 budget, adopted an auroral proton flux of 4×10^5 cm^{-2} sec^{-1} (global mean). From the consistency obtained for He^3, they suggested that this flux might indeed come from the solar wind. In any case, it must be close to an upper limit. A more relaxed limit can be obtained from the neon abundance on the Earth (Junge, Oldenberg & Wasson 1962). Even in the unlikely event that all this neon came from the solar wind, its influx would not exceed 3000 atoms cm^{-2} sec^{-1}. With a Ne/H ratio of 5×10^{-5} (Cameron 1973), the hydrogen flux must be less than 6×10^7 cm^{-2} sec, and probably is much less. Even this figure, however, is significantly smaller than the estimated loss due to charge transfer.

Diffusion and Mixing

If hydrogen is escaping from the top of an atmosphere, it must be flowing up from its source at or near the surface. The details have only recently come under scrutiny, and it has become clear that the flow plays an important, sometimes dominating, role in setting the total escape flux (Hunten 1973a,b; Hunten & Strobel 1974; Liu & Donahue 1974a–c). Earlier work directed at the Earth's thermosphere came close to the issue but did not make the final connection because of the practice of taking a lower boundary at 100–120 km (Bates 1959; Bates & Patterson 1961; Mange 1961; Kockarts & Nicolet 1962, 1963).

It is convenient to divide an atmosphere into two regions, the homosphere below and the heterosphere above; the boundary is the homopause. The homosphere is uniform in composition, except for condensable gases and some minor constituents involved in photochemistry. The stirring of the homosphere takes place by vertical motions on all sorts of scales, often loosely termed turbulence; correspondingly, the homopause is often called the turbopause. In the heterosphere, the prevailing situation is diffusive equilibrium; each stable constituent takes up its individual scale height corresponding to its own mass. Unstable or escaping constituents, flowing upward or downward, take up a dynamic distribution that can be obtained from the diffusion equation (as discussed below). This process is appreciably simplified if the constituent can be treated as minor, that is, as making a negligible contribution to the total density. This restriction, though adopted here, is not necessary (Hunten 1973a,b).

FLOW EQUATION A convenient flow equation can be formulated to include both (molecular) diffusion and mixing (or eddy diffusion) (Colegrove, Hanson & Johnson 1965, 1966):

$$\phi_i = -D_i n_i \left(\frac{1}{n_i}\frac{dn_i}{dz} + \frac{m_i g}{kT} + \frac{1+\alpha_i}{T}\frac{dT}{dz} \right) - K n_i \left(\frac{1}{n_i}\frac{dn_i}{dz} + \frac{m_a g}{kT} + \frac{1}{T}\frac{dT}{dz} \right). \quad (10)$$

Subscript i denotes the flowing constituent (usually H or H_2 in this paper), and a the background atmosphere. The thermal-diffusion factor is α_i; D_i is the diffusion coefficient and K is the eddy diffusion coefficient. Hunten (1973b) gives a collection of values for D_i, which is conveniently written $D_i = b_i/n_a$; b_i is a binary collision parameter, equal to 2.73×10^{19} cm^{-1} sec^{-1} for H in air, and 1.46×10^{19} for H_2 in air (at 208°K). Typically, b (and D along with it) varies as $T^{0.7}$.

K is best described as a parameter to be determined empirically by fitting a solution of Equation 10 to observations of a suitable tracer. Hunten & Strobel (1974) recommend $K = 3 \times 10^5$ cm^2 sec^{-1} near 100 km on Earth, giving high weight to observations of argon, but including helium and oxygen as well. For Venus and Mars, K seems to be close to 10^8 cm^2 sec^{-1}, as discussed below.

D_i varies inversely as atmospheric density and therefore grows rapidly with height. The homopause is conveniently and consistently defined as the height where $D_i = K$. The corresponding density is $n_{ah} = b_i/K$. At any appreciable distance from this level, one or the other term in Equation 10 is dominant and the other usually neglected. Useful analytic solutions exist, some of which Hunten (1975) discusses. Hydrostatic equilibrium is obtained when $\phi_i = 0$. In the homosphere, the second term is dominant, and setting the parenthesis equal to zero gives a distribution of n_i parallel to that of n_a. In the heterosphere, the converse is true, and vanishing of the first parenthesis gives the state of diffusive equilibrium.

LIMITING FLOW For the present purpose, a useful transformation of Equation 10 was introduced by Hunten (1973a,b). The mixing ratio $f_i = n_i/n_a$ is used, and a limiting flux, ϕ_l, is defined as follows:

$$\phi_l = \frac{b_i f_i}{1+f_i}\left[(m_a - m_i)\frac{g}{kT} - \frac{\alpha_i}{T}\frac{dT}{dz}\right]$$

$$\simeq \frac{b_i f_i}{H_a}\left(1 - \frac{m_i}{m_a}\right)$$

$$\simeq \frac{D_i n_i}{H_a}\left(1 - \frac{m_i}{m_a}\right). \tag{11}$$

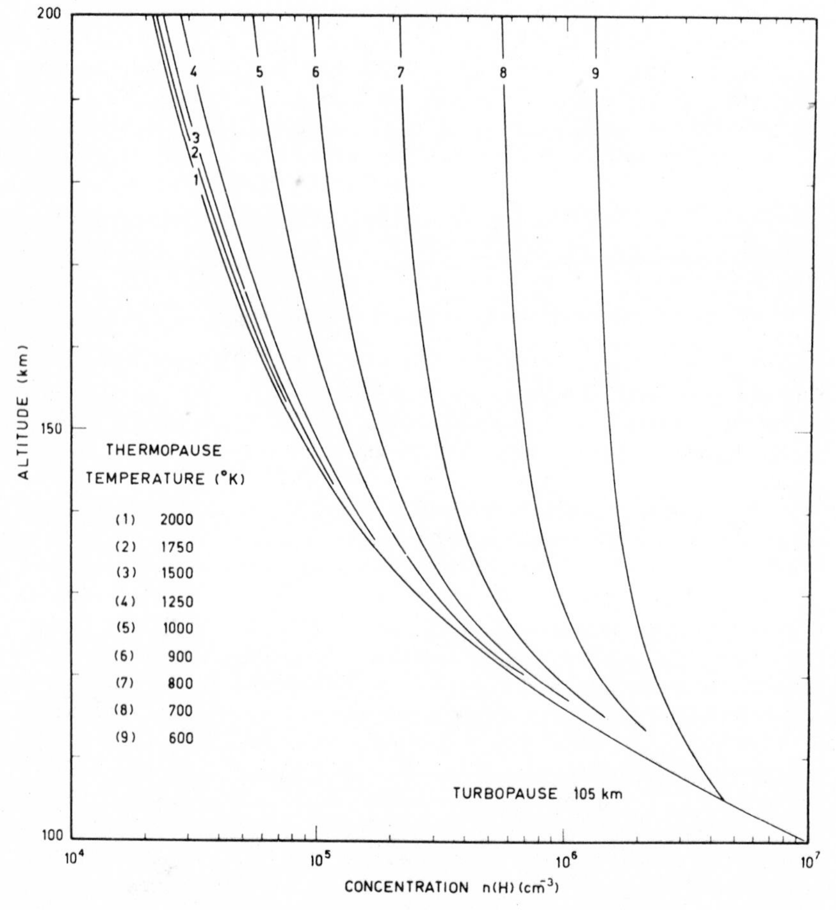

Figure 2 An early model (Kockarts & Nicolet 1963) showing limiting flow of atomic hydrogen. The envelope at the left is the limiting-flow line, and the various curves go into diffusive equilibrium at different points. Only thermal escape is considered.

The approximations hold for a small temperature gradient. A further approximation for a light gas such as H or H_2 is

$$\phi_l \simeq b_i f_i / H_a, \tag{12}$$

and the corresponding limiting velocity is

$$w_l = \phi_l / n_i \simeq b_i / n_a H_a = D_i / H_a.$$

After some manipulation, Equation 11 in Equation 10 gives

$$\phi_i = \phi_l - (K + D_i) n_a \frac{\mathrm{d}f_i}{\mathrm{d}z}. \tag{13}$$

Consequently, there must be a gradient of mixing ratio unless $\phi_i = \phi_l$, although in many cases the gradient is negligibly small. If $\phi_i > \phi_l$, the gradient is negative and the mixing ratio f_i decreases upwards, tending to choke off the flow. Thus, ϕ_l is really a limiting flux, to be exceeded appreciably only where K or n_a is large enough to keep $\mathrm{d}f_i/\mathrm{d}z$ close to zero. These ideas are mainly applied just above the homopause, where both quantities are small. If escape from the top of the atmosphere is easy (normally true for the Earth), the upward flux tends to be only slightly smaller than ϕ_l. Equation 13 then implies $\mathrm{d}f_i/\mathrm{d}z \simeq 0$: the hydrogen takes up almost the same vertical distribution as the background atmosphere. Physically this effect arises because the limiting velocity w_l is inversely proportional to n_a. However, there remains a small upward increase of f_i, and a corresponding increase of ϕ_l by Equation 12. At some height the difference $\phi_l - \phi_i$ becomes large and so does $\mathrm{d}f_i/\mathrm{d}z$: the hydrogen reverts to diffusive equilibrium with a very large scale height. This behavior is illustrated in Figure 2, calculated by Kockarts & Nicolet (1963). The lower envelope of the curves represents limiting diffusion. (These computations include only the Jeans escape process.)

The consequences of the limiting-flow principle are far-reaching. Given an efficient escape process (e.g. a high-enough exospheric temperature), the maximum possible escape flux can be obtained immediately if the mixing ratio of available hydrogen at the homopause is known (Equations 11 or 12). For hydrogen on Earth (as discussed below and illustrated in Figure 2), the appropriate conditions are usually realized.

The meaning of limiting flow can be illustrated with a transformed heterospheric model of Figure 3 (Hunten 1971a). The whole region has uniform density n_{ah}, its height is therefore H_{ah}, and the critical level is slightly below the top. The diffusion coefficient is $D_h = b/n_{ah} = K$ (from the definition of the homopause); the subscript

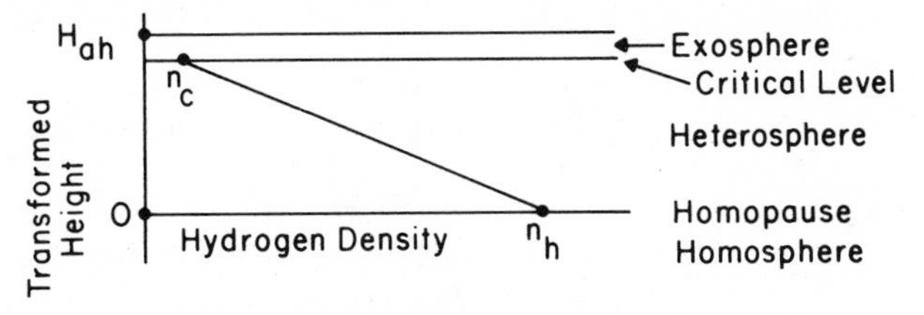

Figure 3 Constant-density model of the heterosphere.

i has been dropped for simplicity. The straight line shows the hydrogen distribution, still with the assumption of easy escape, so that $n_c \ll n_h$. The density drop across the slab is therefore n_h, and the elementary diffusion equation gives $\phi = -D_h \, dn/dz = D_h n_h/H_{ah} = bf_h/H_{ah}$, which is Equation 12. Diffusive equilibrium, on the other hand, corresponds to $n_c \simeq n_h$ and a much smaller flux.

CRITERION FOR LIMITING FLOW We now give an approximate criterion to decide whether or not limiting flow is likely. If the result is borderline, a more detailed examination is necessary. In the limiting-flow situation, $n_c \ll n_h$, as just discussed; in the classical or Spitzer (1952) situation, $n_c \simeq n_h$. We use the limiting velocity at the homopause, $w_h = D_i/H_a = K/H_a$. For the Earth, $w_h = 3 \times 10^5/6 \times 10^5 = 0.5$ cm/sec. Correspondingly, there is an effusion velocity w_c at the critical level; it includes the Jeans velocity w_J of Equation 2 and contributions from any other significant escape processes. The smaller of these two velocities indicates the controlling, or bottleneck, process, i.e. diffusion or escape (Hunten 1973b). In this simple discussion, it is assumed that the critical level and homopause are not very far apart, so that inverse-square effects can be neglected and the scale height of hydrogen taken as infinity.

ION FLOW Large upward and downward fluxes of protons take place on the Earth as ionospheric and magnetospheric conditions change, and the high-latitude flux has already been mentioned as an escape process. All these flows are limited in a manner closely analogous to the discussion above (Hanson & Ortenburger 1961, Banks & Holzer 1969, Banks & Kockarts 1973), The diffusion coefficient for protons is much smaller than for atoms because of the Coulomb interaction with ions and the efficient charge-transfer process (Equation 8).

TOTAL HYDROGEN For the homosphere, a useful flow equation can be written to describe "total hydrogen": the sum of all hydrogenous constituents weighted by the number of hydrogen atoms (Thomas 1973, Hunten & Strobel 1974). At heights where $D_i \ll K$, Equation 13 becomes approximately $\phi_i \simeq -Kn_a(\mathrm{d}f_i/\mathrm{d}z)$. We write one such equation for each component of total-H, weight as described above, and add. The result is

$$\phi_t = -Kn_a \frac{\mathrm{d}f_t}{\mathrm{d}z} = \text{constant}. \tag{14}$$

The total flux must be constant because there can be no sources or sinks of total-H. This result neglects molecular diffusion but nonetheless is accurate enough to indicate the true situation. Processes such as precipitation are not included, so this result is not valid in the troposphere.

Equations 12 and 14, applied to Earth, make predictions that detailed computation confirms (Hunten & Strobel 1974; Liu & Donahue 1974a,b,c). The total-H flux at 30 km is equal to the escape flux (apart from a small r^2 factor). This flux is small enough that $\mathrm{d}f_t/\mathrm{d}z$ is essentially zero throughout the homosphere, that is, f_t is constant from 30 to 100 km. If the chemical form of hydrogen is known at 100 km (the homopause), the limiting flux can be obtained by the expression

$\phi_l = b_i f_t / H$; this flux is therefore almost completely determined by the value of f_t in the stratosphere. The dominant form at 100 km is H_2, which is converted to atoms in the heterosphere (Process 15 below).

Hydrogen Sources

At first sight, the only significant source of escaping hydrogen would seem to be H_2O, which is so abundant in our own environment. In fact, this is far from clear, even for the Earth, where CH_4 seems to be more important, perhaps dominant. On Venus an important source is HCl. On all planets H_2 must be considered, if only as an intermediate, and for primitive atmospheres NH_3 (as well as CH_4) must be included.

DISSOCIATION The stable molecules listed above must be broken up, either by solar photons or free radicals. Photolysis occurs mainly above 40–50 km on the Earth and all the way to the cloud tops on Venus and the surface on Mars. Computation of the rate for a given molecule requires integration over the wavelength of the cross-section multiplied by the attenuated solar flux. The result is a photodissociation coefficient J, as introduced above for ion production. Information on cross-sections is found in reviews by McNesby & Okabe (1964) and Hudson (1971). J coefficients are given by Kockarts (1971) for H_2O in air, by Nicolet (1971) for CH_4 in air, and by Hunten & McElroy (1970) for H_2O in CO_2. Photolysis of CH_4 commonly produces H_2 directly; some of the events with H_2O at short wavelengths do the same. Dissociation rates (J multiplied by concentration) for H_2O and HCl on Venus are shown by McElroy, Sze & Yung (1973).

FREE RADICALS Methane is destroyed by reaction with OH and metastable $O(^1D)$, and H_2 and H_2O by the latter. These processes become dominant below about 50 km on Earth, where the $O(^1D)$ is produced by photolysis of O_3. Densities are given (as diurnal means) by McElroy et al (1974). For the present purpose, the subsequent reactions after methane is first attacked are considered instantaneous, leading finally to two water molecules per methane molecule.

Hunten & McElroy (1970), considering the $O(^1D)$ processes for Mars, found that H_2O destruction is probably much less than direct photolysis. They found that a considerable amount of H_2 is produced in subsequent reactions, and its reaction with $O(^1D)$ is important, particularly at high latitudes where H_2O is frozen out and there would otherwise be no odd hydrogen.

Ozone is rarer on Venus, and the production rate of $O(^1D)$ correspondingly slower; it probably does not play a significant role. More likely, the chlorine atoms from HCl photolysis split up H_2 into atoms. This effectively accomplishes photolysis of H_2, which is very slow by direct absorption. The reverse reaction also occurs (McElroy, Sze & Yung 1973; Sze & McElroy 1975).

At high altitudes on all three planets, an important question is how quickly H_2 can be broken up into atoms. As just mentioned, photolysis is very slow, and in any case attenuation by other gases inhibits it. On the Earth, a very efficient process operates at the high temperatures above 130 km (e.g. Hunten & Strobel 1974):

$$H_2 + O \rightarrow OH + H$$
$$OH + O \rightarrow O_2 + H. \quad (15)$$

An H_2 molecule is thus split by formation of an O_2 molecule. High temperatures are required because the first step is slightly endothermic. Reaction with $O(^1D)$ is also significant below 130 km.

The temperatures on Mars and Venus are now regarded as too low for Process 15 to operate, and the O densities are small as well. Instead, there is an ionospheric process (McElroy & Hunten 1969b):

$$X^+ + H_2 \rightarrow XH^+ + H$$
$$XH^+ + e \rightarrow X + H. \quad (16)$$

The ion X^+ can be CO_2^+, CO^+, O^+, or N_2^+, but not O_2^+ or NO^+. O_2^+ is currently favored over CO_2^+ as the major ion on both planets, and on Mars at least the conversion of H_2 may be incomplete (Liu & Donahue 1975b). One model of Venus shows considerable, but not complete, conversion (Kumar & Hunten 1974).

ODD AND EVEN HYDROGEN The constituents discussed above are long-lived, and discussions of their behavior must include eddy and diffusive transport. Their unstable products, however, have short lifetimes and are conveniently lumped together as "odd hydrogen": they are H, OH, and HO_2. H_2O_2, which is sometimes included, is effectively a dimer of OH and rapidly splits into 2(OH) through photolysis. Conversion of odd back to even hydrogen (H_2 and H_2O) occurs through the reactions

$$OH + HO_2 \rightarrow H_2O + O_2 \quad (17)$$

$$H + HO_2 \rightarrow H_2 + O_2 \quad (18)$$

(or $OH + OH$, or $H_2O + O$)

The importance of the radical HO_2 is obvious, but unfortunately its reaction rates are difficult to measure. The rate coefficient of Reaction 17 has been measured as 2×10^{-10} cm^3 sec^{-1}, but there are strong theoretical reasons for expecting it to be an order of magnitude smaller (Kaufman 1975). The branching ratio of Reaction 18 has not been measured. The first branch shown is particularly important because it is a major source of H_2, which dominates the flow on all three terrestrial planets. Estimates for the partial rate coefficient range from 1.5×10^{-12} cm^3 sec^{-1} (Hunten & McElroy 1970) to 1×10^{-11} cm^3 sec^{-1} (McElroy & Donahue 1972). Liu & Donahue (1974c) find that simultaneously low values for both reaction rates give predictions for the Earth's mesosphere that disagree with observations (cf Figures 5–7 below).

These uncertainties are tolerable for Earth because the escape is controlled to a first approximation by limiting flow of total hydrogen. For other problems and other planets the consequences are more serious.

CONDENSATION The availability of H_2O at heights where it can be converted to odd hydrogen and to H_2 is very much limited by condensation on all three terrestrial planets. For Earth and Mars, the process is freezing out at low temperatures; for Venus, it seems to be absorption in the clouds of concentrated H_2SO_4.

An important but subtle question is the existence and behavior of cold traps, which are named for their counterparts in laboratory vacuum systems but are not nearly as well understood. Air, in passing through a cold region, loses its vapor by condensation and precipitation and subsequently contains only that amount corresponding to the vapor pressure at the minimum temperature. For many problems, including that of escape, the important quantity is the mixing ratio, rather than the partial pressure. A cold trap is therefore most significant if it operates at a high total pressure. For example, the Earth's mesopause is much colder than the tropopause, but the latter is much more important as a trap, and condensation at the mesopause, which forms noctilucent clouds, occurs only under special conditions.

At balloon heights on the Earth, the volume mixing ratio of H_2O is about 4×10^{-6} or 4 ppm (Mastenbrook 1968, Hunten & Strobel 1974). The reason for this remarkably low value is not really understood, although a cold trap at the tropical tropopause may be involved (Urey 1959). Perhaps even more remarkable is that the stratosphere is almost certainly a source of H_2O by oxidation of CH_4 (Bates & Nicolet 1965), and even probably by oxidation of H_2. Recent models agree in finding a large downward flux of H_2O (around 5×10^8 cm^{-2} sec^{-1}) out of the stratosphere (Hunten & Strobel 1974, Liu & Donahue 1974c). Since the methane is biological in origin, and perhaps also the H_2, the present escape rate from Earth is apparently controlled by biological processes. This control is only possible, however, because of the efficient cold trap for H_2O.

For Mars, the observed amount of water vapor fluctuates greatly, but a typical value is a few precipitable microns, or 10^{19} molecules cm^{-2} (Hunten 1971b). The vertical distribution is unknown, but Hunten & McElroy (1970) argued on the basis of typical temperature profiles that the vapor must be largely confined to the bottom 5 km. They estimated the production rate of odd hydrogen, suggesting a mean of 2×10^9 cm^{-2} sec^{-1}. Recently Liu & Donahue (1975b) arrived at a similar picture after first trying a scheme in which water vapor was abundant as high as 40 km; they could not tolerate the large OH production at this height.

Parkinson & Hunten (1972) pointed out that surface condensation is probably important for other components of the hydrogen system, notably H_2O_2 and HO_2. Hunten (1974) has further discussed this suggestion. The vapor pressure of H_2O_2 is similar to that of H_2O; the fact that it freezes out at high Martian latitudes is likely to be an important drain on odd hydrogen during the winter and perhaps an important source in spring. HO_2 also freezes out, though information is mainly confined to temperatures lower than those reached on Mars. When the deposit is warmed, gases released at O_2, H_2O_2, and H_2O.

The visible clouds of Venus are almost certainly concentrated sulfuric acid, around 85% H_2SO_4 by weight or 50% molar (Young 1975). This material's strong affinity for water explains the extreme dryness of the Venus stratosphere (around 1 ppm H_2O); this had previously been puzzling because Venus seems to lack any sort of cold trap (McElroy & Hunten 1969a). A lower limit to total-H on Venus is the HCl abundance, 0.6 ppm (Connes et al 1967). The H_2 abundance is unknown and could conceivably raise this figure by a factor of more than 10.

THE PLANETS

With the background developed above, we now discuss the Earth, Venus, and Mars in turn.

Earth

As outlined earlier, three principal mechanisms contribute significantly to hydrogen escape. Before discussing the details, we summarize our present views of their relative importance, using an exospheric temperature of 1000°K. The classical Jeans process is responsible for only 20% of the loss rate. Charge exchange with plasmaspheric protons dominates, accounting for over 70%. Polar ion flow accounts for 7%. The total flux is governed fairly accurately by the limiting-flow principle of Hunten (1973b). Each component of the flux is very nearly proportional to the H density n_c at the critical level. But the Jeans flux is also very sensitive to temperature (Equation 2). An increase of exospheric temperature increases this component and depresses n_c. The other two components are therefore reduced, and the same total flux will be maintained. Its value is $(2\text{–}3) \times 10^8$ atoms cm^{-2} sec^{-1}.

RECENT PROGRESS As mentioned above, the principle of limiting flow for thermospheric hydrogen was implicit in the work of Bates, Patterson, Mange, Kockarts, and Nicolet around 1960 and had been formulated for ion flow by Banks and his collaborators. But a formal statement came out of work on Titan by Hunten (1973a), then expanded to other planets (Hunten 1973b). Somewhat oversimplified, it reads as follows:

> "The mixing ratio f_t of total-hydrogen is the same at the homopause and in the lower stratosphere. The escape flux is then closely approximated by $\phi \simeq bf_t/H_a$, (Equations 11 and 12), where b is the collision parameter for H_2 molecules."

Thus, once f_t has been set in the stratosphere, the escape flux is determined by the mix of H and H_2 at the homopause. For the principle to work, all forms of hydrogen must be convertible to H and H_2 at the homopause, H_2 must convert to H below the critical level, and H must escape freely. All these conditions are normally satisfied on the Earth, with a possible exception at low exospheric temperatures (discussed below).

A detailed aeronomical model of the region from 50 km to the critical level was next studied by Hunten & Strobel (1974). The only escape was by the Jeans process (which had been omitted in all previous models of this kind). The general ideas were rather accurately verified: the escape flux was independent of variations in reaction rates and solar ultraviolet flux and nearly independent of eddy diffusion coefficient. H_2 was found to outweigh H at the homopause; variations in the ratio were the main contributor to variations in escape flux. Almost simultaneously, a similar, independent study by Liu & Donahue (1974a), who also varied the exospheric temperature, found the expected reduction of flow for temperatures below 1000°K. Otherwise, the flux was 1.8×10^7 cm^{-2} sec^{-1} for each 1 ppm of stratospheric hydrogen.

In different ways, both papers noted a discrepancy between stratospheric and exospheric observations. In the stratosphere, f_t is 9–14 ppm; the Jeans flux at 1100°K is close to 5×10^7 cm^{-2} sec^{-1} from observations of H density and temperature. Hunten & Strobel took $f_t = 9$ ppm and computed a flux of 1.5×10^8 cm^{-2} sec^{-1}. They noted the discrepancy but concluded that it was probably within the combined errors. Liu & Donahue started with the escape flux and computed a total-H mixing ratio; with an upper limit of 1×10^8 cm^{-2} sec^{-1} for the flux, f_t could not exceed 5 ppm. The agreement between the two investigations is excellent. Liu & Donahue, however, concluded that the discrepancy was real. The stratospheric measurements could have been faulty, but more probably other escape mechanisms were at work.

Another discrepancy discussed in both papers relates to an observed variation of Jeans flux with solar activity (Vidal-Madjar, Blamont & Phissamay 1973; Blamont, Cazes & Emerich 1975). Such a variation seemed natural on the old Jeans-Spitzer picture, but it does not accord with the computations when thermal escape is the only loss process.

Liu & Donahue (1974b) therefore investigated polar ion flow and charge exchange as supplementary processes, showing that charge exchange in particular seemed to be large enough to resolve the discrepancies. Shortly thereafter and independently, Bertaux (1974, 1975) made the same suggestion, but without quantitative estimates. He was influenced by his satellite results, which showed a variation of Jeans flux with solar activity, and he realized the conflict with the limiting-flow principle.

THE PRESENT POSITION A consistent picture, including all three loss processes, was given by Liu & Donahue (1974c). The partitioning of the constant total flux among the three processes was done, somewhat arbitrarily, as follows.

The total flux was taken as $\phi_t = 2.68 \times 10^8$ cm^{-2} sec^{-1}, corresponding to $f_t = 14.8$ ppm at 50 km: 0.5 ppm of H_2, 0.25 ppm of CH_4, and 6.4 ppm of H_2O. All three fluxes are proportional to n_c, which must be determined consistently. The factor was taken as $n_c w_J = \phi_J$, calculated from Equation 2 with a Jeans ratio of $B = 0.73$. The charge-exchange flux ϕ_E varies as the column amount of hydrogen and therefore as $n_c H_c$. Tinsley's estimate, $\phi_E = 2 \times 10^8$ cm^{-2} sec^{-1}, refers to a temperature of 1000°K, correspondingly gives $\phi_J = 5 \times 10^7$ cm^{-2} sec^{-1}. These assumptions produce

$$\phi_E = 2 \times 10^8 \frac{n_c H_c}{(n_c H_c)_{1000}} \cdot \frac{\phi_J}{5 \times 10^7} \text{ cm}^{-2} \text{ sec}^{-1}. \tag{19}$$

The polar wind component is proportional to a slowly varying function of temperature, ϕ_{Po}, adapted from computations of Banks & Holzer (1969), thus giving

$$\phi_P = \phi_{Po} \frac{\phi_J}{5 \times 10^7} \simeq 0.36 \phi_J, \tag{20}$$

where $\phi_{P_0} \simeq 1.8 \times 10^7$ cm^{-2} sec^{-1} is approximately valid over the temperature range considered. The sum $\phi_J + \phi_E + \phi_P$ is set equal to ϕ_t, and ϕ_J is found (Equation 5 of Liu & Donahue 1974b). The three components are shown in Table 1, and ϕ_J is plotted as the solid line in Figure 4. The other lines are deduced from observations. The agreement is excellent, perhaps better than could be expected in view of the rather crude nature of the theoretical estimates.

As mentioned above, Bertaux (1974, 1975) independently conceived a similar explanation for his OGO-5 data, shown (with a correction by the Jeans ratio) by the dotted line in Figure 4. Previously Vidal-Madjar, Blamont & Phissamay (1974) had remarked on a variability of ϕ_J during the lifetime of OSO-5, ascribing it to a variation with solar activity in the hydrogen density at 100 km. This, as we now know, is not a viable explanation for variations in ϕ_J. In fact, Vidal-Madjar and his colleagues were probably observing the same type of phenomenon found by Bertaux. Their results are shown by the dashed line in Figure 4.

The treatment outlined here is rather intuitive and needs considerable sharpening. Tinsley's estimate of ϕ_E is also crude—a better treatment and a careful assessment of the hot-proton densities in the plasmasphere would be highly desirable. But the basic result seems fairly secure, namely, the total H-atom escape flux is $(2\text{–}3) \times 10^8$ cm^{-2} sec^{-1} and accords with measured abundances of hydrogen compounds in the stratosphere. The escape rate accepted in much of the literature since Donahue's (1966) review was too small by a factor of 5–10.

It is still unclear how low the exospheric temperature must drop to produce a substantial deviation from the limiting flux. Liu & Donahue (1974a) note that brief diurnal decreases have little effect because of a substantial hydrogen reservoir between the diffusion region and the escape region. The most appropriate diurnal-mean temperature is near the maximum. The charge exchange process, which dominates in any case at the lower temperatures, is most likely adequate to maintain limiting flow under all common circumstances.

The polar wind and charge-exchange mechanisms are not individually worldwide but have nearly complementary distributions: they correspond respectively to

Table 1 Partitioning of the total escape flux among the three mechanisms at various exospheric temperatures T_c.

T_c (°K)	ϕ_J	ϕ_E	ϕ_P
900	2.6	22.2	1.4
950	3.67	21.2	1.9
1000	5	20	1.8
1100	7.05	16.5	2.9
1200	8.7	13.5	3.6
1500	13.5	8.9	4.4
1900	18.2	5.6	2.9

J stands for Jeans, E for charge exchange, P for polar wind. Units are 10^7 cm^{-2} sec^{-1}. From Liu & Donahue (1974b).

regions of open and closed magnetic-field lines. In their respective regions, the two fluxes are comparable, with the polar component being probably somewhat larger. Some lateral flow from lower latitudes may therefore be required. The site of the flow is the lower exosphere, and it has been studied considerably (e.g. McAfee 1967, Vidal-Madjar & Bertaux 1972, Tinsley 1974). As summarized by Liu & Donahue (1974b), most of the flow into the polar cap comes from a ring about 20° wide just outside it. The cap can thus draw on the limiting flow through a much larger region. The previous focus in lateral-flow studies has been the smoothing of the expected large diurnal variation due to the Jeans flux alone. This variation now

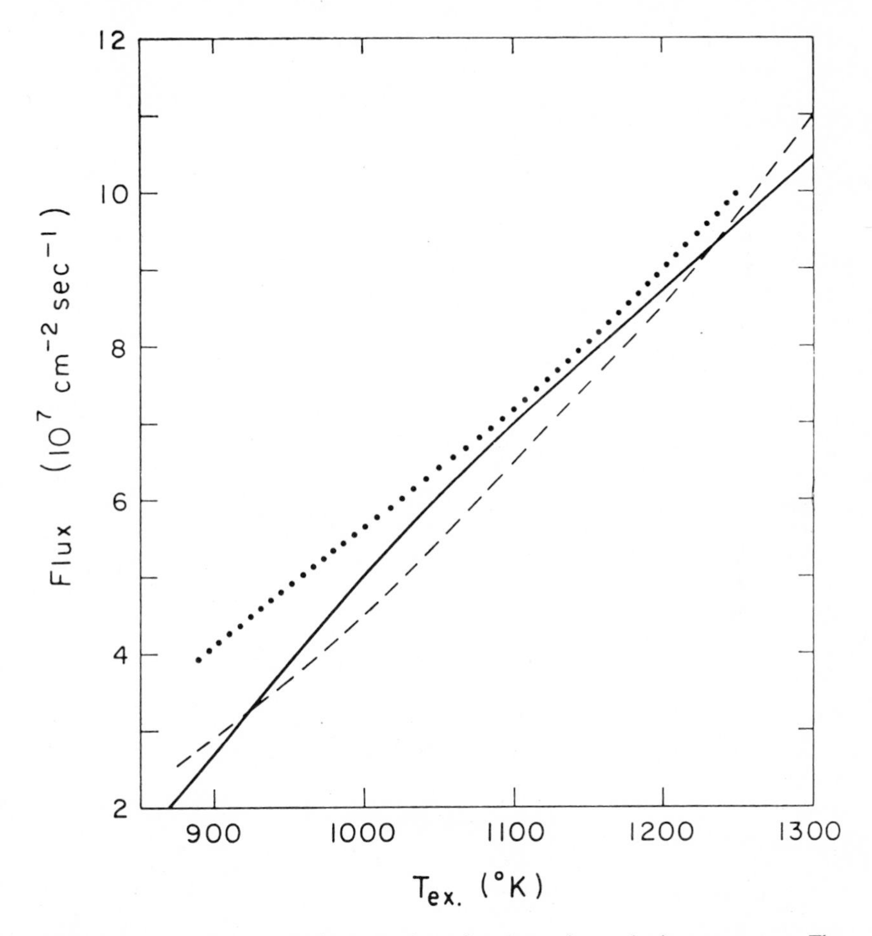

Figure 4 Jeans escape fluxes from Earth as functions of exospheric temperature. The dotted line was observed by Bertaux (1975), the dashed line was observed by Vidal-Madjar, Blamont & Phissamay (1974), and the solid line was calculated. From Liu & Donahue (1974c), with permission of the American Meteorological Society.

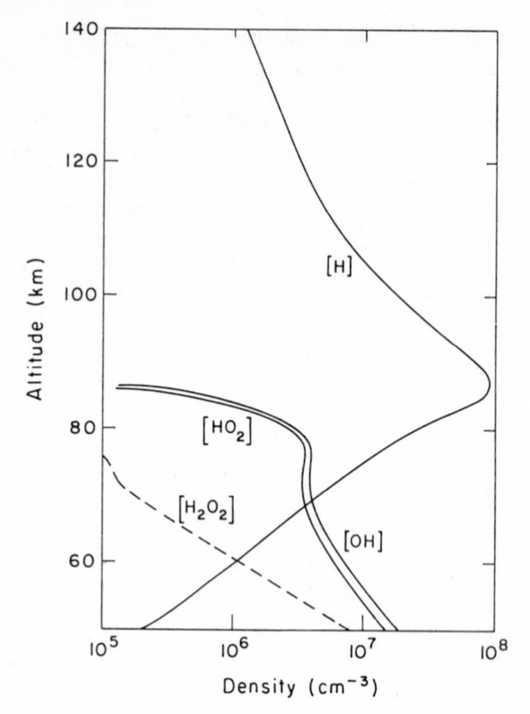

Figure 5 Densities of several hydrogenous constituents in the model of Liu & Donahue (1974c). Reprinted with permission of the American Meteorological Society.

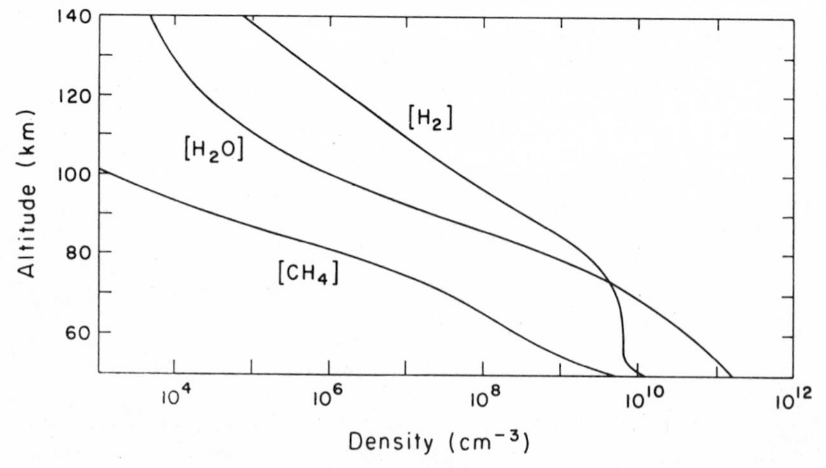

Figure 6 Densities of the other hydrogenous constituents in the model of Liu & Donahue (1974c). Reprinted with permission of the American Meteorological Society.

seems less important, and flow towards the polar regions has to be considered, along with the diurnal variation of the charge-exchange flux.

Figures 5–7 show the densities of hydrogen species and the fluxes for the model of Liu & Donahue (1974c). Particularly noteworthy near the top of Figure 7 is the conversion of the H_2 flux into an H flux by Process 15.

These developments have very much depended on research directed to other planetary bodies. Limiting flow of H in the thermosphere has been known for over 15 years, but the generality of the principle was brought out in Hunten's (1973a) study of Titan. This work was stimulated by observations interpreted in terms of substantial amounts of H_2 in the methane atmosphere of that body. The detailed models of Hunten & Strobel and of Liu & Donahue were both developed out of earlier studies of H_2O photolysis and hydrogen escape on Mars (Hunten & McElroy 1970, McElroy & Donahue 1972). Again, these papers dealt with observed water vapor absorptions and Lyman-α resonance scattering. The influence of planetary studies on our understanding of the Earth is real and important.

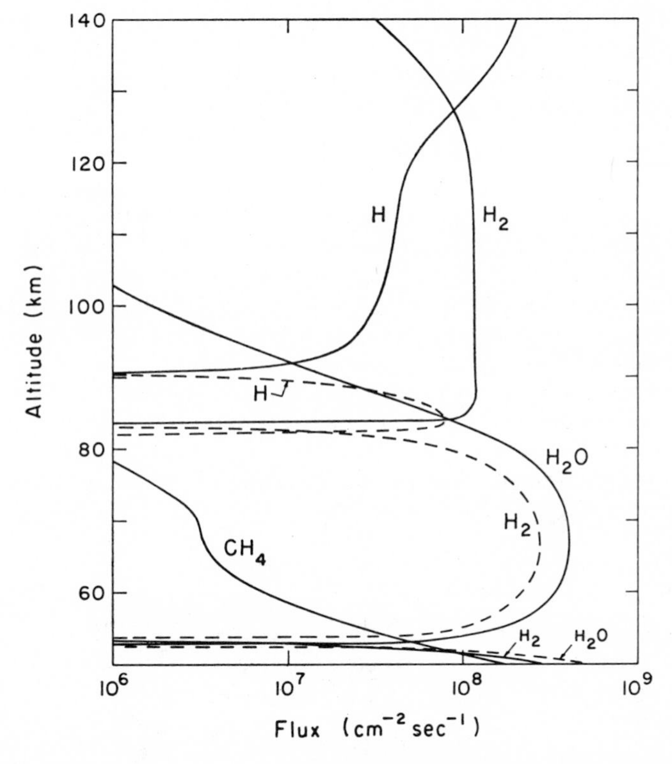

Figure 7 Fluxes in the model of Liu & Donahue (1974c). Solid lines represent upward fluxes and dashed lines represent downward fluxes. Reprinted with permission of the American Meteorological Society.

THE EARLY ATMOSPHERE The revised escape rate of hydrogen has interesting consequences for the Earth's oxygen budget. If the present rate is representative, the corresponding production of free oxygen in 4.5×10^9 years would be 500 g cm^{-2}, or about twice what is now in the atmosphere. Other components of the budget are indicated in Table 2, from Hunten (1973a). The amount of oxygen to produce the Fe_3O_4 now in the crust is nearly 60 times the atmospheric amount. If the oxygen in carbonates also came from the atmosphere, the factor is 270. Hydrogen escape would have had no particular difficulty producing even these amounts in the distant past if the stratosphere had been sufficiently richer in hydrogenous molecules. For example, a flux 10^3 greater than at present would result from a mixing ratio $f_t \simeq 0.4\%$, and would produce all the oxygen in Table 2 in 10^9 years. Loss of hydrogen from CH_4 and NH_3, if required, is equally plausible.

More conservatively, if the escape rate had been somehow maintained at no more than its present value in the early atmosphere, there would still be around 5% of the present O_2 in the atmosphere by 500–1000 million years ago. At that time, presumably, life was still in a glycolytic stage in the ocean, and respiration with photosynthetic conversion of CO_2 to O_2 had not begun. We can now assume considerably more O_2 than was previously available for the transition to respiratory forms of life. The timetable for the development of life and the conversion of CO_2 to O_2 becomes much more comfortable.

Venus

The basis of nearly all ideas about hydrogen on Venus is the measurements by the Lyman-alpha (Lα) photometer on Mariner 5 (Barth et al 1967). These results are shown, along with the recent ones of Mariner 10, in Figure 8 (Broadfoot et al 1974). These data are remarkable in two ways: they imply a small H density, and they cannot be represented by a normal exospheric density distribution. They require at least two components, either two masses or two temperatures, in a ratio of 2–3 (Wallace 1969). Suggestions, never very credible, that the inner component might result from stray light were thoroughly refuted by Mariner 10. This spacecraft, however, was not operated in such a way as to be sensitive to the outer component, which was much better observed by Mariner 5.

The first interpretations relied on a computed exospheric temperature of 700°K by McElroy (1968), which gave a good fit to the outer component of Lα. The inner

Table 2 Oxygen and hydrogen budgets for the primitive Earth

	CO_2	Fe_2O_3	Atmosphere	Totals
Material now present (kg cm^{-2})	72	137	1.035	
O_2 required (kg cm^{-2})	52	13.7	0.24	
H_2O mass (kg cm^{-2})	59	15.4	0.27	
H_2O liquid depth (m)	590	154	2.7	750
H_2 from H_2O (cm^{-2})	2.0×10^{27}	5.2×10^{26}	9×10^{24}	2.5×10^{27}
H_2 from CH_4, NH_3 (cm^{-2})	2.0×10^{27}	—	5.2×10^{25}	2.5×10^{27}

component would then have a mass of 2. Barth, Wallace & Pearce (1968) identified it with H_2 and suggested that the Lα arose by photodissociation at wavelengths short enough to excite one of the atoms. This proposal, also presented (Barth 1968) at the Second Arizona Conference on Planetary Atmospheres, was vigorously attacked by Donahue (1968) in his conference review because of the problem it presented of disposing of the very large number of hydrogen atoms created from H_2 in the model. Later McElroy & Hunten (1969a) showed that the upper limit to the H_2 density in such a model was about 10^7 cm^{-3} at 150 km compared to about 10^{10} cm^{-3} in Barth's model. The obvious alternative seemed to be deuterium atoms (Donahue 1968, 1969; McElroy & Hunten 1969a), which could in principle be greatly enriched in the upper atmosphere. In present-day language, H was to be in an extreme limiting-flow situation and D close to diffusive equilibrium. For this situation to be possible, the eddy-diffusion coefficient K would have to be small, less than 10^5 cm^{-2} sec^{-1}, to give a small homopause velocity w_h. The special assumptions about K and T_c caused no little uneasiness, but the real deathblow to the deuterium hypothesis came from a rocket flight by Wallace et al (1971). Their spectrometer, able to resolve the isotope splitting of Lα, showed no sign of the required D component.

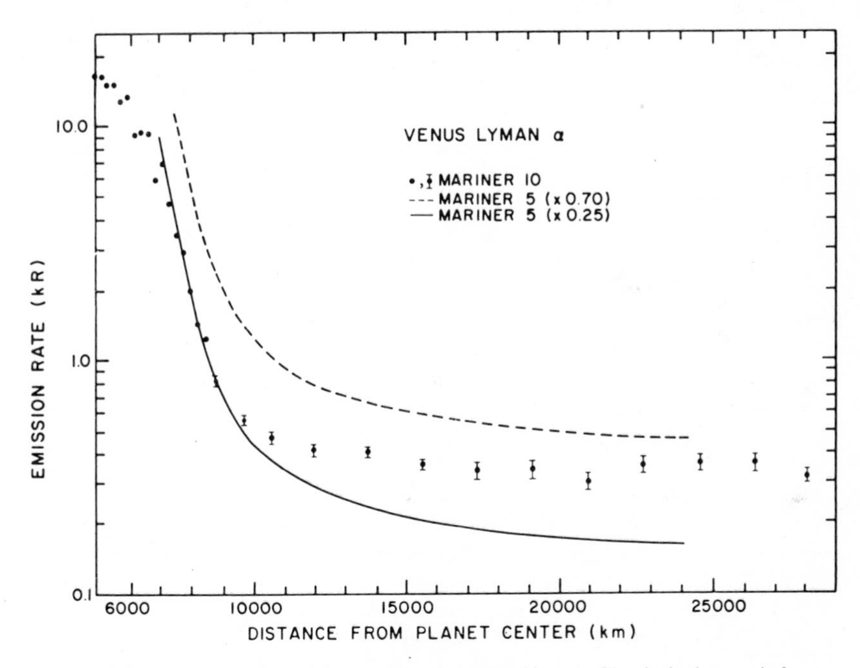

Figure 8 Lyman alpha observations on the day side of Venus. The dashed curve is from a recalibration of Mariner 5, and the solid curve represents an arbitrary further shift needed to match Mariner 10. From Broadfoot et al (1974), with permission of the American Association for the Advancement of Science.

Another problem with Venus was pointed out by J. C. G. Walker (unpublished work) and emphasized by Hunten (1973b) in terms of his limiting-flow principle: a considerable amount of H_2O and HCl exists near the cloud tops of Venus, in addition to an unknown amount of H_2. The mixing ratio of HCl is 0.6 ppm (Connes et al 1967) and about 1 ppm for H_2O (Young 1972, Fink et al 1972). Thus, there are at least 2.5 ppm of total hydrogen near the cloud tops. The HCl photochemical system of Prinn (1971) would add a mixing ratio of H_2 equal to 10 ppm. Then, according to limiting diffusion, if the limiting flow applied to H, the escape flux of H could not be less than 3×10^8 cm^{-2} sec^{-1} and might need to be as large as 25×10^8 cm^{-2} sec^{-1}. The small H densities permitted by the Lα data allowed only 10^6 cm^{-2} sec^{-1} for the Jeans process. Hunten (1973b) discussed this dilemma and concluded that the most likely solution was that the exospheric temperature is 350°K. In this case, the component of low scale height would be hydrogen, limiting diffusion would not apply because of the low exospheric temperature, and the destruction of H_2 by O would be inhibited. Then, with an eddy coefficient, K, of 5×10^6 cm^2 sec^{-1}, the H_2 density would be at its upper limit of 10^7 cm^{-3} in the thermosphere when the mixing ratio in the lower atmosphere is 4 ppm. Fast atoms from dissociation of H_2 might then provide the "hot" component of Lα.

This suggestion was taken up by Kumar & Hunten (1974) in work that was completed just before Mariner 10 encounter. Their model ionosphere, at 350°K exospheric temperature, gave a fair fit to Mariner 5 data, although it is considerably poorer with Mariner 10. They did, however, identify some possible sources of hot H atoms with strengths of the required order. They are ion-H_2 reactions of the form of Equation 16, where X^+ is O^+ or $CO_2{}^+$. The corresponding destruction of H_2 limited its density at the critical level to well under 10^5 cm^{-3}. Previously suggested sources of hot H had been judged inadequate by McElroy & Hunten (1969a), including photodissociation of H_2, which had been advocated by Barth (1971). Sze & McElroy (1975) list many additional possibilities.

Both the ionospheric model and the semiquantitative discussion of H required an eddy coefficient K of 1×10^8 cm^{-2} sec^{-1}. Its effect is to maintain small densities of H and O atoms in the thermosphere and exosphere. The escape flux can only be guessed, because the velocity distribution of the nonthermal atoms is unknown. (The escape rate of the thermal component is negligible.) Kumar & Hunten suggested that 10–30% of the hot atoms might escape, but the figure could be 50% or possibly even higher. The flux would therefore be $(1\text{–}5) \times 10^6$ cm^{-2} sec^{-1}, still far below the limiting value. Most of the H atoms produced in the ionosphere must flow back down to be chemically combined again at deeper levels.

Mariner 10 confirmed that the exospheric temperature in 1974 was around 400°K, both from the Lα data reproduced in Figure 8, and from a similar analysis of helium radiation (Broadfoot et al 1974). These results inspired Liu & Donahue (1975a) and Sze & McElroy (1975) to take up the problem.

For the purposes of this review the two treatments are essentially equivalent. Sze & McElroy include chlorine photochemistry and obtain results that apply all the way down to the cloud tops, but chemical details matter little above 90 km, where transport is dominant. Liu & Donahue (1975a), for example, assumed $f_t = 2.6$ ppm for total-H at the cloud tops. To keep the exospheric H density down at the observed value then required either a large K, a large escape flux, or some combination. The escape flux was simply an arbitrary parameter, assumed to result from a nonthermal mechanism of unknown nature. In the two extreme cases, the nonthermal escape gives limiting flux over a considerable height range; the large K keeps H_2 and H to a small scale height, in a constant mixing ratio with the CO_2, to high altitudes. It also can be thought of as sweeping downwards the H atoms from H_2 destruction. Depending on the specific assumptions, K had to be up to a few times 10^8 cm^2 sec^{-1}. For substantially smaller K, the nonthermal flux had to be nearly 10^8 cm^2 sec^{-1}.

Both papers concluded that the high K was the more likely answer and drew the corollary that the O abundance in the upper atmosphere is therefore small, because of the same downward sweeping that acts on H.

The probable weakness of any nonthermal escape process is due to the small observed H densities in the exosphere. The assessment by Walker, Turekian & Hunten (1970) is still valid for the most part, and many of the ideas have been discussed above under the individual processes. Ion production is simply too slow in the absence of the large O^+ amounts that are so effective for charge exchange on Earth. Venus has no plasmasphere because it lacks a magnetosphere. Solar wind sweeping of $H_2{}^+$ ions was the strongest potential process turned up by Walker et al; their limit was 1.3×10^7 H atoms cm^{-2} sec^{-1}, too small by an order of magnitude to be significant in the present context. In any case, it is inconsistent with the Michel limit, Equation 9, by at least a factor of 3, and it ignores the corresponding accretion.

We should remember as a cautionary warning that the dominant escape process from Earth has only recently been worked out after many decades of belief in Jeans escape. Neverthless, it seems very improbable that Venus has a nonthermal escape flux anything like 10^8 cm^{-2} sec^{-1}. A total flux of order 10^6 cm^{-2} sec^{-1} is much more likely.

Mars

Lyman-alpha emission from the Martian corona was observed by Mariners 6 and 7 (Barth et al 1969). Subsequent analysis established the Jeans escape flux as 1.8×10^8 cm^{-2} sec^{-1}, with $n_c = 3 \times 10^4$ cm^{-3} and $T_c = 350°$K (Anderson & Hord 1971). The Jeans ratio of 0.50 (Chamberlain & Smith 1971) was not included in this result, which therefore becomes 9×10^7 cm^{-2} sec^{-1}. An aeronomical model by Hunten & McElroy (1970) showed that such fluxes could plausibly be accounted for, with photolysis of the observed H_2O amounts as the source. (Barth has told us that, years before his observation, he had been advised not to waste his time—the gravity of Mars was too small to hold appreciable amounts of atomic hydrogen.) As Hunten & McElroy pointed out, the corresponding O_2 production would give the amount in the atmosphere ($\sim 0.1\%$) in only 10^5 years. This mixing ratio had to be assumed, but was later confirmed (Carleton & Traub 1972, Barker 1972, review by Hunten 1974). Hunten (1973b) showed that the model results were in excellent accord with the limiting-flow principle based on the H_2 mixing ratio. The relation of the H_2 to the H_2O is subtle, however, and the evaluation of the total-H mixing ratio requires the H_2O near the surface to be excluded. Also, the model assumed an ionosphere rich enough in CO_2^+ to convert nearly all the H_2 to escaping H.

Considerable interest in the topic was generated when McElroy (1972) noticed an important nonthermal escape process for oxygen that should be roughly in balance with the hydrogen loss. The escape energy from a height of 200 km is $\frac{1}{8}$ eV per atomic mass unit, or 2 eV for an oxygen atom. More than this energy is given in the dissociative recombination of any likely oxygenated molecular ion, CO_2^+, CO^+, or O_2^+. This is true even if the atom is electronically excited within the ground configuration. (For Earth and Venus, the escape energy is too large for the process to work.) Thus essentially half of all ions created above the exobase of Mars produce an oxygen atom that escapes—that is, all the recombination events that produce an atom that travels outward. Hence, it is easy to calculate the nonthermal escape flux simply by calculating the appropriate ionization rate or resorting to a model for the ionosphere. The result is that 7×10^7 O atoms escape per square centimeter per second. Within the errors, the O flux is equal to half the hydrogen thermal escape flux of 9×10^8 atoms cm^{-2} sec^{-1}.

It is difficult to accept as an accident such a coincidence of escape fluxes. McElroy (1972) and McElroy & Donahue (1972) suggested that a connection through the chemistry governs the recombination of CO_2 on Mars. They argued that the escape of oxygen regulates the production of H_2 and hence of atomic hydrogen and that the escape flux of hydrogen is constrained to follow that of oxygen. Qualitatively, if the two were unbalanced—for example, if the O escape rate were too low—the O_2, and hence the O_3 density, would be larger than normal, and the excess $O(^1D)$ would attack H_2, thus increasing its destruction rate. At the same time, the reaction of H with O_2 would increase the amount of HO_2 and OH while decreasing that of H. The overall result would be a decrease in the product of H and HO_2 and therefore a decrease in the rate of production of H_2. Hence H_2 production would decrease, its destruction increase, and the rate of H escape decrease. Two photochemical systems are available; each differs in some details but either one is able to operate as described (McElroy & Donahue 1972, Parkinson & Hunten 1972, Hunten 1974).

However, we are faced here with a paradox in view of Hunten's limiting-flow principle. The H escape from Mars is copious and one would expect the state to be one of limiting diffusion in which the concentration of total-H in the lower atmosphere controls escape. It is difficult to understand how the escape rate of atomic oxygen would determine the total mixing ratio of hydrogen in the lower atmosphere. A possible way out is defining total-H to exclude H_2O near the surface, as suggested above. The rationale would seem to be the very loose coupling between H_2O and the H_2 that actually is the buffer and the carrier of hydrogen to high altitudes.

Liu & Donahue (1975b) have recently investigated the question in detail. They solved the coupled diffusion equations from the surface to the exobase for all constituents expected to be important: CO_2, CO, O_2, H_2O, odd hydrogen (H, OH, HO_2), H_2O_2, and odd oxygen ($O + O_3$). Water was assumed to issue from the surface at a rate equal to the oxygen escape by dissociative recombination. The ionospheric structure was also computed. Too little CO_2^+ was found in the ionosphere to give complete conversion of H_2 to H; most of the ions were converted to O_2^+, which does not attack H_2. Near the exobase, the number density of H_2 remained 4 times larger than that of H. The ratio of escape fluxes (at 320°K) was a factor of 8 the other way, or a factor of 4 in terms of total atoms being lost. The total-H escape flux of 1.1×10^8 cm^{-2} sec^{-1} is nicely in balance with the O flux, 0.53×10^8 cm^{-2} sec^{-1}. The uncorrected Jeans equation was used, but the same effusion velocity could be restored by a small increase of the adopted exospheric temperature. The essential point is that balanced fluxes were found, even with an arbitrary choice of effusion velocity.

Because H_2 remains abundant at the exobase and is too heavy to escape freely, the requirement for limiting flow is not satisfied and the feedback from the O escape can readily operate; the apparent paradox is resolved.

Clearly, this result is sensitive to the computed positive-ion composition of the ionosphere. It should therefore be accepted with some caution until this composition has been confirmed. Possibly the feedback could still work under limiting-flow conditions by controlling the abundance of H_2. As suggested above, total-hydrogen would have to be defined as excluding H_2O.

Liu & Donahue (1975b) also ran two models with the atmospheric O_2 abundance increased by factors of 1.5 and 2. The H_2 abundance was found to decrease by somewhat larger factors. These results verify the most important step in the postulated feedback chain and strongly support its validity. The general conclusion is that on Mars it is not just hydrogen, but rather water, that escapes. These processes may be important in keeping the Martian surface as dry as it is.

ACKNOWLEDGMENT

Kitt Peak National Observatory is operated by the Association of Universities for Research in Astronomy, Inc., under contract with the National Science Foundation.

Literature Cited

Anderson, D. E. Jr., Hord, C. W. 1971. Mariner 6 and 7 ultraviolet spectrometer experiment: analysis of hydrogen Lyman alpha data. *J. Geophys. Res.* 28:6666–73

Banks, P. M., Holzer, T. E. 1969. High-latitude plasma transport: the polar wind. *J. Geophys. Res.* 74:6317–32

Banks, P. M., Kockarts, G. 1973. *Aeronomy*, Parts A and B. New York: Academic

Barker, E. S. 1972. Detection of molecular oxygen in the Martian atmosphere. *Nature* 238:447–48

Barth, C. A. 1968. Interpretation of the Mariner 5 Lyman Alpha Measurements. *J. Atmos. Sci.* 25:564–67

Barth, C. A. 1971. Exospheric temperature of Venus from Mariner 5. *Planetary Atmospheres*, ed. C. Sagan, T. C. Owen, H. J. Smith, pp. 17–22. Dordrecht: Reidel

Barth, C. A., Fastie, W. G., Hord, C. W., Pearce, J. B., Kelly, K. K., Stewart, A. I., Thomas, G. E., Anderson, G. P., Raper, O. F. 1969. Mariner 6: ultraviolet spectrum of Mars upper atmosphere. *Science* 165:1004–5

Barth, C. A., Pearce, J. B., Kelly, K. K., Wallace, L., Fastie, W. G. 1967. Ultraviolet emissions observed near Venus from Mariner 5. *Science* 158:1675–78

Barth, C. A., Wallace, L., Pearce, J. B. 1968. Mariner 5 measurement of Lyman alpha radiation near Venus. *J. Geophys. Res.* 73:2541–45

Bates, D. R. 1959. Some problems concerning the terrestrial atmosphere above about the 100 km level. *Proc. R. Soc.*

London, Ser. A 253:451–62
Bates, D. R., Nicolet, M. 1965. Atmospheric hydrogen. *Planet. Space Sci.* 13:905–9
Bates, D. R., Patterson, T. N. L. 1961. Hydrogen atoms and ions in the thermosphere and exosphere. *Planet. Space Sci.* 5:257–73
Bertaux, J. L. 1974. *L'hydrogène atomique dans l'exosphère terrestre.* PhD thesis, Univ. Paris
Bertaux, J. L. 1975. Observed variations of the exospheric hydrogen density with the exospheric temperature. *J. Geophys. Res.* 80:639–42
Biermann, L., Brosowski, B., Schmidt, H. U. 1967. The interaction of the solar wind with a comet. *Sol. Phys.* 1:254–84
Blamont, J. E., Cazes, S., Emerich, C. 1975. Direct measurement of hydrogen density. *J. Geophys. Res.* 80:2247–65
Brinkmann, R. T. 1971. More comments on the validity of Jeans' escape rate. *Planet. Space Sci.* 19:791–94
Broadfoot, A. L., Kumar, S., Belton, M. J. S., McElroy, M. B. 1974. Ultraviolet observations of Venus from Mariner 10: preliminary results. *Science* 183:1315–18
Cameron, A. G. W. 1973. Elemental and isotopic abundances of the volatile elements in the outer planets. *Space Sci. Rev.* 14:392–400
Carleton, N. P., Traub, W. A. 1972. Detection of molecular oxygen on Mars. *Science* 177:988–92
Chamberlain, J. W. 1963. Planetary coronae and atmospheric evaporation. *Planet. Space Sci.* 11:901–60
Chamberlain, J. W., Smith, G. R. 1971. Comments on the rate of evaporation of a non-Maxwellian atmosphere. *Planet. Space Sci.* 19:675–84
Cloutier, P. A., Daniell, R. E. Jr., Butler, D. M. 1974. Atmospheric ion wakes of Venus and Mars in the solar wind. *Planet. Space Sci.* 22:967–90
Cloutier, P. A., McElroy, M. B., Michel, F. C. 1969. Modification of the Martian ionosphere by the solar wind. *J. Geophys. Res.* 74:6215–28
Cole, K. D. 1966. Theory of some quiet magnetospheric phenomena related to the geomagnetic tail. *Nature* 211:1385–87
Colegrove, F. D., Hanson, W. B., Johnson, F. S. 1965. Eddy diffusion and oxygen transport in the lower thermosphere. *J. Geophys. Res.* 70:4931–41
Colegrove, F. D., Johnson, F. S., Hanson, W. B. 1966. Atmospheric composition in the lower thermosphere. *J. Geophys. Res.* 71:2227–36
Connes, P., Connes, J., Benedict, W. S., Kaplan, L. D. 1967. Traces of HCl and HF in the atmosphere of Venus. *Ap. J.* 147:1230–37
Dalgarno, A. 1960. Low energy stopping power of atomic hydrogen. *Proc. Phys. Soc. London* 75:374–77
Donahue, T. M. 1966. The problem of atomic hydrogen. *Ann. Géophys.* 22:175–88
Donahue, T. M. 1968. The upper atmosphere of Venus: a review. *J. Atmos. Sci.* 25:568–73
Donahue, T. M. 1969. Deuterium in the upper atmospheres of Venus and Earth. *J. Geophys. Res.* 74:1128–37
Fehsenfeld, F. C., Ferguson, E. E. 1972. Thermal energy reaction rate constants for H^+ and CO^+ with O and NO. *J. Chem. Phys.* 56:3066–70
Fink, J., Larsen, H. P., Kuiper, G. P., Poppen, R. F. 1972. Water vapor in the atmosphere of Venus. *Icarus* 17:617–31
Hanson, W. B., Ortenburger, I. B. 1961. The coupling between the protonosphere and the normal F region. *J. Geophys. Res.* 66:1425–35
Herring, J., Kyle, L. 1961. Density in a planetary exosphere. *J. Geophys. Res.* 66:1980–82
Hudson, R. D. 1971. Critical review of ultraviolet photoabsorption cross sections for molecules of astrophysical and aeronomic interest. *Rev. Geophys.* 9:305–406
Hundhausen, A. J. 1972. Dynamics of the outer solar atmosphere. *Physics of the Solar System,* ed. S. I. Rasool, pp. 89–153. Washington: NASA SP-300
Hunten, D. M. 1971a. Hydrogen isotopes around the planets. *Comments Astrophys. Space Phys.* 3:1–6
Hunten, D. M. 1971b. Composition and structure of planetary atmospheres. *Space Sci. Rev.* 12:539–99
Hunten, D. M. 1973a. The escape of H_2 from Titan. *J. Atmos. Sci.* 30:726–32
Hunten, D. M. 1973b. The escape of light gases from planetary atmospheres. *J. Atmos. Sci.* 30:1481–94
Hunten, D. M. 1974. Aeronomy of the lower atmosphere of Mars. *Rev. Geophys. Space Phys.* 12:529–35
Hunten, D. M. 1975. Vertical transport in atmospheres. *Atmospheres of Earth and the planets,* ed. B. M McCormac, pp. 59–72. Dordrecht: Reidel
Hunten, D. M., McElroy, M. B. 1970. Production and escape of hydrogen on Mars. *J. Geophys. Res.* 31:5989–6001
Hunten, D. M., Strobel, D. F. 1974. Production and escape of terrestrial hydrogen. *J. Atmos. Sci.* 31:305–17
Ingersoll, A. P., Leovy, C. B. 1971. The

atmospheres of Mars and Venus. *Ann. Rev. Astron. Astrophys.* 9:147–82

Jeans, J. H. 1925. *The Dynamical Theory of Gases.* Cambridge, Engl: Cambridge Univ. Press. 444 pp.

Johnson, H. E., Axford, W. I. 1969. Production and loss of He^3 in the Earth's atmosphere. *J. Geophys. Res.* 74:2433–38

Junge, C. E., Oldenberg, O., Wasson, J. T. 1962. On the origin of the sodium present in the upper atmosphere. *J. Geophys. Res.* 67:1027–34

Kaufman, F. 1975. Hydrogen chemistry: perspective on experiment and theory. *Atmospheres of Earth and the planets,* ed. B. M. McCormac, pp. 219–32. Dordrecht: Reidel

Kockarts, G. 1971. Penetration of solar radiation in the Schumann-Runge bands of molecular oxygen. *Mesospheric models and related experiments,* ed. G. Fiocco, pp. 160–76. Dordrecht: Reidel

Kockarts, G., Nicolet, M. 1962. Le problème aéronomique de l'hélium et de l'hydrogène neutres. *Ann. Géophys.* 18:269–90

Kockarts, G., Nicolet, M. 1963. L'hélium et l'hydrogène atomique au cours d'un minimum d'activité solaire. *Ann. Géophys.* 19:370–85

Kumar, S., Hunten, D. M. 1974. Venus: an ionospheric model with an exospheric temperature of 350°K. *J. Geophys. Res.* 79:2529–32

Liu, S. C., Donahue, T. M. 1974a. The aeronomy of hydrogen in the atmosphere of the Earth. *J. Atmos. Sci.* 31:1118–36

Liu, S. C., Donahue, T. M. 1974b. Mesospheric hydrogen related to exospheric escape mechanisms. *J. Atmos. Sci.* 31:1466–70

Liu, S. C., Donahue, T. M. 1974c. Realistic model of hydrogen constituents in the lower atmosphere and escape flux from the upper atmosphere. *J. Atmos. Sci.* 31:2238–42

Liu, S. C., Donahue, T. M. 1975a. The aeronomy of the upper atmosphere of Venus. *Icarus* 24:148–56

Liu, S. C., Donahue, T. M. 1975b. The regulation of hydrogen and oxygen escape from Mars. *Icarus.* In press

Mange, P. 1961. Diffusion in the thermosphere. *Ann. Géophys.* 17:277–91

Mastenbrook, H. J. 1968. Water vapor distribution in the stratosphere and high troposphere. *J. Atmos. Sci.* 25:299–311

McAfee, J. R. 1967. Lateral flow in the exosphere. *Planet. Space Sci.* 15:599–609

McElroy, M. B. 1968. The upper atmosphere of Venus. *J. Geophys. Res.* 73:1513–21

McElroy, M. B. 1972. Mars: an evolving atmosphere. *Science* 175:443–45

McElroy, M. B., Donahue, T. M. 1972. Stability of the Martian atmosphere. *Science* 177:986–88

McElroy, M. B., Hunten, D. M. 1969a. The ratio of deuterium to hydrogen in the Venus atmosphere. *J. Geophys. Res.* 74:1720–39

McElroy, M. B., Hunten, D. M. 1969b. Molecular hydrogen in the atmosphere of Mars. *J. Geophys. Res.* 74:5807–9

McElroy, M. B., Sze, N. D., Yung, Y. L. 1973. Photochemistry of the Venus atmosphere. *J. Atmos. Sci.* 30:1437–47

McElroy, M. B., Wofsy, S. C., Penner, J. E., McConnell, J. C. 1974. Atmospheric ozone: possible impact of stratospheric aviation. *J. Atmos. Sci.* 31:287–303

McNesby, J. R., Okabe, H. 1964. Vacuum ultraviolet photochemistry. *Adv. Photochem.* 3:157–240

Meier, R. R., Mange, P. 1973. Spatial and temporal variations of the Lyman-alpha airglow and related atomic hydrogen distributions. *Planet. Space Sci.* 21:309–27

Michel, F. C. 1971. Solar-wind-induced mass loss from magnetic field-free planets. *Planet. Space Sci.* 19:1580–83

Nicolet, M. 1971. Aeronomic reactions of hydrogen and ozone. *Mesospheric models and related experiments,* ed. G. Fiocco, pp. 1–51. Dordrecht: Reidel

Öpik, E. J., Singer, S. F. 1961. Distribution of density in a planetary exosphere. *Phys. Fluids* 4:221–33

Parkinson, T. D., Hunten, D. M. 1972. Spectroscopy and aeronomy of O_2 on Mars. *J. Atmos. Sci.* 29:1380–90

Prinn, R. G. 1971. Photochemistry of HCl and other minor constituents in the atmosphere of Venus. *J. Atmos. Sci.* 28:1058–68

Serbu, G. P., Maier, E. J. R. 1970. Observation from OGO-5 of the thermal ion density and temperature within the magnetosphere. *J. Geophys. Res.* 75:6102–13

Spitzer, L. Jr. 1952. The terrestrial atmosphere above 300 km. *The Atmospheres of the Earth and Planets,* ed. G. P. Kuiper, pp. 211–47. Chicago: Univ. Chicago Press

Sze, N. D., McElroy, M. B. 1975. Some problems in Venus' aeronomy. *Planet. Space Sci.* 23:763–86

Thomas, R. J. 1973. Total mixing ratios. *Planet. Space Sci.* 22:175–77

Tinsley, B. A. 1973. The diurnal variation of atomic hydrogen. *Planet. Space Sci.* 21:686–91

Tinsley, B. A. 1974. Hydrogen in the upper atmosphere. *Fundamentals of Cosmic Physics* 1:201–300. New York: Gordon

& Breach
Tinsley, B. A., Hodges, R. R. Jr., Strobel, D. F. 1975. Diurnal variations of atomic hydrogen: observations and calculations. *J. Geophys. Res.* 80:626–34
Urey, H. C. 1959. The atmospheres of the planets. *Handb. Phys.* 52:363–418
Vidal-Madjar, A., Bertaux, J. L. 1972. A calculated hydrogen distribution in the exosphere. *Planet. Space Sci.* 20:1147–62
Vidal-Madjar, A., Blamont, J. E., Phissamay, B. 1974. Evolution with solar activity of the atomic hydrogen density at 100 kilometers of altitude. *J. Geophys. Res.* 79: 233–41
Walker, J. C. G., Turekian, K. K., Hunten, D. M. 1970 An estimate of the present-day deep-mantle degassing rate from data on the atmosphere of Venus. *J. Geophys. Res.* 75:3558–61
Wallace, L. 1969. Analysis of the Lyman alpha observations of Venus made from Mariner 5. *J. Geophys. Res.* 74:115–31
Wallace, L., Stuart, F. E., Nagel, R. H., Larson, M. D. 1971. A search for deuterium on Venus. *Astrophys. J.* 168:L29–31
Wallis, M. K. 1972. Comet-like interaction of Venus with the solar wind, I. *Cosmic Electrodynamics* 3:45–59
Young, L. D. G. 1972. High resolution spectra of Venus—a review. *Icarus* 17: 632–58
Young, A. T. 1975. The clouds of Venus. *J. Atmos. Sci.* 32:1125–32

ARGILLACEOUS SEDIMENT DEWATERING

John F. Burst
Dresser Minerals Division, Dresser Industries, Inc.,
P.O. Box 6504, Houston, Texas 77005

INTRODUCTION

Sediment dewatering, once thought to be a simple gravitational displacement, turns out to be a complicated interplay of kinetic, thermodynamic, and electrochemical mechanisms. Gravity is certainly the dominant compaction-dewatering mechanism in shallow (up to 3000 ft) sediments, but mineral diagenesis (particularly of clays) and osmotic and aquathermal pressuring are contributing agencies at depth. The requirement for a conduit to provide physical access to the normal hydrodynamic gradient remains.

Conditions that reduce permeability and thereby inhibit dewatering are structured water on fine particle surfaces and in clay interlayers, salinity reversals, excessively thick shale sections, and pore attrition. Better understanding of sediment dewatering awaits more precise thermodynamic data and more accurate measurement of the structured water envelope under various conditions of temperature, pressure, and humidity.

COMPACTION AND FLUID MIGRATION

Modern sediment compaction studies began with Hedberg's (1926) work and Athy's (1930) paper, which established a shale density curve for Permian shales in the midcontinental United States. These evaluations were followed by Hedberg's (1936) curve obtained from Venezuelan sediments and Dickinson's (1953) compendium of Gulf of Mexico Coast data, among others. A comparison of the three curves, which did not overlay, shows that burial depth was not the controlling dewatering mechanism. Dickinson noted that geologic age was definitely a factor. A general review of shale density versus geological time (Manger 1963) indicates that shale compacts continually from its initial deposition to its ultimate lithification as a different rock type (Weaver & Beck 1971).

In the Athy (1930) model, compaction is described as a reduction in volume through closer spacing of grains, recrystallization, and possibly decomposition.

Essentially this theory has remained the same, except that much more is known about the nature of mineral alteration, recrystallization, and decomposition, particularly with respect to clay, than during Athy's era. The 1930 model did describe the development of mica by recrystallization even though the montmorillonite-illite transition was unknown at that time. However, the modifying effects of abnormal pressuring in construction of density-depth, porosity-depth, and compaction-depth curves were not then recognized.

The abnormal pressure modification in sediment dewatering first gained widespread recognition with the publication of Dickinson's (1953) analysis of reservoir pressures. Recognizing that the expulsion of subsurface fluid becomes more difficult as permeability decreases, he first expressed the counterforce relationship between compacting forces and expulsion resistance. Dickinson also extended the Athy-Hedberg continuous compaction curves to considerably greater depths but apparently did not recognize the effects of clay diagenesis; this was left to Powers (1967) and Burst (1969).

Hedberg's (1936) chief contribution to the early dewatering literature was to codify the compaction continuum into a three-stage model:

Stage 1 Zero to 800 lb/in^2 overpressure
- (*a*) Porosity 90–75%; free water loss and first order solid particle rearrangement.
- (*b*) Porosity 75–35%; loss of some absorbed water.

Stage 2 800–6000 lb/in^2 overpressure, porosity 35–10%; mechanical deformation of sediments and continued loss of absorbed water. Some incipient recrystallization.

Stage 3 $>$6000 lb/in^2 overpressure, porosity $<$10%; high pressure recrystallization.

According to this classification, clayey sediments were judged to be essentially dewatered at 6000 lb/in^2 or at about 6000 ft of lithostatic overburden.

Weller (1959) calibrated the various porosity-depth curves (Athy 1930, Hedberg 1936, Terzaghi 1925) into an articulated reflection of total section compaction. He extended the porosity curve limits to 30,000 ft, at which depth he predicted zero porosity. He adjusted the Hedberg limits moderately by converting Hedberg's observations to values based on Atterberg limits and the measured porosities of loose and compacted sand, but in essence he retained Hedberg's tripartite classification.

Weller also recognized the disproportionate effect of clay components in a compacting system by acknowledging that surface forces are enormous where a very large surface area is concentrated in a very small volume. Continuing this line of reasoning, Weller also recognized the complicated clay-water relationship and the possibility that water, tightly bound to clay surfaces, could exhibit solid-state properties.

As may be inferred from this brief historical review, the dewatering of argillaceous sediments and their consequent reduction in volume, usually described as compaction, is a multifaceted process composed of several mechanisms. Principal among these are gravity, pressure, and temperature; ancillary are capillarity, osmosis, ionic-filtration, and perhaps a few other subsurface manifestations of well-known generators or modifiers of liquid movement. The matrix components on which these forces operate are also several fold—pore water, van der Waals milieu, external surface-structured water, clay interlayer water, structurally bound hydroxyl groups, and compositionally bound water molecules such as those structured in the gypsum lattice. With this number of parameters, to say nothing of the modifying effects of fluid chemistry, permeability barriers, nonhydrous mineral alteration, etc, it is not surprising that the full understanding of sediment dewatering is still elusive.

The first sorting out—perhaps arbitrary but simple and useful—is the semantic division of gravitation and compaction. This is proposed as the boundary between that point in burial at which solid sediment particles cease to fall essentially free in a liquid medium and begin instead to compress the liquid as a simple consequence of having formed a contiguous network through the process of nearest neighbor, physical contact (Hedberg, Stage 1a). This boundary is also characterized by a change of migrating water direction from vertical to lateral. Ostensibly, this change occurs quite early in the compaction of clastic sediments and assumes importance in that the gradual closure of vertical permeability reorients the path of least resistance to liquid flow. No doubt incipient closure occurs immediately below the surface in new, clayey sediments, but it may not be definitive until considerable competency has been established, perhaps as deep as 2000 ft (Hedberg, Stage 1b).

This transition is important because it marks the boundary between vertical and lateral fluid migration, thus ushering in a completely new set of parameters for interpreting sediment dewatering. After a certain stage of compaction has been reached, the flow of water must be almost exclusively parallel to bedding planes. At this point, the sediment dewatering process perforce changes from the wave front continuum of essentially vertical rising water to one in which aquifers must discharge laterally into fissures and subsequently to the surface. Deep subsurface brines are locked in until structural movement or perhaps mineral diagenesis can develop conduits of egress.

The mobility variation caused by this migration stricture is important also in that it permits the upward percolation of shallow-formed hydrocarbons such as methane and proscribes directly vertical petroleum migration because liquid hydrocarbon maturation normally occurs deeper in the section than 2000 ft. This merely illustrates the simple physical principal that subsurface liquids (like all others) move under pressure, in the direction of least resistance. This point has been treated extensively by Cordell (1972, 1973, 1974) and Chapman (1974).

The flow of liquids through porous media is normally characterized by the Darcy equation, $Q = K/N \cdot dp/dx$, which states simply that flow is equal to pressure drop modified by the viscosity of the liquid and the permeability of the porous media. Unfortunately, the most common liquid, water (both fresh and saline) does not obey this relationship because water configures itself upon the capillary interfaces, particularly fine-grained particulates, and develops a semisolid flow barrier. The dimensions of the barrier, and thus its barrier potential, are controlled by the nature of the solid surface and the velocity and ionic content of the traversing water. According to von Engelhardt & Tunn (1954), "one can imagine that the fluid becomes fixed on the solid surface of the rock boundary and forms an immobile layer which is no longer part of the flowing fluid."

Orientation of the flat, platey clay particles in sediments is thought to significantly influence the compaction of argillaceous sediments. As reviewed and illustrated by many, including Lambe (1958), Meade (1966), and, most recently, Rieke & Chilingarian (1974), clay platelets respond to confining pressures in terms of their geometrical and electrical characteristics. Inherently repelled by like-ion forces, clays eventually develop a loosely, articulated, "house of cards" configuration in thick, deflocculated suspensions. (See Figure 1*A*, *B*) This is thought to result from the attraction of positively charged plate edges to negatively charged plate surfaces. Upon flocculation (see Figure 1*C*, *D*) repulsive surface forces are neutralized and plate-to-plate floccules begin to form structured domains.

As structural integrity develops, overburden pressure forces become more effective and orientations such as illustrated in Figure 2*A* develop. This type of orientation preference can be documented in many shaley sediments. Another suggested geometry of compacting orientation, as represented by Figure 2*B*, accomodates the many observed instances of clay deposits in which random domains are obvious, such as in the massive kaolin beds of Georgia. As might be expected, orientation effects are less pronounced in finer particle sizes, wherein the platelets assume a more spherical shape.

Actually, the literature is not very instructive in the matter of sedimenting particle interrelationship. For instance, an illustration virtually identical to Figure 1*B*, which is described by Rieke & Chilingarian as deflocculated, is labeled by Meade (1966), as representing the flocculated state. Another aspect of this phase of compaction, also needing more precision, concerns the expressions of sediment concentration. The description by Rieke & Chilingarian (1974, p. 9) of a clay-in-water suspension as "thick mud" begs for quantification. The early evaluations of clay-in-water suspensions were made by colloid chemists who characterized gel structures containing as little as 1–2% montmorillonitic solids. This learning has been transferred to the sedimentological rationale governing argillite petrology

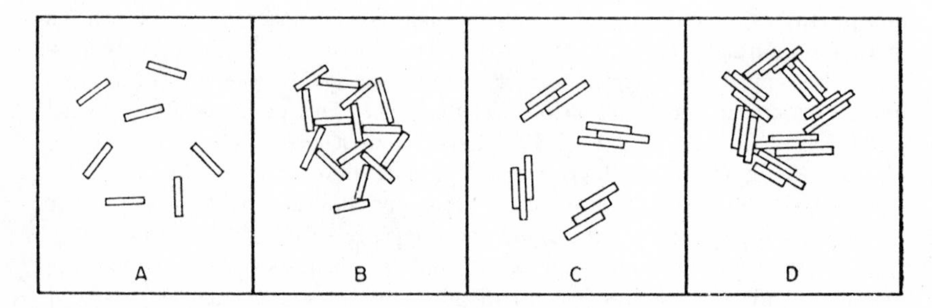

Figure 1 Possible arrangements of clay platelets in thick clay-in-water suspensions. After Rieke & Chilingarian (1974).

with little or no justification. Dilute gel structures such as those used to construct the sedimentation models probably do not exist in nature except as rare anomalies.

Generally, recent interpretations of sedimentary fabrics have recycled decades-old data, which were obtained on not-necessarily pertinent systems and have been modified to interpret newly recognized rock lithologies. Research on clay particle sedimentation and consequent sediment orientation by modern (cryogenic, SEM, and NMR) methods would be fruitful. Keller & Hanson (1975) have been instrumental in generating fabric studies in clays through SEM studies. O'Brien (1971) reviews the present state of prelithification clay floccule work and presents evidence on the gross fabric of kaolinite and illite floccules. Further work is needed in this important aspect of sedimentation.

Magara (1974b) adds a new dimension to the interpretation of subsurface fluid migration. Working from the pertinent thermal expansion coefficients, he calculates that the increase in shale pore water volume generated by geothermal effects is in itself a driving force sufficient to induce fluid movement from deep to shallow sections, or from basin centers to their margins, provided a normal hydrostatic continuum is maintained. The modifying special effect of rock expension is inoperative as long as at least 6% porosity remains in the shale, and Magara, interpreting from Dickinson (1953), estimates that it should not be evidenced above 24,000 ft burial.

The dehydration of clayey sediments is apparently promoted by many physico-

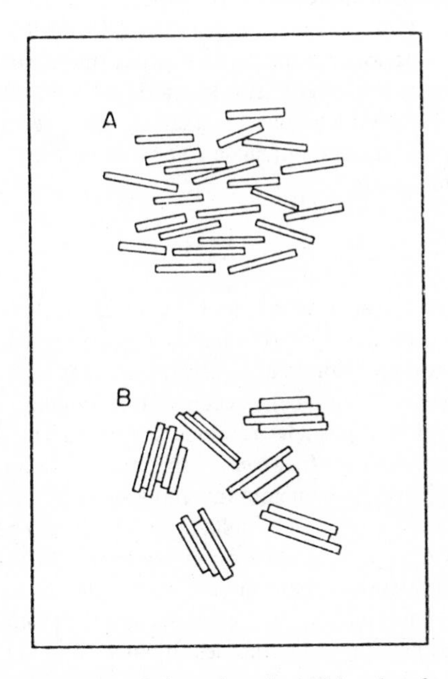

Figure 2 Idealized arrangement of clay mineral particles that form during compaction. After Meade (1966).

chemical relationships, many of which have been investigated and reported upon as independent agencies of dehydration. Most likely, they are highly dependent, representing multifaceted reactions to the impingement of fundamental energy forces.

Classically, shale dehydration has been treated as a reaction to pressure gradients, caused by overburden loading. Magara treats it as temperature dependent. Undoubtedly, it is a function of both and is conditioned by such modifiers as access to normal hydrostatic gradients, chemical composition of the fluid, mineralogical composition of the sediment, and so on.

Magara's aquathermal approach is interesting because it provides a mechanism for the continued dewatering of sediments after clay diagenesis is complete. The internal, geothermally induced pressure, which increases with depth, constitutes a logical force to drive pore water in deeply buried sediments through the ever narrowing porosity channels.

In a later article, Magara (1975) makes another significant contribution to sediment dewatering literature through the introduction of the term *compaction disequilibrium* to describe such phenomena as undercompacted shales. Magara demonstrates the difference in porosity/density characteristics of shale sections in which undercompaction results from compaction disequilibrium versus those in which it derives from montmorillonite dehydration. He concludes that montmorillonite dehydration alone cannot effect abnormal pressuring and that compaction disequilibrium is a more likely principal agent. In these conclusions he differs somewhat from Powers (1967). Magara's main contribution is to emphasize the difference in releasing water from the clay lattice (montmorillonite dewatering) and removing it from the shale (sediment dewatering). In a footnote, Magara notes what is probably the biggest obstacle to assessing the effect of montmorillonite dewatering on the overall question of shale compaction: "the imprecise knowledge of clay interlayer water density."

STRUCTURED WATER

Measurements by von Engelhardt & Tunn (1954), show that the fluid volumes absorbed by clays exceed the pore volumes of the dry powder. In addition, the fluid absorption of montmorillonite considerably exceeds its intracrystalline swelling; therefore, it is concluded that the accumulation of densified, water layers on the outer surface of mineral particles represents a significant portion of the total immobilized water component. In a sense, the water envelope around the clay particle acts as a semipermeable membrane allowing catonic exchanges to occur. The osmotic action of the membrane results in water (fresh?) being extracted from the external electrolyte solution and becoming strongly attached to the clay. The amount of water fixed by this mechanism becomes greater as the degree of dissociation becomes greater. Inasmuch as Na^{+} clays are known to hold more water than Ca^{2+} clays, it follows that Ca^{2+} clays are less dissociated and that their resistance to flow within porous media is less.

A discrepancy in the density estimate of water layers associated with clay lattices is apparent. The coincident hexagonal configurations of water and the silicate lattice have prompted logical conclusions that a systematic fit occurs between the two and that a room-temperature, icelike structure may exist in the bound water. The water hexagon is slightly smaller than its clay counterpart, however, and a certain bond stretching is necessary to effect a proper fit. This implies, of course, that the density of this stretched-bond water is less than the normally configured water which surrounds it. Low & Anderson (1958) showed, for instance, that the density of absorbed water associated with montmorillonite is less than that of liquid water to a distance of 60 Å from the orienting surface. Grim (1968) also suggests that low-density structured water is evident at a considerable distance from the base of its electrical origin.

Another theory (Bradley & Serratosa 1960) proposes that the stretching is so great that unstretched water molecules can infiltrate the expanded hexagonal lattice and raise the overall water layer density above that of normal water. Using this approach, Burst (1969) calculated interlayer water densities as high as 1.15; Williamson (1951) and Martin (1962) had previously suggested interlayer water densities as high as 1.4, a figure accepted by Hanshaw & Bredehoeft (1968) in their authoritative discussion of anomalous fluid pressures.

Martin (1962), however, takes exception to the hexagonal theory of structured water and to the ice analogy. He freely admits that the density and behavior of absorbed water is different from normal, liquid water on the basis of virtually every comparative test, but prefers the description "a two-dimensional fluid." He bases his conclusion on several sets of data, including dielectric and magnetic measurements and the integral entropy data on kaolinite. The data, though meagre, are in Martin's judgment definitive. Martin defines absorbed (structured) water as that occupying the space between the oxygen or hydroxyl surface of a clay crystal and bulk liquid water. In so doing, he also defines the thickness of the structured area as variable from fractionally monomolecular to many tens of molecular layers, depending upon the water content and the type of clay.

Martin's rejection of the ice structure includes the consideration that absorbed water density has not been shown to approach the value for ice (0.90) or for the hexagonal net (0.92). To the contrary, Walker (1956) cites a density of 0.92 for a 14.36 Å unit of Mg-vermiculite, and Burst (1969) calculates 0.94 for a complete monolayer and even less for an unfilled layer. The key to this particular data discrepancy may be that the Walker and Burst figures were derived from Mg-vermiculite whereas the Martin quote derives from an assessment of Na-montmorillonite. Mg-vermiculite possesses more easily measurable and more satisfactorily interpretable surface water relationships than montmorillonite. Further work on resolving the true configuration and consequent density of water structures probably should concentrate on vermiculite.

Frank (1970) supports Martin's position by saying that, although the most stable arrangement of atoms in a system of two H_2O molecules resembles the trans-configuration of ice, this structure is a very inefficient way of filling space. That water formed from melting ice is denser than the ice from whence it came indicates that structural voids either (*a*) no longer exist, (*b*) have been partially filled, (*c*) have been reduced in relative size by bond bending or reorganization. Perhaps the description as icelike has been used by authors in a descriptive rather than a crystallographic sense to illustrate the water net overlaying the hexagonal silicate lattice by relating it to the only commonly observed water-based structure, the snowflake.

Nevertheless, the ice configuration in structured water is valid, density variations and all, if the concept of void filling within the structure can be accepted. Evidenced not only in modular form but also in the shrinkage phenomenon observed on melting, structural vacancies in certain forms of water, such as the four-coordination model of ice, are well accepted. The phenomenon of maximum ice density also confirms water's structural variability. Interstitial filling of the water structure has been proposed several times with the conclusion that fractional, temperature- and pressure-dependent, interstitial vacancy filling can account for density variations (Frank 1970). It should be pointed out that interstitial in this context means water within water—not water within clay lattice, which is the usual interpretation by clay mineralogists.

Unfortunately, the Bradley & Serratosa (1960) work has not been fully exploited in the search for answers to fluid flow problems in fine-grained sediments. Work on water structures has usually concerned liquid water and its normal solid counterpart, ice. Further work on the solid or semisolid, surface-oriented monolayers peculiar to clay lattices is now needed.

The case for room-temperature solid water in clay systems should have been evident from Van Olphen's (1963) work with desorption isotherms. He notes that

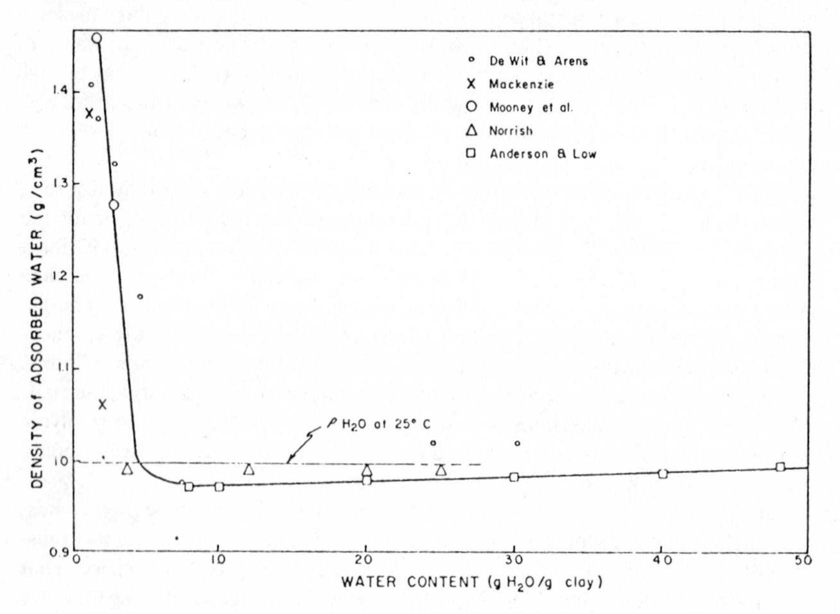

Figure 3 Adsorbed water density on Na-montmorillonite. After Martin (1962).

the work necessary to mechanically shear the last monolayer of water from a calcium montmorillonite lattice is 240 ergs/cm^2 which equates to a force of 79,000 lb/in^2. Values for the second last monolayer are 65 ergs/cm^2 and 37,000 lb/in^2. Gurikov (1963) in reviewing studies on cold neutron ($\sim$0.01 eV) scattering by water, notes that in this respect water is much closer to the solid state than are simple liquids.

Russian literature continued to explore the structured water field until the unfortunate polemics of polywater. Derjaguin's (1970) review of superdense water contains a mention of joint work with the late R. Green-Kelly, which illustrates that the abnormality of clay-associated water has been recognized by the Russian school (also see Kartsev et al 1969). According to Derjaguin, thermal conductivity is extremely sensitive to structural configuration and its measurement can differentiate liquid and crystalline states. He cites the work of Metsik (Derjaguin 1970, p. 54) on multilayered mica sandwiches to show that the "liquid" layers between the mica sheets possess thermal conductivities several tens of times larger than the normal bulk value (of water). Upon heating to 60–70°C, the thermal conductivity value falls, indicating a disruption (liquification) of the layer structure.

Derjaguin's work with Green-Kelly determined that the birefringence of interlayered water could extrapolate the findings of Metsik to extremely fine particles such as montmorillonite. They determined that the difference in refactive index between the two optical directions of montmorillonite water interlayers must be the result of structuring under the specific influence of the surfaces of the clay lattice.

All of these density measurements and calculations, regardless of the relationship of their resultant value with respect to the unity value of ordinary water, depend upon the recognized but not-too-well-understood phenomenon of structured water. A listing of the various densities proposed for the water molecule is of little avail because probably they are all correct for the conditions of evaluation. Apparently, ordinary water is a misnomer in density determinations since it is far more affected by and responsive to its environment than is usually admitted.

To illustrate several of the density measurements in the literature, Martin (1962) constructed a graph (see Figure 3). Adsorbed (structured) water density has been measured at values from slightly less than 1 to greater than 1.4. The only significant variation in measurement from the trends indicated are the traversing of the unity values at low water contents, as suggested by Walker (1956) and Burst (1969). This phenomenon is further treated in the next section.

The working hypotheses and demonstrated theories concerning the nature of structured water are summarized below.

The Hexagonal Net

One theory describes structured water as forming a hexagonal net wherein loosely articulated water structures are distributed across the regular geometry of clay surfaces in conformance with the hexagonal electrical charge distribution of the structural lattice. This theory accommodates the low density observations of Low & Anderson (1958) as well as (by in-filling) the high density calculations of many workers in the field. This explanation of clay-surface, oriented, structural water is the most adequate. First suggested by Hendricks & Jefferson (1938), this interpretation has consistently gained credibility as a succession of investigators conclude that the effect of water film epitaxis is three-dimensional both within the lattice of expandable clays and around the gross particles in all instances of fine-grained argillaceous material. The most significant recent contribution to understanding of the hexagonal net theory is Bradley & Serratosa's (1960) modification of the Hendricks water net to accommodate two additional water molecules within an epitactic supercell. Armed with this interpretive option, virtually all of the present analytical data on clay surface–water net relationships can be rationalized.

The Two-Dimensional Fluid Theory

The two-dimensional fluid theory of clay-surface, structured water appears to have been first proposed by MacEwan (1948) who considered the Hendricks net to be less rigid and essentially mobile. Although most analytical data argue that a considerably less-than-liquid barrier is present on clay surfaces, Martin (1962) also adopted the two-dimensional fluid theory, expanding it to include a three-dimensional spacial arrangement that includes the entire interfacial region between the oxygen or hydroxyl surface of the clay crystal and bulk liquid water. The physical attributes of this space form the basis for the various theories of structural water. Proponents of the hexagonal net theory conclude that this intermediate area resembles the clay solid whereas supporters of the two-dimensional liquid interpretation conclude that it more resembles bulk liquid. Martin's data on differential entropy favors the latter.

Ice Structure Interpretations

Ice structure interpretations of adsorbed water have been popular, no doubt prompted by the similarity of the hexagonal appearance of snowflakes and the hexagonal representations of surface-structured water. The analogy is useful in visualizing the concept of solid water, but the hypothesis (rather than theory) has not stood up consistently to analytical probing. Martin, for instance, suggests that both dielectric and magnetic (NMR) properties indicate that sorbed water does not have an icelike structure. The original and, to some extent, subsequent attempts to relate clay surface water and the ice structure apparently stem from the relatively low, similar densities of each. Although some forms of ice may be capable of fitting onto the silicate lattice, the observed density dichotomy of adsorbed water (both high and low density) necessitates the existence of both high and low density room-temperature ice structures, which is unlikely. The water in-filling suggestion of Bradley & Serratosa (1960) seems more reasonable.

CLAY MINERAL DIAGENESIS

Each early investigator of sediment compaction, by confining study to a restricted geologic province and a narrow depth interval, inadvertently plotted a time-dependent member of a curve series [see Figure 4, adapted from Magara (1968)] that is now known to operate through the entire sedimentary section. More recent data from deep drill holes show that the overriding control of sediment dewatering is exercised by the kinetic integration of pressure, temperature, and time, although an interesting thermodynamic interjection (clay mineral diagenesis) at one critical point in a specific but pervasive clay lattice dehydration, pointed out by Keller (1963), Powers (1967), and Burst (1969), modifies the kinetic concept somewhat.

Clay diagenesis, as a temperature-dependent function responding to the impress of geothermal gradient, is now a fixture in geologic interpretation. Specifically, the progressive conversion of montmorillonite to illite, with its attendant water expulsion, has been repeatedly documented (see Dunoyer de Segonzac 1969, 1970). The dehydration sequence, as represented by the disposition of expressed lattice water, and its consequences are still being investigated, however. Powers (1967), Burst (1969), and Perry & Hower (1972) have treated clay dehydration in turn, and with each analysis the interpretations became more complex as the intricacies of the mechanism were better appreciated. Powers, noting the difference between consolidation resulting from initial compaction and compaction attributable to the montmorillonite conversion, designated the two effects as stage I and stage II (see Figure 5). Burst used approximately the same boundary conditions in his interpretation of the dehydration sequence, but reduced stage II (the montmorillonite-illite reaction) to a relatively narrow, temperature-dependent band and added a deeper, third stage to accommodate deep burial water loss subsequent to the major diagenetic conversion. Perry & Hower refined the interpretation by splitting Burst's stage II into two parts representing (*a*) an initial high rate of water expulsion engendered by the relatively rapid collapse of approximately 65% of the montmorillonite layers at the top of the diagenetic zone, and (*b*) a second high rate of water expulsion resulting from the quick collapse of that fraction of the montmorillonite lattices representing the interval between 65% and 80% of total collapsed lattices. Perry & Hower added a fourth dehydration stage that corresponds to Burst's stage III.

Although the montmorillonite-illite transformation was recognized as a laboratory phenomenon and was deduced in the field, principally by agronomists, from apparent potassium fixation, it was not until the late 1950s that Powers (1959), Burst (1959), and the assorted papers of Weaver (1958a) (1958b) (1959) brought to

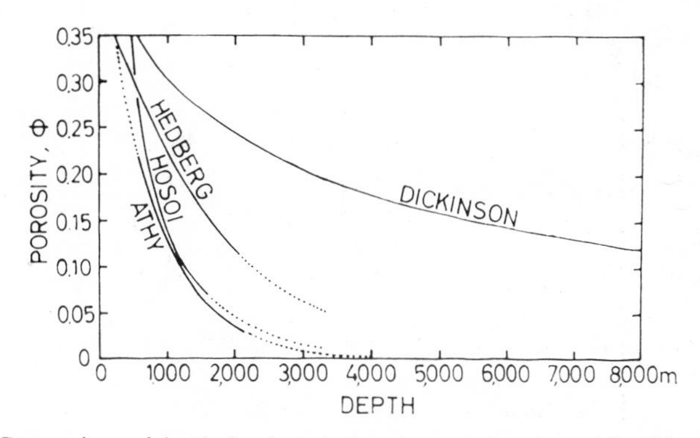

Figure 4 Comparison of depth-density relations in several regions. After Magara (1968).

light the persistent alteration of montmorillonite with burial depth. Each of the authors cited above advanced different theories to explain the observations. Weaver's interpretation, which was supported by many classical stratigraphers, was that clays received their physical and chemical characteristics from parent material, carried these relationships into superposed stratigraphic sequences, and retained them in essentially unaltered form. Powers, noticing a consistent difference between shallow clays and those at depth, proposed the equivalence level, below which K^+ is adsorbed preferentially to Mg^{2+} by clays (to form illite) and above which Mg^{2+} is the preferred adsorbate producing chlorite and vermiculite.

Burst (1959) also favored chemical alteration, but after first trying to relate down hole clay mineralogy to ancient sedimentary environment, as an extension of Power's (1954) work, he finally concluded that the clays were responding to geothermally induced chemical alterations and that Tertiary (at least) argillaceous sediments truly reflected their geothermal history (Burst 1969). The key observation was that the montmorillonite-illite transition always progressed toward the collapsed (dewatered) lattice and that it never subsequently reversed. This theory dominates in interpretations since 1969, with later work contributing variations, improvisions, and data ancillary to the central theme.

The essential element of Burst's (1969) work is the interpretation of Bradley & Serratosa's (1960) in-filled water structure theory in terms of subsurface thermodynamics. By calculating the density of interlayer water and fitting the required water contents to the available void spaces, Burst was able to divide clay mineral dewatering into shallow and deep kinetic releases of interlayer water and a mid-depth release, which is apparently controlled by thermodynamic forces.

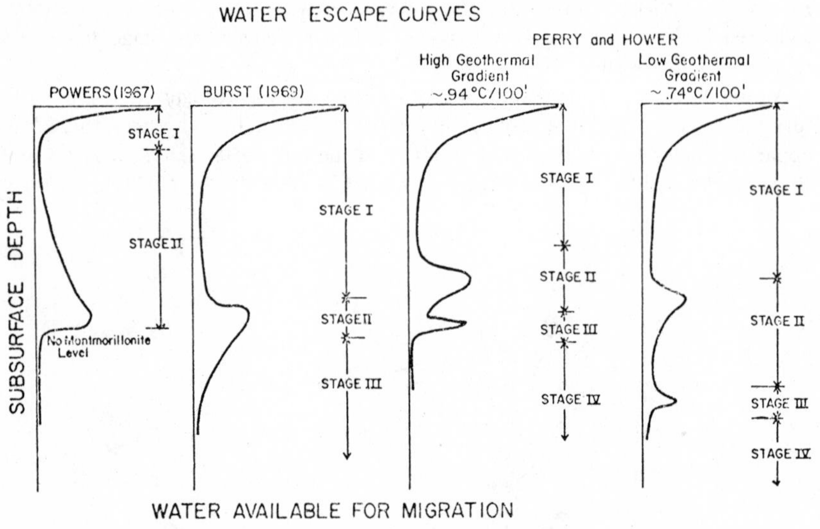

Figure 5 Water escape curves. After Perry & Hower (1972).

The continuous diagenetic effects of time, temperature, and pressure therefore experience a discontinuity with respect to sediment dewatering at the stage where the clay lattices in shale sections contain only two water interlayers (Burst 1970). At this stage, the attractive forces generated by exchange sites within swelling clay lattices cause the last two water layers to be (*a*) spread slightly to accommodate the charge distribution, (*b*) infilled with extra water molecules at the rate of one additional water for each eight molecules normally present, and (*c*) compacted in an overlapped geometry due to the hexagonal close-pack rule.

This results in a configuration in which the water trapped within the clay mineral lattice occupies less space inside the clay lattice than it would in the pore space (see Figure 6). Pressure, therefore, is no longer an effective dewatering agent and the shaley interval resists the normal diagenetic progression as it waits for forma-

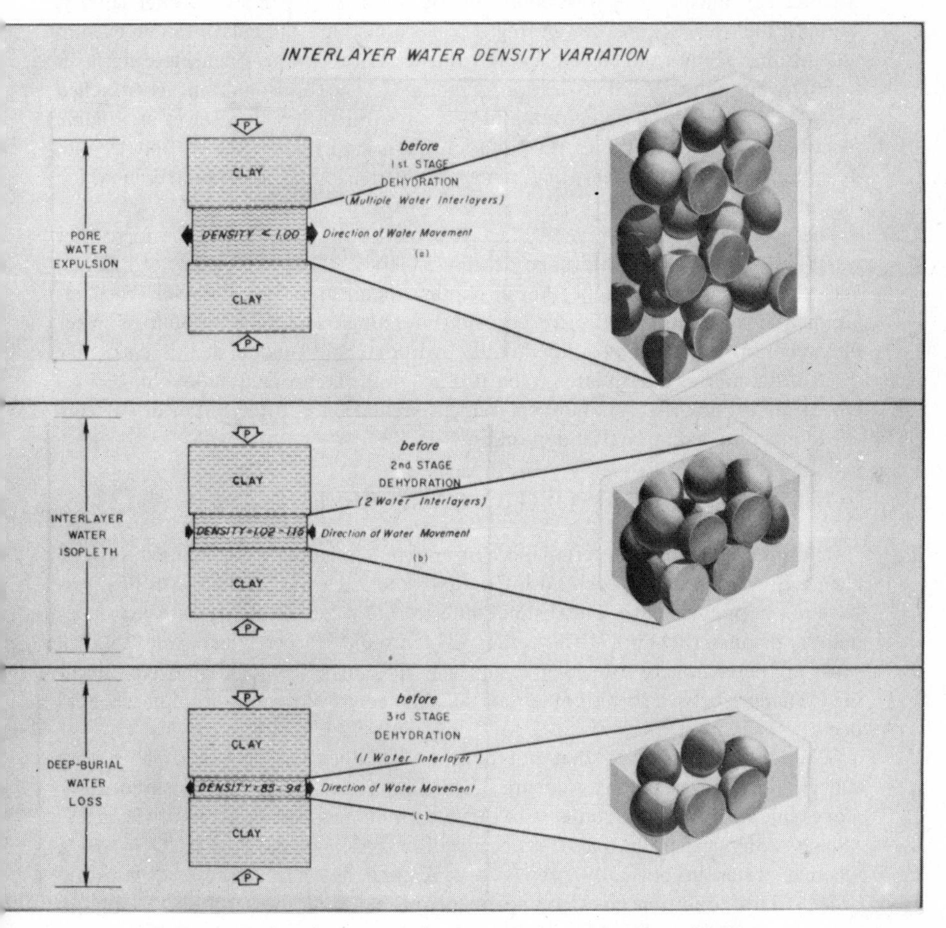

Figure 6 Interlayer water density variations. After Burst (1969).

tional temperatures to increase sufficiently to force the stabilized interlayer water into the pores. The vertical extent of this stabilized zone depends essentially upon the rates of sedimentation and pore water expulsion. In rapidly depositing sections, heat accumulation is slow because of excessive entrapped water. As a result, the geothermal gradient is lowered and the interlayer dewatering level is deepened. Normally the sediment temperature eventually rises enough to mobilize the interlayer water which then drains out through the pore system. If the system is closed, however, perhaps by the self-sealing action of monolayer immobility in thick shale sections, the pore space cannot accommodate the mobilized water, which is attempting to increase its volume by 10–15% upon liquefaction, and the condition of overpressuring develops.

When only two water layers are present, dewatering cannot take place without thermal assistance even against pore water densities of three times seawater salinity. Without the overpacking, dewatering is considerably less difficult although even in this instance the normal close packing tendencies of the lattices densifies interlayer water to the point that dehydration is inhibited. The semisolid, surface-attached water molecules inhibit dehydration in two ways. First, they themselves are unable to flow in response to the Darcy equation because of high viscosity, and second, by stabilizing as a semisolid, they reduce the permeability of the matrix in which they are a component.

The net effect of clay mineral diagenesis on sediment dewatering is to restrict water movement in the midrange depths (3,000–12,000 ft in Tertiary sediments) and to encourage it at deeper burial depths where interlayer water has been reliquified and acts as an effective porosity breeching agent. The amount of water moving from interlayer positions to the hydrodynamic conduit at this particular stage of dewatering is estimated to be 10–15% of the compacted bulk volume. This represents a thickness/volume displacement second only to the initial dewatering of sediment immediately after deposition.

PORE WATER: COMPOSITION AND EFFECT

Insight into sediment dewatering mechanisms could seemingly be obtained through chemical evaluation of pore and interstitial waters. However, studies in this area have not been conclusive. The distinct difference in water chemistry between shales and sandstones (Schmidt 1973) seems to contradict earlier conclusions that all rocks are permeable to some degree and that, in terms of geologic time, equilibrating exchanges between ionic content of the pore waters of various lithologies would occur.

Schmidt (1973) showed that not only does abnormally pressured shale pore water have a lower salinity than the water in adjacent normally pressured sandstones but also the anion contents of the two lithologies were consistantly inverted. "Shale water generally has a concentration order of $SO_4^{=} > HCO_3^{-} > Cl^{-}$, whereas water in normally pressured sandstone has a reversed concentration order." This condition could obtain in permeable units, however, if the ions in question possessed different mobilities, as suggested by White (1965). Ionic mobility, in this case, is controlled by the semipermeable membrane effect of clay layers separating aquifers.

It is somewhat difficult to apply the differential mobility theory to Schmidt's observations because he worked with both normally and abnormally pressured formations—abnormally pressured units, particularly shales, are supposedly impermeable (Dickey, Shriram, & Paine 1968). The theory appears valid, however, as long as shale members maintain some permeability. Compaction must control permeability. At one time, all shales must have displayed permeability to water, for, as Jones (1969) points out, "How else could compaction occur?"

The shale permeability question is of utmost importance in sediment dewatering because it dictates whether Darcy flow or non-Newtonian factors control water movement and whether reverse osmosis is a tenable dehydration mechanism, and because it conditions, to some great extent, the location of water-deposited mineral accumulations.

Pore water chemistry in itself is not too important in sediment dewatering except as it reflects the ionic chemistry of the shale matrix and the adjustments in salinity caused by mineral diagenesis. Both of these properties can be monitored by observing changes in cation exchange capacity with depth (Schmidt 1973). As montmorillonite changes to illite, the CEC falls because the exchange capacity of illite is considerably lower than that of montmorillonite. Also, because Na^+ is the major exchangeable ion on montmorillonite, the population of exchangeable Na^+ decreases with depth and, presumably, the number of Na^+ ions in the pore water increases commensurately. Tracing the paths of these newly liberated ions on the basis of concentration changes is complicated by the ion-free interlayer water which enters the pore water system at the same time. Exchangeable K^+ increases relatively as the Na^+ is removed, but subsequently decreases as the structural lattices sequester these ions by fixation. The role of calcium in this type of study is not known; however, because of its hydrated ion radius 9.6 A (vs 5.6 Å for Na^+) it is presumed to be less mobile. According to Posokhov (1966; see Rieke & Chilingarian 1974, p. 224) the mobility of hydrated ions has been ordered in the following relative sequence:

(*a*) Highest $NO_3^- > Cl^- > SO_4^{2-} > CO_3^{2-}$ lowest
(*b*) Highest $K^+ > Na^+ > Ca^{2+} > Mg^{2+}$ lowest

On the same page, same publication, Nightingale (1959) is quoted as listing hydrated ionic radii (Å) as follows:

K^+	Na^+	Ca^{2+}	Mg^{2+}
3.31	3.58	4.12	4.28

Although it is seen that numerical values can differ somewhat (probably depending on the nature of the medium), the sequential relativity is preserved.

The effect of these various changes with respect to sediment dewatering centers around the relative impermeability of sodium based clays to aqueous flow. A fluid flow balance develops between pore volume and chemical barriers, in which pore space, although relatively abundant in shallow sediments, is flow inhibited by sodium-complexed clays, and although relatively restricted at depth, is enclosed by argillites considerably more amenable to aqueous flow. Field observations indicate that flow follows available pore space rather than ionic environment except in the case of abnormally pressured shales wherein compaction may be retarded by ionic barriers to dewatering.

Sodium in formational waters has long been known to restrict water flow through the mechanism of clay blocking (von Engelhardt & Gaida 1963, F. Jones 1964). The effect is most noticeable when fresh water is introduced into a sodium dominant pore milieu such as may occur when drilling mud filtrate enters formational porosity. It is also in evidence at the freshwater–saltwater interface in the shallow subsurface (Manheim & Horn 1968), at salinity reversal interfaces, and at the clay diagenesis isopleth. Inasmuch as the newly liquified water in this last-named interval is essentially fresh, its entry into the surrounding saline-saturated formational matrix could cause blocking by any sodium-based clays and thus may represent the missing component in the oft-proposed association between clay mineral diagenesis and abnormal subsurface pressures (P. Jones 1969). The activator of clay blocking appears to be the sodium ion both in pore water and on the clay surface, rather than the presence of salinity. Specifically, the swelling of expandable (sodium) clay mineral particles upon contact with relatively fresh water causes the most severe water sensitivity problems. Swollen particles restrict flow in rock pores, and minute, expanded lamellae break away and further restrict flow by lodging in pore constrictions (Dodd et al 1955). Some relief from this condition is cited by F. O. Jones (1964), who shows that clay (Wyoming bentonite) preferentially absorbs calcium over sodium at exchange sites when at least 10% of the total ions in the liquid exchange medium are calcium. This probably means that sodium saturated clays are rare in nature and that clay blocking through completely natural processes is infrequent.

Membrane action, also known as ion filtration, salt filtration, and reverse osmosis may appear in the section at that point where true consolidation begins and pore water is no longer able to traverse unimpeded through the various lithologies. Membranes inhibit hydraulic continuity and are probably responsible for most of the variable chemistries displayed by stratigraphic neighbors, although Manheim & Horn (1968) contend that these played no important role in concentrating brines in the Atlantic continental margins. In this case, however, simple subsurface leaching (Rittenhouse 1964) and osmotic distribution are probably overriding factors. At any rate, there is general agreement that compositional changes are evident between adjoining subsurface strata, and reverse osmosis is high on the list of possible causative mechanisms.

Osmosis tends to cause a pressure increase in the formation containing a higher concentration of ions. This osmotic pressure could conceivably change the pattern of flow (Bredehoeft et al 1963). A way in which this can happen is detailed by Olsen (1972), who shows that the combination of osmotic and electroosmotic gradients can reduce or even eliminate the effects of hydraulic pressure, thereby altering or eliminating hydraulic flow. According to Olsen, the magnitude of hydraulic flow decreases more rapidly with depth than do the osmotic and electroosmotic flows; thus the latter are increasingly important at greater depths of burial.

Magara (1974a) recently presented an ion filtration explanation for salinity concentrations in Gulf Coast formational fluids. He argues convincingly that relatively fresh water is expressed from shales leaving a salt build-up in the shale pore water. He also suggests that, because migrating water has a tendency to move toward sandy strata, the porosity of clays tends to fall as the sand is approached and the salinity of the remaining water will increase. He infers directed water flow by noting that osmotic flow from a less concentrated point to a more concentrated means (in the context of the geometry he adopts) flow from the centers to the margins of basins. In his example, osmotic pressure is small compared to compaction pressure—probably a universal condition.

The effects of pore water chemistry on sediment dewatering appear to be confined to (*a*) blocking of the hydraulic gradient by the combined presence of fresh water and high sodium content, and (*b*) osmotic pressure which moves dilute clay expressed water toward more concentrated fluids and, in the reverse capacity, causes the separation of this dilute fluid from shales, resulting in a build-up of salt within the shale.

SALINITY PROFILES

The amount of water transferred from lattice to pore during clay diagenesis has been estimated as between 10–15% of the compacted sediment volume. Inasmuch as clay mineral diagenesis represents not only a relocation of water molecules but also a change in their mobility potential as they change from semisolid, lattice-affixed crystal components to a liquid, significant reorientation of the entire subsurface pore component system can be expected. For example, the interjections of "fresh" interlayer water into the saline pore water of deeply buried sediments probably causes the phenomenological salinity reversals experienced in young sediments of the Gulf Coast Tertiary.

Classically, deeper penetration into the subsurface was understood to be accompanied by rising pore water salinity values. Occasionally encountered freshets were considered anomalies in the normal linear salinity-depth relationship. It now appears, insofar as Tertiary sections are concerned, that this judgment was premature. Deeper drilling through the diagenetic isopleth (Burst 1970) has developed new evidence. Relatively recent work by Timm & Maricelli (1953) and others show that salinity reversals are common to Gulf Coast geology and indicate that a rational, systematic mechanism to explain their frequency is required. Clay mineral diagenesis seems to meet the requirement.

Dickey (1969) has reviewed the question of paleosalinity in considerable detail. Although his work was published before the potential impact of clay diagenesis was fully appreciated, his illustrations and descriptions are extremely useful. According to his analysis subsurface brines are normally more concentrated than seawater (35,000 ppm), ranging between 50,000 ppm and 200,000 ppm total solids. The salinity of meteoric waters usually increases linearly with depth at rates of increase varying between 10,000 ppm and 100,000 ppm per thousand feet (50–300 mg/liter^{-1} m^{1}. In some instances, however, such as in Tertiary sediments of the Gulf

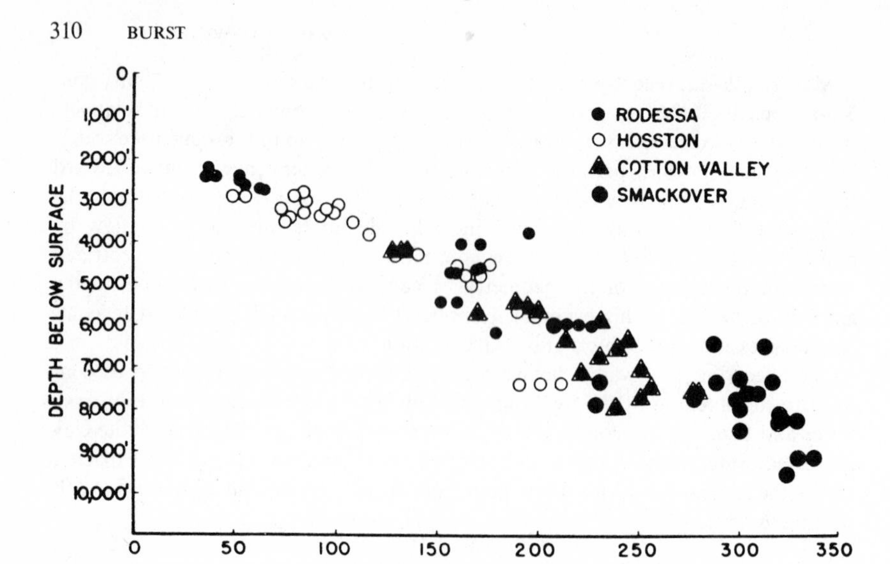

Figure 7 Concentration vs depth, Jurassic to Cretaceous. After Dickey (1969).

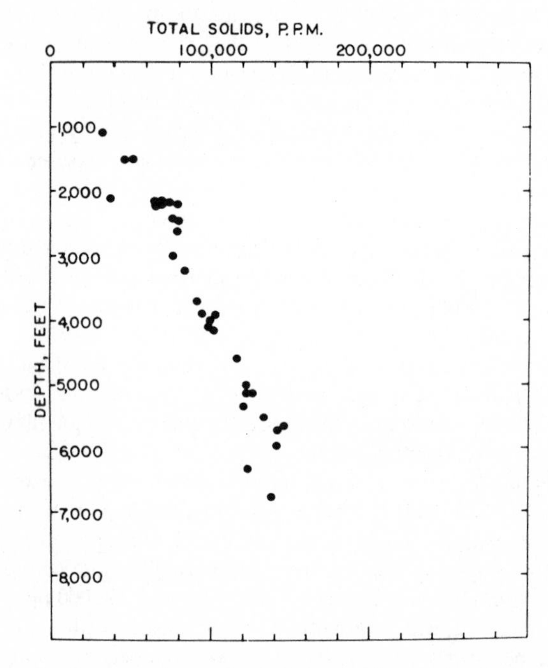

Figure 8 Concentration vs depth, Eocene Wilcox. After Dickey (1969).

Coast, the linear salinity increase no longer holds. Compared to older sediments from the Jurassic and Cretaceous (see Figure 7) wherein linearity is precise, salinity levels often stabilize at depths like those in the Eocene Wilcox (see Figure 8) and are increasingly recognized to reverse direction entirely at depth (see Figure 9), thus displaying values more suggestive of shallow aquifers than of deeply buried sediments.

Overton & Zanier (1970), working in the Gulf Coast, have proposed a salinity profile wherein the deep zone water freshening is first apparent at about 10,000 ft and lasts to about 15,000 ft. Their prescribed cause for this fresh water interval is shale mineralizing in conformance with the previous observations of Powers (1967) and Burst (1969). Another possible explanation (Beall & Fisher 1969) is that at overburden pressures of 8,000–12,500 lb/in^2, pore sizes become so small ($<1\mu$) that ionic filtration obtains and fresh water is exuded into permeable sands whereas saline components are trapped within the shales. Louden (1971) characterizes the salinity gradient, beginning with the top 3000 ft of section, which he describes as actively compacting from 50% to 18% porosity. Aquifers, receiving essentially fresh water from the contracting pores, transport it in response to hydrodynamic gradient. Salinity builds up between 3000 ft and 9,000–12,000 ft, where ionic filtration (and clay diagenesis) again supply fresh water. "Normal" salinity

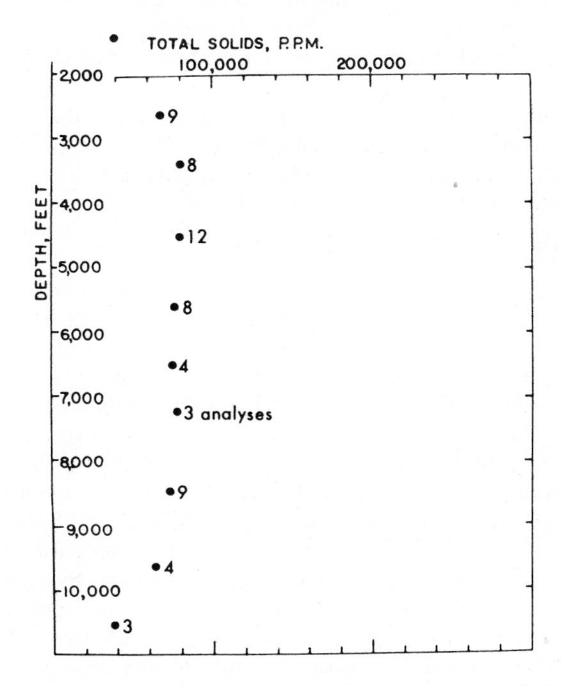

Figure 9 Concentration vs depth, illustrating salinity reversal, South Louisiana. After Dickey (1969).

gradients resume below 16,000 ft. In this model, therefore, three salinity reversals are viewed as normal in a sedimentary section.

Salinity reversals are important from an academic point of view because they lend interpretive value to the effects of pressure-temperature relationships in the subsurface (Magara 1974a) and of clay mineral diagenesis (Powers 1967, Burst 1969). There may also be an incredibly significant practical aspect to this phenomenon. Deep strata salinity reversals of sufficient magnitude to result in brackish water point up the possibility of fresh water production in arid, land-locked countries through reverse osmotic processes that can efficiently convert brackish concentrations.

Overton & Timko (1969) speculated that deep drinking water could possibly have been derived from compacted sands and shales; in these instances shales appear to be useful ionic filters. In light of present knowledge, fresh water accumulations through clay lattice dewatering presents an alternative interpretation, depending on the depth of low salinity water generation. If salinity reversals continue to be found coincident with the diagenetic isopleth, logic seems to dictate that clay lattice dewatering is the responsible agent. If this relationship is not readily apparent, the agencies of ionic filtration and aquathermal flow should be considered.

Perhaps the theories of ionic filtration and clay diagenesis can be joined by a brief but possibly significant reference in the Overton & Timko paper (p. 38), which points out that at 190°F, filtration of salt water through a red mud cake produced potable drinking water. Of course, 190°F is within striking range of the 220°±F required by Burst's (1969) lattice dehydration and suggests that a deep salinity reversal is a combination, or overlapping, of these two desalination phenomena. The significance of the Overton & Timko experiments should not be overlooked by potable water explorationists. The temperature 190°F is well within reach of solar energy collectors, and the prospect of a simple, subdistillation temperature, fresh water generator is always exciting.

The salinity profile problem reverts to the more basic (and previously discussed) water density and structuring problem. As salinity builds up in shale sections, the fresh structured water associated with clay lattices apparently remains intact until diagenesis ruptures the structural configurations. Pore water reaching aquifers up to this time carries the full budget of dissolved salts without dilution, which accounts for the linearity of salinity gradients (no modification due to clay desorption). As the diagenetic isopleth is crossed, the fresh water envelope around clay particles becomes unstructured and salinity dilution results. Ultimate dilution is attained at that point where the combination of lattice collapse and ionic filtration are both operative. In the very deepest section, diagenesis is completed and pore water salinites once again build in accordance with solution chemistry reactions.

Other possible mechanisms for subsurface fresh water generation include the gypsum-anhydrite conversion, reviewed extensively by Hanshaw & Bredehoeft (1968), and the conversion of anhydrite to native sulphur, detailed by F. T. Manheim (1970). The gypsum-anhydrite water source stems from the shallow burial (600–900 m) dehydration reaction $CaSO_4 \cdot 2H_2O = CaSO_4 + 2H_2O$. Heard & Rubey

(1966) calculate that a 50-ft layer of gypsum would generate a layer of water 24.25 ft thick, which according to Hanshaw & Bredehoeft (1968) is sufficient to consider this mechanism as a constant subsurface fluid generator (and a prodigious source of fresh water).

Manheim (1970) considers a further step in this diagenetic change in which anhydrite is eventually reduced to native sulphur through the reactions:

$$3CaSO_4 + C_{10}H_{22} \rightarrow 3H_2S + C_6H_{14} + 3CaCO_3 + CO_2 + H_2O$$

and

$$3H_2S + SO_4^{=} \rightarrow 4S + 2H_2O + 2OH^-.$$

These produce what Manheim describes as important sources of fresh water in domal cap rock.

At least five sources of fresh water are therefore available (one ion filtration source and four mineral diagenesis sources) in the deep subsurface. These reservoirs of potentially potable water for irrigation and human consumption in areas where meteoric waters cannot satisfy regional demand have not yet been exploited.

Sediment dewatering is conditioned by salinity at any saltwater–freshwater interface where the fluid flow characteristics of clay-bearing sediments may be altered. Salinity also controls another significant but often overlooked dewatering mechanism—the mechanical withdrawal of subsurface water for manufacturing, agricultural, and human uses. This has led to surface subsidence, saltwater invasion, water table lowering, and other environmental modifiers.

ABNORMAL FLUID PRESSURES

The question of abnormal or anomalous shale pressures (also known as geopressures and often described in terms of low density or undercompacted shales) must be addressed in any discussion of sediment dewatering. Although abnormal pressuring represents the exact opposite of sediment dewatering, it is in the dewatering concept because it displaces fluids anomalously in the sedimentary column and consequently is responsible for anomalous dewatering sequences. The best estimate of circumstances surrounding the phenomenon of abnormal subsurface pressures (although it may not cover all instances) is that they derive from pockets of sediment that were cut off from the prevailing hydrologic gradient early in their depositional history and subsequently lowered into the sedimentary section as isolated, low density anomalies. Unable to discharge their fluid load in response to increasing overburden, they build up abnormal internal pressures.

The physical properties of the abnormal pockets retain many of the characteristics of the shallower level (i.e. low salinity), although the pressure relationship between the isolated anomaly and the surrounding matrix changes in accordance with the densities of their respective pore water components. This is adequately explained by Barker (1972). Describing a mechanism which he calls aquathermal pressuring, Barker reduces the phenomenological concept to simple physical chemistry via the pressure-temperature-density diagram for water. He shows quite logically that water entrapped in a sediment without access to a hydrologic gradient, develops enormous excess pressures (see Figure 10). Magara (1974b, see above) uses the same force to transport deeply buried hydrocarbons. The force becomes effective as a dewatering mechanism when it can breech the confining sediment matrix and induce leakage to an aquifer.

Apparently, as early as 1953, Dickinson recognized the function of isolation as a requisite to abnormal pressuring, but in an interpretation normal to that time, he attributed the pressure build-up to overburden compaction rather than to geothermal temperature.

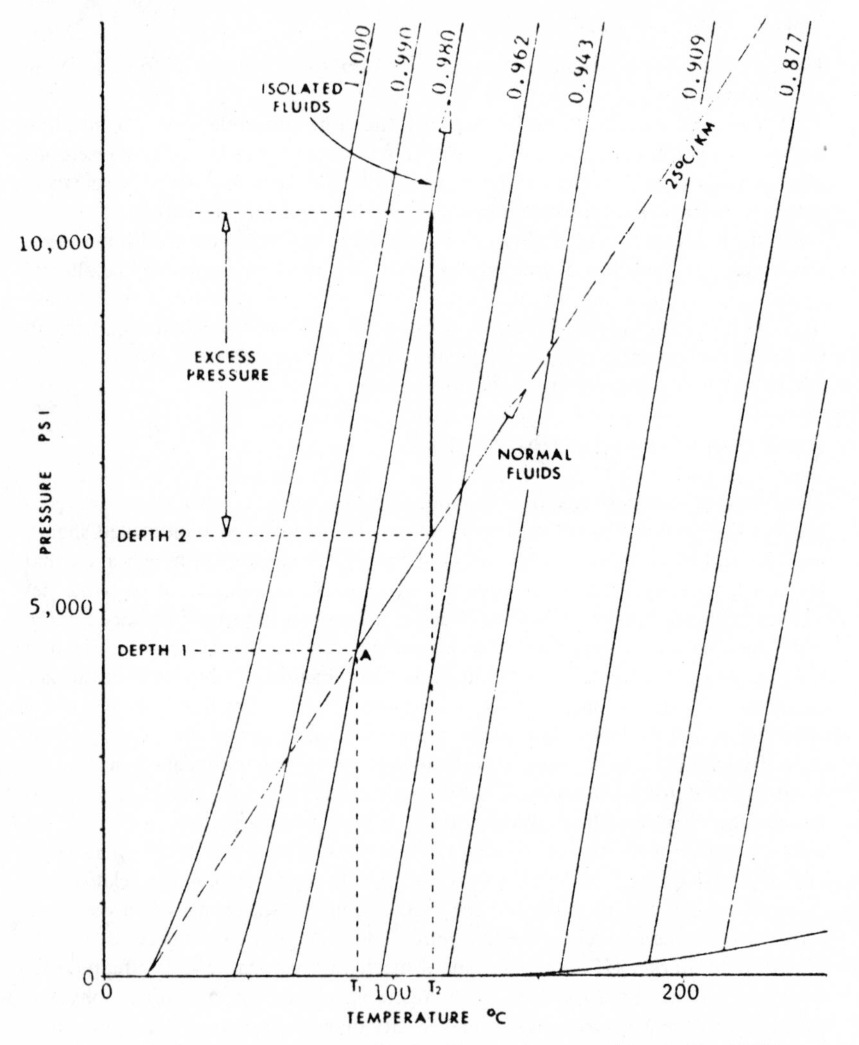

Figure 10 Pressure-temperature-density diagram for water. After Barker (1972).

Barker's deduction that abnormal pressuring of 4400 lb/in^2 results from a 50°C temperature rise translates into the sequestering of pore fluids at, for instance, about 7000 ft burial, producing such a pressure abnormality at 13,000 ft under a typical Gulf Coast geothermal gradient of 1.5°F per 100 ft of buried depth. In turn, it means a combination of overburden weight and compacted fine-grained sediment thickness must have coexisted at 7000 ft to effectively block normal hydrological drainage. By estimating the average solid content of the shale seal over abnormally pressured pockets, converting to the volume that this seal occupied during the early stages of compaction and integrating its preseal compaction rate, an estimate of the minimum burial depth for the inception of abnormal shale pressuring should be possible. Chapman (1972) suggests a minimum depth of 2000 ft but allows that it may be even shallower.

Relieving abnormal pressuring and reestablishing normal sediment dehydration apparently entails the straightforward and simple processes of eventual formational leakage, faulting, and diapirism. Apparently, the regulators operate frequently; most abnormal pressuring is located in young (Tertiary and Recent) formations. Therefore, inasmuch as abnormal pressures are eventually equilibrated, this phenomenon can be considered a transitory inhibitor to sediment dewatering; however, it is academically interesting as well as economically significant when considered in its role of spoiler in petroleum exploration.

CONCLUSIONS

Sediment dewatering is controlled by a balance of mechanisms that tends to promote fluid flow from solids (compaction, mineral diagenesis, ion filtration, and aquathermal pressures) and various modifiers that inhibit flow (clay blocking, structured water envelopes and progressive porosity reduction). Gravity, once virtually the sole agency cited for compaction and fluid flow, has two new allies to help continue dewatering in the deeper portions of the sedimentary section:

1. Clay (and other) mineral diagenesis, activated by elevated matrix temperatures that overcome chemisorption and in some instances molecular bonds, releases liquid water from solid and semisolid sediment components. This process generates fluid flow and continues dewatering at depths where purely gravitational effects are largely negated by sediment consolidation and incipient lithification.

2. Deeper still in the section, the internal pressure of pore water expanding (more rapidly than its enclosing rock matrix) in response to geothermal heat maintains flow despite rapidly deteriorating pore interconnections. For the first time, therefore, an adequate theory is available to explain sediment dewatering from slurry to slate.

Application of these processes (as well as gravitational compaction itself) in natural environments is not completely understood, principally because of "the imprecise knowledge of clay interlayer water density" (Magara 1975). Often reported, its value has been measured or calculated by various workers as being both above and below the unity value of common, pure, liquid water.

This variation, although small numerically, generates completely different dewatering explanations. The imprecision presents a challenge that must be met before diagenetic sediment dehydration and other, equally significant fluid migration processes can be adequately deciphered. Presently needed are sophisticated thermodynamic experimentation to reduce the variables known to contribute to density value variations and more precise fluid mechanic interpretations to articulate the various fluid flow theories into a comprehensive, total column, dewatering mechanism.

Literature Cited

Athy, L. F. 1930. Density, porosity and compaction of sedimentary rocks. *Bull. Am. Assoc. Petrol. Geol.* 14(1): 1–24

Barker, C. 1972. Aquathermal pressuring—role of temperature in development of abnormal-pressure zones. *Bull. Am. Assoc. Petrol. Geol.* 56(10): 2068–71

Beall, A. O. Jr., Fisher, A. G. 1969. Sedimentology. In *Initial Reports of the Deep Sea Drilling Project,* Vol. 1, ed. M. Ewing et al. Washington DC: GPO. 672 pp.

Bradley, W. F., Serratosa, J. M. 1960. A discussion of the water content of vermiculite. In *Clays and Clay Minerals. Proc. Nat. Clays Clay Miner. Conf., 7th,* ed. A. Swinford. Oxford: Pergamon. 369 pp.

Bredehoeft, J. D., Blyth, C. R., White, W. A., Maxey, G. B. 1963. Possible mechanism for concentration of brines in sub-surface formations. *Bull. Am. Assoc. Petrol. Geol.* 47(2): 257–69

Burst, J. F. 1959. Post diagenetic clay mineral environmental relationships in the Gulf Coast Eocene. In *Clays and Clay Minerals. Proc. Nat. Clay Miner. Conf., 6th,* ed. A. Swinford. Oxford: Pergamon. 411 pp.

Burst, J. F. 1969. Diagenesis of Gulf Coast clayey sediments and its possible relation to petroleum migration. *Bull. Am. Assoc. Petrol. Geol.* 53(1): 73–93

Burst, J. F. 1970. The diagenetic isopleth. *Proc. Symp. Abnormal Pressure, 2nd,* pp. 121–30. Louisiana State Univ., Batton Rouge

Chapman, R. E. 1972. Clays with abnormal interstitial fluid pressures. *Bull. Am. Assoc. Petrol. Geol.* 56(4): 790–95

Chapman, R. E. 1974. Depth of oil origin and primary migration: geologist's discussion. *Bull. Am. Assoc. Petrol. Geol.* 58(10): 2029–67

Cordell, R. J. 1972. Depths of oil origin and primary migration: review and critique. *Bull. Am. Assoc. Petrol. Geol.* 56(10): 2029–67

Cordell, R. J. 1973. Colloidal soap as proposed primary migration medium for hydrocarbons. *Bull. Am. Assoc. Petrol. Geol.* 57(9): 1618–43

Cordell, R. J. 1974. Depths of oil origin and primary migration: geologist's discussion—reply. *Bull. Am. Assoc. Petrol. Geol.* 58(9): 1857–61

Derjaguin, B. V. 1970. Superdense water. *Sci. Am.* 223(5): 53–70

Dickey, P. A. 1969. Increasing concentration of subsurface brines with depth. *Chem. Geol.* 4: 361–70

Dickey, P. A., Shriram, C. R., Paine, W. R. 1968. Abnormal pressure in deep wells in southwestern Louisiana. *Science* 160 (3828): 609–15

Dickinson, G. 1953. Reservoir pressures in Gulf Coast Louisiana. *Bull. Am. Assoc. Petrol. Geol.* 37(2): 410–31

Dodd, C. G., Conley, F. R., Barnes, P. M. 1955. Clay minerals in petroleum reservoir sands and water sensitivity effects. In *Clays and Clay Minerals. Proc. Nat. Clays Clay Miner. Conf. Proc. 3rd,* ed. A. Swinford. Oxford: Pergamon. 573 pp.

Dunoyer de Segonzac, G. 1969. Les mineraux argileux dans la diagenese, passage metamorphisme (thesis, Univ. Strasbourg). *Mem. Serv. Carte Geol. Alsace Lorraine,* Vol. 29. 320 pp.

Dunoyer de Segonzac, G. 1970. The transformation of clay minerals during diagenesis and low grade metamorphism: review. *Sedimentology* 15: 281–346

Frank, H. S. 1970. The structure of ordinary water. *Science* 169(3946): 635–41

Grim, R. E. 1968. *Clay Mineralogy.* New York: McGraw-Hill. 596 pp.

Gurikov, Y. V. 1963. The similarity in the structures of water and ice. *Zh. Strukt. Khim.* 4(6): 824–29

Hanshaw, B. B., Bredehoeft, J. D. 1968. On the maintenance of anomalous fluid pressures. *Geol. Soc. Am. Bull.* 79(9): 1097–1106

Heard, H. C., Rubey, W. W. 1966. Tectonic implications of gypsum dehydration. *Geol. Soc. Am. Bull.* 77: 741–60

Hedberg, H. D. 1926. The effect of gravitational compaction on the structure of sedimentary rocks. *Bull. Am. Assoc. Petrol. Geol.* 10: 1035–73

Hedberg, H. D. 1936. Gravitational compaction of clays and shales. *Am. J. Sci.* 31: 241–87

Hendricks, S. B., Jefferson, M. E. 1938. Crystal structure of magnesium vermiculite—chlorites. *Am. Mineral.* 23: 851–62

Jones, F. O. Jr. 1964. Influence of composition of water on clay blocking of permeability. *J. Petrol. Technol., April,* pp. 441–46

Jones, P. H. 1969. Hydrodynamics of geopressure in the northern Gulf of Mexico basin. *J. Petrol. Technol., July,* pp. 803–10

Kartsev, A. A., Vagin, S. B., Baskvo, E. A. 1969. Paleohydrogeology. *Nedra Moscow.* 150 pp.

Keller, W. D. 1963. Diagenesis in clay minerals—a review. In *Clays and Clay Minerals. Proc. Nat. Clays Clay Miner. Conf., 11th,* ed. W. B. Bradley, 13: 136–57. Oxford: Pergamon. 368 pp.

Keller, W. D., Hanson, R. F. 1975. Dissimilar fabrics by scan electron microscopy of sedimentary versus hydrothermal kaolins in Mexico. *Clays Clay Miner.* 23(3): 201–5

Lambe, T. W. 1958. The structure of compacted clay. *J. Soil Mech. Found. Div. Am. Soc. Civil Eng.* 84: 1654–1711

Louden, L. R. 1971. "Chemical Caps" can cause pressure buildup. *Oil Gas J.,* Nov. 15, pp. 144–46

Low, P. F., Anderson, D. M. 1958. The partial specific volume of water in bentonite suspensions. *Soil Sci. Soc. Am. Proc.* 22: 22–24

MacEwan, D. M. C. 1948. Complex formation between montmorillonite, halloysite and certain organic liquids. *Trans. Faraday Soc.* 44: 349–67

Magara, K. 1968. Compaction and migration of fluids in miocene mudstone, Nagaoka plain, Japan. *Bull. Am. Assoc. Petrol. Geol.* 2466–2501

Magara, K. 1974a. Compaction, ion formation, and osmosis in shale and their significance in primary migration. *Bull. Am. Assoc. Petrol. Geol.* 58(2): 283–90

Magara, K. 1974b. Aquathermal fluid migration. *Bull. Am. Assoc. Petrol. Geol.* 58(2): 2513–21

Magara, K. 1975. Re-evaluation of montmorillonite de-hydration as cause of abnormal pressure and hydrocarbon migration. *Bull. Am. Assoc. Petrol. Geol.* 59(2): 291–302

Manger, G. E. 1963. Porosity and bulk density of sedimentary rocks. *US Geol. Surv. Bull.* 1144-E: 34–40 and Table 3

Manheim, F. T. 1970. Brines and interstitial brackish water in drill cores from the deep Gulf of Mexico. *Science* 170(3953): 57–61

Manheim, F. T., Horn, M. K. 1968. Composition of deeper subsurface waters along the Atlantic continental margin. *Southeast. Geol.* 9(4): 215–36

Martin, R. T. 1962. Adsorbed water on clay: a review. In *Clays and Clay Minerals. Proc. Nat. Clays Clay Miner. Conf., 9th,* pp. 28–70. Oxford: Pergamon

Meade, R. H. 1966. Factors influencing the early stages of compaction of clays and sands. Review. *J. Sediment. Petrol.* 36: 1085–1101

Nightingale, E. R. 1959. Phenomenological theory of ion solution effective radii of hydrated ions. *J. Phys. Chem.* 63(9): 1381–87

O'Brien, N. R. 1971. Fabric of kaolinite and illite floccules. *Clays Clay Miner.* 19: 353–59

Olsen, H. W. 1972. Liquid movement through kaolinite under hydraulic, electric and osmotic gradients. *Bull. Am. Assoc. Petrol. Geol.* 54(10): 2022–27

Overton, H. L., Timko, D. J. 1969. The salinity principle, a tectonic stress indicator in marine sands. *Log Anal.* 10(3): 34–43

Overton, H. L., Zanier, A. M. 1970. Hydratable shales and the salinity high enigma. *Soc. Pet. Engr. Pap. No. 2989* Presented at Am. Inst. Min. Metall. Eng., 45th Ann. Fall Meet., Houston, Texas. 9 pp.

Perry, E. A. Jr., Hower, J. 1972. Late-stage dehydration in deeply buried pelitic sediments. *Bull. Am. Assoc. Petrol. Geol.* 56(10): 2013–21

Posokhov, E. V. 1966. Formation of chemical composition of underground water (basic factors). *Gidrometeorol. Izd. Leningrad.* 285 pp.

Powers, M. C. 1954. Clay diagenesis in the Chesapeake Bay area. In *Clays and Clay Minerals. NAS-NRC Publ.* 327: 68–80

Powers, M. C. 1959. Adjustment of clays to chemical change and the concept of the equivalence level. In *Clays and Clay Minerals. Proc. Nat. Clays Clay Miner. Conf.,* ed. A. Swinford. Oxford: Pergamon. 411 pp.

Powers, M. C. 1967. Fluid release mechanisms in compacting marine mud rocks and their importance in oil exploration. *Bull. Am. Assoc. Petrol. Geol.* 51(7): 1240–54

Rieke, H. H. III, Chilingarian, G. V. 1974. Compaction of argillaceous sediments. In *Developments in Sedimentology,* Vol. 16. New York: Elsevier. 424 pp.

Rittenhouse, G. R. 1964. Possible mechanism for concentration of brines in subsurface formations. *Bull. Am. Assoc. Petrol. Geol.* 48:234–36

Schmidt, G. W. 1973. Interstitial water composition and geochemistry of deep Gulf Coast shales and sandstone. *Bull. Am. Assoc. Petrol. Geol.* 57(2):321–37

Terzaghi, K. 1925. Principles of soil mechanics II, compressive strength of clays. *Eng. News. Rec.* 95:796–800

Timm, B. C., Maricelli, J. J. 1953. Formation waters in southwest Louisiana. *Bull. Am. Assoc. Petrol. Geol.* 37:394–409

Van Olphen, H. 1963. Compaction of clay sediments in the range of molecular particle distances. In *Clays and Clay Minerals. Proc. Nat. Clays Clay Miner. Conf., 11th,* pp. 178–87. New York: Macmillan. 368 pp.

Von Engelhardt, W., Gaida, K. H. 1963. Concentration changes of pore solutions during the compaction of clay sediments. *J. Sediment. Petrol.* 33:919–30

Von Engelhardt, W., Tunn, W. L. M. 1954. The flow of fluids through sandstones. *Ill. State Geol. Surv. Circ. 194.* 17 pp.

Walker, G. F. 1956. The mechanism of dehydration of mg-vermiculite. In *Clays and Clay Minerals. NAS-NRC Publ.* 456: 101–15

Weaver, C. E. 1958a. Geologic interpretation of argillaceous sediments. *Bull. Am. Assoc. Petrol. Geol.* 42:254–309

Weaver, C. E. 1958b. The effects and geological significance of potassium "fixation" by expandable clay minerals derived from muscovite, biotite, chlorite, and volcanic materials. *Am. Mineral.* 43: 839–61

Weaver, C. E. 1959. The clay petrology of sediments. In *Clays and Clay Minerals. Proc. Nat. Clays Clay Miner. Conf., 6th,* pp. 154–87

Weaver, C. E., Beck, K. C. 1971. Clay water diagenesis during burial: How mud becomes gneiss. *Geol. Soc. Am. Spec. Pap. 134.* 96 pp.

Weller, J. M. 1959. Compaction of sediments. *Bull. Am. Assoc. Petrol. Geol.* 43(2): 273–310

White, D. E. 1965. Saline waters of sedimentary rocks. In *Fluids in Subsurface Environments. Am. Assoc. Petrol. Geol. Mem.*, ed. A. Young, J. E. Galley, 4:342–66

Williamson, W. O. 1951. The physical relationships between clay and water. *Brit. Ceram. Soc. Trans.* 50:10–34

THE IONOSPHERES OF MARS AND VENUS

John C. McConnell
Department of Physics and Centre for Research in Experimental Space Science,
York University, Toronto, Canada

INTRODUCTION

With the advent of deep space probes and new developments in ground-based spectroscopy, the field of planetary studies has truly become a science. We have seen our early terrapocentric theories of Mars and Venus become altered beyond recognition as data from planetary flybys, orbiters, probes, and ground-based studies have yielded new insights into the structures of the lower atmospheres. Undoubtedly, as still more data are returned, the views presented here on the upper atmospheres and ionospheres of Mars and Venus will be modified further. This review is not a compendium of references; rather it deals mostly with how our understanding of the ionospheres has changed since the review by Whitten & Colin (1974). The atmospheres of Mars and Venus above 100 km are primary concerns: although our concept of the Martian upper atmosphere has changed little since it was reviewed by McConnell (1973), our picture of the upper atmosphere of Venus, based mainly on airglow and radio occultation data, has become somewhat confused since the Whitten & Colin (1974), review.

The lower atmospheres of Mars and Venus are composed mainly of CO_2 (cf Hunten 1971). However, a substantial percentage (up to 25%) of the Martian atmosphere could consist of Ar or some other heavy inert gas (cf McConnell 1973, Levine & Riegler 1974, Owen 1974). On Venus, up to 5% of the lower atmosphere could consist of N_2 or some other relatively inert constituents (e.g. Hunten 1971, Kumar 1975). Based on our knowledge of the Martian and Cytherian temperature structures, there is no reason to suggest that CO_2 should not be a relatively major component in the upper atmospheres. However, CO_2 rapidly photodissociates to CO and O in the upper atmosphere. Under mixing conditions, characterized by a one-dimensional eddy diffusion coefficient K similar to those of the Earth i.e. $K \cong 1-5 \times 10^6\ cm^2\ sec^{-1}$, one would also expect substantial (10–50%) amounts of O and CO in the upper atmosphere.

THE UPPER ATMOSPHERE OF MARS

Information pertaining to the structure of the neutral atmosphere is derived primarily from the spectacularly successful limb-scanning ultraviolet spectrometer experiment of Barth and co-workers (e.g. Barth et al 1972), which scanned the spectral region 1200–4500 Å during the Mariner 6 and 7 (M6, M7) flybys and the 1200–3500 Å region during the Mariner 9 (M9) orbit of Mars. A spectrum from M6 and M7 (see Figure 1) shows many band and line features: CO(4+), CO Cameron bands, CO_2^+ UV doublet, F-D-B bands, and atomic multiplets of H, O, and C. The major features and the principal excitation mechanisms are listed in Table 1. A detailed

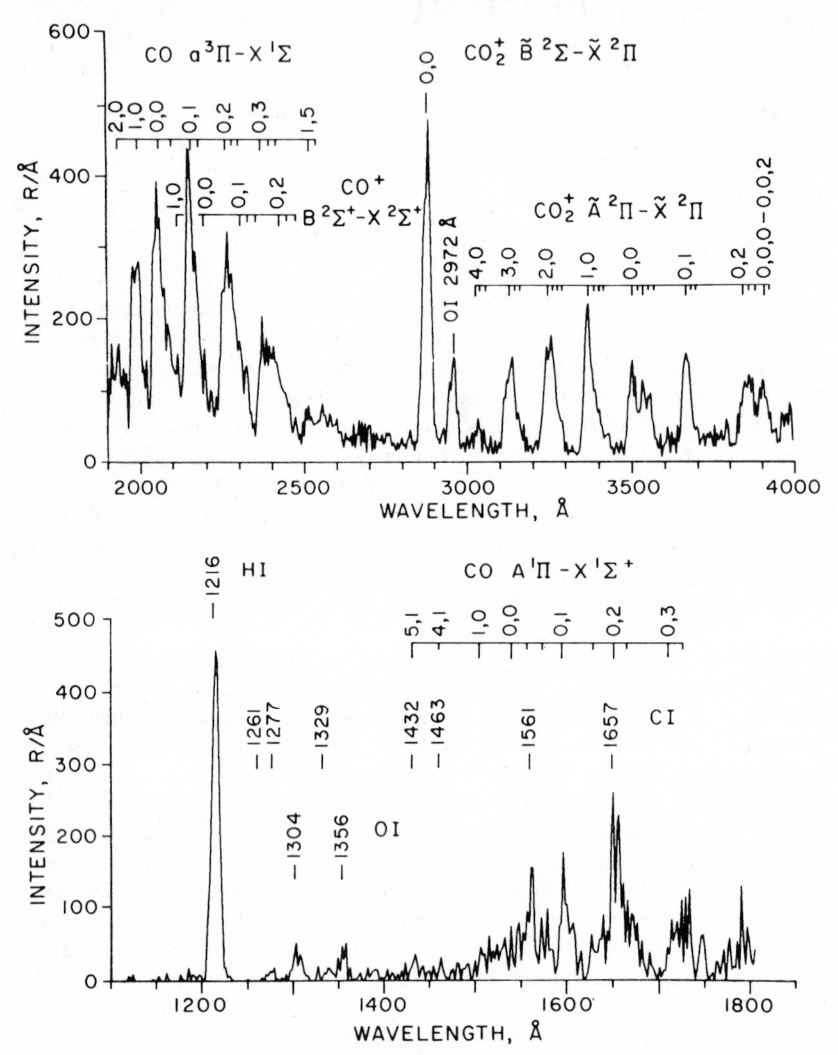

Figure 1 Ultraviolet spectra of the upper atmosphere of Mars, 1100–1800 at 10 Å resolution and 1900–4000 at 20 Å resolution. The spectra shown are the result of the summation of four individual observations (Barth et al 1971).

Table 1 UV Emission features in the Martian and Cytherian Upper Atmospheres

Observed emissions	Excitation mechanisms[a]
$CO(a^3\pi - X^1\Sigma)$ Cameron bands 1800–2600 Å	(1) $h\nu + CO_2 \rightarrow CO^* + O'$; (2) $e + CO_2 \rightarrow CO^* + O' + e$; (3) $CO_2^+ + e \rightarrow CO^* + O'$
$CO(A^1\pi - X^1\Sigma)$ fourth positive bands 1400–1800 Å	(1), (2), (3)
$CO^+(B^2\Sigma - X^2\Sigma)$ first negative bands 2100–2800 Å	(4) $h\nu + CO_2 \rightarrow CO^{+*} + O + e$; (5) $e + CO_2 \rightarrow CO^{+*} + O + 2e$
$CO_2^+(B^2\Sigma_u - X^2\pi_g)$ ultraviolet doublet 2890 Å	(6) $h\nu + CO_2 \rightarrow CO_2^{+*} + e$; (7) $e + CO_2 \rightarrow CO_2^{+*} + e$
$CO_2^+(A^2\pi_u - X^2\pi_g)$ Fox-Duffendack-Barker bands 3000–4500 Å	(6), (7), (8) $h\nu + CO_2^+ \rightarrow CO_2^{+*}$
$O(^1S - {}^1D)$ 2972 Å	(1), (3)
$O(^5S - {}^3P)$ 1356 Å	(2), (9) $e + O \rightarrow O^* + e$
$O(^3S - {}^3P)$ 1304 Å	(10) $h\nu + O \rightarrow O^*$, (9), (2), (1)
$H(^2P - {}^2S)$ 1216 Å	(11) $h\nu + H \rightarrow H^*$
$C(^3S - {}^3P)$ 1657 Å	(12) $h\nu + CO_2 \rightarrow C^* + O' + O''$; (13) $e + CO_2 \rightarrow C^* + O' + O'' + e$; (14) $h\nu + C \rightarrow C^*$

[a] The symbols *, ′, and ″ indicate the excited state.

analysis of these airglow features can yield important information on the composition and temperature structure of the upper atmosphere (McConnell & McElroy 1970, Stewart 1972, McConnell 1973). One main conclusion from an analysis of the CO Cameron bands (one of the brightest features) is that the exospheric temperature[1] varies within the range 270–400°K (Stewart et al 1972). The variability and magnitude of the exospheric temperature is also borne out by the interpretation of the O(1304) multiplet (Strickland et al 1973[2]), and the H-Lyman-α (HLα) emission (Anderson 1974). An analysis of the O(1304) emission and CO fourth positive system indicates that the mixing ratios of O and CO in the upper atmosphere are small: $\sim 1\%$ at a total number density of $\sim 2 \times 10^{10}$ cm^{-3} (McElroy & McConnell 1971, Strickland et al 1972, 1973, Thomas 1971, Mumma et al 1975a,b). The Lα data, analyzed by Anderson & Hord (1971) and Anderson (1974), yield H densities at the exobase (see footnote 1) of $\sim 2 \times 10^4$ cm^{-3} and exospheric temperatures of $\sim$300–350°K. The thermal escape rate of H, calculated using the Jeans escape formula, is $\sim 10^8$ H atoms cm^{-2} sec^{-1}. Emissions from N_2, N_2^+, NO, N, and H_2 have not been observed by the UV spectrometer. Dalgarno & McElroy (1970) have used the absence of N_2 and N_2^+ emissions to put a limit of 5% on the N_2 mixing ratio in the lower atmosphere, whereas McElroy & McConnell (1971) noted $NO/CO_2 \leqq 10^{-4}$ at 120 km is true in the absence of the (1, 0) NOγ-band. Moos (1974a) has recently used the lack of H_2 Lyman-band emissions in the M9 UV data to derive an upper limit of $H_2 \leqq 3 \times 10^5$ cm^{-3} at 250 km. Figure 2 shows a typical model atmosphere for Mars (Barth et al 1972). The CO densities are somewhat low and the O densities somewhat high (cf McElroy & McConnell 1971) compared to those mentioned above; however, the values are representative.

The above low densities of O and CO were surprise results from M6 and M7. Because CO and O are rapidly produced in the upper atmosphere, such low densities imply either very rapid in situ chemistry or dynamics for removal. Chemical sinks have been discounted, leaving dynamics as the main removal mechanism for the products of CO_2 dissociation. A value of $K \simeq 10^8$ cm^2 sec^{-1} is required to produce agreement with the UV airglow data. As noted earlier on earth, a typical K in the thermosphere is from 1×10^6 to 5×10^6 cm^2 sec^{-1}. The rapid mixing is probably effected by waves propagated from the lower atmosphere (C. B. Leovy, 1974, private communication) or by a planetary-wide circulation such as occurs on Venus[3] (Dickinson 1971).

The energy radiating from the upper atmosphere of Mars that was observed by the UV spectrometer is a relatively small fraction of the total energy deposited by the

[1] The exospheric temperature is the temperature on the region of the atmosphere where collisions are sufficiently infrequent to invalidate the normal approximations of continuum mechanics. The exobase is taken as the level where the collision mean free path is equal to the local scale height.

[2] Figure 8 of Strickland et al (1973) clearly indicates the variability of the exospheric temperature of O; see, for example, Figure 2 of McElroy & McConnell (1971).

[3] However, as Dickinson (1971) noted, the driving conditions for the upper atmospheric circulations are different.

solar flux in this region of the atmosphere. The rest of the energy is used to ionize and dissociate CO_2. Some of this energy subsequently reappears locally as heat, and some, stored as chemical energy, is lost lower down in the atmosphere when CO and O recombine. Calculation of the local heating efficiency of solar radiation in a Martian type atmosphere is difficult (Henry & McElroy 1968, Stewart 1972). However, present studies of the thermospheric heat budget indicate a problem, namely, the amount of energy calculated to be deposited locally, leads to overly high values for exospheric temperatures (Stewart 1972, Stewart et al 1972). Normally in these calculations the $O(^1D)$ produced from photodissociation of CO_2 (Table 1), is quenched by CO_2 and the electronic energy is assumed to reappear locally as heat. Stewart et al (1972) have suggested that the problem may be somewhat alleviated if the $O(^1D)$ quenching by CO_2 yields vibrationally excited CO_2, which then radiates the energy to space. However, control of upper atmospheric temperatures by solar energy implies a correlation of temperature with changing solar energy in the extreme ultraviolet. Although the airglow intensity shows such a correlation, the exospheric temperature does not (Stewart et al 1972). This could be because the upper atmosphere is under the control of lower atmospheric dynamics. The rapid mixing which flows O and CO from the upper atmosphere could also remove heat faster than thermal conduction (e.g. McElroy 1967, Stewart 1972; but see also Hunten 1974).

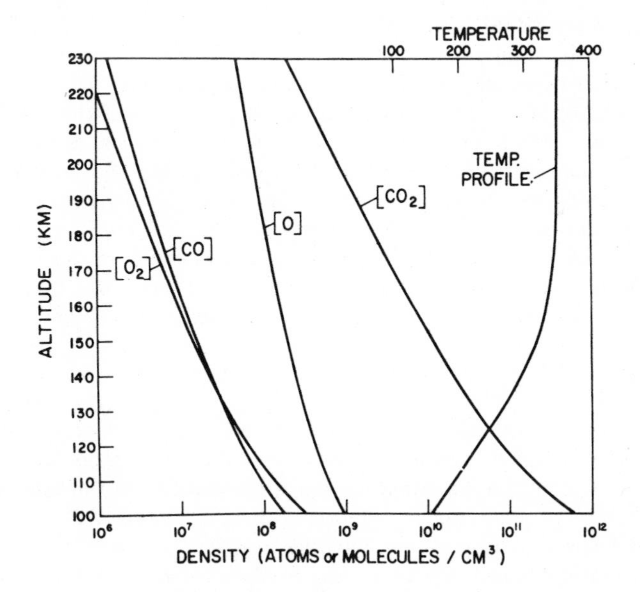

Figure 2 Mars model atmosphere. This model atmosphere is based on Mariner 6 and 7 ultraviolet spectrometer and radio occultation results, as well as on ground-based telescope observations (Barth et al 1972).

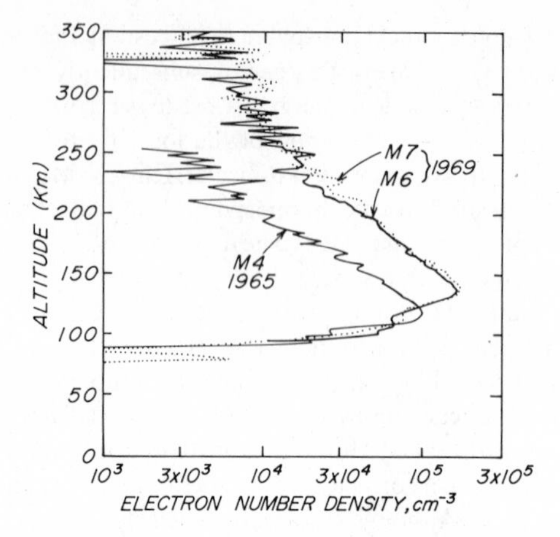

Figure 3 The Martian ionosphere as observed during the Mariner 4, 6, and 7 occultation experiments (Rasool & Stewart 1971).

THE IONOSPHERE OF MARS

The ionosphere of Mars was first detected by the S-band occultation experiment on board Mariner 4 (M4). The initial interpretations of the ionosphere data (which was the sole direct information on the upper atmosphere) were controversial. The range of views held at that time are well summarized by the articles in *The Atmospheres of Mars and Venus* (Brandt & McElroy 1968). With the data from M6, M7, and M9, it appears that the type of ionosphere observed by M4 was more the exception than the rule. Results comparing the M4, M6, and M7 ionospheres are shown in Figure 3.

Current theories of the Mars ionosphere hold that it is an F_1-type ionosphere. In the Earth's ionosphere, the F_1 region is characterized by cross-sections for ionization and absorption of solar radiation of $\sim 10^{-17}$ cm^2. Also, the rate-limiting steps that lead to disappearance of the ions and electrons are the following reactions:

$$O^+ + N_2 \rightarrow NO^+ + N$$

$$O^+ + O_2 \rightarrow O_2^+ + O,$$

which convert the atomic ions (the main ions produced at these altitudes) to molecular ions. The F_2 region arises because (*a*) the main ion produced is atomic, and (*b*) the loss rate for the atomic ion increases exponentially upwards. Above the F_2 peak, loss of ions through diffusion is more important than chemical loss, whereas below the F_2 peak, ion molecule reactions followed by electronic recombination are the main loss processes. On Mars, the main source of ions appears to be CO_2[4] photoionization (Table 2), and the main loss of ions is electronic

[4] Ar^+ is possibly important; however, the reaction $Ar^+ + CO_2$ is so rapid and the Ar scale height is sufficiently similar to that of CO_2 that large densities of Ar will not seriously change the above arguments.

recombination of CO_2^+ and O_2^+. The most important ion reactions are shown in Table 2. The principal source of O_2^+ ions is the reaction

$$O + CO_2^+ \rightarrow O_2^+ + CO$$
$$\rightarrow O^+ + CO_2$$

followed by

$$O^+ + CO_2 \rightarrow O_2^+ + CO.$$

Even with such low densities of O as mentioned earlier, O_2^+ is thought to be the major ion in the Martian ionosphere. This is certainly consistent with the UV airglow data for CO_2^+. The CO_2^+ A state can be excited by fluorescence (Dalgarno et al 1970) as well as by direct excitation (Table 1). Thus observation of the F-D-B bands can put strong constraints on the presence of CO_2^+ in the ionosphere (Stewart 1972, McConnell 1973). Estimates of CO_2^+ densities yield abundances that are 10–20% of the total measured electron density. Because reaction 4 (Table 2) is the main sink for CO_2^+, knowledge of CO_2^+ densities can yield estimates of O densities. Such estimates are in reasonable agreement with the interpretation of the O(1304) data.

The F_1 theory assumes that ion flow is unimportant, and therefore steady state

Table 2 Ionospheric reactions in the upper atmospheres of Mars and Venus[a]

No.	Reaction	Rate[b]
1.	$CO_2 + h\nu \rightarrow CO_2^+ + e$ $CO^+ + O + e$ $CO + O^+ + e$	7.5×10^{-7} sec^{-1} [c]
2.	$O + h\nu \rightarrow O^+ + e$	4×10^{-7} sec^{-1}
3.	$He + h\nu \rightarrow He^+ + e$	1.0×10^{-7} sec^{-1}
4.	$O + CO_2^+ \rightarrow O^+ + CO_2$ $\rightarrow O_2^+ + CO$	9.6 (−11) 1.6 (−10)
5.	$O^+ + CO_2 \rightarrow O_2^+ + CO$	1.2 (−9)
6.	$H_2 + CO_2^+ \rightarrow CO_2H^+ + H$	1.4 (−9)
7.	$He^+ + CO_2 \rightarrow CO^+ + O + He$ $\rightarrow CO_2^+ + He$ $\rightarrow CO + O^+ + He$	1.2 (−9)
8.	$O^+ + H_2 \rightarrow HO^+ + H$	2.0 (−9)
9.	$HO^+ + e \rightarrow HO^+ + H$	2.0 (−7) estimate
10.	$CO_2H^+ + e \rightarrow CO_2 + H$	3.5 (−7) estimate
11.	$CO_2^+ + e \rightarrow CO + O$	3.8 (−7)
12.	$O_2^+ + e \rightarrow O + O$	2.2 (−7)
13.	$CO^+ + e \rightarrow C + O$	6.0 (−7)
14.	$N_2^+ + e \rightarrow N + N$	1.8 (−7)

[a] In cubic centimeters per second at 300°K unless otherwise stated.
[b] See Liu & Donahue (1975), Kumar & Hunten (1974) for references.
[c] Ionization rates are for zero optical depth at Venus.

is assumed in the calculations. Figure 4 shows a fit to a few of the observations. Kliore et al (1973) have also shown that the F_1 theory produces agreement in the variation of the height and peak electron density of the F_1 layer with changing solar zenith angle. In addition, the absolute electron densities in an F_1 model should show a correlation with solar activity, as Kliore et al (1972) have shown.

A recurring problem in the interpretation of the ionospheric profiles is that the absolute densities calculated using known solar fluxes and electron recombination coefficients are too small by about 50–70%. This could imply that the solar fluxes used are too small by a factor of 2–3, that an extra source of ionization such as the solar wind is present, or that the electronic recombination coefficients used are too small. Recent measurements of solar fluxes on Atmospheric Explorer C indicate that they are not the problem (M. B. McElroy, private communication, 1975). Not enough is known concerning the interaction of the solar wind with a planetary ionosphere such as that of Mars to reliably judge whether such interaction could possibly give rise to additional ionization. Unfortunately, in both solutions, additional ionization probably means additional airglow; one would exchange an ionization problem with one of excess airglow! Most likely at present is that the recombination coefficient of O_2^+ is smaller because O_2^+ ions are produced in vibrationally excited states, via reactions 4 and 5 (Table 2). It has been speculated that vibrationally excited diatomic ions recombine much more slowly than those in the ground vibrational states (O'Malley 1969). More definitive laboratory data on ion molecule and recombination rates are required to resolve this problem.

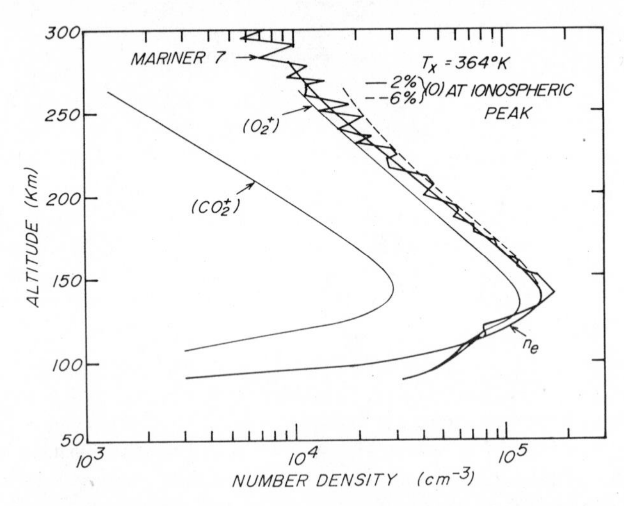

Figure 4 Calculated Martian electron density profiles for model atmospheres similar to that shown in Figure 3. The solar fluxes have been arbitrarily increased by a factor of 2.7. The Mariner 7 results are from Figure 3. Also shown are the O_2^+ and CO_2^+ concentrations for 2% of O at 135 km, and the electron density calculated for 6% of O at 135 km (McConnell 1973).

Using the simple theory of the F_1-type region, the topside plasma scale height *Hp* should reflect the scale height *Hn* of the species that is the major ionization source. In fact, if the electron, ion, and neutral temperatures are equal, the plasma scale height is just twice the neutral scale height. Results from M6, M7, and M9 showed that this is approximately true, but presently impossible to demonstrate in detail because the occultation and UV experiments derive atmospheric parameters at different locations and *Hp* and *Hn* vary considerably: $Hp = 38.5 \pm 5$ and $Hn = 17.8^{+7}_{-3}$ km.

The *Hp* values measured for the even-numbered orbits up to 60 on M9 (Kliore et al 1972) were all taken at approximately the same local time (solar zenith angle $\sim 50°$) but at different longitudes. The latitudes varied between 10° and 40° for the even-numbered orbits, and the longitude varied from 150° to 0° to 240°. When *Hp* is plotted against even-orbit number (effectively longitude), it is possible (with a little imagination) to see a wave structure in the plasma scale heights.[5] It would be interesting to see if the wavelengths predicted from tidal motions correlated with the wavelength (possibly ethereal) in *Hp*.

ATMOSPHERIC ESCAPE

As mentioned earlier, H escapes at a rate of $\sim 10^8$ atoms cm^{-2} sec^{-1}. This H is supplied ultimately by photolysis of H_2O in the lower atmosphere to yield stoichiometrically $H_2 + \frac{1}{2}O_2$, followed by diffusion of H_2 to the upper atmosphere where the ionospheric reactions 6 and 10 occur (Table 2). The H escape by thermal evaporation of high energy atoms in the tail of the Maxwellian distribution occurs in regions where the densities are low enough that atoms with sufficient energy may escape without collisions. As noted by Brinkmann (1971) and McElroy (1972), atmospheric escape may also occur via processes 11–14 in Table 1, in which the excess recombination energy appears as kinetic energy from the fragments. This subject has been treated in detail by McElroy (1972), whose main conclusion is that large amounts of O, C, and N may escape via nonthermal processes. Indeed, escaping H and O appear to be in balance with photolysis of H_2O in the lower atmosphere. If the CO_2 presently observed in the atmosphere is representative of the total amount of outgassed CO_2, then McElroy (1972) finds that an equal amount of CO_2 could have escaped. In addition, a two-meter-thick layer of H_2O and possibly all the N_2 outgassed to date could have left the planet. However, Owen (1974) and Levine & Riegler (1974) have suggested that 25% of the Martian atmosphere is Ar and that the outgassing rates of the Earth and Mars are similar; if true, implies an amount of outgassed CO_2 some thousand times the amount observed with concomitantly large amounts of N_2 and H_2O (Owen 1974). If this is the case, the disappearance of such copious amounts of CO_2, H_2O, and N_2 presents an interesting challenge in understanding the evolution of the Martian atmosphere.

[5] The data from the extended emission are much less certain than those from the main mission (Kliore et al 1973).

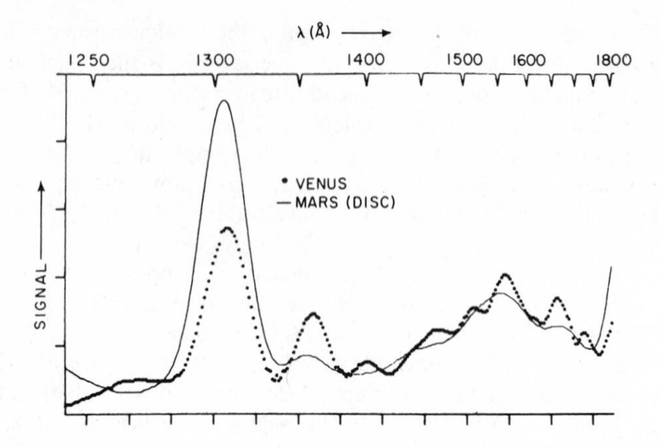

Figure 5 Disc spectra of Mars and Venus. The spectrum of Venus has been divided by 28 for the comparison. The wavelength scale is nonlinear. A sliding sum of 10 adjacent points has been made. Intensities are given in Table 3. From Moos (1974b).

THE UPPER ATMOSPHERE OF VENUS

Most information pertaining to the Cytherian upper atmosphere is derived from airglow data in the spectral region 300–1900 Å (Barth et al 1967, Moos et al 1969, Rottman & Moos 1973, Broadfoot et al 1974). Because the Cytherian atmosphere is mainly CO_2, one expects the Martian and Cytherian airglow spectra to be similar. To the first order, a similarity is observed, that is, the emissions shown in Table 1 for the spectral range 1200–1900 Å are observed in the Venus spectra. However, even allowing for the difference in the sun-planet distance ratio (2.1), the intensities of the various features are very different (see Figure 5). Table 3 lists the ratios of the O(1304, 1356) and 1375–1800 Å brightnesses for Mars and Venus (Moos 1974b). The simplest explanation of the ratios in Table 3 is that the mixing ratios of O and CO in the Cytherian upper atmosphere are much greater than those for Mars by about 10–20 times. However, one does not necessarily expect 10–20 times more emission. Because the O(1304) triplet is an optically thick emission, the increase in intensity is approximately logarithmic rather than linear. The situation is more complicated because, as O increases, the $e + O$ source (No. 9 in Table 1) of 1304 Å becomes more important. The O(1356) emission is excited by photoelectrons on O and CO_2 (see Table 1), and therefore the fraction of 1356 Å emission due to O is approximately linear with O. Direct excitation of CO on Mars is not the main source of the CO fourth positive bands. Increased CO amounts in the Cytherian atmosphere seem required to explain the brightness in the 1400–1900 Å region, as observed by Rottman & Moos (1973). If this is the case, the main excitation mechanism would appear to be fluorescence of CO. If one allows for various uncertainties in the excitation parameters, the arguments of Rottman &

Table 3 Ratios of Venus and Mars disc brightness at selected wavelength regions

Wavelength Å	Venus disc/(Mars disc times 4.4)
1216	1.3
1304	3.8
1356	12
1375–1800	5.6

Moos (1973) and Strickland (1973) seem to indicate mixing ratios for O and CO at the 2×10^{10} cm^{-3} density level of ~ 5–20%.

The M5 ultraviolet photometer data from the 1250–2250 Å and 1350–2250 Å channels have recently been reanalyzed by Anderson (1975a). He attributes the difference in signal between the two channels at the terminator to O(1304 Å) emission from the upper atmosphere. The intensities he derives, extrapolated to the planet's day side using the Strickland's (1973) results, agree with the rocket results of Rottman & Moos (1973).

The Lα measurements of Rottman & Moos (1973), obtained using a rocket-borne prism spectrometer, are in reasonable agreement with the M5 UV photometer measurements Barth et al (1968). The rocket measurements indicate Lα intensities less bright than the M5 data by about a factor of 2 but contain no information on the altitude distribution of constituents that cause the airglow. As indicated in the sections on Mars, limb-grazing measurements of airglow provide altitude information on the emitting molecule. Interpretation of the M5 Lα data indicated two components to the emission (cf Figure 6). It has been suggested that the outer component was H at 700°K and the inner component D (Donahue 1969, McElroy & Hunten 1969). An alternative suggestion was that the inner component was thermal H at 350°K and the outer component "hot" H, arising possibly from the photodissociation of H_2 (Barth 1968).

To date, nothing has been said of the M10 ultraviolet objective-grating spectrometer results (Broadfoot et al 1974). Radiation was detected in 10 channels in the spectral range 200–1700 Å, with a spectral resolution of 20 Å. In addition, the spectrometer had two zero-order channels to monitor radiation in the 200–1500 and 1150–1700 Å regions. The list of channels used is shown in Table 4. Only preliminary results of the Venus data are available (Broadfoot et al 1974, Kumar 1975, Kumar & Broadfoot 1975).

The Lα measurement (shown in Figure 6) implies an exospheric temperature of ~ 400°K. The integrated disc brightness of 19 kR is in good agreement with the rocket measurements of Rottman & Moos (1973). A second component was not observed, but the data are preliminary (McElroy 1974). Anderson (1975b) has reanalyzed the M5 Lα data in detail, using radiative transfer calculations. His conclusions differ significantly from earlier interpretations: the bright limb data exhibited two components with exospheric temperatures of 275 ± 50°K and 1020 ± 100°K, with densities at the exobase of $2 \pm 1 \times 10^5$ cm^{-3} and 1.3×10^3 cm^{-3} respectively. The dark disc data were best fitted with a dual component density model. On the night side, the cool temperature component had an exospheric temperature of 150°K, whereas the hot component was characterized by a temperature of 1500°K. The exobase densities were similar to those of the day side. From these data, the D hypothesis is no longer tenable (see also Wallace et al 1971).

The puzzling feature of the previous two temperature models was why one component exhibited exactly twice the scale height of the other. With this new information on Lα, the problem no longer exists. As noted above, Barth (1968) suggested the two-temperature model, but his suggestions for the source of the hot H—photodissociation of H_2—was shown to be untenable (McElroy & Hunten 1969). Kumar & Hunten (1974), recently reviving the two-temperature model, suggest reactions such as reactions 6, 8, 9, and 10 (Table 2) as the source of hot H. Sze & McElroy (1975) also note that, because the reaction

$$H^+ + O \rightleftarrows O^+ + H$$

is very fast, the H temperature may reflect the ion temperature. If high ion temperatures obtain in the exosphere, then the temperature of the hot H component is related to the ion temperature. The thermal escape rates on the day side for

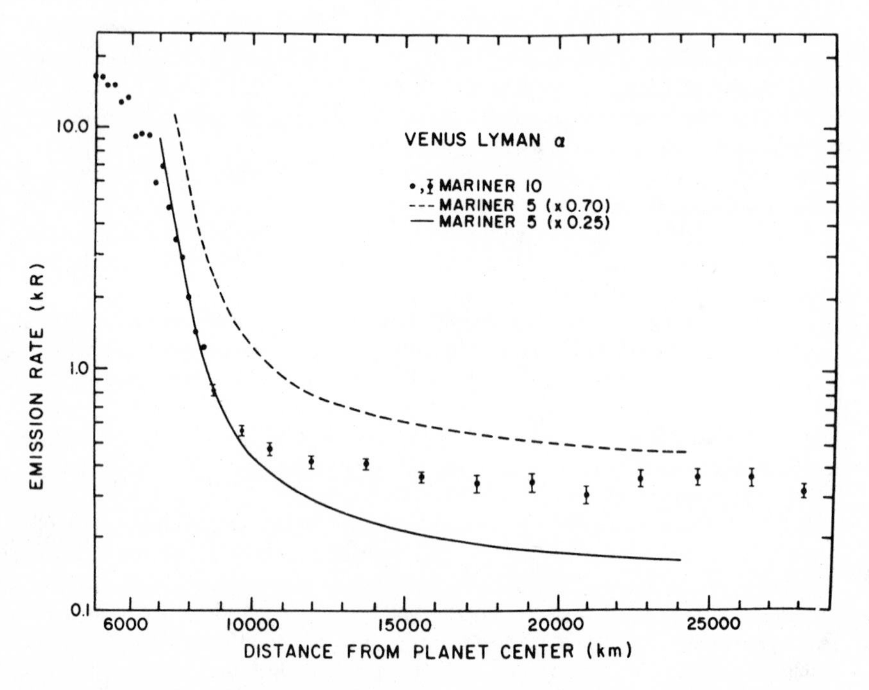

Figure 6 Hydrogen Lα emission rate at Venus vs minimum distance of the line of sight from the center of the planet. For comparison, the data obtained by Mariner 5, scaled down by factors of 4 and 0.7, are also shown (Broadfoot et al 1974).

Table 4 Ultraviolet emissions detected at Venus by the Mariner 10 grating spectrometer

Probable emitting species	Channel (Å)	Count rate (sec^{-1})	Intensity (kR) A[a]	Intensity (kR) B[b]
Zero order	1150–1700	26,200	4000	
Zero order	200–1500	12,160		
He^+	304	100		
Background	430	67		
He	584	233	0.61	
Ne	740	87		
A	867	100		
A	1048	147		
H	1216	693	19	27
O	1304	267	17	5.5
CO, fourth-positive	1480	987	55	~2–3
C	1657	260	30	4

[a] Approximate intensity at 13,000 km, as given by Broadfoot et al (1974).
[b] Intensity of features given by Rottman & Moos (1973) which are uncertain by a factor of ~ 1.7.

each component are 40 cm^{-2} sec^{-1} and 2×10^6 cm^{-2} sec^{-1} for the 275°K and 1020°K components, respectively. We return to the implications of the Lα airglow later.

The M10 spectrometer observations at 584° Å provide the first positive detection of He in the Cytherian atmosphere. The 600 Rayleighs of emission is most probably caused by resonance scattering of sunlight. The estimate of He densities by Kumar & Broadfoot (1975) indicate a density of 4×10^5 cm^{-3} at 200 km. The exospheric temperature of the He is also inferred to be ~400°K (Kumar 1975, Kumar & Broadfoot 1975), which is similar to the Mariner 10 results for H.

The O, C, and CO emissions at longer wavelengths (Table 4) measured by M10 appear much more intense than those measured by Rottman & Moos (1973). At present no reasonable excitation mechanism is available. In addition, both zero-order channels show very intense signals. As for the previously mentioned emissions, no explanation of such bright emissions is yet available, other than scattered sunlight off the spacecraft.

The main deductions from the airglow data could be summarized as follows: (*a*) an exospheric temperature $\gtrsim$400°K is typical of the day side upper atmosphere; (*b*) mixing ratios of O and CO at 145 km are ~10–20%; (*c*) H density is variable but is ~10^5 cm^{-3} at exobase; (*d*) D is not present; and (*e*) He density is 4×10^5 cm^{-3} at exobase. The number of observations is very limited, however, and higher temperatures may occur. With temperatures $\gtrsim$400°K in the upper atmosphere of Venus, the problem of the heat budget that seems to occur on Mars would raise its head on Venus.

The above conclusions from the airglow data are the most obvious. In a personal communication, J. C. G. Walker (cited in Hunten 1973) noted that the low densities of thermospheric H are not immediately compatible with the lower atmospheric mixing ratio of hydrogen-containing compounds where $(2H_2O + HCl + 2H_2)/CO_2 \geqq 10^{-6}$ is true.

In the lower atmosphere H_2O, HCl, and H_2 are subject to dissociation, and so these relatively stable forms of H will eventually be converted to atomic H. As H atoms are conserved, the mixing ratio of H in the upper part of the mixed atmosphere must still be $\geqq 10^{-6}$. Above the turbopause, where the time constant for eddy diffusion is approximately equal to the time constant for molecular diffusion, H (in diffusive equilibrium) takes on its own scale height, which is so large that we can treat it as infinite. For a K $\sim 10^6$ cm^2 sec^{-1}, the background number density at the turbopause is $\sim 2 \times 10^{13}$ cm^{-3}, giving an H density $\geqq 2 \times 10^7$ cm^{-3}. Thus, because the H scale height is so large, this should be approximately the exospheric density. As noted above, the observed density is $\sim 10^5$ cm^{-3} at the exobase. The first order implication is that K should be $\sim 10^8$ cm^2 sec^{-1} in order to maintain low H mixing ratios at the exobase. The above arguments are modified by two features: (*a*) H_2O etc may not be fully dissociated at the turbopause and (*b*) H is actually escaping. The main effect of item *a* is because of the change in the diffusion coefficient. Bearing in mind the magnitude of the alteration of K, this is not a serious limitation to the argument. Probably, a subsidiary effect is to dissociate H_2 by CO_2^+ (reactions 6 and 10 in Table 2) near the exobase, making it somewhat easier for the H to diffuse to the exobase. This would only exacerbate

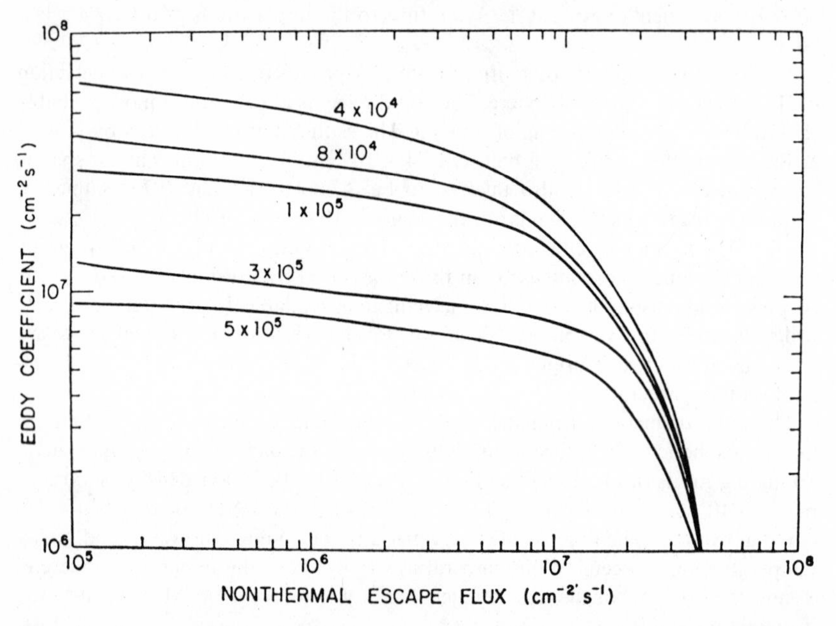

Figure 7 Contours of H density at the exobase level on Venus as a function eddy diffusion coefficient and nonthermal escape flux. The computations were carried out for a model atmosphere of exospheric temperature 398°K. The H_2 and H_2O mixing ratios at the cloud tops were 2×10^{-7} and 8×10^{-7}, respectively (Sze & McElroy 1975).

the problem. Item *b* only modifies the arguments seriously if the escape flux is of order of the diffusion-limited flux (Mange 1955, Hunten 1973), which is $\sim 2 \times 10^7$ cm^{-2} sec^{-1} for an H mixing ratio of 10^{-6}. If this occurs, H remains mixed for some distance above the turbopause. However, in order to obtain the diffusion limited flux for an H density of 10^5 cm^{-3} at the exobase, one requires exospheric temperatures of $\sim 1000°K$, much greater than those observed. Also the hot H component only supplies a flux of $\sim 10^6$ cm^{-2} sec^{-1}. Again this does not significantly alter the arguments. Thus the Lα data imply that either a value of $K \simeq 10^8$ cm^2 sec^{-1} obtains in the upper atmosphere of Venus, or low K values obtain and there is a nonthermal escape flux of $\sim 5 \times 10^7$ cm^{-2} sec^{-1}. This large escape flux would force diffusion-limited conditions on H and, as noted above, result in H being mixed above the turbopause, yielding low thermospheric H densities. Liu & Donahue (1975) and Sze & McElroy (1975) have performed detailed computations on this problem. Figure 7 shows the detailed behavior of H densities at the exobase as a function of K and the nonthermal escape flux. Although at present nonthermal escape processes seem efficient in the Earth's exosphere (Tinsley et al 1975, Liu & Donahue 1974), no mechanism sufficient to provide H escape fluxes of $\sim 5 \times 10^7$ cm^{-2} sec^{-1} on Venus has been uncovered (Kumar & Hunten 1974, Liu & Donahue 1975, Sze & McElroy 1975).

Large values of K have import on the problem of the O and CO abundances in the upper atmosphere. Large K values (as for Mars) imply low O and CO abundances, i.e. $\leqq 1\%$, contrary to the implications of the airglow data. A possible solution to the dilemma is to invoke esoteric airglow excitation sources, such as inflowing solar wind, which incidentally would also help the ionospheric problem (discussed later). Figure 8 shows a typical model atmosphere for Venus for an

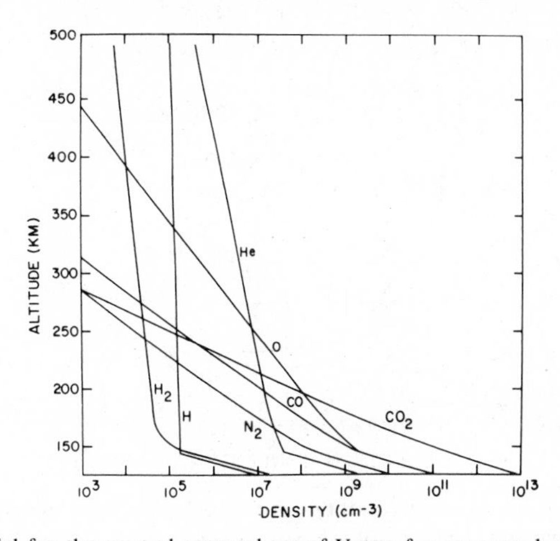

Figure 8 Model for the neutral atmosphere of Venus for an exospheric temperature of 350°K. The mixing ratios are CO_2 (0.98), CO(0.01), O(0.01), $N_2(10^{-3})$, $H_2(2 \times 10^{-6})$, and $He(2 \times 10^{-4})$. Total density is $\simeq 1.0 \times 10^{13}$ cm^{-3} at 125 km. Eddy diffusion coefficient $K = 10^8$ cm^2 sec^{-1} and height of the turbopause is 145 km (Kumar & Hunten 1974).

exospheric temperature of 380°K and a K of 10^8 cm^2 sec^{-1}. The densities for CO, O, N_2, and H_2 must remain a matter for debate, however, until the airglow-transport problem is resolved.

THE IONOSPHERE OF VENUS

The ionospheric profiles measured by the M5 and M10 radio occultation experiments are shown in Figures 9, 10, and 11. On the night side, electron densities of 10^4 cm^{-3} seem typical in the upper atmosphere. A notable difference in the night side electron density profiles of M5 and M10 is the much less extensive top side ionosphere at the time of M10 (Fjeldbo et al 1975). Differences in solar wind conditions, could account for this. The night side ionosphere of Mariner 5 has been discussed by McElroy & Strobel (1969). The wind speeds required by McElroy &

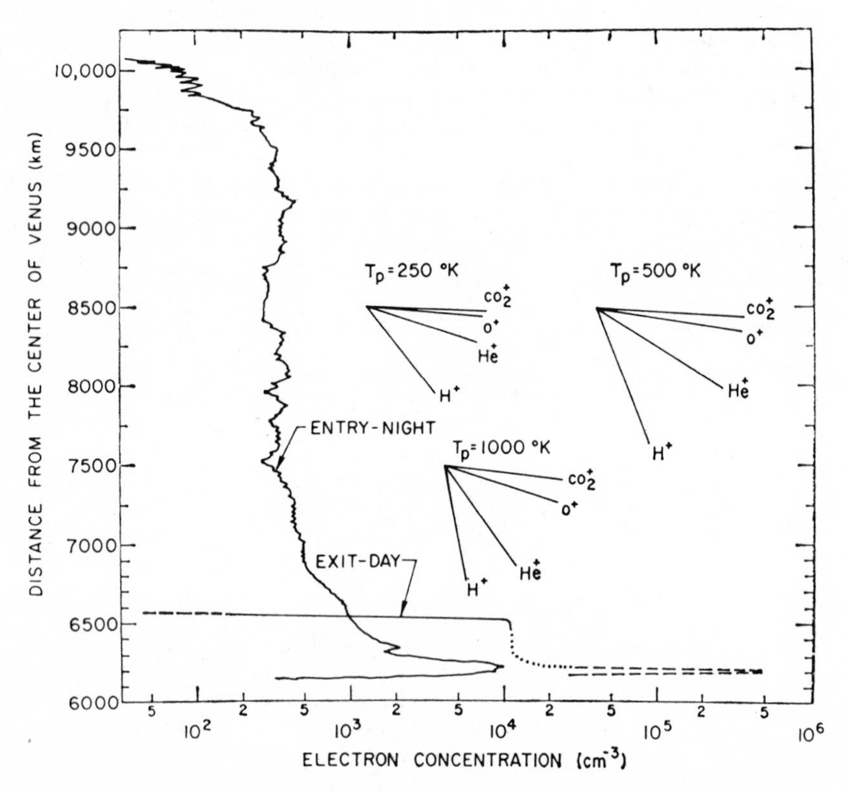

Figure 9 Profiles of electron concentrations in the night and day ionospheres of Venus. The night profile and the solid part of the day profile are from integral inversion of the phase data. The dashed day peak is from very preliminary model-fitting to amplitude data. The dotted line is in an area where no direct measurements were made because of the formation of caustics (Mariner Stanford Group 1967).

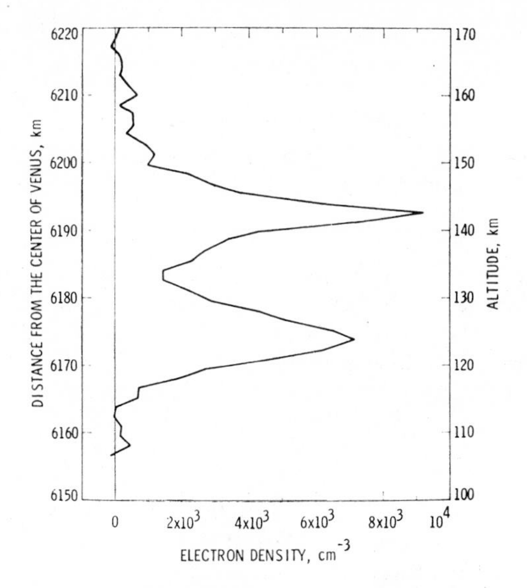

Figure 10 Electron density profile obtained with Mariner 10 during ingress on the night side of Venus. The measurements were made near longitude 69.3°E and latitude 1.3°N. The solar zenith angle was 117.7° (Fjeldbo et al 1975).

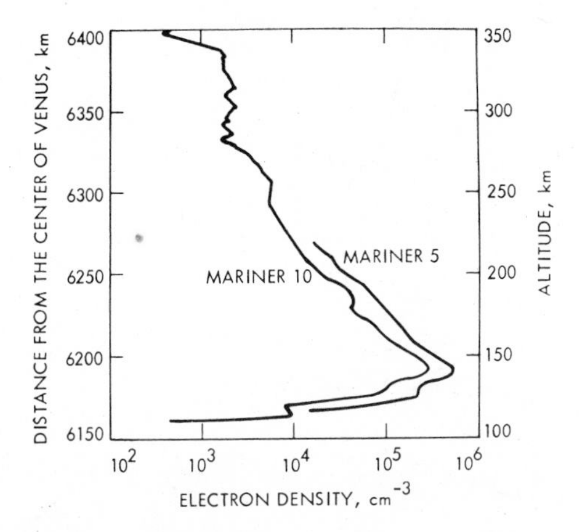

Figure 11 Electron density profiles obtained on the day side of Venus. The Mariner 10 egress measurements were made near longitude 236.6°E and latitude 56.0°S where the solar zenith angle was 67.0°. Also shown are the results of a refined analysis of the 2.3 GHz tracking data from Mariner 5 obtained on October 19, 1967. The latter measurements were made during Mariner 5's egress near 32.5°S at a solar zenith angle of about 33° (Fjeldbo et al 1975).

Strobel (1969) to carry day side ions such as He^+, H_2^+ to the night side seem reasonable, based on the results of the thermospheric circulation suggested by Dickinson (see Dickinson & Ridley 1975 and references therein). The ions transported to the night side react with CO_2 (reaction 7, Table 2, for example). As a result of the ion molecule and recombination reactions (reactions 11 and 12 of Table 2), night glow occurs (McElroy & Strobel 1969).

The day side electron density profiles of M5 and M10 are shown in Figure 11 for the region of the electron density maximum. Modeling of the ionosphere based on a predominantly CO_2 upper atmosphere suggests that in the region of the electron density maximum the ionosphere is of the F_1 type, as for Mars, with the main ions being CO_2^+ and O_2^+ (McElroy 1968, McElroy & McConnell 1971, Kumar & Hunten 1974, Bauer & Hartle 1974, Nagy et al 1975). The M5 results are consistent with an exospheric temperature of 700°K and O mixing ratios at the ion peak $\leqq 5\%$ (McElroy 1968). However, in these models the thermospheric temperature at 145 km is quite low, ~200–300°K (McElroy 1968), and the predominant ions are CO_2^+ and O_2^+ (McElroy & McConnell 1971). Photochemical equilibrium is assumed. As is true for Mars, the reproduction of the magnitude of the observed densities presents a problem (cf Stewart 1971). More recent attempts (Kumar & Hunten 1974, Nagy et al 1975) to model the Cytherian atmosphere use exospheric temperatures of about 380°K (e.g. see Figure 12). In these low temperature

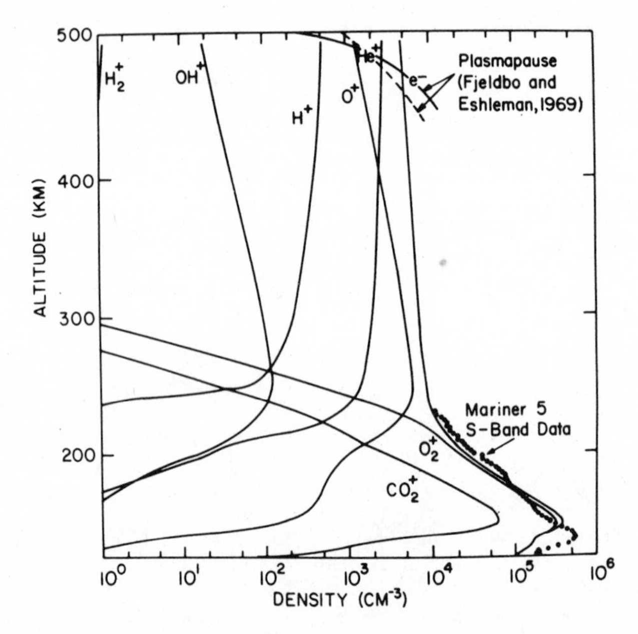

Figure 12 Ionosphere model for the neutral atmosphere shown in Figure 8. Also shown are the electron density profile obtained from the Mariner 5 S-band experiment and the plasmapause profile deduced from Mariner 5 49.8 MHz data (Kumar & Hunten 1974).

models, the temperature in the region of the ion peak is unchanged from earlier models, and reasonable agreement with the M5 and M10 data is obtained. In the region of 160–220 km, the O/CO_2 ratio is increased over that of earlier models (cf McElroy & McConnell 1971) because lower exospheric temperatures are used and photoionization of O (reaction 2, Table 2) is an important source of ions. As noted earlier for Mars, in the F_1 models the plasma scale height reflects the scale height of the ion source, not the terminal molecular ion. Thus, in order to fit the M5 data in the 160–220 km region, an $O + CO_2$ ion source at a temperature of $\sim 380°K$, is required instead of a CO_2 ion source at $\sim 700°K$. The low temperature models have difficulty reproducing the ion densities measured at 500 km by M5. This could indicate high ion and electron temperatures in the 200–500 km region resulting from magnetospheric heating and thus would relate to solar wind interaction.

A notable feature of the Cytherian day side ionospheres is the existence of an ionopause where a sharp decrease in electron density is observed. The ionopause electron density measured by M5 was $\sim 10^4$ cm^{-3} and its altitude was about 500 km (Figure 9). As shown in Figure 11, the ionopause electron density and height measured by M10 were $\sim 10^3$ and 350 km respectively (Howard et al 1974, Fjeldbo et al 1975). The ionopause is presumably the mark of the interaction with the solar wind (see following section). Depending on the details of the solar wind interaction (e.g. Michel 1971) the apparent variability of the ionopause could be due to changing solar wind characteristics or to varying thermospheric conditions, which alter the densities of neutral species at great heights and thus change the ion-mass loading of the solar wind.

The major ion above 200 km in the models of Kumar & Hunten (1974) and Nagy et al (1975) is different, despite similar reaction schemes. According to Kumar & Hunten (1974), O^+ is the major ion, but Nagy et al (1975) estimated that O_2^+ is the major ion. In both models, the ions are in diffusive equilibrium. The difference is attributable to the allowance that Nagy et al (1975) made for the temperature dependence of the recombination coefficients 11 and 12 (Table 2). With electron temperatures typical of both models, the rates of 11 and 12 are some 4–6 times slower than with the constant rates Kumar & Hunten (1974) used. This allows the molecular ions to dominate to heights where diffusion takes over. Kumar & Hunten noted that 1% O gives their best agreement with the M5 ionosphere, but Nagy et al (1975) found that 5% O will still fit the results. The difference is attributable mainly to the mixing ratio being referred to different levels. The mixing ratio of O in Figure 8 is more like 4% when referred to a total density of 2×10^{10} cm^{-3}, similar to that used by Nagy et al (1975).

It is difficult in the models of Kumar & Hunten (1974) and Nagy et al (1975) to fit the M5 data above 220 km. It is also difficult to model the M10 electron density profile above 220 km. At these altitudes, the ions are in diffusive equilibrium and photochemical steady state is no longer appropriate. To suppress the topside plasma scale height, Nagy et al (1975) require an upward ion wind of the order of 10 km sec^{-1}. An alternative suggestion (Bauer & Hartle 1974) revitalized the solar wind interaction theory of Cloutier et al (1969) by suggesting that photoions in the topside ionosphere are swept up by the penetrating solar wind and flow to the lower ionosphere. Bauer & Hartle (1974) tentatively state that the topmost ion is He^+ (Figure 13). However, the amounts of He required are some 10 times those inferred from M10 results (Kumar & Broadfoot 1975). This could be a discrepancy, but more likely the detailed physics of the solar wind—ionospheric interaction is inadequate. In the 180–240 km region, the major ion is O^+, whereas below this the major ion is probably O_2^+. Bauer & Hartle (1974) explained the region below 160 km as an F_1 region. Thus, as noted earlier, the plasma scale height must reflect that of the ion source. If the thermospheric temperature is of the order of 380°K, the atmosphere must be mainly CO_2 (i.e. $O \leqq 20\%$). However, if the thermospheric temperature profiles resemble models, the temperature is more like 200–300°K, and a larger abundance of O can be allowed without violating the plasma scale height constraints. O and CO_2 densities are restricted in that, for the F_1 electron density peak to occur at the correct altitude, the column density of absorbers should be $\sim 10^{17}$ cm^{-2} above 145 km (cf Hunten 1968). Based on these arguments, one expects a CO_2 density of the order of 2×10^7 to 4×10^7 cm^{-3} at 200 km. If the region is suppressed by a solar wind of velocity v, then the time constant for flow of O^+ must be less than the time constant for reaction with CO_2 (reaction 5 in Table 2). Using the CO_2 densities quoted above, $v \simeq 5 \times 10^4$ cm sec^{-1}. Also in this region

$$n(\mathrm{O}^+) = \mathrm{H(O)}J_2 n(\mathrm{O})/v,$$

where n is the species density, J_2 is the photoionization rate given in Table 2 and H(O) is the scale height for atomic oxygen. Using the above data and ion

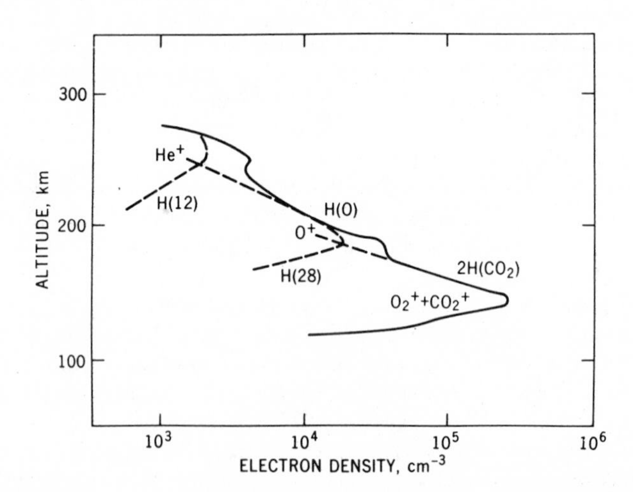

Figure 13 Dayside electron density profile (solid curve) of the Venus ionosphere observed by the Mariner 10 radio occultation experiment. Low-density features above 260 km have been omitted because they are of the order of the experiment noise level (10^3 cm^{-3}) (however, see Figure 11). Dashed curves represent the ion density distributions proposed by Bauer & Hartle (1974) to fit the observed electron density profile (Bauer & Hartle 1974).

densities from Figure 13, this gives $n(O) = 5 \times 10^8$ cm^{-3}, which in turn implies an O density of 5×10^9 cm^{-3} or a mixing ratio of 20% at 145 km—the F_1 peak. In the region of photochemical steady state, time constant arguments indicate that $v \leqq 10^4$ cm sec^{-1}. The numbers indicated here are somewhat different in magnitude from those of Bauer & Hartle (1974). The conclusions remain the same, however, and the implication of large O densities in the upper atmosphere still holds. Thus this ionospheric model seems to agree with the airglow studies. It should be noted that the solar wind model suggested by Bauer & Hartle (1974) is qualitative in nature and does not appear to have been verified by calculation.

THE INTERACTION OF THE SOLAR WIND WITH THE ATMOSPHERES OF MARS AND VENUS

A discussion of the upper atmospheres and ionospheres of Mars and Venus would be incomplete without some mention of our present knowledge of solar wind interaction with Mars and Venus. The most recent papers on this subject are by Bogdanov & Vaisberg (1975) on Mars 2 and 3, Bridge et al (1974), and Ness et al (1974). Hill & Michel (1975) outlined the current understanding of the solar wind planetary interaction theory.

On the Earth a bow shock forms when the solar wind interacts with the Earth's magnetic field. However, the magnetic moments of Mars and Venus are less than 10^{-4} times the Earth's magnetic moment (cf Whitten & Colin 1974, Bogdanov & Vaisberg 1975, Ness et al 1974) and some other process must account for shocks observed on these planets. Cloutier et al (1969), noting that photoionization of neutral atmosphere would add more ion mass to the impinging solar wind than it could carry, suggest that photoionization of the upper neutral atmosphere provides a sufficient obstacle to make the solar wind go subsonic, thereby generating a standing bow wave (cf Michel 1971).

The existence of a bow shock on Mars is indicated by a discontinuity observed in the nearby magnetic field measured by M4 (Smith et al 1965); it has also been inferred from the experiments on the Mars 2 and 3 orbiters. Figure 14 shows various samples of the bow shock crossings as measured by the changes in the ion flux energy spectra (Bogdanov & Vaisberg 1975), as well as the M4 results and a calculated "mean" shock. The measurements indicate clearly that the bow shock location is variable. The estimated stagnation point, which can range from approximately the Martian radius, 3400 km, to 6000 km, is not correlated with variations in solar wind velocity. Apparently the observations preclude a Martian dipole moment with properties estimated by Dolginov et al (1973). As the bow shock was crossed and the orbiters Mars 2 and 3 approached Mars, the relatively hard magnetosheath spectrum softened considerably, becoming similar to that for hot ions with temperatures of ~ 20 eV. The magnetic field simultaneously increased. The region of the hot ions, as measured by Mars 2 and 3, is shown in Figure 15. The results are consistent with ions forming the solar wind obstacle. The evidence also suggests that a solar wind interaction model similar to that of Johnson & Midgley (1969) is appropriate for Mars. In their model, the maintenance of an electrical connection between the induced magnetosphere and ionosphere is intimately related to the neutral atmosphere. Thus the variations in exospheric temperature noted earlier for the M9 encounter could yield density variations in H (possibly He), with a time scale and magnitude that could account for the fluctuating obstacle and bow shock boundaries.

The evidence for a bow shock on Venus is less extensive than for Mars. Its presence seems well documented by Mariner 5 and 10 and Venera 4 and 6 (cf Bridge et al 1974, Ness et al 1974). Figure 16 indicates the paths taken by the M5 and M10 flybys. The bow shock and the ionopause appear to have been at different locations for M5 and M10. The bow shock was located at position 1 and between positions 4 and 5 on the M5 flyby. The trajectory of M10 meant that it crossed the shock so far downstream that it was not detectable. The plasma science experiment and dual magnetometer experiments indicate a bow shock located between positions *D* and *E* (Figure 16). If the solar wind interaction

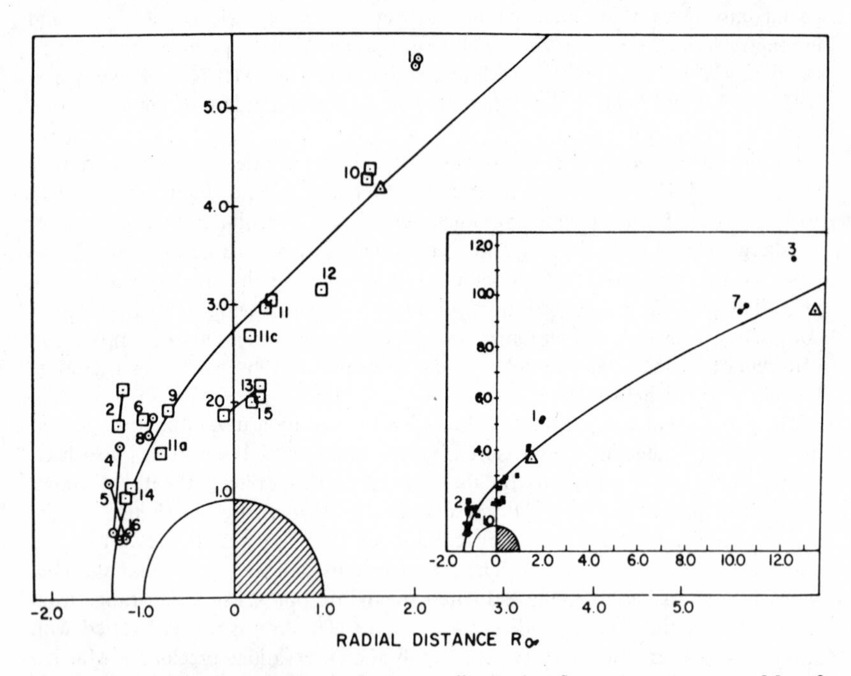

Figure 14 Martian bow shock crossings according to ion flux measurements on Mars 2 (circles) and Mars 3 (square) from December 14, 1971, to May 12, 1972. The short connecting lines show the uncertainty in determination of shock location because of telemetry sequence. The curve shows the mean shape of the bow shock calculated from the observed crossings by using the approximation of the parabolic cross-section in solar-oriented cylindrical coordinates. In the lower righthand corner, the small-scale figure is given to show the distant crossings of the bow shock. Two bow shock crossings by Mariner 4 in 1965 are shown by triangles (Bogdanov & Vaisberg 1975).

mechanism on Venus is similar to that on Mars, then the Mars 2 and 3 results indicate that the difference between the M5 and M10 results could be expected.

With regard to the solar wind as a potential ionization source in the ionosphere (e.g. Stewart 1971), the models of Cloutier et al (1969), Dessler (1968), and Johnson & Midgley (1969) yield very different estimates for the amount of the solar wind that actually impinges on the ionosphere (Michel 1971). For example, Johnson & Midgley (1969) estimated that about 1 part in 10^4 of the incident solar wind impacts the ionosphere.

SUMMARY AND CONCLUSIONS

Inevitably, exploration of new fields to solve first-order problems uncovers much more complex problems; such has been the case with the exploration of Mars and Venus. The first-order problems regarding the composition and thermal structure have been addressed. A comparison of actual measurements of these basic properties with models based on common assumptions about the terrestrial atmosphere has revealed serious gaps in our knowledge of the physical and dynamical processes occurring in planetary atmospheres. The experiments on board the Viking 1

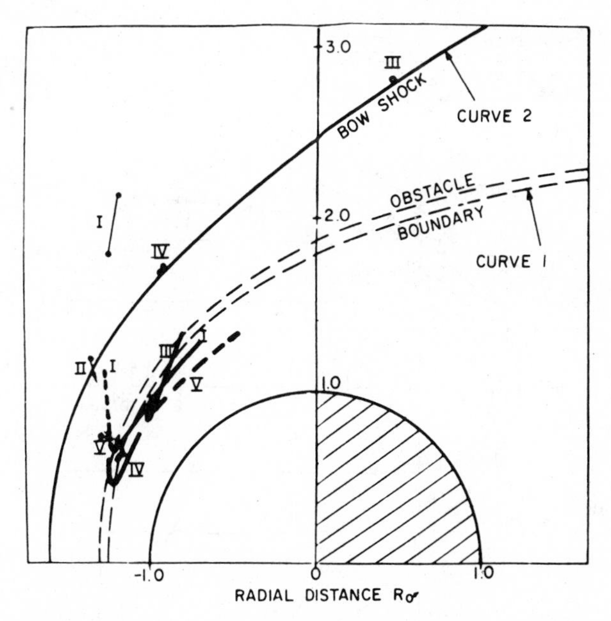

Figure 15 Observations of the region of soft ions (cushion) on Mars 2 and Mars 3 satellites (thick lines). Bow shock crossings are shown by circles with the same numbers as the corresponding soft ion region observations. The solid line is one of the bow shocks from Spreiter & Rizzi (1972) taken to fit the observed crossings, and the double-dashed curve represents the boundary of the obstacle (Bogdanov & Vaisberg 1975).

and 2 landers and orbiters, which are presently on course for Mars and due to arrive around mid-June 1976, should yield much-needed information. The next planned U.S. missions to Venus are the Venus Pioneer bus, carrying several probes to explore various regions of the Cytherian atmosphere, and the Venus Pioneer orbiter.

Our understanding of the thermospheric heat budget on Mars and Venus appears deficient. Is it controlled by lower atmospheric coupling, solar wind coupling, or a complex interaction of both? If coupling is with the lower atmosphere, is it caused by wave propogation or planetary-wide circulation? On Mars, measurements of night side ionospheric profiles could prove useful in estimating the strength of ionospheric winds. The dual frequency radio occultation experiment on board Vikings 1 and 2 are much more sensitive than the single frequency experiments employed on M4, M6, M7, and M9. The Viking experiment results should enhance understanding of the thermospheric circulation and heat budget of Mars.

The problem of excess ionization on Mars and Venus is puzzling. Is it simply caused by vibrationally excited ions or solar wind ionization? If it is the former,

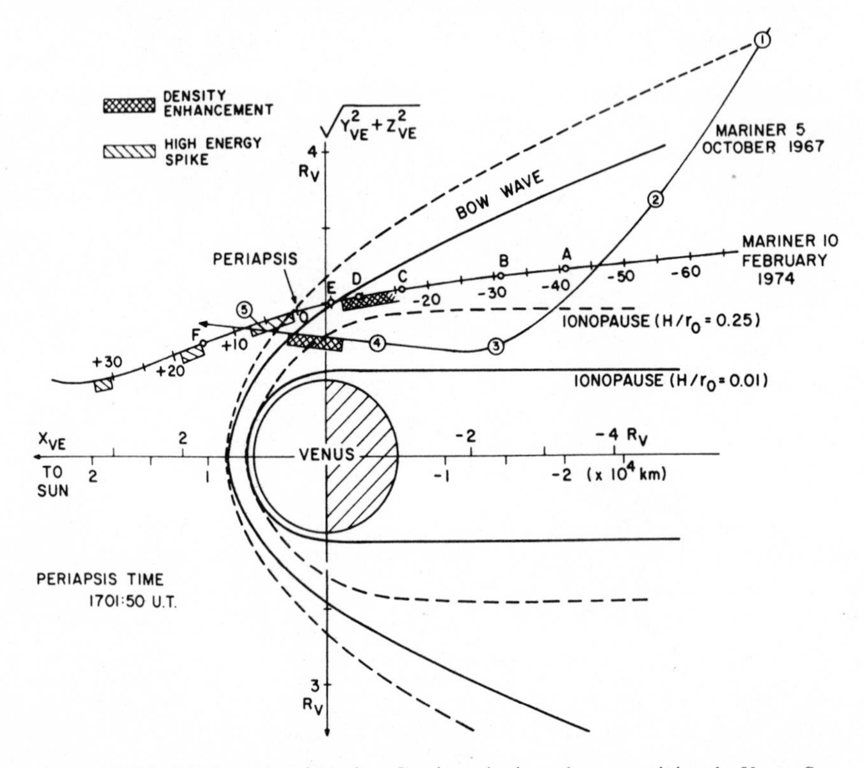

Figure 16 The Mariner 10 and Mariner 5 trajectories in a plane containing the Venus-Sun line. The planet and two predictions of fluid theory for the case of flow along the Venus-Sun line are also shown. The letters refer to events along the track of Mariner 10, and the circled numbers refer to events along the track of Mariner 5 (Bridge et al 1974).

then laboratory work should confirm the hypothesis. A laboratory analysis is enormously complicated, but would be worthwhile, especially because excited molecular ions may be present in laboratory plasmas in a quantity sufficient to cast doubt on some of the present recombination work results (cf Gutcheck & Zipf 1973). If the excess ionization is due to solar wind penetration, a considerable fraction of the impinging solar wind is probably involved (Stewart 1971). We noted above that the Mars 2 and 3 measurements of the hot ions in the magnetosphere of Mars indicated the plausibility of the Johnson & Midgley (1969) solar wind interaction model. In this model, only about 1 part in 10^4 of the solar wind impinges the ionosphere directly. Thus, for Mars at least, the solar wind ionization source is unlikely. The Viking entry retarding potential analyzer (RPA) experiment (Nier et al 1972) should provide additional evidence on this mechanism.

The physics of the interaction of the solar wind with the ionospheres of Mars and Venus is still not well understood. Further theoretical work is required. The data base for comparison is so far rather limited, and the Viking RPA experiment will provide data on solar wind electrons, ionospheric photoelectrons, electron temperature in the ionosphere, and ion temperature, composition, and concentration (Nier et al 1972). Such information should elucidate the physical processes occuring in the magnetosphere-ionosphere region as well as in the bow shock magnetosheath region. Experiments planned for the Pioneer Venus orbiter and probe include magnetometer, plasma probe, AC electrical field sensor, and RPA (cf Whitten & Colin 1974). If Pioneer Venus eventually returns the anticipated data, our understanding of the Venus–solar wind interaction should increase considerably.

The data on the mixing conditions in the upper atmosphere of Venus are contradictory, at present. Further reduction of the M10 spectrometer data may help to resolve the problems raised in this review, but the search for escape mechanisms for H and/or airglow excitation mechanisms should continue. At worst we may have to wait for the UV spectrometer and the ion and neutral mass spectrometer data from the Pioneer Venus missions to adequately resolve this matter.

The issue of D on Venus seems settled by the M10 data and the reanalysis of the M5 data,[6] but the problem of the two-temperature model has yet to be addressed seriously. Its solution may yield information on the solar wind interaction.

Almost all our experimental information on the upper atmospheres of Mars and Venus is from planetary flybys and orbiting spacecraft. The data represent the dedicated efforts of planners, engineers, experimenters, and the people who track the spacecrafts. The exciting new information on planetary atmospheres has opened new vistas in planetary studies. This review presents implications of just a small fraction of data returned from the planetary missions; the larger canvas remains to be painted.

ACKNOWLEDGMENTS

This work was supported by a grant from the National Research Council of Canada.

[6] However, McElroy (1974) has noted that H/D ratio rates in the lower atmosphere could yield information of the evolutionary history of H_2O on Venus and possibly on the importance of the solar wind as a source or sink of planetary H.

Literature Cited

Anderson, D. E. Jr. 1974. Mariner 6, 7, and 9 ultraviolet spectrometer experiment: analysis of Hydrogen Lyman alpha data. *J. Geophys. Res.* 79:1513–18

Anderson, D. E. Jr. 1975a. The Mariner 5 ultraviolet photometer experiment: analysis of Rayleigh-scattered and 1304-Å radiation from Venus. *J. Geophys. Res.* 80:3063–67

Anderson, D. E. Jr. 1975b. The mariner 5 ultraviolet photometer experiment: analysis of the Hydrogen Lyman-alpha data. *J. Geophys. Res.* 80: In press

Anderson, D. E. Jr., Hord, C. W. 1971. Mariner 6 and 7 ultraviolet spectrometer experiment: analysis of hydrogen Lyman alpha data. *J. Geophys. Res.* 76:6666–73

Barth, C. A. 1968. Interpretation of the Mariner 5 Lyman-α measurements. *J. Atmos. Sci.* 25:564–67

Barth, C. A. et al 1971. Mariner 6 and 7 ultraviolet spectrometer experiment: upper atmospheric data. *J. Geophys. Res.* 76:2213–27

Barth, C. A., Pearce, J. B., Kelly, K. K., Wallace, L., Fastie, W. G. 1967. Ultraviolet emissions observed near Venus from Mariner 5. *Science* 158:1675–78

Barth, C. A., Stewart, A. I., Hord, C. W., Lane, A. L. 1972. Mariner 9 ultraviolet spectrometer experiment: Mars airglow spectroscopy and variations in Lyman alpha. *Icarus* 17:457–68

Barth, C. A., Wallace, L., Pearce, J. B. 1968. Mariner 5 measurement of Lyman-alpha radiation near Venus. *J. Geophys. Res.* 73:2541–45

Bauer, S. J., Hartle, R. E. 1974. Venus ionosphere: an interpretation of Mariner 10 observations. *Geophys. Res. Lett.* 1:7–9

Bogdanov, A. V., Vaisberg, O. L. 1975. Structure and variations of solar wind—Mars interaction region. *J. Geophys. Res.* 80:487–94

Brandt, J. C., McElroy, M. B., eds. 1968. *The Atmosphere of Mars and Venus.* New York: Gordon & Breach. 288 pp.

Bridge, H. S. et al 1974. Observations at Venus encounter by the Plasma Science Experiment on Mariner 10. *Science* 183: 1293–96

Brinkmann, R. T. 1971. Mars: has nitrogen escaped? *Science* 174:944–46

Broadfoot, A. L., Kumar, S., Belton, M. J. S., McElroy, M. B. 1974. Ultraviolet observations of Venus from Mariner 10: preliminary results. *Science* 183:1315–18

Cloutier, P. A., Michel, F. C., McElroy, M. B. 1969. Modification of the Martian ionosphere by the solar wind. *J. Geophys. Res.* 74:6215–28

Dalgarno, A., Degges, T. C., Stewart, A. I. 1970. Mariner 6: origin of Mars ionized carbon dioxide ultraviolet spectrum. *Science* 167:1490–91

Dalgarno, A., McElroy, M. B. 1970. Mars: is nitrogen present? *Science* 170:167–68

Dessler, A. J. 1968. Ionizing plasma flux in the Martian upper atmosphere. In *The Atmospheres of Mars and Venus,* ed. J. C. Brandt, M. B. McElroy, pp. 241–50. New York: Gordon & Breach. 288 pp.

Dickinson, R. E. 1971. Circulation and thermal structure of the Venus thermosphere. *J. Atmos. Sci.* 28:885–94

Dickinson, R. E., Ridley, E. C. 1975. A numerical model for the dynamics and composition of the Venusian thermosphere. *J. Atmos. Sci.* 32:1219–31

Dolginov, S. S., Yeroshenko, Y. G., Zhuzgov, L. N. 1973. Magnetic field in the very close neighborhood of Mars according to data from the Mars 2 and Mars 3 spacecraft. *J. Geophys. Res.* 78:4779–86

Donahue, T. M. 1969. Deuterium in the upper atmospheres of Venus and Earth. *J. Geophys. Res.* 74:1128–37

Fjeldbo, G., Seidel, B., Sweetnam, D., Howard, T. 1975. The Mariner 10 radio occultation measurements of the ionosphere of Venus. *J. Atmos. Sci.* 32:1232–36

Gutcheck, R. A., Zipf, E. C. 1973. Excitation of the CO fourth positive system by the dissociative recombination of CO_2^+ ions. *J. Geophys. Res.* 78:5429–36

Hansen, J. E., ed. 1974. *The Atmosphere of Venus. Proc. Conf. Goddard Inst. Space Studies, New York, Oct. 15–17.* 198 pp.

Henry, R., McElroy, M. B. 1968. Photoelectrons in planetary atmospheres. In *The Atmospheres of Mars and Venus,* ed. J. . Brandt, M. B. McElroy, pp. 251–85. New York: Gordon & Breach. 288 pp.

Hill, T. W., Michel, F. C. 1975. Planetary magnetospheres. *Rev. Geophys. Space Sci.* 13:967–74

Howard, H. T. et al. 1974. Venus: mass, gravity field, atmosphere and ionosphere as measured by the Mariner 10 dual-frequency radio system. *Science* 183:1297–1301

Hunten, D. M. 1968. The ionosphere and upper atmosphere of Mars. In *The Atmospheres of Mars and Venus,* ed. J. C. Brandt, M. B. McElroy, pp. 147–80. New York: Gordon & Breach. 288 pp.

Hunten, D. M. 1973. The escape of light gases from planetary atmospheres. *J. Atmos. Sci.* 30:1481–94

Hunten, D. M. 1974. Energetics of thermospheric eddy transport. *J. Geophys. Res.* 79:2533–34

Johnson, F. S., Midgley, J. E. 1969. Induced magnetosphere of Venus. *Space Res.* 9:760–63. Amsterdam: North-Holland. 770 pp.

Kliore, A. J. et al 1972. The atmosphere of Mars from Mariner 9 radio occultation measurements. *Icarus* 17:484–516

Kliore, A. J., Fjelbo, G., Seidel, B. L., Sykes, M. J., Woiceshyn, P. M. 1973. S band radio occultation measurements of the atmosphere and topography of Mars with Mariner 9: extended mission coverage of polar and intermediate latitudes. *J. Geophys. Res.* 78:4331–51

Kumar, S. 1975. The ionosphere and upper atmosphere of Venus. In *Atmospheres of Earth and the Planets,* ed. B. M. McCormac, pp. 375–84. Dordrecht: Reidel. 458 pp.

Kumar, S., Broadfoot, A. L. 1975. He 584 Å airglow emission from Venus: Mariner 10 observations. *Geophys. Res. Lett.* 2:357–60

Kumar, S., Hunten, D. M. 1974. Venus: an ionospheric model with an exospheric temperature of 350°K. *J. Geophys. Res.* 79:2529–32

Levine, J. S., Riegler, G. R. 1974. Argon in the Martian atmosphere. *Geophys. Res. Lett.* 1:285–87

Liu, S. C., Donahue, T. M. 1974. Mesospheric hydrogen related to exospheric escape mechanisms. *J. Atmos. Sci.* 31: 1466–70

Liu, S. C., Donahue, T. M. 1975. The aeronomy of the upper atmosphere of Venus. *Icarus* 24:148–56

McConnell, J. C. 1973. The atmosphere of Mars. In *Physics and Chemistry of Upper Atmospheres,* ed. B. M. McCormac, pp. 309–34. Dordrecht-Holland: Reidel. 389 pp.

McConnell, J. C., McElroy, M. B. 1970. Excitation processes for Martian dayglow. *J. Geophys. Res.* 75:7290–93

McElroy, M. B. 1967. The upper atmosphere of Mars. *Astrophys. J.* 150:1125–38

McElroy, M. B. 1968. The upper atmosphere of Venus in light of the Mariner 5 measurements. *J. Atmos. Sci.* 25:574–77

McElroy, M. B. 1972. Mars: an evolving atmosphere. *Science* 175:443–45

McElroy, M. B. 1974. Aeronomy of Venus. See Hansen 1974, pp. 141–54

McElroy, M. B., Hunten, D. M. 1969. The ratio of deuterium to hydrogen in the Venus atmosphere. *J. Geophys. Res.* 74: 1720–39

McElroy, M. B., McConnell, J. C. 1971. Atomic carbon in the atmospheres of Mars and Venus. *J. Geophys. Res.* 76: 6674–90

McElroy, M. B., McConnell, J. C. 1971. Dissociation of CO_2 in the Martian atmosphere. *J. Atmos. Sci.* 28:879–84

McElroy, M. B., Strobel, D. F. 1969. Models for the night time Venus ionosphere. *J. Geophys. Res.* 74:1118–27

Mange, P. 1955. Diffusion processes in the thermosphere. *Ann. Geophys.* 11:153–68

Mariner Stanford Group 1967. Venus: ionosphere and atmosphere as measured by dual frequency radio occultation of Mariner 5. *Science* 158:1678–83

Michel, F. C. 1971. Solar wind interaction with planetary atmospheres. *Rev. Geophys. Space Phys.* 9:427–35

Moos, H. W. 1974a. An upper limit on H_2 ultraviolet emissions from the Martian exosphere. *J. Geophys. Res.* 79:2887–89

Moos, H. W. 1974b. Comparison of the far ultraviolet spectra of Venus and Mars. *J. Geophys. Res.* 79:685–87

Moos, H. W., Fastie, W. G., Bottema, M. 1969. Rocket measurement of ultraviolet spectra of Venus and Jupiter between 1200 Å and 1800 Å. *Astrophys. J.* 155:887–97

Mumma, M. J., Morgan, H. D., Mentall, J. E. 1975a. Reduced absorption of the nonthermal $CO(A^1\pi - X^1\Sigma^+)$ fourth positive group by thermal CO and implications for the Mars upper atmosphere. *J. Geophys. Res.* 80:168–72

Mumma, M. J., Stone, E. J., Zipf, E. C. 1975b. Nonthermal rotational distribution of $CO(A^1\pi)$ fragments produced by dissociative excitation of CO_2 by electron impact. *J. Geophys. Res.* 80:161–67

Nagy, A. F., Liu, S. C., Donahue, T. M., Atreya, S. K., Banks, P. M. 1975. A model of the Venus ionosphere. *Geophys. Res. Lett.* 2:83–86

Ness, N. F., Behannon, K. W., Lepping, R. P., Whang, Y. C., Schatten, K. H. 1974. Magnetic field observations near Venus: preliminary results from Mariner 10. *Science* 183:1301–6

Nier, A. O., Hanson, W. B., McElroy, M. B., Seiff, A., Spencer, N. W. 1972. Entry science experiments for Viking 1975. *Icarus* 16:74–91

O'Malley, T. F. 1969. Theory of dissociative recombination in vibrationally excited gases. *Phys. Rev.* 185:101–85

Owen, T. 1974. Martian climate: an empirical test of possible gross variations. *Science* 183:763–64

Rasool, S. I., Stewart, R. W. 1971. Results and interpretation of the S-band occultation experiments on Mars and Venus. *J. Atmos. Sci.* 28:869–78

Rottman, G. J., Moos, H. W. 1973. The ultraviolet (1200–1900 angstrom) spectrum of Venus. *J. Geophys. Res.* 78: 8033–48

Smith, E. J., Davis, L. Jr., Coleman, P. J. Jr., Jones, D. E. 1965. Magnetic field measurements near Mars. *Science* 149:1241–42

Spreiter, J. R., Rizzi, A. W. 1972. The Martian bow wave-theory and observation. *Planet. Space Sci.* 20:205–8

Stewart, A. I. 1972. Mariner 6 and 7 ultraviolet spectrometer experiment: implications of CO_2^+, CO and O airglow. *J. Geophys. Res.* 77:54–68

Stewart, A. I., Barth, C. A., Hord, C. W., Lane, A. I. 1972. Mariner 9 ultraviolet spectrometer experiment: structure of Mars' upper atmosphere. *Icarus* 17:469–74

Stewart, R. W. 1971. The electron distributions in the Mars and Venus upper atmospheres. *J. Atmos. Sci.* 28:1069–73

Strickland, D. J. 1973. The OI 1304- and 1356-Å emission from the atmosphere of Venus. *J. Geophys. Res.* 78:2827–36

Strickland, D. J., Stewart, A. I., Barth, C. A., Hord, C. W., Lane, A. L. 1973. Mariner 9 ultraviolet spectrometer experiment: Mars atomic oxygen 1304-Å emission. *J. Geophys. Res.* 78:4547–59

Strickland, D. J., Thomas, G. E., Sparks, P. R. 1972. Mariner 6 and 7 ultraviolet spectrometer experiment: analysis of the OI 1304 Å and OI 1356 Å emissions. *J. Geophys. Res.* 77:4052–68

Sze, N. D., McElroy, M. B. 1975. Some problems in Venus' aeronomy. *Planet. Space Sci.* 23:763–86

Thomas, G. E. 1971. Neutral composition of the upper atmosphere of Mars as determined from the Mariner UV spectrometer experiments. *J. Atmos. Sci.* 28:859–68

Tinsley, B. A., Hodges, R. R., Strobel, D. F. 1975. Diurnal variations of atomic hydrogen: observations and calculations. *J. Geophys. Res.* 80:626–34

Wallace, L., Stuart, F., Nagel, R., Larson, M. 1971. A search for deuterium on Venus. *Astrophys. J.* 168:L29–L31

Whitten, R. C., Colin, L. 1974. The ionospheres of Mars and Venus. *Rev. Geophys. Space Phys.* 12:155–92

THE URANIUM-SERIES METHODS OF AGE DETERMINATION[1]

Teh-Lung Ku
Department of Geological Sciences, University of Southern California, Los Angeles, California 90007

INTRODUCTION

Of the many nuclear clocks commonly used in geology, those based on the radioactive daughter products of uranium and thorium are perhaps the least known. Extensive applications of this collectively termed uranium-series (disequilibrium) methods are relatively recent, and much developmental work pertaining to their strengths and weaknesses are still underway. Nevertheless, the past decade has seen a steady increase of literature on the subject. Such a trend is almost certain to continue, particularly in view of enhanced interest in deciphering the timing of many profound changes in the physical and biological nature of the Earth that are embodied in the Pleistocene record.

No ideal dating technique applicable to the entire Pleistocene epoch is yet available. For instance, radiocarbon ages of more than 40,000 yr are normally considered unreliable and the dating of the Quaternary with U-He and K-Ar is limited by the scarcity of suitable preserved and datable deposits (Broecker 1965). In principle, the various uranium-series methods cover ages ranging from several years to more than a million years and are applicable to a variety of materials. Thus far these methods have contributed to several areas of study, in particular to establishing the chronology of the late Pleistocene climatic fluctuations. They are also potentially valuable to the fields of anthropology and archaeology.

The three naturally occurring uranium and thorium decay series each include nuclides of different nuclear and chemical properties, with a wide range of decay half-lives (Figure 1). These nuclides can be separated through the processes of weathering, transportation, and deposition. Once separated, they tend to restore radioactive equilibria with one another, with rates dictated by their respective decay constants. Such rather unique relationships render those nuclides useful as tracers

[1] Contribution No. 355 from the Department of Geological Sciences, University of Southern California.

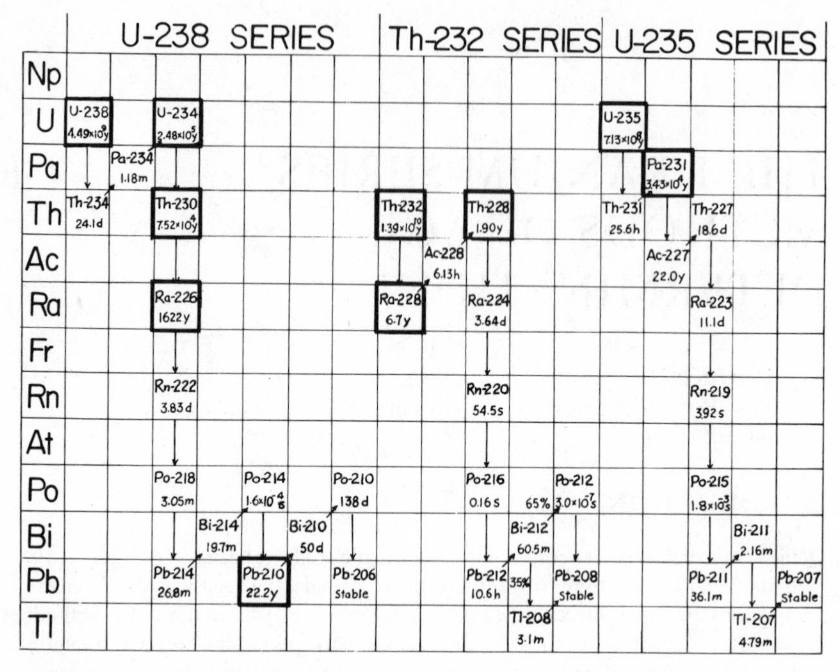

Figure 1 The three naturally occurring radioactive decay series. Nuclides framed with heavy lines are those of geochronological interest discussed in this paper.

and/or time-measuring tools for the study of a number of oceanographic, hydrologic, meteorologic, and geologic problems. In this review, we confine our discussion mostly to their use in age determination of geologic materials.

The uranium-series methods encompass a spectrum of techniques involving many isotopes (Figure 1; Goldberg & Bruland 1974). However, they can be divided into two categories: (*a*) methods based on decay of unsupported intermediate members of the series, and (*b*) methods based on accumulation of decay products of uranium. Either of these situations may be encountered in the dating of igneous rocks. The order of presentation will come under these groupings with emphasis on the application of the isotope ^{230}Th (ionium). Such an emphasis is both inevitable and justified in that ^{230}Th dating of deep-sea sediments, coral, and mollusks constitutes the bulk of the data in this field of study; it bears on principles, assumptions, and limitations associated with the use of other uranium-series isotopes.

NATURE OF RADIOACTIVE DISEQUILIBRIUM

Because of the relatively short half-lives of the intermediate members of the decay series (Figure 1), normally secular equilibrium among them should have long since been established on the Earth. This is, however, not the case. Take the marine hydrosphere as an example. Activities of the ^{238}U series nuclides in sea water exhibit the following relationships:[2] $^{234}U = 1.14(^{238}U)$, $^{230}Th < 0.01(^{234}U)$, $^{226}Ra > 30(^{230}Th)$, $^{210}Pb < {}^{226}Ra$ (except in surface ocean), and so forth. The primary cause for these disequilibria is difference in solubility of the isotopes involved.

During weathering, uranium is oxidized to form UO_2^{2+} which is quite soluble because anionic complexes, such as $UO_2(CO_3)_3^{4-}$, are formed (McKelvey et al 1955, Starik & Kolyadin 1957). Because of the α- and β-decays accompanying the transformation of ^{238}U to ^{234}U, it is more likely that ^{234}U rather than ^{238}U will occupy metastable lattice positions induced by α-recoil dislocation (Cherdyntsev et al 1961, Dooley et al 1966), or that ^{234}U will be in the soluble, +6 valence state (Rosholt et al 1963, 1966; Dooley et al 1966). Either leads to preferential leaching of ^{234}U over ^{238}U from weathered minerals (Cherdyntsev 1955), and thus to $^{234}U/^{238}U$ ratios of larger than unity in natural solutions, including seawater. Kigoshi (1971) observed the dissolution of α-recoil ^{234}Th into the solution in contact with mineral grains. Hence, a certain fraction of the ^{234}U atoms could have originally been produced in the interstitial fluid of a sediment.

As the oceanic ^{234}U decays to ^{230}Th, the latter, because of the high ionic potential of the Th^{4+} ion, is readily adsorbed or precipitated as a hydrolysate. Thus ^{230}Th is greatly depleted in the water column (Koczy et al 1957) and enriched in the surface bottom deposits (Pettersson 1937, Isaac & Picciotto 1953). The excess ^{226}Ra over ^{230}Th in seawater is a result of the diffusion of ^{226}Ra across the sediment-water interface, which is generated by the decay of ^{230}Th in sediments (Kröll 1953; Koczy 1954, 1958). As ^{226}Ra in the ocean further decays to ^{210}Pb, fractionation again occurs. The residence time of ^{210}Pb is considerably shorter than that of ^{226}Ra, so that in most parts of the ocean, ^{210}Pb is depleted with respect to ^{226}Ra (Goldberg 1963, Craig et al 1973, Bruland et al 1974b) and enriched in surface sediments (Koide et al 1972).

The kind of fractionation pattern outlined above is also seen in the decay series of ^{235}U and ^{232}Th. The ^{235}U-^{231}Pa relationship resembles that of ^{234}U-^{230}Th, in that ^{231}Pa, being insoluble, is hence impoverished relative to ^{235}U in seawater and enriched in sediments (Sackett 1960). In the ^{232}Th series, the diffusional input of ^{228}Ra from sediments results in the excess of ^{228}Ra and ^{228}Th over their progenitor ^{232}Th in the sea (Koczy et al 1957, Moore & Sackett 1964, Moore 1969).

In summary, because of diversities in the chemical and nuclear properties, disequilibria among members of the decay series exist. In seawater and other surface solutions, the activities of those isotopes of geochronological interest generally decrease in the following order:

$$^{234}U > {}^{238}U > {}^{226}Ra > {}^{210}Pb > {}^{230}Th$$
$$^{235}U > {}^{231}Pa$$
$$^{228}Ra > {}^{228}Th > {}^{232}Th$$

[2] Hereafter, symbols for the isotopic species, when used in equations or figures or when expressed as ratios, all denote activities, i.e. $N \cdot \lambda$ (N is the number of atoms of the species and λ its decay constant).

Understandably, fractionations such as these will have made their imprints on the associated sediments and other solids precipitated from the solutions. Restoration to radioactive equilibrium of the isotopes in the solid phase is the basis of various geochronological yardsticks discussed below. Several sources that provide further general references and background information on the subject include Koczy (1963), Broecker (1965), Wetherill & Tilton (1967), Komura & Sakanoue (1967), Bernat (1969), Szabo (1969), Cherdyntsev (1971), and Goldberg & Bruland (1974).

METHODS BASED ON DECAY OF UNSUPPORTED INTERMEDIATE DAUGHTER NUCLIDES

As mentioned, natural fractionations taking place in the hydrological cycle lead to the preferential precipitation of certain insoluble daughter products in the decay series over their more soluble parents. In bottom deposits, these daughter nuclides could therefore be largely unsupported (or in excess) and will diminish with time through decay. If the original amount of excess daughter nuclides in a freshly deposited sediment is known, the extent to which this excess has been reduced gives a measure of the time since the settling of that sediment. In this way, the age or accumulation rate of a deposit is determined. Mathematically (the parent nuclides, having much longer half-lives than their respective daughters in excess, can be treated as being stable in the formulation),

$$c = c_0 \exp(-\lambda t), \quad \text{or} \quad t = -(1/\lambda) \ln (c/c_0), \tag{1}$$

where λ is the decay constant of the unsupported radiogenic daughter, c_0 is the initial excess, and c is the excess found t years later. The crucial problem is the assignment of c_0.

Dating Deep-Sea Deposits

Unsupported ^{230}Th ($t_{1/2} = 75{,}200$ years) and ^{231}Pa ($t_{1/2} = 34{,}300$ years) have been used for dating deep-sea sediments and ferromanganese nodules. The precipitation of the two isotopes as they are formed from the oceanic uranium is essentially quantitative, with less than 0.05% of the ^{230}Th and less than 0.2% of the ^{231}Pa remaining in solution (Moore & Sackett 1964). Their concentrations in the deposits, which depend to first order on the accumulation rate of the deposits, and measurement techniques will determine the applicable age range. Currently, sediments of up to 350,000 years old are datable by ^{230}Th, and up to 150,000 years old by ^{231}Pa; for the slowly accreted Mn nodules, the corresponding ranges are longer.

^{230}Th METHOD Attempts to utilize ^{230}Th for sediment dating began with Piggot & Urry (1939, 1942), who measured ^{226}Ra as an index of ^{230}Th. They expressed c and c_0 of Equation 1 in concentration units (i.e. amount of ^{230}Th per gram of total sediment) and assumed c_0 to be constant.

Let t be the age of the sediment layer at depth x, then $t = x/s$, where s is the average sedimentation rate (in centimetres per unit time) down to depth x. From Equation 1:

$$\ln c = -(\lambda/s)x + \ln c_0, \quad (2)$$

where λ in the present case is the decay constant of ^{230}Th. When ln c is plotted against x, as customarily done, a straight-line relationship would point to the constancy of both c_0 and s; it not only serves to validate the constant c_0 assumption, but also enables the sedimentation rate, s, to be assessed from the slope of the straight line.

Such an approach has been basically adopted by many of the later investigations on sediment dating. Interestingly, major improvement made on the original work of Piggot & Urry seems to have been only in the analytical realm. The migratory behavior of ^{226}Ra in sediments, first noted by Kröll (1953, 1954), has prompted the direct measurement of ^{230}Th, which was not feasible before World War II. Furthermore, the excess ^{230}Th used in Equation 2 is obtained by subtracting the measured activity of ^{234}U from that of ^{230}Th (i.e. $^{230}Th_{ex} = {}^{230}Th - {}^{234}U$). This correction arises from the consideration that uranium in a typical pelagic sediment is mostly associated with detrital clays (Ku 1965) in which a ^{234}U-equivalent amount of detrital ^{230}Th (U-supported ^{230}Th) is present. The correction becomes important in deeper strata where the quantity of excess ^{230}Th is reduced. Measurement precision of $\sim 2\%$ for $^{230}Th_{ex}$ using α-particle spectrometry can be achieved; however, analytical uncertainties as they now stand may often not be the limiting factor that affects the precision of age determination of sediments.

Shown in Figure 2*a* is an Equation 2 plot for a Caribbean core V12-122. An average sedimentation rate of 2.52 cm 10^{-3} yr^{-1} over the length of the core is obtained from the slope of the regression line best fitting the data points. The linear regression coefficient (hence the rate derived therefrom) has a standard error of $\pm 4.1\%$, comparable to the analytical errors as shown. Within this error limit, constancy of c_0 and s is indicated, implying uniformity in the deposition of both ^{230}Th and total sediments over the past 300,000 years or so. V12-122 is one of the better cores in terms of showing good linearity in plots like Figure 2. Cores exhibiting larger scattering of data (therefore giving larger errors in rate estimates) than V12-122 are not uncommon, especially those cores with varied lithology.

The nonlinearity plots could be explained if the initial concentration of ^{230}Th varied with time. And the suspected culprit is change in the bulk sediment settling rate, because the very long residence time of uranium (10^5–10^6 yr) combined with the very short residence time of ^{230}Th (~ 100 yr) favors the concept of a constant ^{230}Th precipitation over the datable interval. To cope with this problem, two major modifications of the method have been attempted. They are the ^{230}Th/^{232}Th and the ^{231}Pa/^{230}Th methods.

^{230}Th/^{232}Th METHOD Picciotto & Wilgain (1954) suggested using the ratio ^{230}Th/^{232}Th instead of ^{230}Th per total sediment for c and c_0 in Equation 2, reasoning that, because of the chemical identity of the two thorium isotopes, ^{232}Th should provide an estimate of the amount of ^{230}Th initially present in a sediment. In other words, despite variation in the bulk rate of sedimentation, the ratio of the precipitated ^{230}Th/^{232}Th in the freshly deposited sediment remains constant through time at a locality. Several investigators have followed the suggestion and applied it (with

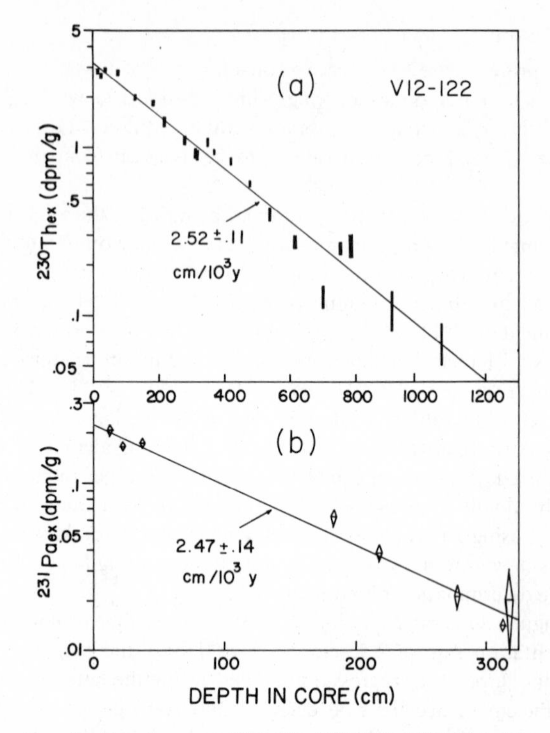

Figure 2 Equation 2 plots for core V12-122 (17°00′N, 74°24′W). The parameter c is expressed as (*a*) excess ^{230}Th concentrations, and (*b*) excess ^{231}Pa concentrations. The sedimentation rates are obtained from slopes of the best-fitting straight lines for the data points; in this case, they are from the linear regression coefficients of the ln $c-x$ correlation matrix. Analytical data are from Ku (1966) and Broecker & van Donk (1970). Broecker & van Donk (1970) have cited an average rate of 2.35 ± 0.10 cm/10^3 yr for the core.

slight operational variations) to cores from many parts of the ocean. They include, among others, Baranov & Kuzimina[3] (1958), Goldberg and his co-workers (Goldberg & Koide 1958, 1962, 1963; Goldberg et al 1964; Goldberg & Griffin 1964), Almodovar (1960), and Miyake & Sugimura (1961). However, these applications have invited criticisms (Rosholt et al 1961; Koczy 1961, 1963; Sarma 1964; Sackett 1964; Broecker 1965). Objections stem primarily from the argument that, although they are chemically similar, ^{230}Th and ^{232}Th (or, for that matter, Fe_2O_3) vastly differ geochemically in their marine cycles. The former is produced in the sea, whereas the latter is mostly an integral part of the continental detritus carried to the sea floor.

The critics' concern is justified. Most ^{232}Th in sediments is detrital in origin. Thorium content of pelagic biogenous (calcareous) components is low (Koczy 1949, Holmes et al 1968); most of the ^{232}Th resides in the nonbiogenous (clay)

[3] These authors used ratios of excess ^{230}Th to MnO and Fe_2O_3, as they noted that ^{232}Th co-varies with MnO and Fe_2O_3 in the cores examined.

components. Thorium content of these clays approximates those of average igneous rocks and shales and does not show any correlative trend with sedimentation rate, water depth, location of occurrence, etc (Ku 1966). Even with the acid-leach technique (Goldberg & Koide 1962), the major fraction of the ^{232}Th analyzed in the ^{230}Th/^{232}Th dating is detrital, not authigenic as intended by the method (Goldberg & Griffin 1964).

The question remains as to the validity of the ^{230}Th/^{232}Th–derived results. Obviously, in cores with nearly constant ^{232}Th distribution, there should be little difference between ^{230}Th expressed as ^{230}Th/^{232}Th ratios and as concentrations. Both are subject to the same uncertainties in case scatter of data in the ln c vs depth plot occurs. The situation is exemplified by Figure 3 for core AII42-13 (Ku et al 1972). In essence, the advantage claimed by the ^{230}Th/^{232}Th method over the Piggot-Urry approach is nullified. For cores of varying ^{232}Th content, however, depth plots of ^{230}Th/^{232}Th and ^{230}Th per total sediment will be different. For the following reasons the difference can be significantly amplified in high $CaCO_3$ cores.

Although knowledge of the precise removal mechanism of ^{230}Th in the ocean is lacking, there is evidence bearing on the ^{230}Th precipitation as being facilitated by settling of particulate material of both biogenic and detrital nature. This is in contrast with the detrital affiliation of ^{232}Th. The implication is that depth plots of ^{230}Th/^{232}Th should resemble those of ^{230}Th per noncarbonate (i.e. ^{230}Th concentration on a carbonate-free basis), as shown in Figure 3*b*. In cores with varying but overall high carbonate content, percentagewise, changes in the non-carbonate (hence ^{232}Th) may become much larger than the corresponding changes in carbonate, leading to depth plots of ^{230}Th/^{232}Th that drastically differ from those plots if ^{230}Th per total sediment is used. A case in point is the data on another Mid-Atlantic Ridge core ZEP-15 (Goldberg et al 1964), located near AII42-13. [Analogous situations can be found in many other cores from the Ridge area, e.g. Goldberg & Griffin (1964) and Ku et al (1972).] Carbonate contents in this core vary from 72% to 84%; hence noncarbonates fluctuate between 16% and 28%—a variation almost spanning a factor of two. Figure 4 shows two ways of plotting (not shown is the plot using ^{230}Th per noncarbonate, which would yield patterns close to those of ^{230}Th/^{232}Th, as has been noted). Although in both plots the mean slope of the linear fit for the data points are similar—thus giving a mean sedimentation rate of about 4 mm 10^{-3} yr^{-1}—details in the scatter are different, especially for the upper portion of the core.

ZEP-15 is one of the Mid-Atlantic Ridge cores whose ^{230}Th/^{232}Th chronology has shown serious discrepancies with the ^{14}C data; ages based on ^{230}Th/^{232}Th can be as much as five to ten times too great (Goldberg & Griffin 1964; Goldberg et al 1964; Broecker 1965). The discrepancies can be attributed chiefly to the manner in which the ^{230}Th data were interpreted (Ku et al 1968, 1972), although viewpoints have differed (Goldberg 1968). In Figure 4*a*, the 1.3 mm 10^{-3} yr^{-1} rate shown is based on the assumption of a constant initial ^{230}Th/^{232}Th. In the light of the above discussions, this is clearly unwarranted. The steep slope for the top 11.5 cm is mainly induced by a downward decrease of carbonate content (hence increase of ^{232}Th) in this interval, rather than by ^{230}Th decay. It is therefore no surprise that the age of 88,000 years assigned at 11.5 cm is too old compared to the ^{14}C age of 17,000 years (Broecker 1965). Figure 4*b* presents the depth plot of ^{230}Th concentration for ZEP-15. A rate of sedimentation of 4.3 mm 10^{-3} yr^{-1} for the top 65 cm is estimated from the best-fitting line, which can be compared with the 5 mm 10^{-3} yr^{-1} rate for the 0–150 cm interval of core AII42-13 (Figure 3). Both are average rates with some ±15% to ±25% uncertainties. They do not specify short-term rate variations within the averaged intervals, because in doing so one has to assume

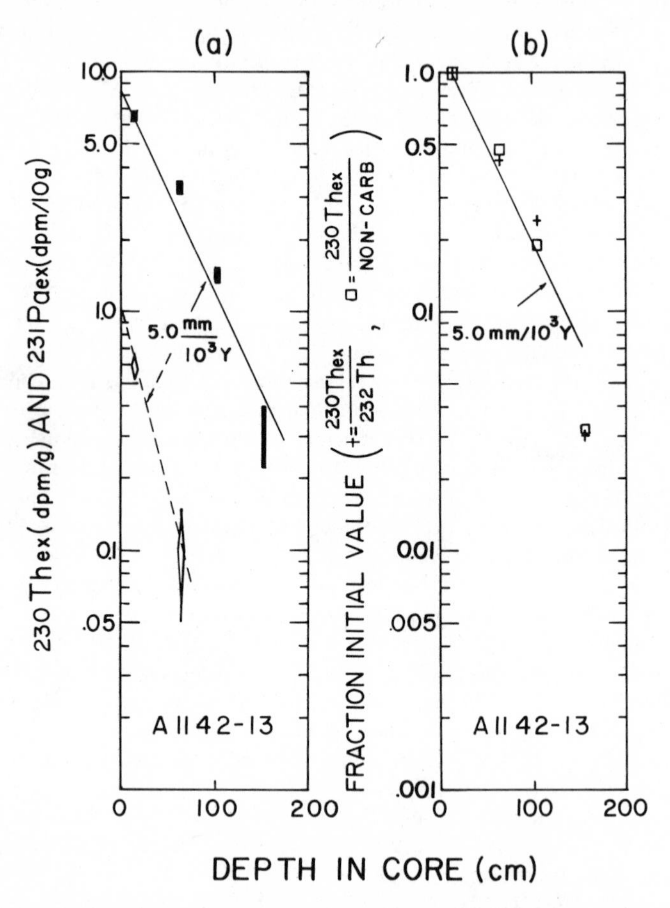

Figure 3 Equation 2 plots for core AII42-13 (19°40′N, 42°44′W): (*a*) *c* is expressed as excess ^{230}Th (solid bars) and excess ^{231}Pa (unfilled diamonds) concentrations. Average sedimentation rates derived from the best-fitting lines for both ^{230}Th and ^{231}Pa agree. (*b*) *c* is expressed as $^{230}Th_{ex}/^{232}Th$ and as $^{230}Th_{ex}$ per noncarbonate (for comparison, the data are normalized against the measured ratios in the uppermost layers). Because carbonate content, and therefore ^{232}Th, in this core is relatively uniform, (*a*) and (*b*) plots show similar results (Ku et al 1972).

a rigid constant c_0 that both the Piggot-Urry and the Picciotto-Wilgain approaches fail to provide.

Briefly, the function of $^{230}Th/^{232}Th$ is similar to that of ^{230}Th per noncarbonate used by Ku & Broecker (1966) and Scott et al (1972). Both methods fail to improve the ^{230}Th per total sediment approach. In fact, they may be more vulnerable to influence by carbonate fluctuations. With this in mind and with cautious data interpretation, the $^{230}Th/^{232}Th$–derived results should be usable. The technique offers some analytical convenience in that only ratios of the thorium isotopes are measured. In recent years, nondestructive γ-ray techniques have been developed to measure ^{230}Th and ^{232}Th by way of simultaneous analyses of ^{214}Bi and ^{208}Tl (Osmond & Pollard 1967, Yokoyama et al 1968, Bhandari et al 1971). They provide rapid, albeit crude, means of surveying oceanic sdimentation rates.

$^{231}Pa/^{230}Th$ METHOD The idea of using ^{231}Pa in conjunction with ^{230}Th was conceived independently by Sackett (1960) and Rosholt et al (1961). Both nuclides are removed from seawater in a time short compared to their radioactive half-lives. The relative abundance of their progenitors, ^{238}U, ^{235}U, and ^{234}U, in the ocean has been rather constant (Hamer & Robbins 1960, Thurber 1962), and river contributions of unsupported ^{231}Pa and ^{230}Th are negligible (Scott 1968). Therefore,

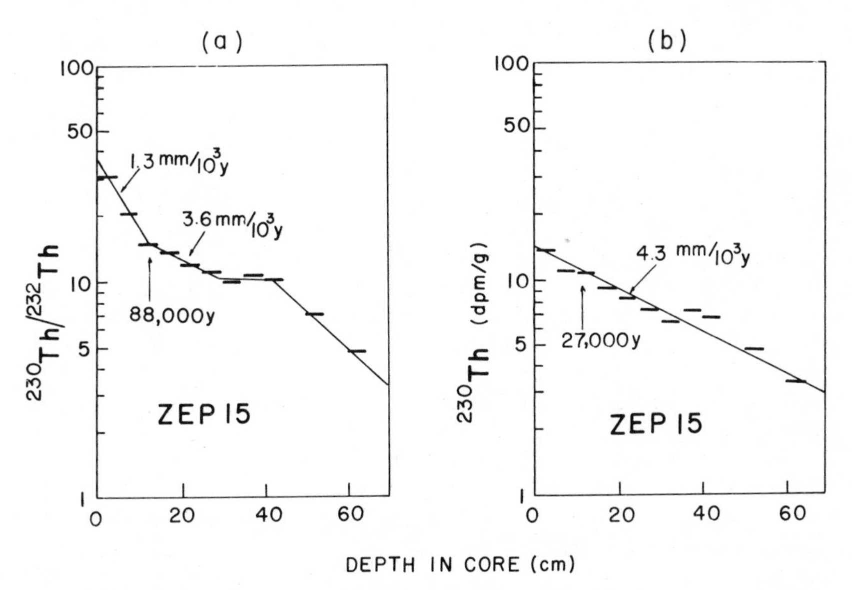

Figure 4 Similar comparison as made in Figure 3 for core ZEP-15 (21°06′N, 44°57′W). Here, because of the nonuniform carbonate distribution in the core, the two ways of expressing *c* in Equation 2 give quite different results. Shown in (*a*) are rates interpreted by Goldberg et al (1964) assuming a constant initial $^{230}Th/^{232}Th$. An alternative and (in the author's opinion) more correct way of deriving the ages is through the plot and interpretation of (*b*).

the $^{231}Pa/^{230}Th$ ratio added to the sediment should be fixed at a value close to their production ratio, i.e. 1/11 or 0.091, if no fractionation occurs during precipitation of the two species. The age (t) of a sediment with a measured ratio $(^{231}Pa_{ex}/^{230}Th_{ex})_t$ is then:

$$(^{231}Pa_{ex}/^{230}Th_{ex})_t = (^{231}Pa_{ex}/^{230}Th_{ex})_0 \exp(\lambda_0 - \lambda_1)t, \tag{3}$$

where λ_0 and λ_1 are decay constants of ^{230}Th and ^{231}Pa, respectively, and $(^{231}Pa_{ex}/^{230}Th_{ex})_0$, the ratio of excess ^{231}Pa to excess ^{230}Th in newly deposited sediment, assumes the theoretical value 0.091. In practice, $^{231}Pa_{ex} = {}^{231}Pa - {}^{235}U$, and $^{230}Th_{ex} = {}^{230}Th - {}^{234}U$.

The $(^{231}Pa_{ex}/^{230}Th_{ex})_t$ ratio in Equation 3 is a function of t only; therefore it is independent of changing oceanic conditions such as fluctuations in bulk sedimentation rate and uranium content of seawater and should decrease with a half-life of about 62,000 years. Depending upon the detectability of the residual excess ^{231}Pa, this so-called $^{231}Pa/^{230}Th$ method is usually applicable over the last 150,000 years.

The first test of the method would be to see if indeed $(^{231}Pa_{ex}/^{230}Th_{ex})_0$ is equal to 0.091. This appears to be borne out by some of the early measurements (Sackett 1960; Rosholt et al 1961, 1962). In three cores from the Caribbean Sea, Rosholt et al (1961, 1962) found a set of dates[4] that are consistent with stratigraphic correlation and with the ^{14}C ages in the upper portion of the cores, within the errors of $\pm(5\text{–}20)\%$. However, subsequent studies have revealed an alarming number of cores showing surface $^{231}Pa_{ex}/^{230}Th_{ex}$ ratios that are much too low (Sackett 1964; Ku 1965, 1966; Ku et al 1972). As an extreme case, surface layer of core AII42-13 (Figure 3*a*) shows this ratio being as low as in the vicinity of 0.01, which according to Equation 3 would have an apparent age of over 150,000 years! The low ratios cannot be explained by suggestions (Rosholt et al 1961, Sackett 1964) such as missing core tops, contamination by reworked sediment, or preferential upward migration of ^{230}Th in the core, as discussed by Ku (1966).

On examining the integrated amounts of unsupported ^{231}Pa ($\Sigma^{231}Pa_{ex}$) and unsupported ^{230}Th ($\Sigma^{230}Th_{ex}$) in cores from various parts of the ocean (Ku 1966, Ku et al 1972), it is evident that the ratio $\Sigma^{231}Pa_{ex}/\Sigma^{230}Th_{ex}$ in sediments is less than the $^{235}U/^{234}U$ ratio of 1/25 in sea water. This inequality could be attributable to a deficiency of ^{231}Pa (Scott 1968, Turekian & Chan 1971, Ku et al 1972) or to a surplus of ^{230}Th (Ku 1966, Ku & Broecker 1967a) in sediments with respect to their oceanic supplies from the ^{235}U and ^{234}U parents. On the other hand, the ratio $\Sigma^{231}Pa_{ex}/\Sigma^{230}Th_{ex}$ in deep-sea manganese nodules has been found to be markedly higher than the seawater $^{235}U/^{234}U$ ratio (Sackett 1966, Ku & Broecker 1969). Clearly fractionation between ^{231}Pa and ^{230}Th must have taken place during their precipitation from the sea.

This unexpected finding negates the use of Equation 3 for age determination as orginally proposed for the $^{231}Pa/^{230}Th$ method. However, concentrations of ^{230}Th and ^{231}Pa when plotted separately according to Equation 2 often give rates consistent

[4] The $^{231}Pa/^{230}Th$ dates were calculated without considering the 14% excess ^{234}U over ^{238}U, because this phenomenon was not known at that time.

with each other, within errors, as illustrated in Figures 2 and 3*a*. It may imply that the initial ratio of the two isotopes at a given depositional site have remained sufficiently constant, even though it shows spatial variations. If so, the $^{231}Pa/^{230}Th$ ratios can be used either via the Equation 2 plot, thus obtaining average sedimentation rates (Sackett 1964, Ku & Broecker 1967a, Broecker & Ku 1969), or via Equation 3 by substituting the measured $^{231}Pa_{ex}/^{230}Th_{ex}$ in surface sediments for the theoretical value of 0.091. The first approach is preferred because strict constancy for the initial $^{231}Pa_{ex}/^{230}Th_{ex}$ ratios cannot be assured. It should be pointed out that ages obtained from the use of the ratio $^{231}Pa/^{230}Th$ are intrinsically more vulnerable to such errors as introduced by sediment mixing (Sarma 1964), analytical blanks (Broecker & van Donk 1970), uranium corrections (Broecker & Ku 1969), and uncertainties in the decay constant of ^{231}Pa or ^{230}Th than from the use of either isotope alone.

^{231}Pa METHOD Similar to the Piggot-Urry method for ^{230}Th, the decrease of unsupported ^{231}Pa with depth in sediment cores and manganese nodules have been used to derive average accumulation rates over the last 150,000 years (Sarma 1964, Ku 1965, Ku & Broecker 1967b). The technique has proven useful as a complement to the ^{230}Th method, though both are subject to the same limitations as discussed. Figures 2 and 3*a* illustrate the point.

Sackett (1965) used a different approach in treating the ^{231}Pa data. His assumption is that the deposition rate of ^{231}Pa onto a given area of ocean bottom remains constant. If the total amount of excess ^{231}Pa in the core and the amount of excess ^{231}Pa below a given level in the core are known, the age at that level can be calculated by a formula similar in form to Equation 1:

$$t = -(1/\lambda_1) \ln (n/n_t) \tag{4}$$

where λ_1 is the decay constant of ^{231}Pa, n is the excess ^{231}Pa below the level with age t, and n_t is the total excess ^{231}Pa in the core. In practice, the core is divided into contiguous, uniform segments; the age at the divisions are to be determined. Each segment is measured of its excess ^{231}Pa. The deepest segment measured should contain no excess ^{231}Pa. Although hinging on the constancy of ^{231}Pa deposition, ages obtained by this approach are independent of variations in the bulk sedimentation rate. Sackett (1965) has shown results in reasonable agreement with the ^{14}C ages obtained on four cores.

The same approach should be applicable to ^{230}Th. Although Broecker (1965) cited three ^{230}Th ages by this way and Schornick (1972) attempted the approach on several manganese nodules, the quality of the data base these workers used does not permit a rigorous evaluation. Considering that ^{230}Th can be measured more precisely than ^{231}Pa and that constant ^{230}Th deposition is indicated at least in some cores, a systematic study of the method using ^{230}Th is warranted.

SYNOPSIS OF DATING DEEP-SEA DEPOSITS The bulk of the data has certainly established the first-order logarithmic decreases of both ^{230}Th and ^{231}Pa in bottom deposits, strengthening the concept that particle-by-particle sedimentation prevails in the deep sea. Mean sedimentation rates can be determined from the slope of the line best fitting the data points in the Equation 2 plot, using either concentrations of ^{230}Th or ^{231}Pa, or ratios such as ^{230}Th/^{232}Th and ^{231}Pa/^{230}Th. Within error limits of the best linear fits, rates estimated in this manner agree with those derived from other dating tools such as ^{14}C, K-Ar, ^{10}Be, and paleomagnetic reversals [for sediments, see e.g. Ku et al (1968), Dymond (1969), Amin (1970); for manganese nodules, see Ku (1976)]. The error limits can be $\pm 4\%$ or better for very uniformly deposited sediments, but are in general about $\pm 20\%$. Short-term sedimentation rate fluctuations (on a time scale of tens of thousands of years or less) are not resolvable by the ratio methods, as had been hoped. Perhaps future investigations into the constant flux of ^{230}Th (as envisaged by Sackett for ^{231}Pa) may shed some light on this aspect.

Studies of elemental flux and budget in the hydrosphere require knowledges of deep ocean rate processes. Such knowledges are made more accessible by the dating tools described. Sharpening of the tools for greater precision is not crucial at present for the geochemical studies, since the general $\pm 20\%$ error is more than outweighed by uncertainties in other parameters, such as sediment and nodule bulk density. To fix the chronology of climate over the past several hundred thousand years recorded in deep-sea cores, further improvement over data such as those from core V12-122 would be desirable, as indicated in discussions in Broecker & van Donk (1970), Broecker & Ku (1969), and Emiliani & Rona (1969), for example. It is a challenging task. However, as oxygen isotopic ratios in marine foraminifera have assumed new value as indices of global ice volume and stratigraphic correlation (Shackleton & Opdyke 1973), the task may be tackled by accurate dating of only a few "ideal" cores, like V12-122.

Dating Nearshore Marine and Lake Deposits

The use of ^{230}Th and ^{231}Pa is generally limited to sediments deposited at rates of 10 cm 10^{-3} yr^{-1} (or 0.1 mm/yr) and less. For deposits such as coastal and lake sediments accumulated at rates of the order of millimeters per year, ^{210}Pb has proven to be a useful geochronometer. It provides chronologies of the past 100–200 years.

^{210}Pb METHOD The isolation of ^{210}Pb, a member of the ^{238}U-series with a half-life of 22 years, can be traced back to its precursor ^{222}Rn. This noble gas nuclide tends to escape from the surface of the earth into the atmosphere. Most of the ^{222}Rn atoms entering the atmosphere remain in the troposphere where they decay to ^{210}Pb through a series of short-lived intermediate nuclides. The residence time for ^{210}Pb in the troposphere is estimated to range from days to about a month before it is removed by precipitation and dry fallout (Francis et al 1970, Pierson et al 1966).

This atmospheric flux of unsupported ^{210}Pb, assumed to have remained constant at a given locality, was initially proposed by Goldberg (1963) to date permanent snow fields, using Equation 2. The general validity of the proposition has since been confirmed (Crozaz et al 1964, Windom 1969). Application of the technique has been extended to lake and coastal marine sediments. Exponential decreases of

$^{210}Pb_{ex}$ ($= {}^{210}Pb - {}^{226}Ra$) with depth in lake cores have given measures of average sedimentation rates that essentially agree with those derived from other lines of evidence such as artificial radioactivity (fission products) and palynology (Krishnaswami et al 1971, Koide et al 1973, Robbins & Edgington 1975). Indications are that input of ^{210}Pb from rivers feeding the lakes may be insignificant and most of the excess ^{210}Pb in sediments originates from overhead precipitation (Goldberg 1963, Krishnaswami et al 1971).

In the nearshore marine sediments, the excess ^{210}Pb has an additional source: ^{226}Ra in seawater. The ^{210}Pb formed from this source, together with that from the atmospheric fallout and terrestrial runoff (if any), is scavenged by sinking particulates of both biogenous and detrital origin throughout the water column. The scavenging in the coastal regions is rather efficient. There, the residence time of ^{210}Pb in the surface mixed layer of the ocean is estimated to range from a few years (Rama et al 1961, Koide et al 1972) to less than a month (Bruland. et al 1974b), depending upon the intensity of biological productivity.

The reliability of the method when applied to the marine environment was first documented in a study of the varved sediment from the Santa Barbara Basin off California (Koide et al 1972). As shown in Figure 5, the ^{210}Pb-derived rate of 3.9 mm/yr, averaged over a 112 year period, compares favorably with the mean annual valve thickness of 3.3 mm in the core. The agreement strengthens the method's critical assumption—relatively constant flux of ^{210}Pb to a deposition site.

Part of the ^{210}Pb in sediments must be supported by ^{226}Ra. This has not been corrected in the Figure 5 data. As has been noted in discussing ^{230}Th, the correction becomes more important in the deeper strata. In order to date deeper sections in the core, hence to extend the range of the method, the ^{226}Ra corrections were made in subsequent studies (Bruland et al 1974a, Bruland 1974, Murray 1975).

Koide et al (1972) postulated postdepositional migration of lead to explain the low surficial concentration of ^{210}Pb in two of the cores they analyzed. Finding no supporting evidence for such a postulation, Robbins & Edgington (1975) suggested sediment mixing as an alternative explanation. Their suggestion, however, cannot account for the observed maximum ^{210}Pb values near the sediment-water interface. A more likely cause would be the very high water content of the surficial strata. Failure to remove the interstitial salts could result in a dilution of ^{210}Pb concentration in the analyzed samples, as noted by Bruland (1974). Since high and variable water contents are not uncommon to the newly deposited coastal sediments, expressing ^{210}Pb concentration on a salt-free basis or normalized to aluminum content of the sediment (Goldberg & Bruland 1974) would be desirable. In the Al-normalization, ^{210}Pb is assumed to be principally associated with the aluminum-silicate phases. This needs to be substantiated.

The method was initially tested on anoxic sediments in which bioturbation is minimum. Recent studies (Murray 1975, Murray & Ku 1974) have also indicated its utility in relatively oxygenated areas with sufficiently rapid sedimentation (e.g. >1 mm/yr); there, disturbances of benthic fauna, if occurring, are commonly limited to a depth of around 5 cm. The applicability of ^{210}Pb to other coastal environments such as estuaries, lagoons, and continental shelves, should be further assessed.

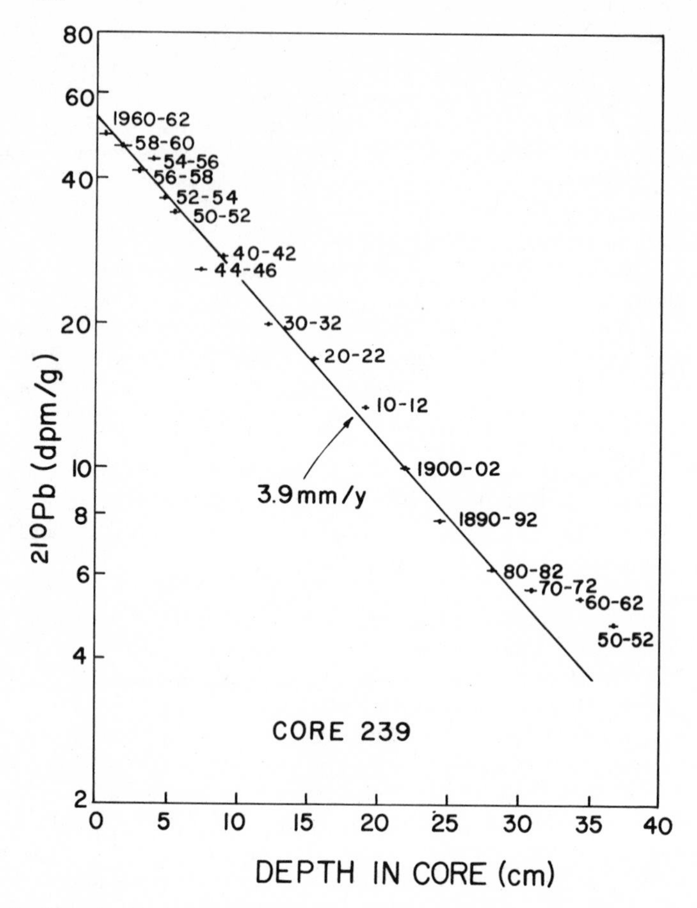

Figure 5 Equation 2 plot for core 239 from the Santa Barbara Basin; c is expressed as ^{210}Pb concentrations. The dates assigned to the analyzed layers are derived from counting varves in the sediments. Koide et al (1972).

$^{228}Th/^{232}Th$ METHOD The activity ratio $^{228}Th/^{232}Th$ in surface and deep waters of the ocean commonly varies between 5 and 30 (averaging about 15) (Nikolayev et al 1962, Moore & Sackett 1964, Somayajulu & Goldberg 1966, Miyake et al 1970, Imai & Sakanoue 1973, Knauss et al 1975). This means that only a small fraction of the ^{228}Th in these waters is generated from ^{232}Th in solution; the remainder comes mainly from the diffusional input of ^{228}Ra from bottom sediments (Koczy et al 1957, Moore 1969). Similar situations presumably exist in lacustrine environments. On precipitation, excess ^{228}Th could be detectable in surface layers of nearshore and lake sediments; with time, it should disappear with a half-life of 1.9 yr. Using an acid-leach technique, Koide et al (1973) measured $^{228}Th/^{232}Th$ ratios in sediments and obtained sedimentation rates for the top 4–8 cm of two cores. The rates were comparable to those based on the ^{210}Pb chronology. These workers suggested that the $^{228}Th/^{232}Th$ ratio may be used for dating sediments deposited around the last decade and as permissive evidence for the recovery of the uppermost levels of the deposit during coring operation.

Recent studies (Murray 1975) on ^{210}Pb-dated sediments of the San Pedro shelf off southern California have shown that surface $^{228}Th/^{232}Th$ ratios in these sediments, measured by both the acid-leach and total dissolution techniques, are significantly less than the equilibrium value of unity. It appears that extensive ^{228}Ra migration has occurred, rendering the suggested $^{228}Th/^{232}Th$ method inapplicable to these sediments.

In addition to the methods discussed above, excess ^{226}Ra has been found in freshwater manganese nodules and used to estimate the accretion rates of these lake deposits (Krishnaswami & Moore 1973).

Use of ^{234}U-^{238}U Disequilibrium

The excess ^{234}U over ^{238}U in natural waters raises the possibility of an age method based on the disappearance of the excess ^{234}U in solids precipitated from such waters. Substituting $c = (^{234}U/^{238}U) - 1$ and $c_0 = (^{234}U/^{238}U)_o - 1$ in equation 1, we obtain

$$(^{234}U/^{238}U) - 1 = [(^{234}U/^{238}U)_0 - 1] \exp(-\lambda_4 t) \qquad (5)$$

where $(^{234}U/^{238}U)_0$ is the initial $^{234}U/^{238}U$ ratio and λ_4 is the decay constant of ^{234}U. The age of a deposit can be calculated if $(^{234}U/^{238}U)_0$ is known. For terrestrial waters, conditions are generally too variable to warrant estimation of $(^{234}U/^{238}U)_0$ (Cherdyntsev et al 1963, Thurber 1963, Chalov et al 1964, Thompson et al 1975). This is not the case for the ocean. Thurber (1962, 1963) analyzed recent marine corals and oolites and consistently found a 15% excess of ^{234}U over ^{238}U. This finding has been confirmed by later analyses and by direct measurements of uranium in seawater from various parts of the ocean (summary in Goldberg & Bruland 1974). Open ocean $^{234}U/^{238}U$ values have been estimated, at 95% confidence limits, to be 1.14 ± 0.03; variations shown are comparable to the analytical precisions. The uniformity in isotopic composition as well as concentration of uranium in the ocean (Ku et al 1974), a reflection of uranium's long open ocean residence time, justifies us to assume $(^{234}U/^{238}U)_0$ to be 1.14 or 1.15 for marine precipitates. As ^{234}U decays with a 248,000 year half-life, the $^{234}U/^{238}U$ method would potentially be useful to date marine deposits over the last million years. The success of its applications, however, has mostly been limited to unaltered fossil corals, based on agreement with ages derived from the $^{230}Th/^{234}U$ (to be discussed later) and He/U methods (Thurber et al 1965, Veeh 1966). Studies of fossil marine mollusks suggest (*a*) that a sizable fraction of their uranium comes from the sediment matrix, which prevents a reliable estimate of the initial ^{234}U excess, and (*b*) that in some cases continuous exchange of uranium with its surroundings may have occurred (Kaufman et al 1971). With few exceptions (Ku & Broecker 1967b, Krishnamoorthy et al 1971), Equation 5 is seldom applicable to pelagic deposits. This is because in a typical pelagic sediment (*a*) most uranium is nonauthigenic, and (*b*) part of the ^{234}U is subject to postdepositional migration (Ku 1965). The second statement was reached by a diffusion-model fit of the $^{234}U/^{238}U$ data observed in three red-clay cores (Figure 6). The fraction of the "mobile" ^{234}U may be related to the oxidation (from +4 to +6 valence state) of ^{234}U accompanying radioactive decay (Kolodny & Kaplan 1970), or to the α-recoil range of ^{234}Th in sediments, which is a function of sediment particle size and density (Kigoshi 1971). In any case, because of their recoil nature the in situ–generated ^{234}U atoms do not appear tightly bound. As the number of such atoms increases with depth in the core, upward diffusion of ^{234}U along its concentration gradient would ensue.

Although models such as those proposed by Ku (1965) and by Kolodny & Kaplan (1970) allow parametric sedimentation rates or ages to be assessed for certain deposits, they provide more geochemical insight than geochronological utility. In this context, one should note that $^{234}U/^{238}U$ ratios have been used extensively as an index of geochemical processes operating in uranium deposits, soils, and groundwaters (Dooley et al 1966, Rosholt et al 1966, Osmond et al 1974), and to distinguish uranium sources in metalliferous sediments associated with ocean ridge processes (Ku 1969, Veeh & Boström 1971, Rydell & Bonatti 1973, Rydell et al 1974).

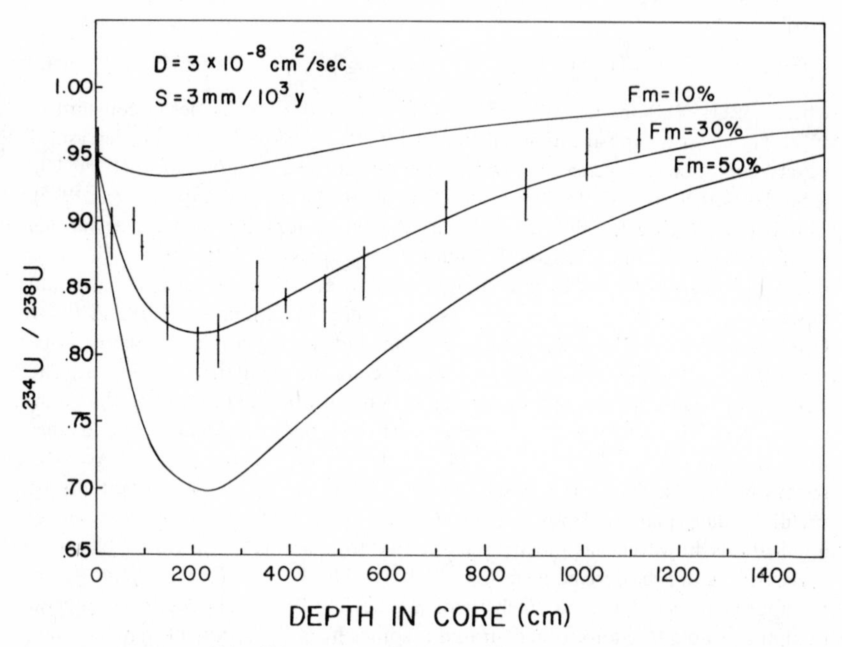

Figure 6 Measured $^{234}U/^{238}U$ ratios (bars) in a red-clay core V10-95 (26°31′N, 51°47′W). The curves refer to the distributional patterns of $^{234}U/^{238}U$ if (*a*) the core has a constant sedimentation rate of 3 mm/10^3 yr, (*b*) a fraction, F_m, of the in situ–produced ^{234}U in sediments is subject to postdepositional diffusion, and (*c*) the diffusion coefficient of this mobile ^{234}U is 3×10^{-8} cm^2/sec Ku (1965).

Chalov et al (1964) proposed the use of $^{234}U/^{238}U$ to date the initiation of closed (no-outflow) lakes. Under the assumption that influx of uranium to the lake is large compared with its removal and bears a constant isotopic composition, $(^{234}U/^{238}U)_0$ as measured in feedwaters, the age of the lake can be calculated from the $^{234}U/^{238}U$ ratio of the present lake water (Chalov et al 1964, see also Komura & Sakanoue 1967):

$$(^{234}U/^{238}U)-1 = [(^{234}U/^{238}U)_0-1][1-\exp(-\lambda_4 t)](\lambda_4 t)^{-1} \quad (6)$$

where λ_4 is the ^{234}U decay constant. Ages of two alpine lakes as determined by this method were consistent with those derived from chloride accumulation and other geological considerations (Chalov et al 1964). Noting that the uranium isotopic composition in the feed streams was more uniform than its concentration over yearly cycles, these workers contended that this method should be more reliable than any constant-influx methods using element concentrations. The constancy of the observed feedwater $^{234}U/^{238}U$ ratios, of course, has to be extrapolated over the entire dating periods of hundreds of thousands of years. The extrapolation involves taking risks.

Other Applications

Because unsupported ^{228}Ra ($t_{1/2} = 6.7$ yr) is present in coastal surface water, its incorporation into coral skeleton is useful for diciphering the chronology of modern corals. The decay of this ^{228}Ra with depth in corals gives mean skeletal growth rates over the last 30 years, assuming that ratios $^{228}Ra/Ca$ or $^{228}Ra/^{226}Ra$ initially incorporated by the corals remain constant in time. Rates thus obtained agreed with the ^{90}Sr and bomb ^{14}C data (Moore et al 1973, Moore & Krishnaswami 1974) and indicated that the density bands as revealed by X-radiography (Knutson et al 1972) are indeed annual (Dodge & Thomson 1974, Moore & Krishnaswami 1974). Similar approach may be attempted using ^{210}Pb. Strong deviations from the ideal $^{210}Pb/^{226}Ra$ decay curve were noted in samples showing relatively smooth $^{228}Ra/^{226}Ra$ decrease (Moore & Krishnaswami 1972, Moore et al 1973), implying a greater variability of ^{210}Pb than the radium isotopes in the coastal surface ocean.

White lead is a pigment in painting and lead metal is an important constituent of other works of art, such as pewter and lead-bearing bronzes. The ^{210}Pb-^{226}Ra disequilibrium engendered by the lead extraction from ores could therefore provide a dating tool in the fields of art authentication and archaeology. Keisch (1968) used excess ^{210}Pb to distinguish eighteenth-century and earlier paintings from forgeries made in the twentieth century.

METHODS BASED ON ACCUMULATION OF DECAY PRODUCTS OF URANIUM

The decay products here refer principally to ^{230}Th and ^{231}Pa. Under this category, most of the work and contributions have so far come from dating carbonate material, chiefly corals, mollusks, and speleothems. This we review at some length.

Dating Carbonates

Uranium is coprecipitated with calcite or aragonite from natural waters that are essentially free of thorium and protactinium. In a fossil carbonate, the amounts of ^{230}Th and ^{231}Pa initially present could be considered negligible in comparison with those subsequently generated by the decay of uranium. If the carbonate remained a closed system, the amounts of ^{230}Th and ^{231}Pa generated therein are functions of time and uranium content. This forms the basis of the two age methods described below.

$^{230}Th/^{234}U$ METHOD Potratz et al (1955) and Barnes et al (1956) first reported the use of ^{230}Th accumulation for dating corals. Concurrently, Cherdyntsev (1956) suggested its application to fossil bones. These and several subsequent workers (e.g. Sackett 1958, Tatsumoto & Goldberg 1959, Blanchard 1963), while exploring the possibility of dating other materials such as oolites and shells, used the relationship:

$$^{230}Th/^{238}U = 1-\exp(-\lambda_0 t) \tag{7}$$

With the discovery of ^{234}U-^{238}U disequilibrium in natural solutions, Equation 7 should be modified to:

$$^{230}Th/^{234}U = (^{238}U/^{234}U)[1-\exp(-\lambda_0 t)] + [1-(^{238}U/^{234}U)][\lambda_0/(\lambda_0-\lambda_4)][1-\exp(\lambda_4-\lambda_0)t] \tag{8}$$

In Equations 7 and 8, λ_0 and λ_4 represent respectively the decay constants of ^{230}Th and ^{234}U; ^{230}Th, ^{234}U, and ^{238}U are measured activities in a sample of t years old. The modification necessitates measurement of $^{234}U/^{238}U$ as well as $^{230}Th/^{234}U$ for each sample (except for corals and oolites). Figure 7 shows a graphic solution for Equation 8 (Kaufman & Broecker 1965). The current techniques permit measurement back up to 350,000 years.

$^{231}Pa/^{235}U$ METHOD The ingrowth of ^{231}Pa from ^{235}U for dating carbonates was proposed by Sackett (1958). The sample age t is calculated from:

$$^{231}Pa/^{235}U = {}^{231}Pa/0.046(^{238}U) = 1-\exp(-\lambda_1 t) \tag{9}$$

where λ_1 is the decay constant of ^{231}Pa. In practice, ^{235}U is usually obtained from measurement of the more abundant ^{238}U, noting a constant activity ratio $^{235}U/^{238}U$ of 0.046 for natural uranium (Hamer & Robbins 1960). This method has a potential range of up to 150,000 years. It has not been as widely used as the $^{230}Th/^{234}U$ method because of the low abundance and analytical tediousness of ^{231}Pa.

EVALUATION OF RESULTS All age-dating methods require that, over the time interval considered, the sample remain closed to gain or loss of nuclides of interest—the closed system assumption. Though implicit in our previous discussion of the decay methods, this assumption has not been singularly stressed. The reason is threefold.

First, notwithstanding inference being made of possible postdepositional migrations of ^{230}Th and ^{210}Pb (Sackett 1965, Koide et al 1972), satisfactory mechanisms or proofs for such migrations have yet to be presented. The diffusion coefficients of ^{234}U and ^{226}Ra in sedimentary columns have been estimated to be of the order of 10^{-8}–10^{-9} cm^2/sec (Ku 1965, Koczy & Bourret 1958, Goldberg & Koide 1963). These estimates should be the upper limits for the less soluble isotopes like ^{230}Th, ^{231}Pa, and ^{210}Pb. Koczy (1965) set diffusion coefficient for ^{230}Th at about 10^{-12} cm^2/sec, which would be small enough to be negligible in age determination of sediments accumulated at rates of millimeters to centimeters per thousand years. Secular equilibrium between ^{230}Th and ^{234}U has been consistently observed in deeper sections of sediment cores (Ku 1965, 1966).

Second, diffusion of ^{234}U and ^{226}Ra deters direct use of these nuclides in age-dating (the $^{234}U/^{238}U$ method and the use of ^{226}Ra as an index of ^{230}Th), but does not affect the use of ^{230}Th and ^{210}Pb because the latter nuclides are plentiful relative to ^{234}U and ^{226}Ra in younger deposits.

Third, the decay methods employ the best-fitting procedure, which is essentially an averaging technique to smooth out perturbations of various kinds, including any small-scale migration of isotopes.

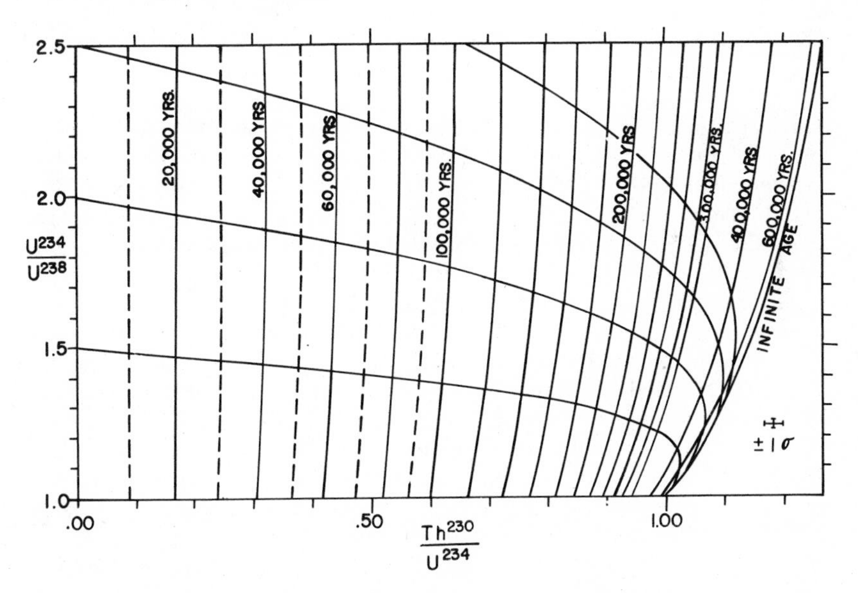

Figure 7 Variation of $^{234}U/^{238}U$ and $^{230}Th/^{234}U$ ratios with time in a closed system free of initial ^{230}Th. The near-vertical lines are isochrons (locus of points for samples of the same age but different initial $^{234}U/^{238}U$ ratios). The horizontal curves denote age paths of samples with initial ratios 1.5, 2.0, and 2.5. Typical error in analysis is also indicated.

For the ingrowth methods discussed here, however, ages are directly related to the parent-daughter ratios, and the closed system requirement becomes rather stringent. Indeed, that stipulation has severely limited the applicability of these methods to certain types of material, like mollusks.

Unaltered fossil corals free of recrystallization or void-filling material have consistently yielded reliable age results. Corals deposit aragonitic skeletons that contain about 3 ppm uranium and negligible thorium and protactinium. Upon death of the organism, the coral skeleton apparently behaves as a closed system until its recrystallization to calcite, during which process preferential loss of U relative to Th and Pa may occur (Sackett 1958). Ample evidence testifies to the reliability of results obtained on the unaltered corals, including (*a*) concordant dates derived from $^{230}Th/^{234}U$, $^{231}Pa/^{235}U$, and $^{234}U/^{238}U$ ratios (assuming an initial 15% excess ^{234}U) in the same sample (e.g. Ku 1968); (*b*) agreement with stratigraphic information (e.g. Mesolella et al 1969); and (*c*) checkout with other geochronometers such as ^{14}C and He/U (e.g. Thurber et al 1965, Fanale & Schaeffer, 1965).

Uranium-series dating of corals associated with former shorelines in different geographic locations have produced valuable information on the chronology of Pleistocene glacioeustatic oscillations and in several places on the rates of tectonic uplift. A number of raised reefs from tectonically active regions have been thoroughly dated, such as Barbados (Broecker et al 1968), New Guinea (Veeh & Chappell 1970, Bloom et al 1974), and the Ryukyu Islands (Sakanoue et al 1967, Konishi et al 1970). Many dates are also available from more stable areas, such as Florida (Osmond et al 1965, Broecker & Thurber 1965), Hawaii (Veeh 1966, Ku et al 1974), the Bahamas (Broecker & Thurber 1965, Neumann & Moore 1975), and several atolls and small islands in the Pacific and Indian Oceans (Thurber et al 1965, Labeyrie et al 1967, Veeh 1966). Chiefly from these studies, a record of the relative sea level changes over the past 150,000 years has emerged (Broecker & van Donk 1970, Steinen et al 1973, Bloom et al 1974). The reef dating appears to have registered major sea level fluctuations with periods of the order of 20,000 years, and the last time the sea stood significantly higher than it does today must have occurred around 115,000–140,000 years ago (centred about 125,000 years ago). As the resolution of the present dating techniques is usually about 10,000–20,000 years (Figure 7), refinement or extension of the sea level record requires further analytical improvements as well as extensive field investigations. The Barbados data have indicated that favorable conditions for coral growth also prevailed 170,000–240,000 years ago (Mesolella et al 1969).

Like corals, the inorganically precipitated marine oolites incorporated a few ppm uranium and negligible thorium in their aragonitic crystals when they formed. Unrecrystallized oolites appear also suitable for U-series dating (Tatsumoto & Goldberg 1959, Osmond et al 1965, Broecker & Thurber 1965), although because of their geological occurrences (often in dunes) they are not ideal markers for former sea stands, unlike reef corals.

Uranium-series dating of speleothems (cave-deposited travertines) has recently been reviewed by Harmon et al (1975). The most common speleothem forms—stalactite, stalagmite, and flowstone—are developed by slow precipitation of $CaCO_3$. Such deposits tend to be macrocrystalline, compact, and not subject to postdepositional leaching. Pure calcite speleothems with well-preserved internal stratigraphy can be reliably dated by the $^{230}Th/^{234}U$ method, provided they contain sufficient uranium (U content may range from 10 ppb to 100 ppm). Internally consistent sequences of $^{230}Th/^{234}U$ ages in a sample were secured, and in a few cases, where comparison among ^{14}C, $^{231}Pa/^{235}U$, and $^{230}Th/^{234}U$ ages was made, the agreement was good (Cherdyntsev et al 1965, Spalding & Mathews 1972, Thompson 1973). However, the prospect for application of $^{234}U/^{238}U$ to dating speleothems is not good, because in most cases the initial excess could not be accurately determined (Duplessy et al 1970, Thompson et al 1975).

Besides answering the intriguing question of speleothem growth rates, dating of cave deposits has great utility in geomorphic studies of karst landforms (e.g. Ford 1973). Perhaps more importantly, it offers a unique means of assessing Pleistocene climatic changes in continental areas when combined with $^{18}O/^{16}O$ and D/H studies. Research in this area, which has just begun, has already yielded promising results (Hendy & Wilson 1968, Duplessy et al 1970, Thompson et al 1974, Schwarcz et al 1975). Two interesting studies on the $^{230}Th/^{234}U$ dating of presently submerged (below sea surface) stalagmites from caves in the Bahamas and Bermuda (Spalding & Mathews 1972, Harmon & Schwarcz 1974) suggested a way to educe heights of the Quaternary low sea stands.

Understandably, the importance of successful dating of molluscan shells, by virtue of their ubiquity and abundance in geologic deposits, cannot be overstated. The subject has indeed commanded much attention. For age dating, the numbers of U-series isotopic analyses done on mollusks probably exceeds that on any other material. Yet, despite these efforts, uncertainties remain about how to distinguish the failures from the successes where ages are not independently known, as noted in a detailed critique by Kaufman et al (1971).

The first glimpse of the complexity in mollusk dating came from the work of Broecker (1963) and Blanchard (1963), who reported that fossil marine mollusks typically have much more uranium than living ones. Fossil shells of various ages show mean values in the vicinity of 1 ppm U, whereas 95% of their modern counterparts contain less than 0.5 ppm U (Kaufman et al 1971). Thus most of the uranium in mollusk is of secondary origin. Uranium content in fossils does not tend to increase with age, however, leading Broecker (1963) to suggest that the uptake of uranium occurs shortly after the death of the organism. Data gathered to date seem to indicate that fixation of most of the uranium, a process probably related to the degradation of organic material, ceases after a few thousand years. This will not cripple the method because its age range of interest often goes beyond 50,000 years. But exceptions to this rule may occur (Kaufman et al 1971), and there is no sure way to sort them out.

The next complication concerns the problem of extraneous ^{230}Th (and ^{231}Pa). Fossil mollusks, unlike their living equivalents and corals, frequently show relatively large amounts of ^{232}Th (Blanchard et al 1967). This is especially true for calcitic shells. The ^{232}Th was apparently acquired from a continental environment in which the shells are often in contact with silicate sedimentary matrix. The question arises as to whether corrections can be made for the "common" ^{230}Th, which is presumably incorporated along with the observed ^{232}Th. The correction should be R^{232}Th exp$(-\lambda_0 t)$, in which R is the ^{230}Th/^{232}Th ratio entering the shell, λ_0 is the decay constant of ^{230}Th, and t is the time elapsed since addition of the common ^{230}Th. R may have a range of values, as reflected by the ^{230}Th/^{232}Th data on natural waters and soils (Szabo 1969, Cherdyntsev 1971). Attempts have been made to estimate R on a regional basis (Kaufman & Broecker 1965, Osmond et al 1970). Even if this could be done in each case, the problem of assigning t in the correction term remains. The age errors incurred from erroneous common ^{230}Th correction have been discussed (Kaufman et al 1971). There is as yet no data base whereupon corrections for extraneous ^{231}Pa (if any) can be formulated. As a working guideline, shells (or corals etc) with ^{230}Th/^{232}Th ratios less than about 20 should be avoided.

The ultimate question lies in whether or not mollusks give reliable uranium-series ages. The answer could be referred to Kaufman et al (1971) who concluded that at least half of the mollusk data compiled show a discordance between ^{230}Th/^{234}U ages and ^{231}Pa/^{235}U ages and between their ^{230}Th/^{234}U ages and age estimates from paleontology, K-Ar, ^{14}C, and coral ^{230}Th/^{234}U data. The picture would change little even if these workers' data compilation included additional and more recent published results on concordancy checks (e.g. Sakanoue et al 1967, Veeh & Valentine 1967, James et al 1971, Szabo & Vedder 1971, Ku & Kern 1974, Szabo & Gard 1975). Aside from the secondary uranium and thorium uptake noted above, the root

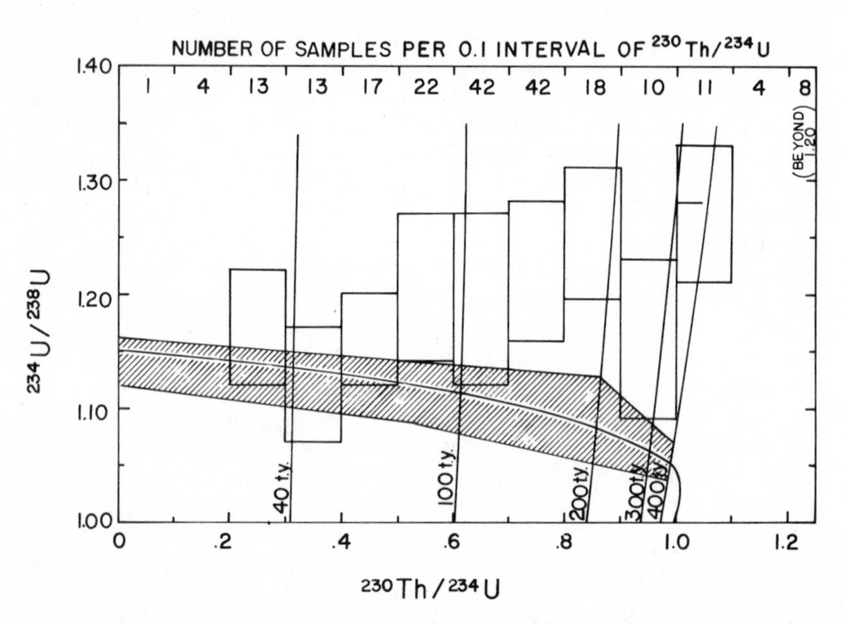

Figure 8 Data on over 200 marine mollusks in a Figure 7 plot. The rectangles are drawn in for each 0.1 interval of ^{230}Th/^{234}U such that they include the median 60% of the mollusk data points for that interval. The shaded area represents the range for corals. Note, whereas corals follow the closed system age path with an initial marine ^{234}U/^{238}U of 1.15, mollusks do not (Kaufman et al (1971).

of the problem can be traced to the observed $^{234}U/^{238}U$ ratios in fossil mollusks of both marine and terrestrial origin: (*a*) they commonly reflect those of the terrestrial environment (i.e., >1.15), and (*b*) contrary to the prediction of Figure 7, they show an increasing trend as $^{230}Th/^{234}U$ increases (Figure 8). Clearly, assimilation or exchange of uranium and possibly its products took place in a continental environment, and at least half the time the process persisted long enough to pose a serious obstacle to age determination.

To correct the effect of isotope migration, an open system model has been constructed (Rosholt 1967, Szabo & Rosholt 1969). The model deals with a certain type of discordant $^{231}Pa/^{235}U$-$^{230}Th/^{234}U$ results and designates a steady-state contamination process with fixed migrational patterns for the isotopes involved. At times the open system dates seem to improve over the closed system ages, but not always. The reality and universality of the model have been questioned (Kaufman et al 1971, Kaufman 1972).

At the present, the best way to distinguish reliable from questionable ages is a concordancy check between $^{230}Th/^{234}U$ and $^{231}Pa/^{235}U$. Without such a check, one places little confidence in isolated dates. Other proposed closed system criteria such as nonrecrystallization, plausible U and Th concentrations, and $^{226}Ra/^{230}Th$ and $^{228}Th/^{232}Th$ ratios (Kaufman & Broecker 1965, Stearns & Thurber 1965, Blanchard et al 1967) are useful but less stringent. There is as yet no effective guideline for selection of suitable samples before analyses. Aside from the preservation appearance, one clue may be that specimens found in carbonate matrices show less signs of radionuclide migration than those found in silicate sediments, as suggested by the results of Stearns & Thurber (1965) and James et al (1971). Although shells of different initial mineralogy (aragonite vs calcite) have different isotopic characteristics, it has not yet been determined which mineralogy is the more suitable (Kaufman et al 1971, Broecker & Bender 1972). Data on the solitary corals and their coexisting aragonitic shells in raised Californian marine terraces (silicate sediments) offer some interesting comparison (Veeh & Valentine 1967, Ku & Kern 1974). The aragonitic corals contain primary marine uranium whereas the shells do not. Also the latter consistently give younger $^{230}Th/^{234}U$ ages than the former. Does this mean that any consideration in mineralogy and sediment matrix is easily overridden by the uranium source problem? Questions and clues such as the foregoing are subjects for future inquiry. Only after much further research into the nature of the disruptive processes at work can one begin attempting to salvage questionable ages via modeling or other measures of a general nature and thereby improve the present dismal success rate of $\leq 50\%$.

Other works on carbonate material include the dating of marls in the Jordan River–Dead Sea Graben (Kaufman 1971) and of the arid zone calcretes (caliche) in the southwestern U.S. (Ku 1975). Both studies involve inorganically precipitated carbonates mixed with some silicate detrital components which require common ^{230}Th corrections. As reflected in their internally consistent $^{230}Th/^{234}U$ ages, materials used in these studies apparently form an undisturbed system. In this regard, the absence of organic matter may play a role. Uranium distribution in molluscan shells is not uniform; higher concentrations are believed to be associated with organic coatings (Lahoud et al 1966, Komura & Sakanoue 1967).

Applications to Other Systems

Following the early work of Cherdyntsev (1956), dating of fossil bones with the uranium-series nuclides has appeared in several recent articles (Szabo et al 1969, 1973; Hansen & Begg 1970; Howell et al 1972; Ward et al 1972; Szabo & Collins 1975). The situation encountered in the bone system resembles that in mollusks. In bones, uranium is largely secondary and shows rather high concentrations (mostly 20–400 ppm, with $^{234}U/^{238}U$ of 1.1–1.5). Thorium contents are usually low—of the order of 0.0 χ ppm. Therefore correction for the extraneous ^{230}Th is hardly required. Cases for open system in fossil bones have been documented by the discordant $^{230}Th/^{234}U$ and $^{231}Pa/^{235}U$ ages. From a survey of the literature, the incidence of such discordancy appears slightly less than 50%. Hence in this respect bones seem to fare better than mollusks; the high uranium content of the former may be a factor. In view of its potential contribution to archaeology, dating of bones deserves continuous efforts. Calibration of uranium-series dates with those of ^{14}C (on collagen) and fission track would be useful.

Application of the ^{230}Th and ^{231}Pa ingrowth methods to the following systems have also yielded meaningful age information: marine phosphorites (Kolodny & Kaplan 1970, Veeh et al 1973, Burnett 1974), shallow-water marine manganese nodules (Ku & Glasby 1972), saline deposits of dry lakes (Goddard 1970), and secondary uranium-bearing minerals (e.g. Allegre 1964).

DATING IGNEOUS ROCKS

Chemical fractionation between different elements of the decay chains also occurs during magmatic processes. In unweathered Quaternary igneous rocks, often radioactive equilibrium exists between ^{234}U and ^{238}U, but not between the uranium parents and their daughters: ^{230}Th, ^{231}Pa, ^{226}Ra, and ^{210}Pb (Anestad-Fruth 1963; Somayajulu et al 1966; Oversby & Gast 1968; Cherdyntsev et al 1967, 1968). Dating of these rocks has been done by utilizing the disequilibrium relationship between ^{230}Th and ^{238}U; the former nuclide can be either in excess or depleted with respect to the latter. As first attempted by Cerrai et al (1965), one assumes that when a magma solidifies, its different minerals or phases may inherit different Th/U (hence $^{230}Th/^{234}U$) but identical $^{230}Th/^{232}Th$ and $^{234}U/^{238}U$ ratios. A given mineral, after having acted as a closed system for t years following crystallization, should have its $^{230}Th/^{232}Th$ ratio as:

$$^{230}Th/^{232}Th = (^{230}Th_0/^{232}Th)\exp(-\lambda_0 t)+(^{238}U/^{232}Th)[1-\exp(-\lambda_0 t)] \quad (10)$$

where $^{230}Th_0$ is the initial ^{230}Th activity and λ_0 the decay constant of ^{230}Th. In this equation, there are two unknowns: $^{230}Th_0/^{232}Th$ and t. Thus the age can be uniquely determined from two or more minerals of common origin. For instance, for a pair of minerals, A and B, from the same rock,

$$t = (1/\lambda_0)\ln[1-(i_A-i_B)/(u_A-u_B)]^{-1} \quad (11)$$

where i stands for $^{230}Th/^{232}Th$ and u stands for $^{238}U/^{232}Th$, with subscripts A

and B refering to minerals A and B respectively. It can be seen from this equation that the precision in estimating t depends heavily on the difference in the U-Th fractionation between the two minerals, as denoted by the term $u_A - u_B$.

The method has been applied in the studies of various acidic rocks, including granite, lavas, pyroclastics, and pumice (Kigoshi 1967, Taddeucci et al 1967, Discendenti et al 1970, Fukuoka 1974). These studies have cited general agreement of the ^{230}Th dates with independent estimates from K-Ar, ^{14}C, fission track, or stratigraphic evidences. Noteworthy is that the ^{230}Th ages thus derived refer to the time of crystallization of minerals, which could take place in the magma chamber, and they are upper limits for the eruption ages.

A graphic presentation of Equation 10 has been used (Kigoshi 1967, Allegre 1968). As shown in Figure 9, minerals of the same formation age from a rock should fall on a single straight line (isochron) with slope $1 - \exp(-\lambda_0 t)$ and y-intercept

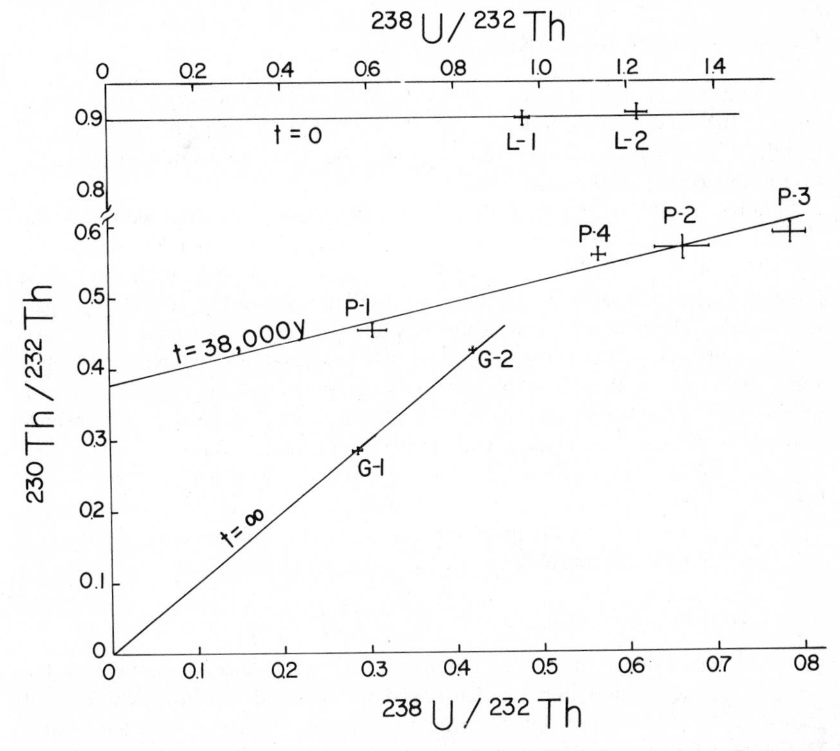

Figure 9 Equation 10 plots for separate fractions isolated from three rocks: a Cretaceous granite (G), a 1783 A.D. lava (L), and a pumice (P) whose age has been inferred at 35,700 years through ^{14}C dating. Data are from Kigoshi (1967) in which the separated fractions were obtained by partial dissolution or extraction. These chemical procedures may involve hot-atom effect. A better procedure would be to use physically separated mineral phases, as in Taddeucci et al (1967).

$({}^{230}Th_0/{}^{232}Th) \exp(-\lambda_0 t)$. Those not plotted on a defined isochron are products of different crystallization time and melt $({}^{230}Th/{}^{232}Th)$ composition domain. Adopting such concepts, Allegre (1968) suggested significant time differentiation between mineral formation at depth and eruption (by noting the glass phase plotted below the mineral isochron), and Fukuoka & Kigoshi (1974) identified the presence of relic zircon grains (by noting certain zircons plotted above the isochron). On the other hand, Kuptsov et al (1969) held that volcanic rocks may involve multiple (more than two) generations of minerals that would defy treatment by the method discussed here. In such cases, these workers suggested the use of Equations 7 and 9 to obtain maximum ages for those minerals showing deficient ${}^{230}Th$ and ${}^{231}Pa$.

Studies of young igneous rocks with the U-series isotopes have apparently just begun. Much more data are needed before their potential contributions to elucidating igneous processes as well as to age dating can be fully appreciated.

CONCLUDING REMARKS

The unique aspect of the uranium-series disequilibrium methods is their diversity in terms of time and material to which they apply, and of the scope of problems they handle. Also, reliability checks can sometimes be made by examining internally several related parent-daughter pairs.

Although Joly (1908) reported the ${}^{226}Ra$ enrichment in deep-sea sediments and manganese nodules retrieved from the Challenger Expedition (1872–1876)—the first observation of the effect of "disequilibrium"—as early as 1908, utilization of the fractionations produced in nature's laboratory did not progress rapidly until the past two decades. The field, still comparatively young and expanding, offers challenges. New datable material is being discovered and new relationships explored. While some areas are in the midst of seeming vacillations pending strokes of fresh approaches, some are so nascent in their developmental stage that the obstacle to progress is simply lack of data. The speculation and controversy in an expanding field such as this can unsettle those who are mainly interested in applying the methods or their results. If this review helps these consumers and those contemplating entry to realize the status, the problems, and some of the prospects of the field, it will more than serve its purpose.

ACKNOWLEDGMENTS

I wish to acknowledge the Oceanography and the Earth Science Sections of the National Science Foundation for their support of my studies in this field (current grants DES72-01557 and DES72-01661). I thank Mu-ching Lin and Carol McClenning for their assistance in the preparation of the manuscript.

Literature Cited

Allegre, C. J. 1964. De l'extension de la methode de calcul graphique Concordia aux mesures d'ages absolus effectues a l'aide du desequilibre radioactif. Cas des mineralisations secondaires d'uranium. *C. R. Acad. Sci. Paris* 259:4086–89

Allegre, C. J. 1968. ^{230}Th dating of volcanic rocks: a comment. *Earth Planet. Sci. Lett.* 5:209–10

Almodovar, I. 1960. *Thorium isotopes method for dating marine sediments.* PhD thesis. Carnegie Inst. Technol., Pittsburgh. 112 pp.

Amin, B. S. 1970. *Dating of ocean sediments by radioactive methods.* MS thesis. Univ. Bombay. 100 pp.

Anestad-Fruth, E. 1963. *Uranium series disequilibrium in recent volcanic rocks.* MA thesis. Columbia Univ., New York. 22 pp.

Baranov, V. I., Kuz'mina, L. A. 1958. The rate of silt deposition in the Indian Ocean. *Geochemistry* 2:131–40

Barnes, J. W., Lang, E. J., Potratz, H. A. 1956. Ratio of ionium to uranium in coral limestone. *Science* 124:175–76

Bernat, M. 1969. Utilisation des methodes basées sur le desequilibre radioactif dans la geologic du Quaternaire. *Cah. ORSTOM, Ser. Geol. I,* 2:3–27

Bhandari, N., Bhat, S. G., Krishnaswami, S., Lal, D. 1971. A rapid beta-gamma coincidence technique for determination of natural radionuclides in marine deposits. *Earth Planet. Sci. Lett.* 11:121–26

Blanchard, R. L. 1963. *Uranium decay series disequilibrium in age determination of marine calcium carbonates.* PhD thesis. Washington Univ., St. Louis. 175 pp.

Blanchard, R. L., Cheng, M. H., Potratz, H. A. 1967. Uranium and thorium series disequilibria in recent and fossil marine molluscan shells. *J. Geophys. Res.* 72: 4745–57

Bloom, A. L., Broecker, W. S., Chappell, J. M. A., Matthews, R. K., Mesolella, K. J. 1974. Quaternary sea level fluctuations on a tectonic coast: new ^{230}Th/^{234}U dates from the Huon Peninsula, New Guinea. *Quat. Res.* 4:185–205

Broecker, W. S. 1963. A preliminary evalua tion of uranium series inequilibrium as a tool for absolute age measurement on marine carbonates. *J. Geophys. Res.* 68: 2817–34

Broecker, W. S. 1965. Isotope geochemistry and the Pleistocene climatic record. In *The Quaternary of the United States,* ed. H. E. Wright Jr., D. G. Frey, pp. 737–53. Princeton, NJ: Princeton Univ. Press

Broecker, W. S., Bender, M. L. 1972. Age determination on marine strandlines. In *Calibration of Hominoid Evolution,* ed. W. W. Bishop, J. A. Miller, pp. 19–35. New York: Scottish Acad. Press

Broecker, W. S., Ku, T. L. 1969. Caribbean cores P6304-8 and P6304-9: new analysis of absolute chronology. *Science* 166:404–6

Broecker, W. S., Thurber, D. L. 1965. Uranium-series dating of corals and oolites from Bahaman and Florida Key limestones. *Science* 149:58–60

Broecker, W. S., Thurber, D. L., Goddard, J., Ku, T. L., Matthews, R. K., Mesolella, K. J. 1968. Milankovich hypothesis supported by precise dating of coral reefs and deep-sea sediments. *Science* 159:297–300

Broecker, W. S., van Donk, J. 1970. Insolation changes, ice volumes, and the O^{18} record in deep-sea cores. *Rev. Geophys. Space Phys.* 8:169–98

Bruland, K. W. 1974. *Pb-210 geochronology in the coastal marine environment.* PhD thesis. Univ. Calif. San Diego, La Jolla, Calif. 106 pp.

Bruland, K. W., Bertine, K., Koide, M., Goldberg, E. D. 1974a. History of metal pollution in southern California coastal zone. *Environ. Sci. Technol.* 8:425–32

Bruland, K. W., Koide, M., Goldberg, E. D. 1974b. The comparative marine geochemistries of lead 210 and radium 226. *J. Geophys. Res.* 79:3083–86

Burnett, W. C. 1974. *Phosphorite deposits from the sea floor off Peru and Chile: radiochemical and geochemical investigations concerning their origin.* PhD thesis. Univ. Hawaii, Honolulu. 164 pp.

Cerrai, E., Dugnani Lonati, R., Gazzarini, F., Tongiorgi, E. 1965. Il methodo ionio-uranio per la determinazione dell'eta dei minerali vulcanici recenti. *Rend. Soc. Mineral. Ital.* 21:47–62

Chalov, P. I., Tuzova, T. V., Musin, Y. A. 1964. The U^{234}/U^{238} ratio in natural waters and its use in nuclear geochronology. *Geochem. Int.* 3:402–8

Cherdyntsev, V. V. (with P. I. Chalov et al) 1955. Ob izotopnom sostave radio elementov-V privodnykh obyektakh V sviazi S voprosami geokhronologii. In *Trudy III Sessii Komissi Opredeleniyu Absolyutnogo Yozrastu,* pp. 175–233. Moscow: Izd. Akad. Nauk SSSR

Cherdyntsev, V. V. 1956. Determination of the absolute age of the Paleolithic. *Sov. Arkheol.* 25:64–86

Cherdyntsev, V. V. 1971. *Uranium-234.* Jerusalem: Israel Program for Scientific Translation. 234 pp.

Cherdyntsev, V. V., Kazachevskii, I. V., Kuz'mina, E. A. 1963. Isotopic composition of uranium and thorium in the supergene zone. *Geochemistry* 3:271–83

Cherdyntsev, V. V., Kazachevskii, I. V., Kuz'mina, E. A. 1965. Dating of Pleistocene carbonate formations by the thorium and uranium isotopes. *Geochem. Int.* 794–801

Cherdyntsev, V. V., Kislitsina, G. I., Kuptsov,

V. M., Kuz'mina, E. A., Zverev, V. L. 1967. Radioactivity and absolute age of young volcanic rocks. *Geochem. Int.* 4: 639–46

Cherdyntsev, V. V., Kuptsov, V. M., Kuz'mina, E. A., Zverev, V. L. 1968. Radioisotopes and protactinium age of neovolcanic rocks of the Caucasus. *Geochem. Int.* 5:56–64

Cherdyntsev, V. V., Orlov, D. P., Isabaev, E. A., Ivanov, V. I. 1961. Isotopic composition of uranium in minerals. *Geochemistry* 10:927–36

Craig, H., Krishnaswami, S., Somayajulu, B. L. K. 1973. ^{210}Pb-^{226}Ra: radioactive disequilibrium in the deep sea. *Earth Planet. Sci. Lett.* 17:295–305

Crozaz, G., Picciotto, E., De Brueck, W. 1964. Antarctic snow chronology with Pb-210. *J. Geophys. Res.* 69:2597

Discendenti, A., Nicoletti, M., Taddeucci, A. 1970. Datazione K-Ar e ^{230}Th di alcuni prodotti vel Vulcano di Latera (Monti Vulsini). *Period. Mineral.* (*Roma*) 39:461–68

Dodge, R. E., Thomson, J. 1974. The natural radiochemical and growth records in contemporary hermatypic corals from the Atlantic and Caribbean. *Earth Planet. Sci. Lett.* 23:313–22

Dooley, J. R., Granger, H. C., Rosholt, J. N. 1966. Uranium-234 fractionation in the sandstone type uranium deposits of the Ambrosia Lake District, New Mexico. *Econ. Geol.* 61:1362–82

Duplessy, J. C., Labeyrie, J., Lalou, C., Nguyen, H. V. 1970. Continental climatic variations between 130,000 and 90,000 years B.P. *Nature* 226:631–32

Dymond, J. 1969. Age determinations of deep-sea sediments: a comparison of three methods. *Earth Planet. Sci. Lett.* 6:9–14

Emiliani, C., Rona, E. 1969. Caribbean cores P6304-8 and P6304-9: new analysis of absolute chronology. A reply. *Science* 166: 1551–52

Fanale, F. P., Schaeffer, O. A. 1965. The helium-uranium method for dating marine carbonates. *Science* 149:312–17

Ford, D. C. 1973. Development of the canyons of the South Nahanni River, N.W.T. *Can. J. Earth Sci.* 10:366–78

Francis, C. W., Chester, G., Haskin, L. A. 1970. The determination of Pb-210 mean residence time in the atmosphere. *Environ. Sci. Technol.* 4:586–89

Fukuoka, T. 1974. Ionium dating of acidic volcanic rocks. *Geochem. J.* 8:109–16

Fukuoka, T., Kigoshi, K. 1974. Discordant Io-ages and the uranium and thorium distribution between zircon and host rocks. *Geochem. J.* 8:117–22

Goddard, J. G. 1970. *Th^{230}/U^{234} dating of saline deposits from Seales Lake, California.* MS thesis. Queens College, New York. 50 pp.

Goldberg, E. D. 1963. Geochronology with Pb-210. In *Radioactive Dating,* pp. 121–31. Vienna: IAEA

Goldberg, E. D. 1968. Ionium/thorium geochronologies. *Earth Planet. Sci. Lett.* 4: 17–21

Goldberg, E. D., Bruland, K. 1974. Radioactive geochronologies. *The Sea,* ed. E. D. Goldberg, 5:451–89. New York: Wiley-Interscience

Goldberg, E. D., Griffin, J. 1964. Sedimentation rates and mineralogy in the South Atlantic. *J. Geophys. Res.* 69:4293–4309

Goldberg, E. D., Koide, M. 1958. Ionium-thorium chronology in deep-sea sediments of the Pacific. *Science* 128:1003

Goldberg, E. D., Koide, M. 1962. Geochronological studies of deep-sea sediments by the ionium-thorium method. *Geochim. Cosmochim. Acta* 26:417–43

Goldberg, E. D., Koide, M. 1963. Rates of sediment accumulation in the Indian Ocean. In *Earth Science and Meteoritics,* ed. J. Geiss, E. D. Goldberg, pp. 90–102. Amsterdam: North-Holland

Goldberg, E. D., Koide, M., Griffin, J. 1964. A geochronological and sedimentary profile across the North Atlantic Ocean. *Isotopic and Cosmic Chemistry,* ed. H. Craig, S. L. Miller, G. J. Wasserburg, pp. 211–32. Amsterdam: North-Holland

Hamer, A. N., Robbins, E. J. 1960. A search for variations in the natural abundance of uranium-235. *Geochim. Cosmochim. Acta* 19:143–45

Hansen, R. O., Begg, E. L. 1970. Age of Quaternary sediments and soils in the Sacramento area, California, by uranium and actinium series dating of vertebrate fossils. *Earth Planet. Sci. Lett.* 8:411–19

Harmon, R. S., Schwarcz, H. P. 1974. Late Pleistocene history of Bermuda as recorded by presently submerged speleothems. *Geol. Soc. Am. Abstr.* 6:778

Harmon, R. S., Thompson, P., Schwarcz, H. P., Ford, D. C. 1975. Uranium-series dating of speleothems. *Nat. Speleol. Soc. Bull.* 37:21–33

Hendy, C. H., Wilson, A. T. 1968. Paleoclimatic data from speleothems. *Nature* 219:48–51

Holmes, C. W., Osmond, J. K., Goodell, H. G. 1968. The geochronology of foraminiferal ooze deposits in the "Southern Ocean." *Earth Planet. Sci. Lett.* 4:368–74

Howell, F. C., Cole, G. H., Kleindienst, M. R.,

Szabo, B. J., Oakley, K. P. 1972. Uranium-series dating of bone from the Isimila prehistoric site, Tanzania. *Nature* 237 : 51–52

Imai, T., Sakanoue, M. 1973. Content of plutonium, thorium and protactinium in sea water and recent coral in the north Pacific. *J. Oceanogr. Soc. Jpn* 29 : 76–82

Isaac, N., Picciotto, E. 1953. Ionium determination in deep-sea sediments. *Nature* 171 : 742–43

James, N. P., Mountjoy, E. W., Omura, A. 1971. An early Wisconsin reef terrace at Barbados, West Indies, and its climatic implications. *Bull. Geol. Soc. Am.* 82 : 2011–18

Joly, J. J. 1908. On the radium content of deep-sea sediments. *Philos. Mag.* 16 : 190–97

Kaufman, A. 1971. U-series dating of Dead Sea Basin carbonates. *Geochem. Cosmochim. Acta* 35 : 1269–81

Kaufman, A. 1972. Comments on the U-series dating of molluscs from southern California. *Earth Planet. Sci. Lett.* 14 : 447–50

Kaufman, A., Broecker, W. S. 1965. Comparison of Th^{230} and C^{14} ages for carbonate materials from Lakes Lahontan and Bonneville. *J. Geophys. Res.* 70 : 4039–54

Kaufman, A., Broecker, W. S., Ku, T. L., Thurber, D. L. 1971. The status of U-series methods of mollusk dating. *Geochim. Cosmochim. Acta* 35 : 1155–83

Keisch, B. 1968. Dating works of art through their natural radioactivity : improvements and applications. *Science* 160 : 413–16

Kigoshi, K. 1967. Ionium dating of igneous rocks. *Science* 156 : 932–34

Kigoshi, K. 1971. Alpha-recoil thorium-234 : dissolution into water and the uranium-234/uranium-238 disequilibrium in nature *Science* 173 : 47–48

Knauss, K. G., Ku, T. L., Moore, W. S. 1975. Radium and thorium isotopes in surface waters of the eastern equatorial Pacific. *Fall Ann. Meet. Am. Geophys. Union, San Francisco, 1975* (Abstr.)

Knutson, D. W., Buddemeier, R. W., Smith, S. V. 1972. Coral chronometers : seasonal growth bands in reef corals. *Science* 177 : 270–72

Koczy, F. F. 1949. Thorium in sea water and marine sediments. *Geol. Foren. Stockholm Forhandl.* 71 : 238–42

Koczy, F. F. 1954. Geochemical balance in hydrosphere. In *Nuclear Geology,* ed. H. Faul, pp. 120–27. New York : Wiley

Koczy, F. F. 1958. Natural radium as a tracer in the ocean. *Proc. U.N. Int. Conf. Peaceful Uses of Atomic Energy, 2nd* 18 : 351–57

Koczy, F. F. 1961. Ratio of thorium-230 to thorium-232 in deep-sea sediments. *Science* 133 : 978–79

Koczy, F. F. 1963. Age determination in sediments by natural radioactivity. *The Sea,* ed. M. N. Hill, 3 : 816–31. New York : Wiley-Interscience

Koczy, F. F. 1965. Remarks on age determination in deep-sea sediments. *Prog. Oceanogr.* 3 : 155–71

Koczy, F. F., Bourret, R. 1958. *Radioactive Nuclides in Ocean Water and Sediments.* Prog. Rep., Mar. Lab., Univ. Miami, Florida

Koczy, F. F., Picciotto, E., Poulaert, G., Wilgain, S. 1957. Mesure des isotopes du thorium dans l'eau de mar. *Geochim. Cosmochim. Acta* 11 : 103–29

Koide, M., Bruland, K. W., Goldberg, E. D. 1973. Th-228/Th-232 and Pb-210 geochronologies in marine and lake sediments. *Geochim. Cosmochim. Acta* 37 : 1171–87

Koide, M., Soutar, A., Goldberg, E. D. 1972. Marine geochronology with ^{210}Pb. *Earth Planet. Sci. Lett.* 14 : 442–46

Kolodny, Y., Kaplan, I. R. 1970. Uranium isotopes in sea-floor phosphorites. *Geochim. Cosmochim. Acta* 34 : 3–24

Komura, K., Sakanoue, M. 1967. Studies on the dating methods for Quaternary samples by natural alpha-radioactive nuclides. *Sci. Rep. Kanazawa Univ.* 12 : 21–66

Konishi, K., Schlanger, S. O., Omura, A. 1970. Neotectonic rates in the central Ryukyu Islands derived from ^{230}Th coral ages. *Mar. Geol.* 9 : 225–40

Krishnamoorthy, T. M., Sastry, V. N., Sarma, T. P. 1971. U-234/U-238 as a tool for dating marine sediments. *Cur. Sci.* 40 : 279–81

Krishnaswami, S., Lal, D., Martin, J. M., Meybeck, M. 1971. Geochronology of lake sediments. *Earth Planet. Sci. Lett.* 11 : 407–14

Krishnaswami, S., Moore, W. S. 1973. Accretion rates of fresh-water manganese deposits. *Nature* 243 : 114–16

Kröll, V. St. 1953. Vertical distribution of radium in deep-sea sediments. *Nature* 171 : 742

Kröll, V. St. 1954. On the age determination in deep-sea sediments by radium measurements. *Deep-Sea Res.* 1 : 211–15

Ku, T. L. 1965. An evaluation of the U^{234}/U^{238} method as a tool for dating pelagic sediments. *J. Geophys. Res.* 70 : 3457–74

Ku, T. L. 1966. *Uranium series disequilibrium in deep-sea sediments.* PhD thesis. Columbia Univ., New York. 157 pp.

Ku, T. L. 1968. Protactinium-231 method of dating coral from Barbados Island. *J. Geophys. Res.* 73:2271–76

Ku, T. L. 1969. Uranium series isotopes in sediments from the Red Sea hot-brine area. In *Hot Brines and Recent Heavy Metal Deposits in the Red Sea*, ed. E. T. Degens, D. A. Ross, pp. 521–24. New York: Springer-Verlag

Ku, T. L. 1975. *^{230}Th/^{234}U Dating of Desert Caliche.* Presented at Geol. Soc. Am. Penrose Conf., June 8–13, Mammoth Lakes, Calif.

Ku, T. L. 1976. Rates of manganese accretion. In *Marine Manganese Deposits*, ed. G. P. Glasby. Amsterdam: Elsevier. In press

Ku, T. L., Bischoff, J. L., Boersma, A. 1972. Age studies of Mid-Atlantic Ridge sediments near 42°N and 20°N. *Deep-Sea Res.* 19:233–47

Ku, T. L., Broecker, W. S. 1966. Atlantic deep-sea stratigraphy: extension of absolute chronology to 320,000 years. *Science* 151:448–50

Ku, T. L., Broecker, W. S. 1967a. Rates of sedimentation in the Arctic Ocean. *Prog. Oceanogr.* 4:95–104

Ku, T. L., Broecker, W. S. 1967b. Uranium, thorium and protactinium in a manganese nodule. *Earth Planet. Sci. Lett.* 2:317–20

Ku, T. L., Broecker, W. S. 1969. Radiochemical studies on manganese nodules of deep-sea origin. *Deep-Sea Res.* 16:625–37

Ku, T. L., Broecker, W. S., Opdyke, N. 1968. Comparison of sedimentation rates measured by paleomagnetic and the ionium methods of age determination. *Earth Planet. Sci. Lett.* 4:1–16

Ku, T. L., Glasby, G. P. 1972. Radiometric evidence for the rapid growth rate of shallow-water continental margin manganese nodules. *Geochim. Cosmochim. Acta* 36:699–703

Ku, T. L., Kern, J. P. 1974. Uranium-series age of the upper Pleistocene Nester Terrace, San Diego, California. *Bull. Geol. Soc. Am.* 85:1713–16

Ku, T. L., Kimmel, M. A., Easton, W. H., O'Neil, T. J. 1974. Eustatic sea level 120,000 years ago on Oahu, Hawaii. *Science* 183:959–62

Ku, T. L., Knauss, K. G., Mathieu, G. G. 1974. Uranium in open ocean: concentration and isotopic composition. *EOS Tran. Am. Geophys. Union* 55:314

Kuptsov, V. M., Cherdyntsev, V. V., Kuz'mina, E. A., Sulerzhitskii, L. D. 1969. Ionium ages and the process of formation of volcanic rocks. *Geochem. Int.* 6:671–75

Labeyrie, J., Lalou, C., Delibrias, G. 1967. Étude des transgressions marines sur un atoll du Pacifique par les methodes du carbone-14 et du rapport uranium-234/thorium-230. In *Radioactive Dating and Methods of Low-level Counting*, pp. 349–58. Vienna: IAEA

Lahoud, J. A., Miller, D. S., Friedman, G. M. 1966. Relationship between depositional environment and uranium concentrations of molluskan shells. *J. Sediment. Petrol.* 36:541–47

McKelvey, V. E., Everhart, D. L., Garrels, R. M. 1955. Origin of uranium deposits. *Economic Geology*, 15th Anniv. Vol., Part I, ed. A. M. Bateman, pp. 464–533. Lancaster, Penn: Lancaster Press

Mesolella, K. J., Matthews, R. K., Broecker, W. S., Thurber, D. L. 1969. The astronomical theory of climatic change: Barbados data. *J. Geol.* 77:250–74

Miyake, Y., Sugimura, Y. 1961. Ionium-thorium chronology of deep-sea sediments of the western Pacific Ocean. *Science* 133: 1823

Miyake, Y., Sugimura, Y., Yasujima, T. 1970. Thorium concentration and the activity ratios ^{230}Th/^{232}Th and ^{228}Th/^{232}Th in sea water in the western north Pacific. *J. Oceanogr. Soc. Jpn* 26:130–26

Moore, W. S. 1969. Oceanic concentrations of ^{228}Radium. *Earth Planet. Sci. Lett.* 6: 437–66

Moore, W. S., Krishnaswami, S. 1972. Coral growth rates using Ra-228 and Pb-210. *Earth Planet. Sci. Lett.* 15:187–90

Moore, W. S., Krishnaswami, S. 1974. Correlation of x-radiography revealed banding in corals with radiometric growth rates. *Proc. Int. Symp. Coral Reefs, 2nd*, ed. Great Barrier Reef Committee, Brisbane, 2:269–76

Moore, W. S., Krishnaswami, S., Bhat, S. G. 1973. Radiometric determinations of coral growth rates. *Bull. Mar. Sci.* 23:157–76

Moore, W. S., Sackett, W. M. 1964. Uranium and thorium series inequilibrium in sea water. *J. Geophys. Res.* 69:5401–5

Murray, S. M. 1975. *Accumulation rates of sediments and metals off southern California as determined by Pb^{210} method.* PhD thesis. Univ. South. Calif., Los Angeles. 146 pp.

Murray, S. M., Ku, T. L. 1974. Trace metal accumulation in recent sediments off southern California. *Geol. Soc. Am. Abstr.* 6:884

Neumann, A. C., Moore, W. S. 1975. Sea level events and Pleistocene coral ages in the northern Bahamas. *Quat. Res.* 5:215–24

Nikolayev, P. S., Lazarev, K. F., Grashchenko,

S. M. 1962. The concentration of thorium isotopes in the water of the sea of Azov. *Dokl. Akad. Nauk. SSSR* 138:489–90

Osmond, J. K., Carpenter, J. R., Windom, H. L. 1965. Th^{230}/U^{234} age of the Pleistocene corals and oolites of Florida. *J. Geophys. Res.* 70:1843–47

Osmond, J. K., Kaufman, M. I., Cowart, J. B. 1974. Mixing volume calculations, sources and aging trends of Floridan aquifer water by uranium isotope methods. *Geochim. Cosmochim. Acta* 38:1083–1100

Osmond, J. K., May, J. P., Tanner, W. F. 1970. Age of the Cape Kennedy barrier-and-lagoon complex. *J. Geophys. Res.* 75: 469–79

Osmond, J. K., Pollard, L. D. 1967. Sedimentation rate determination in deep sea cores by gamma-ray spectrometry. *Earth Planet. Sci. Lett.* 3:476–80

Oversby, V. M., Gast, P. W. 1968. Lead isotope compositions and uranium decay series disequilibrium in recent volcanic rocks. *Earth Planet. Sci. Lett.* 5:199–206

Pettersson, H. 1937. Der Verhältnis Thorium zu Uran in den Gesteinen und un Meer. *Anz. Akad. Wiss. Wien, Math-Naturwiss. Kl.*: 127–28

Picciotto, E., Wilgain, S. 1954. Thorium determination in deep-sea sediments. *Nature* 173:632–33

Peirson, D. H., Cambray, R. S., Spicer, G. S. 1966. Lead-210, bismuth-210, and polonium-210 in the atmosphere: accurate ratio measurement and application to aerosol residence time determination. *J. Geophys. Res.* 77:6515–27

Piggot, C. S., Urry, W. D. 1939. The radium content of an ocean bottom core. *J. Wash. Acad. Sci.* 29:405–15

Piggot, C. S., Urry, W. D. 1942. Time relations in ocean sediments. *Bull. Geol. Soc. Am.* 53:1187–1210

Potratz, H. A., Barnes, J. W., Lang, E. J. 1955. A radiochemical procedure for thorium and its application to the determination of ionium in coral limestone. *Los Alamos Sci. Lab. Pub. LA-1845*. 19 pp.

Rama, Koide, M., Goldberg, E. D. 1961. Lead-210 in natural waters. *Science* 134: 98–99

Robbins, J. A., Edgington, D. N. 1975. Determination of recent sedimentation rates in Lake Michigan using Pb-210 and Cs-137. *Geochim. Cosmochim. Acta* 39: 285–304

Rosholt, J. N. 1967. Open system model for uranium-series dating of Pleistocene samples. In *Radioactive Dating and Methods of Low-Level Counting*, pp. 299–311. Vienna: IAEA

Rosholt, J. N., Doe, B. R., Tatsumoto, M. 1966. Evolution of the isotopic composition of uranium and thorium in soil profiles. *Geol. Soc. Am. Bull.* 77:987–1003

Rosholt, J. N., Emiliani, C., Geiss, J. Koczy, F. F., Wangersky, P. J. 1961. Absolute dating of deep-sea cores by the Pa^{231}/Th^{230} method. *J. Geol.* 69:162–85

Rosholt, J. N., Emiliani, C., Geiss, J., Koczy, F. F., Wangersky, P. J. 1962. Pa^{231}/Th^{230} dating and O^{18}/O^{16} temperature analysis of core A254-BR-C. *J. Geophys. Res.* 67: 2907–11

Rosholt, J. N., Shields, W. R., Garner, E. L. 1963. Isotopic fractionation of uranium in sandstone. *Science* 139:224–26

Rydell, H. S., Bonatti, E. 1973. Uranium in submarine metalliferous deposits. *Geochim. Cosmochim. Acta* 37:2557–65

Rydell, H. S., Kraemer, T., Boström, K., Joensuu, O. 1974. Post-depositional injections of uranium-rich solutions into East Pacific Rise Sediments. *Mar. Geol.* 17:151–64

Sackett, W. M. 1958. *Ionium-uranium ratios in marine deposited calcium carbonates and related materials.* PhD thesis. Washington Univ., St. Louis. 106 pp.

Sackett, W. M. 1960. The protactinium-231 content of ocean water and sediments. *Science* 132:1761–62

Sackett, W. M. 1964. Measured deposition rates of marine sediments and implications for accumulation rates of extraterrestrial dust. *Ann. N.Y. Acad. Sci.* 119:339–46

Sackett, W. M. 1965. Deposition rates by the protactinium method. *Symposium on Marine Geochemistry,* ed. D. R. Schink, J. T. Corless, pp. 29–40. Univ. Rhode Island

Sackett, W. M. 1966. Manganese nodules: thorium-230: protactinium-231 ratios. *Science* 154:646–47

Sakanoue, M., Konishi, K., Komura, K. 1967. Stepwise determinations of thorium, protactinium and uranium isotopes and their applications in geochronological studies. In *Radioactive Dating and Methods of Low-Level Counting,* pp. 313–29. Vienna: IAEA

Sarma, T. P. 1964. *Dating of marine sediments by ionium and protactinium methods.* PhD thesis. Carnegie Inst. Technol., Pittsburgh. 107 pp.

Schornick, J. C. Jr. 1972. *Uranium and thorium isotope geochemistry in ferromanganese concretions from the southern ocean.* PhD thesis. Florida State Univ., Tallahassee. 161 pp.

Schwarcz, H. P., Harmon, R. S., Ford, D. C. 1975. North American climate during the last 200,000 years inferred from stable isotope studies of speleothem. *Geol. Soc. Am. Abstr.* 7:851

Scott, M. R. 1968. Thorium and uranium concentrations and isotope ratios in river sediments. *Earth Planet. Sci. Lett.* 4:245–52

Scott, M. R., Osmond, J. K., Cochran, J. K. 1972. Sedimentation rates and sediment chemistry in the southern Indian Basin. In *Antarctic Oceanology II: The Australian-New Zealand Sector. Antarctic Res. Ser.*, ed. D. E. Hayes, 19:317–34. Washington DC: AGU

Shackleton, N. J., Opdyke, N. D. 1973. Oxygen isotope and paleomagnetic stratigraphy of equatorial Pacific core V28-238: oxygen isotope temperatures and ice volumes on a 10^5 and 10^6 year scale. *Quat. Res.* 3:39–55

Somayajulu, B. L. K., Goldberg, E. D. 1966. Thorium and uranium isotopes in sea water and sediments. *Earth Planet. Sci. Lett.* 1:102–6

Somayajulu, B. L. K., Tatsumoto, M., Rosholt, J. N., Knight, R. J. 1966. Disequilibrium of the ^{238}U series in basalt. *Earth Planet. Sci. Lett.* 1:387–91

Spalding, R. F., Mathews, T. D. 1972. Stalagmites from caves in the Bahamas: indicators of low sea level stand. *Quat. Res.* 2:470–72

Starik, I. E., Kolyadin, L. B. 1957. The occurrence of uranium in ocean water. *Geochemistry* 3:245–56

Stearns, C. E., Thurber, D. L. 1965. Th^{230}/U^{234} dates of late Pleistocene marine fossils from the Mediterranean and Morrocan littorals. *Quaternaria* 7:29–42

Steinen, R. P., Harrison, R. S., Matthews, R. K. 1973. Eustatic low stand of sea level between 125,000 and 105,000 B.P.: evidence from the sub-surface of Barbados, West Indies. *Bull. Geol. Soc. Am.* 84:63–70

Szabo, B. J. 1969. Uranium-series dating of Quaternary successions. *Études sur le Quaternaire dans le Monde,* pp. 941–49. Congress INQUA, Paris, 8th

Szabo, B. J., Collins, D. 1975. Ages of fossil bones from British interglacial sites. Nature 254:680–82

Szabo, B. J., Gard, L. M. 1975. Age of the South Bight II marine transgression at Amchitka Island, Aleutians. *Geology* 2: 457–60

Szabo, B. J., Malde, H. E., Irwin-Williams, C. 1969. Dilemma posed by uranium-series dates on archaeologically significant bones from Valsequillo, Puebla, Mexico. *Earth Planet. Sci. Lett.* 6:237–44

Szabo, B. J., Rosholt, J. N. 1969. Uranium-series dating of Pleistocene molluscan shells from southern California—an open system model. *J. Geophys. Res.* 74:3253–60

Szabo, B. J., Stalker, M., Churcher, C. S. 1973. Uranium-series ages of some Quaternary deposits near Medicine Hat, Alberta, Canada. *Can. J. Earth Sci.* 10:1464–69

Szabo, B. J., Vedder, J. G. 1971. Uranium-series dating of some Pleistocene marine deposits in southern California. *Earth Planet. Sci. Lett.* 11:283–90

Taddeucci, A., Broecker, W. S., Thurber, D. L. 1967. ^{230}Th dating of volcanic rocks. *Earth Planet. Sci. Lett.* 3:338–42

Tatsumoto, M., Goldberg, E. D. 1959. Some aspects of the marine geochemistry of uranium. *Geochim. Cosmochim. Acta* 17: 201–8

Thompson, P. 1973. *Spelechronology and late Pleistocene climates inferred from O, C, H, U and Th isotopic abundances in speleothems.* PhD thesis. McMaster Univ., Hamilton, Ontario. 340 pp.

Thompson, P., Ford, D. C., Schwarcz, H. P. 1975. U^{234}/U^{238} ratios in limestone cave seepage waters and speleothem from West Virginia. *Geochim. Cosmochim. Acta* 39: 661–69

Thompson, P., Schwarcz, H. P., Ford, D. C. 1974. Continental Pleistocene climate variations from speleothem age and isotopic data. *Science* 184:893–95

Thurber, D. L. 1962. Anomalous U^{234}/U^{238} in nature. *J. Geophys. Res.* 67:4518–20

Thurber, D. L. 1963. *Anomalous U^{234}/U^{238} and an investigation of the potential of U^{234} for Pleistocene chronology.* PhD thesis. Columbia Univ., New York

Thurber, D. L., Broecker, W. S., Blanchard, R. L., Potratz, H. A. 1965. Uranium-series ages of Pacific atoll coral. *Science* 149:55–58

Turekian, K. K., Chan, L. H. 1971. The marine geochemistry of the uranium isotopes, ^{230}Th and ^{231}Pa. *Activation Analysis in Geochemistry and Cosmochemistry,* ed. A. O. Brunfelt, E. Steinnes, pp. 311–20. Oslo: Universitetsforlaget

Veeh, H. H. 1966. Th-230/U-238 and U-234/U-238 ages of Pleistocene high sea level stand. *J. Geophys. Res.* 71:3379–86

Veeh, H. H., Boström, K. 1971. Anomalous U-234/U-238 on the East Pacific Rise. *Earth Planet. Sci. Lett.* 10:372–74

Veeh, H. H., Burnett, W. C., Soutar, A. 1973. Contemporary phosphorites on the continental margin of Peru. *Science* 181: 844–45

Veeh, H. H., Chappell, J. M. A. 1970. Astronomical theory of climatic change: support from New Guinea. *Science* 167:

862–65

Veeh, H. H., Valentine, J. W. 1967. Radiometric ages of Pleistocene fossils from Cayucos, California. *Bull. Geol. Soc. Am.* 78 : 547–50

Ward, L. M. Jr., Lewis, G. E., Whitemore, F. C. Jr., 1972. Steller's Sea Cow in Pleistocene interglacial beach deposits on Amchitka, Aleutian Islands. *Bull. Geol. Soc. Am.* 83 : 867–70

Wetherill, G. W., Tilton, G. R. 1967. Geochronology. *Research in Geochemistry,* ed. P. H. Abelson, Vol II, 1–28. New York: Wiley

Windom, H. L. 1969. Atmospheric dust records in permanent snow fields : implications to marine sedimentation. *Bull. Geol. Soc. Am.* 80 : 761–82

Yokoyama, Y., Tobailem, J., Grjebine, T., Labeyrie, J. 1968. Determination de la vitesse de sedimentation oceanique par une methode non destructive de spectrometrie gamma. *Geochim. Cosmochim. Acta* 32 : 374–52

THE TOPSIDE IONOSPHERE: A REGION OF DYNAMIC TRANSITION[1]

P. M. Banks and R. W. Schunk
Department of Applied Physics and Information Science,
University of California at San Diego, La Jolla, California 92093

W. J. Raitt
Mullard Space Science Laboratory, University College, London,
Holmbury St. Mary, Dorking, Surrey, England

INTRODUCTION

The Earth is known to be surrounded by an envelope of relatively cool plasma (ion and electron temperatures from 300° to 10,000°K). Its origin is in the absorption of solar ultraviolet radiation by neutral gases at 60–300 km altitude in the upper atmosphere. In the decades prior to in situ measurements with rockets and spacecraft, exploration of the complex behavior of this medium was based on its reflection of high frequency radio waves. From this work it was apparent that the radio reflection process itself implied the presence of free electrons and ions; consequently, the term ionosphere was coined as part of the general nomenclature developed to describe various global features of the atmosphere.

Early radio experiments (1925–1930) disclosed that the ionosphere is stratified into various layers, the highest being the F_2 layer with a peak concentration of 10^5 to 10^6 electrons cm^{-3} (see Figure 1). Although the radio methods could not identify the ion composition of the ionosphere, later direct probes found that NO^+ was the dominant molecular ion of the E region and O^+ the major F region ion. Before 1957 and the age of space measurements, many of the relevant parameters related to the solar radiation spectrum and the structure of the upper atmosphere could only be speculated. The early theoretical models of Bradbury (1938) and in particular Ferraro (1945)—which described ionospheric behavior in terms of ionization production, loss, diffusion in the neutral atmosphere, plasma pressure gradients, and gravity—were adequate to explain the F_2 layer. The increase in

[1] This research was supported in part through NASA grant NGR 05-009-075 and NSF grant DES74-22343.

electron (and ion) density below the F layer peak was found to result from competition between ion production (O^+) and loss (to N_2 and O_2), whereas above the peak the density decreased in accordance with a gravity-dominated electrostatic equilibrium similar to the exponential decrease found for the neutral atmosphere.

Owing to the properties of high frequency radio wave reflections, the ionosphere above the peak of the F_2 layer could not be probed by ground stations. In the absence of measurements it was generally assumed that the topside ionosphere lying above the peak of the F_2 layer gradually decreased in concentration in accordance with electrostatic equilibrium and eventually merged with the interplanetary medium.

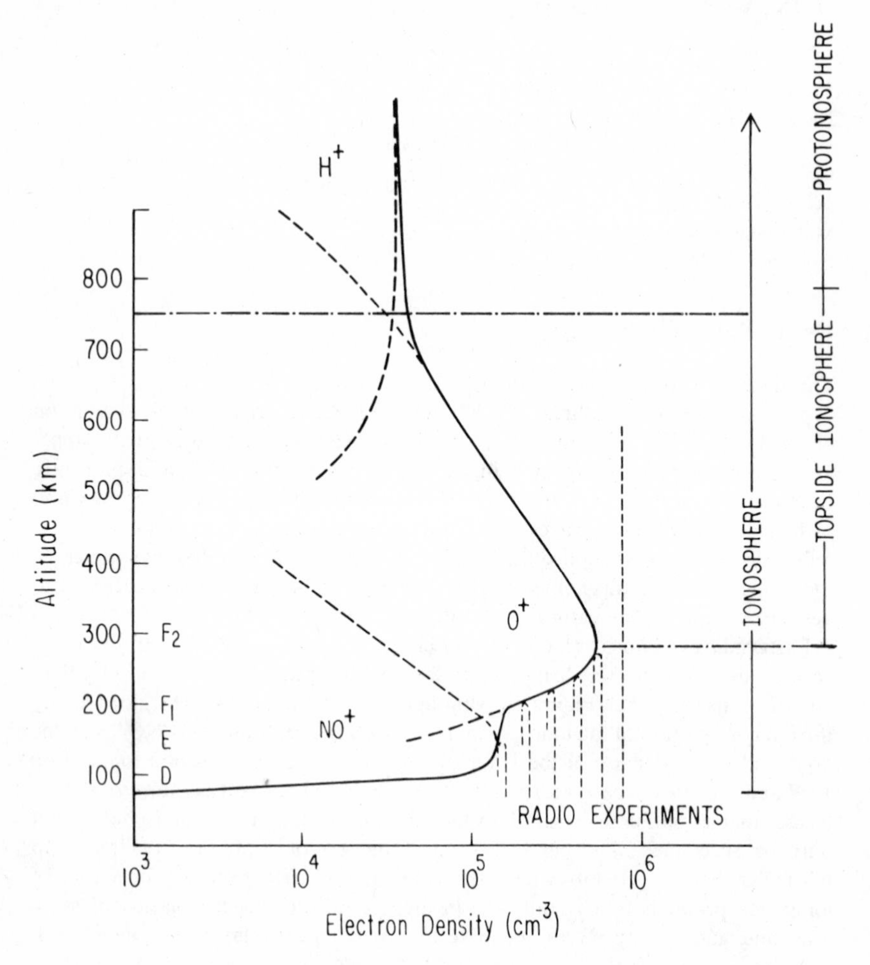

Figure 1 Schematic illustration of the ionosphere showing the various layers D, E, F_1, and F_2.

This view prevailed until 1953, when Storey (1953), using lightning-generated, low frequency radio waves (whistlers), was able to deduce the presence of appreciable electron concentrations (~ 1000 electrons cm^{-3}) at altitudes of 3–4 earth radii. Such densities were inconsistent with then-current ideas of the way the topside ionosphere should behave, and it was speculated that the plasma was composed of protons and electrons of solar origin. This conflicted with the observed interaction of the solar wind with comets, however, and the source of the plasma remained undetermined until 1960. Johnson (1960) suggested that the simple reversible charge exchange reaction $O^+ + H \leftrightarrows H^+ + O$, proposed earlier by Dungey (1955), could produce copious quantities of H^+ that could readily diffuse along magnetic field lines into regions far from the earth.

Recent progress in our understanding of the behavior of cool plasma within the magnetosphere (see Figure 2) has been rapid. In essence, the F_2 region of the ionosphere is both a source and a sink of ionization for the regions far from the earth. When the plasma pressure is sufficiently low in the regions far from the Earth, H^+ is created in the topside ionosphere by charge exchange with O^+ and a net outward flow of H^+ is established with a density, temperature, and flow velocity consistent with various external influences. Likewise, when a sufficiently high plasma pressure is present far from the Earth, the H^+ flow can reverse in

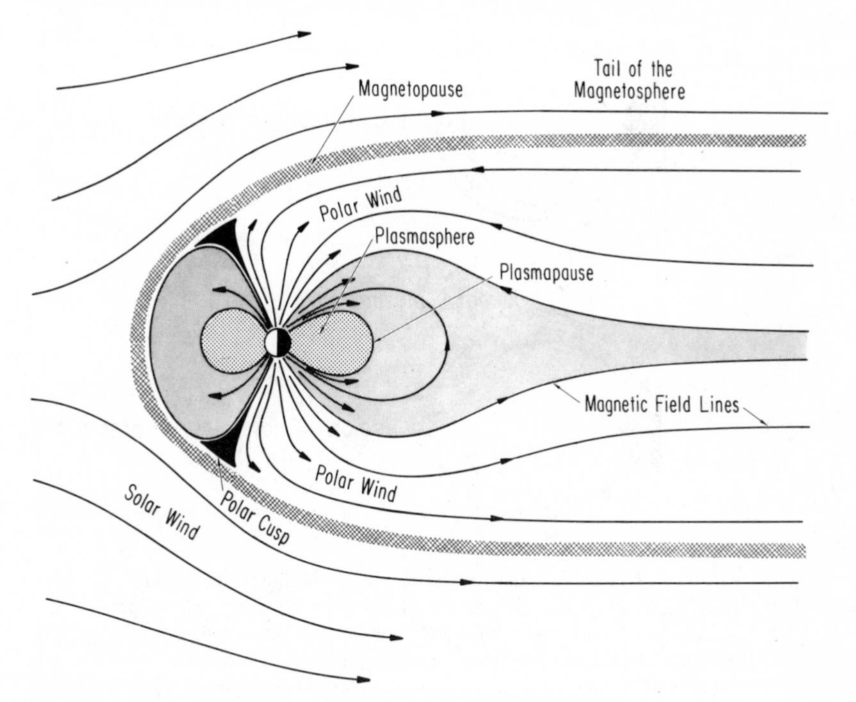

Figure 2 General view of the magnetosphere showing polar wind, plasmasphere, plasmapause, and magnetopause. From Banks (1972).

direction, with H^+ being converted to O^+ within the topside regions above the F_2 peak. Finally, a unique state of diffusive equilibrium can prevail such that gravity and plasma pressure gradients in the electron, O^+, H^+, and other ion gases that may be present precisely balance without flow. These three basic states are illustrated schematically in Figure 3, indicating the usefulness of what is now called the reservoir model; because of the lack of the appreciable collisions with atmospheric particles, the low energy plasma of the magnetosphere is constrained to remain in a particular magnetic flux tube, so that flows up and down the magnetic field lines can increase or deplete the plasma density in the magnetosphere or ionosphere.

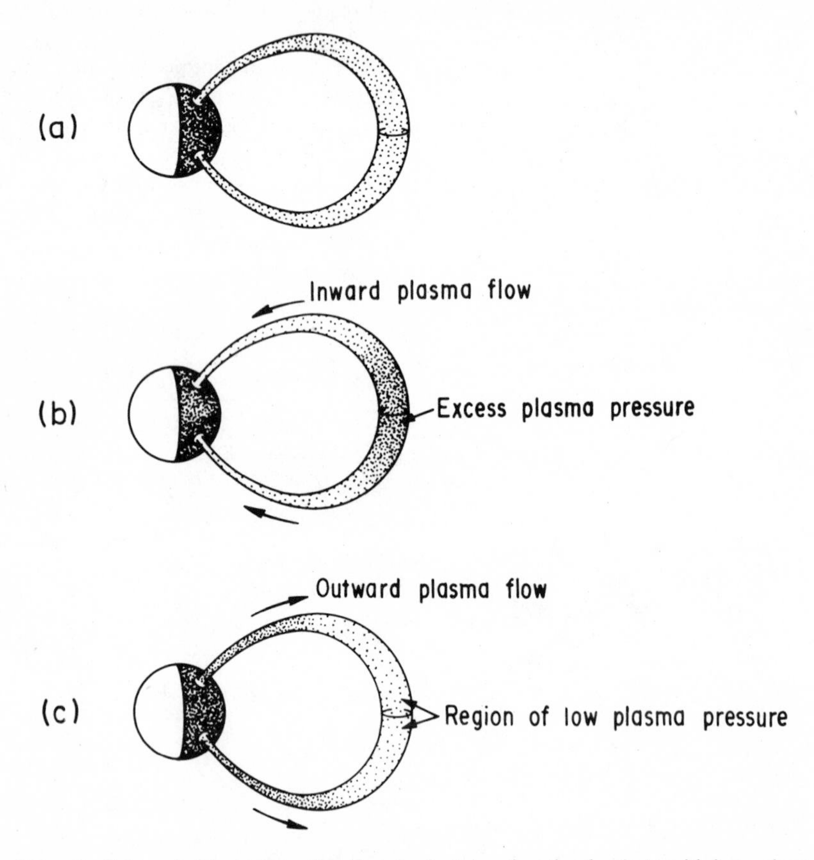

Figure 3 Schematic illustration of the three basic states of motion for the topside ionosphere. (*a*) Diffusive equilibrium: static distribution of plasma with small inward flow at low altitudes to accommodate ion losses in the F_2 region. (*b*) Inward diffusive flow: plasma moves inward and is lost in the F_2 region. Speed of flow may become supersonic if excess plasma pressure is sufficiently large. (*c*) Outward diffusive flow: plasma moves outward to equalize plasma pressure. Speed of flow may become supersonic if pressure deficit is large. From Banks & Kockarts (1973). Reproduced by permission of Academic Press.

Because of gravity, the dominant ion species of the regions more than several thousand kilometers from the Earth is H^+, although, as discussed later, He^+ is also present in measurable amounts.

This review is principally concerned with the various dynamic processes occurring in the topside ionosphere, which lies between the F_2 peak and the regions near 3000 km altitude, where H^+ becomes the dominant ion and where ionic reactions cease to have practical importance. Actually, the nomenclature is difficult here because the term topside ionosphere could justifiably be applied to all cool plasma of terrestrial origin lying above the F_2 peak. However, through processes described later, the great bulk of cool plasma around the Earth is contained within the plasmasphere, a toruslike volume surrounding the Earth with an outer boundary layer (the plasmapause) similar in shape to the surface of revolution created by geomagnetic field lines at 60° magnetic latitude. For magnetospheric processes it is often convenient to speak in terms of the regions away from the Earth that are inside or outside the plasmasphere. Then the ionosphere is left as a (relatively) low altitude collision-dominated plasma having some limited topside extension. However, moving away from the Earth at latitudes poleward of the plasmaspause, there is still a transition from the O^+ of the F_2 layer to the very low concentration of H^+ in what may reasonably be called the outer ionosphere. Previously, Geisler & Bowhill (1965) proposed the term protonosphere to describe middle latitude regions corresponding to the plasmasphere where H^+ rapidly becomes the dominant ion species. If, however, we extend the term protonosphere to include all regions above 2000–3000 km altitude, where H^+ is the dominant ion of thermal energy and ion reaction processes are negligible, a consistent nomenclature is possible. Thus ionosphere means the plasma regions extending from the lower D region to the base of the protonosphere, the latter surface varying greatly with geographical position and depends strongly upon plasma flow and atmospheric composition. The topside ionosphere is then the transition region lying between the F_2 peak and the protonosphere, and the plasmasphere and the plasmapause become a singular feature of the protonosphere. The regions are indicated schematically in Figure 4 for low, middle, and high latitudes.

There are many dynamic processes occurring in the topside ionosphere. As indicated previously, this region is one of considerable transition, providing the coupling process that link the dynamic protonosphere with the relatively static ionosphere lying below the F_2 peak. Within the topside ionosphere, an important change in ion composition is observed ($O^+ \rightarrow H^+$), along with substantial increases in plasma temperature. Furthermore, in the regions outside the plasmasphere (considering also its extension into the ionosphere) large-scale vertical flows of H^+ and He^+ occur with speeds along **B** of up to 10–20 km sec^{-1}. Such a range of behavior within the topside ionosphere reflects the various boundary conditions exerted by the protonosphere and underlying ionosphere, such as plasma partial-pressure gradients, flow velocity, and ion concentration, as well as factors associated with the neutral atmosphere and solar radiation.

In many areas, theoretical studies of the topside ionosphere have progressed more rapidly than experimental studies, largely because it is difficult to measure vertical (or parallel-to-**B**) profiles of ion composition, velocity, and temperature in the regions between 300 and 3000 km altitude. By far the largest body of direct data has been gathered from various spacecraft; as a result of their rapid tangential motion with respect to the Earth, these craft emphasize latitudinal changes in parameters that result from the more dramatic vertical processes. Thus the details of the vertical structure tend to result from theoretical models, whereas the geographical behavior tends to remain experimental.

The mathematical basis for recent theoretical models of the topside is discussed in the section following immediately. Computational results illustrating the response of the topside to a variety of dynamic conditions are presented in the third section (on models of the topside ionosphere). The fourth section (on experimental

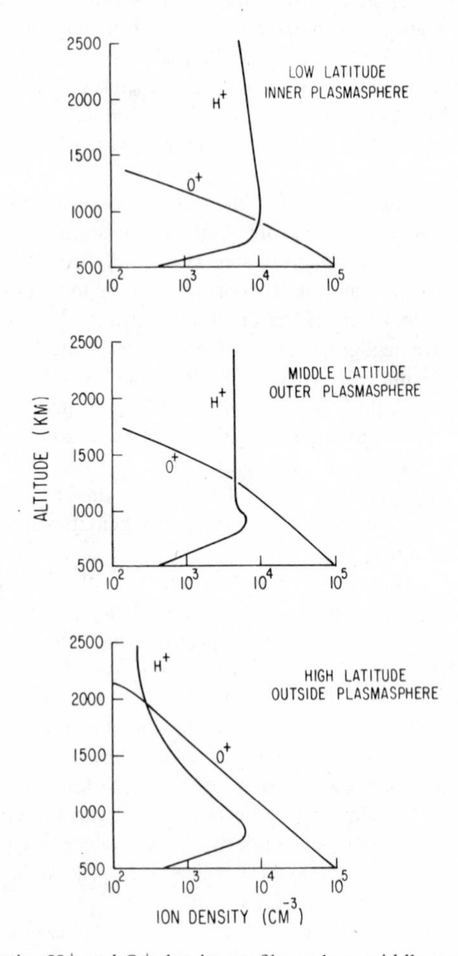

Figure 4 Illustrative H^+ and O^+ density profiles at low, middle, and high latitudes.

observations) is devoted to recent ground-based and spacecraft experimental results that show the broad range of global behavior. Finally, the summary comments on the future directions of research in this field.

THEORY OF MULTISPECIES ION FLOW

Early studies of the topside ionosphere by Johnson (1960), Mange (1960), Kockarts & Nicolet (1963), Bauer (1966), and others gave special emphasis to a static, nonflow model where the principal topside ions (O^+, He^+, H^+) arranged themselves in stratified layers according to ion mass. Later studies (e.g. see Banks 1972) made clear that a more dynamical coupling was actually present with large ($\sim 10^8$ ions cm^{-2} sec^{-1}) fluxes of ions passing through the topside ionosphere in response to diurnal variations of plasma pressure in the F_2 region and the magnetosphere (Hanson & Ortenburger 1961, Hanson & Patterson 1964, Geisler & Bowhill 1965, Banks et al 1971, Schunk & Walker 1970b).

Neglecting for the moment minor ions such as He^+, N^+, and O^{2+}, the basic model for O^+ and H^+ flow in the topside ionosphere can be summarized as follows: For plasma flow away from the Earth, O^+ created in the vicinity of the F_2 peak and above diffuses slowly (~ 10 m sec^{-1}) upward in response to its partial pressure gradient and the electron pressure gradient. In the regions below ~ 550 km, H^+ is in chemical equilibrium with O^+ as a consequence of charge exchange, giving an expression for the H^+ density as (Banks & Kockarts 1973, p. 193) $n(H^+) = n(O^+)n(H)/n(O)(T_n/T_i)^{\frac{1}{2}}$ where n represents the indicated species concentration and T_n and T_i are the neutral gas and ion temperatures, respectively.

From this relation and the known altitude distributions of atomic hydrogen and oxygen in the upper atmosphere, it follows that in chemical equilibrium the H^+ density will increase with altitude. Eventually, however, diffusion dominates chemistry, and plasma flow (H^+) can be established in response to H^+ pressure gradients. As H^+ flows outward, it experiences a frictional drag owing to H^+–O^+ Coulomb collisions (collisions with the neutral gases have no appreciable momentum effect, although they do affect the chemistry via the above relation). This drag between ion species is very important because it acts to limit the outward H^+ flux to some maximum value, which is of the order 10^8 ions cm^{-2} sec^{-1}, depending upon a number of factors including ion temperature and the neutral atomic hydrogen, atomic oxygen, and atomic oxygen ion concentrations (Geisler 1967). The frictional drag also affects the H^+ and O^+ temperature, through a dissipative process, which tends to increase the H^+ temperature above that of O^+ (Maier & Hoffman 1974, Banks et al 1974, Raitt et al 1975). When the H^+ pressure of the overlying protonosphere is sufficiently low, the H^+ flow speed parallel to $\mathbf{B}$ can be as large as 10–20 km sec^{-1}, and above the critical point the flow within the topside ionosphere becomes supersonic (Banks & Holzer 1968, 1969a,b; Banks et al 1971; Marubashi 1970).

With regard to inward plasma flow, the situation is more subdued. For sufficiently low F_2 region O^+ pressure, H^+ can flow inwards from the protonosphere to act as a source of O^+ (via charge exchange), which tends to maintain the F_2 region.

Unlike the result for plasma outflow, H^+–O^+ collisions impose no fundamental limit to the inward flux of H^+. However, since the total amount of H^+ contained in a magnetic flux tube inside the plasmasphere is of the order of 10^{13} ions cm^{-2} sec^{-1}, very large inward plasma flows can be sustained for only brief periods until the plasmasphere is depleted. In contrast with the outward flow case, the ion gas heating associated with inward plasma flows is very small because large inward H^+ fluxes correspond to large H^+ concentrations and consequent low H^+ velocities.

The majority of theoretical studies of the topside ionosphere have been based upon numerical solutions to the density, momentum, and energy conservation equations derived from moments of Boltzmann's equation. It is only recently, however, that a consistent approach to this problem has been taken in a format suitable for atmospheric dynamics. Schunk (1975) has derived a 13-moment approximation, based on the work of Grad (1949, 1958), which gives internally consistent expressions for density, velocity, temperature, heat conduction, and viscous stress for each species of a gas mixture.

The 13-Moment Approximation

The starting point for the derivation of the 13-moment system of transport equations is Boltzmann's equation. In a gas mixture, the motion of each species is described by a separate Boltzmann equation, which takes the form

$$\frac{\partial f_s}{\partial t} + \mathbf{v}_s \cdot \nabla f_s + \left[\mathbf{G} + \frac{e_s}{m_s}\left(\mathbf{E} + \frac{1}{c}\mathbf{v}_s \times \mathbf{B}\right)\right] \cdot \nabla_{v_s} f_s = \frac{\delta f_s}{\delta t} \tag{1}$$

where $f_s(\mathbf{r}, \mathbf{v}_s, t)$ is the distribution function of species s, $\mathbf{v}_s$ is the velocity, e_s is the charge, m_s is the mass, $\mathbf{G}$ is the acceleration due to gravity, $\mathbf{E}$ is the electric field, $\mathbf{B}$ is the magnetic field, c is the speed of light, $\partial/\partial t$ is the time derivative, ∇ is the coordinate space gradient, and ∇_{v_s} is the velocity space gradient.

In the following analysis, it is convenient to express Boltzmann's equation in terms of the random velocity, $\mathbf{c}_s$, defined as

$$\mathbf{c}_s = \mathbf{v}_s - \mathbf{u}_s \tag{2}$$

where $\mathbf{u}_s$ is the average species velocity, defined below. In terms of the random velocity, Equation 1 becomes

$$\frac{\partial f_s}{\partial t} + (\mathbf{c}_s + \mathbf{u}_s) \cdot \nabla f_s - \frac{D_s \mathbf{u}_s}{Dt} \cdot \nabla_{c_s} f_s - \mathbf{c}_s \cdot \nabla \mathbf{u}_s \cdot \nabla_{c_s} f_s$$
$$+ \left[\mathbf{G} + \frac{e_s}{m_s}\left(\mathbf{E} + \frac{1}{c}\mathbf{u}_s \times \mathbf{B}\right)\right] \cdot \nabla_{c_s} f_s + \frac{e_s}{m_s c}(\mathbf{c}_s \times \mathbf{B}) \cdot \nabla_{c_s} f_s = \frac{\delta f_s}{\delta t} \tag{3}$$

where D_s/Dt is the convective derivative,

$$\frac{D_s}{Dt} = \frac{\partial}{\partial t} + \mathbf{u}_s \cdot \nabla. \tag{4}$$

The collision term, $\delta f_s/\delta t$, in Boltzmann's equation represents the rate of change of f_s in a given region of phase space as a result of collisions. For binary elastic collisions, the appropriate collision term is the Boltzmann collision integral

$$\frac{\delta f_s}{\delta t} = \int \mathrm{d}\mathbf{c}_t \, \mathrm{d}\Omega g_{st} \sigma_{st}(g_{st}, \theta)[f'_s f'_t - f_s f_t] \tag{5}$$

where $\mathrm{d}\mathbf{c}_t$ is the volume element in velocity space, $\mathrm{d}\Omega$ is an element of solid angle in the center-of-mass reference frame, θ is the center-of-mass scattering angle, g_{st} is the relative velocity of the colliding particles s and t, $\sigma_{st}(g_{st}, \theta)$ is the differential scattering cross section, and the primes denote quantities evaluated after a collision.

The 13-moment system of transport equations is obtained by taking velocity moments of Boltzmann's equation. However, before taking moments, it is useful to introduce the following physically significant moments of the distribution function:

$\frac{3}{2}kT_s = \frac{1}{2}m_s\langle c_s^2\rangle$, species temperature;

$\mathbf{q}_s = \frac{1}{2}n_s m_s\langle c_s^2\mathbf{c}_s\rangle$, heat flow vector;

$\mathbf{P}_s = n_s m_s\langle \mathbf{c}_s \mathbf{c}_s\rangle$, pressure tensor;

$\boldsymbol{\tau}_s = \mathbf{P}_s - p_s\mathbf{I}$; stress tensor;

$\boldsymbol{\mu}_s = \frac{1}{2}n_s m_s\langle c_s^2\mathbf{c}_s \mathbf{c}_s\rangle$, higher-order pressure tensor; and

$\mathbf{Q}_s = n_s m_s\langle \mathbf{c}_s \mathbf{c}_s \mathbf{c}_s\rangle$, heat flow tensor;

where n_s is the density of species s, $p_s = n_s kT_s$ is the partial pressure, k is Boltzmann's constant, $\mathbf{I}$ is the unit dyadic, and the bracket symbol denotes the average

$$\langle \mathbf{A}\rangle = \frac{1}{n_s}\int \mathrm{d}\mathbf{c}_s \, f_s \mathbf{A}. \tag{6}$$

Multipling Equation 3 by 1, $m_s \mathbf{c}_s$, $\frac{1}{2}m_s c_s^2$, $m_s \mathbf{c}_s \mathbf{c}_s$, and $\frac{1}{2}m_s c_s^2\mathbf{c}_s$, and integrating over velocity space, we obtain, respectively, the continuity, momentum, energy, pressure tensor, and heat flow equations for species s:

continuity

$$\frac{\partial n_s}{\partial t} + \nabla \cdot (n_s \mathbf{u}_s) = \frac{\delta n_s}{\delta t}, \tag{7a}$$

momentum

$$n_s m_s \frac{D_s \mathbf{u}_s}{Dt} + \nabla \cdot \mathbf{P}_s - n_s m_s \mathbf{G} - n_s e_s\left(\mathbf{E} + \frac{1}{c}\mathbf{u}_s \times \mathbf{B}\right) = \frac{\delta \mathbf{M}_s}{\delta t}, \tag{7b}$$

energy

$$\frac{D_s}{Dt}(\tfrac{3}{2}p_s) + \tfrac{3}{2}p_s(\nabla \cdot \mathbf{u}_s) + \nabla \cdot \mathbf{q}_s + \mathbf{P}_s : \nabla \mathbf{u}_s = \frac{\delta E_s}{\delta t}, \tag{7c}$$

pressure tensor

$$\frac{D_s \mathbf{P}_s}{Dt} + \nabla \cdot \mathbf{Q}_s + \mathbf{P}_s(\nabla \cdot \mathbf{u}_s) + \frac{e_s}{m_s c}[\mathbf{B} \times \mathbf{P}_s - \mathbf{P}_s \times \mathbf{B}] + \mathbf{P}_s \cdot \nabla \mathbf{u}_s + (\mathbf{P}_s \cdot \nabla \mathbf{u}_s)^T = \frac{\delta \mathbf{P}_s}{\delta t}, \tag{7d}$$

and heat flow

$$\frac{D_s \mathbf{q}_s}{Dt} + \mathbf{q}_s \cdot \nabla \mathbf{u}_s + \mathbf{q}_s(\nabla \cdot \mathbf{u}_s) + \mathbf{Q}_s : \nabla \mathbf{u}_s + \nabla \cdot \boldsymbol{\mu}_s$$

$$+ \left[\frac{D_s \mathbf{u}_s}{Dt} - \mathbf{G} - \frac{e_s}{m_s}\left(\mathbf{E} + \frac{1}{c}\mathbf{u}_s \times \mathbf{B}\right)\right] \cdot (\boldsymbol{\tau}_s + \tfrac{5}{2}p_s \mathbf{I}) - \frac{e_s}{m_s c}\mathbf{q}_s \times \mathbf{B} = \frac{\delta \mathbf{q}_s}{\delta t}, \tag{7e}$$

where the transpose of a tensor $\mathbf{A}$ is denoted by $(\mathbf{A})^T$ and where the operation $\mathbf{Q}_s : \nabla \mathbf{u}_s$ corresponds to the double product of the two tensors $\mathbf{Q}_s$ and $\nabla \mathbf{u}_s$ (cf Chapman & Cowling 1970). The quantities on the right-hand sides of Equations (7a–e) describe, respectively, the rate of change of density, momentum, energy, pressure, and heat flow as a result of collisions. These quantities can be evaluated rigorously only for Maxwell molecule interactions, where the interaction potential varies inversely as the fourth power of the particle separation. For other interaction potentials, it is necessary to adopt approximate expressions for the distribution functions of the colliding particles in order to evaluate the collision terms.

Even with Maxwell molecule collision terms, the general system of moment equations presented above does not constitute a closed set, because the equation governing the moment of order r contains the moment of order $r+1$. Therefore, in order to close the system of moment equations, it is necessary to adopt an approximate expression for the distribution function. The system of equations is closed at any level by using the approximate distribution function to express higher-order moments in terms of lower-order moments. A particularly useful approximation is Grad's 13-moment approximation (Grad 1949, 1958),

$$f_s = f_{so}\left[1 + \frac{m_s}{2kT_s p_s}\boldsymbol{\tau}_s : \mathbf{c}_s \mathbf{c}_s - \left(1 - \frac{m_s c_s^2}{5kT_s}\right)\frac{m_s}{kT_s p_s}\mathbf{q}_s \cdot \mathbf{c}_s\right] \tag{8}$$

where

$$f_{so} = n_s \left(\frac{m_s}{2\pi k T_s}\right)^{3/2} \exp\left(-m_s c_s^2 / 2kT_s\right). \tag{9}$$

In the 13-moment approximation, stress and heat flow are put on an equal footing with density, drift velocity, and temperature, and it is necessary to solve flow equations in order to obtain the individual components of the stress tensor and heat flow vector. Since the stress tensor is symmetric and traceless, there are a total of 13 parameters describing each species in the gas mixture.

By taking the higher-order pressure tensor and heat flow tensor moments of the species distribution function (Equation 8), these moments can be expressed in terms of the 13 lower-order moments:

$$\boldsymbol{\mu}_s = \frac{5}{2}\frac{kT_s}{m_s}\left[p_s \mathbf{I} + \tfrac{7}{5}\boldsymbol{\tau}_s\right] \tag{10}$$

and

$$(Q_s)_{jkl} = \tfrac{2}{5}\left[(q_s)_j \delta_{kl} + (q_s)_l \delta_{jk} + (q_s)_k \delta_{jl}\right] \tag{11}$$

where we have temporarily introduced, index notation in Equation 11. From Equations 10 and 11, we can readily obtain the terms needed to close the system of moment equations:

$$\nabla \cdot \mathbf{Q}_s = \tfrac{2}{5}[\nabla \mathbf{q}_s + (\nabla \mathbf{q}_s)^T + (\nabla \cdot \mathbf{q}_s)\mathbf{I}] \tag{12}$$

$$\mathbf{Q}_s : \nabla \mathbf{u}_s = \tfrac{2}{5}[\mathbf{q}_s(\nabla \cdot \mathbf{u}_s) + (\nabla \mathbf{u}_s) \cdot \mathbf{q}_s + \mathbf{q}_s \cdot \nabla \mathbf{u}_s] \tag{13}$$

$$\nabla \cdot \boldsymbol{\mu}_s = \frac{5}{2}\frac{k}{m_s}[\nabla(T_s p_s) + \tfrac{7}{5}\nabla \cdot (T_s \boldsymbol{\tau}_s)]. \tag{14}$$

Only the pressure tensor and heat flow equations are affected by the closure of the system of moment equations. With allowance for Equations 12 to 14, these equations become

pressure tensor

$$\frac{D_s \mathbf{P}_s}{Dt} + \tfrac{2}{5}[\nabla \mathbf{q}_s + (\nabla \mathbf{q}_s)^T + (\nabla \cdot \mathbf{q}_s)\mathbf{I}] + \mathbf{P}_s(\nabla \cdot \mathbf{u}_s) + \frac{e_s}{m_s c}[\mathbf{B} \times \mathbf{P}_s - \mathbf{P}_s \times \mathbf{B}] + \mathbf{P}_s \cdot \nabla \mathbf{u}_s + (\mathbf{P}_s \cdot \nabla \mathbf{u}_s)^T = \frac{\delta \mathbf{P}_s}{\delta t} \tag{15}$$

and heat flow

$$\frac{D_s \mathbf{q}_s}{Dt} + \tfrac{7}{5}\mathbf{q}_s \cdot \nabla \mathbf{u}_s + \tfrac{7}{5}\mathbf{q}_s(\nabla \cdot \mathbf{u}_s) + \tfrac{2}{5}(\nabla \mathbf{u}_s) \cdot \mathbf{q}_s + \frac{5}{2}\frac{k}{m_s}[\nabla(T_s p_s) + \tfrac{7}{5}\nabla \cdot (T_s \boldsymbol{\tau}_s)] + \left[\frac{D_s \mathbf{u}_s}{Dt} - \mathbf{G} - \frac{e_s}{m_s}\left(\mathbf{E} + \frac{1}{c}\mathbf{u}_s \times \mathbf{B}\right)\right] \cdot (\boldsymbol{\tau}_s + \tfrac{5}{2}p_s \mathbf{I}) - \frac{e_s}{m_s c}\mathbf{q}_s \times \mathbf{B} = \frac{\delta \mathbf{q}_s}{\delta t}. \tag{16}$$

Equations 7a–c, 15, and 16 correspond to the closed system of moment equations for species s. The solution of this system of equations yields information on the species density, drift velocity, temperature, stress tensor, and heat flow. By allowing for the appropriate collision processes, these equations can be applied to a multi-component plasma of arbitrary degree of ionization as well as to a mixture of neutral gases. Furthermore, processes such as photoionization, recombination, and dissociation can also be considered.

With regard to the collision terms for the 13-moment approximation, general expressions have been derived only for low speed flows, where the species drift velocity differences are small compared to the species thermal speeds (Burgers 1969). If we further assume that the species temperature differences are small compared to the average temperature of the gas mixture, these low speed collision terms can be expressed in the form (cf Schunk 1975)

$$\frac{\delta n_s}{\delta t} = 0 \tag{17}$$

$$\frac{\delta \mathbf{M}_s}{\delta t} = -\sum_t n_s m_s \nu_{st}(\mathbf{u}_s - \mathbf{u}_t) + \sum_t \nu_{st}\frac{z_{st}\mu_{st}}{kT_{st}}\left[\mathbf{q}_s - \frac{\rho_s}{\rho_t}\mathbf{q}_t\right] \tag{18}$$

$$\frac{\delta E_s}{\delta t} = -\sum_t \frac{n_s m_s \nu_{st}}{m_s + m_t} 3k(T_s - T_t) \tag{19}$$

$$\frac{\delta \mathbf{P}_s}{\delta t} = -\sum_t \frac{2m_s \nu_{st}}{m_s + m_t} \left\{ \mathbf{P}_s - \frac{n_s}{n_t} \mathbf{P}_t + \tfrac{3}{10} z''_{st} \frac{m_t}{m_s} \left[\boldsymbol{\tau}_s + \frac{\rho_s}{\rho_t} \boldsymbol{\tau}_t \right] \right\} - \tfrac{3}{5} z''_{ss} \nu_{ss} \boldsymbol{\tau}_s \tag{20}$$

$$\frac{\delta \mathbf{q}_s}{\delta t} = -\sum_t \nu_{st} \left\{ D^{(1)}_{st} \mathbf{q}_s - D^{(4)}_{st} \frac{\rho_s}{\rho_t} \mathbf{q}_t + \tfrac{5}{2} p_s (\mathbf{u}_s - \mathbf{u}_t) \cdot \left[1 - \frac{m_t z_{st}}{m_s + m_t} \right] \right\} - \tfrac{2}{5} z''_{ss} \nu_{ss} \mathbf{q}_s, \tag{21}$$

where $\rho_s = n_s m_s$ is the mass density, $\mu_{st} = m_s m_t/(m_s + m_t)$ is the reduced mass, and $T_{st} = (m_t T_s + m_s T_t)/(m_s + m_t)$ is the reduced temperature. The quantity ν_{st} is the momentum transfer collision frequency between gases s and t; values are given in Appendix A for the gas mixture appropriate to the Earth's topside ionosphere. The quantities z_{st}, z'_{st}, z''_{st}, $D^{(1)}_{st}$ and $D^{(4)}_{st}$ are pure numbers that differ for different combinations of species s and t; values are given in Appendix B.

The heat flow terms that appear in the momentum collision term (Equation 18) account for thermal diffusion effects, and the drift velocity terms in the heat flow collision term (Equation 21) account for thermoelectric effects. Thermal diffusion can have an important influence on the charged particle distributions at altitudes above the F region peak electron density when the electron and ion temperature gradients are greater than 1°K km^{-1} (Walker 1967; Schunk & Walker 1969, 1970a,b). For ion and electron temperatures that increase with altitude, thermal diffusion acts to drive the heavy ions toward higher altitudes, i.e. toward hotter regions. Thermoelectric effects can have an important influence on the electron temperature profile when the field aligned electron current is greater than 10^{-5} amp m^{-2} (Schunk & Walker 1970c). For auroral conditions, the electron temperature in the topside ionosphere can be reduced by as much as 1000°K by thermoelectric transport processes (Rees et al 1971).

The 5-Moment Approximation

Although the low speed collision terms are appropriate for a variety of ionospheric problems, they cannot be applied to problems involving large relative flows or large temperature differences between the interacting species. As a consequence, the low speed collision terms are not appropriate for the polar wind where H^+ flows through O^+ with drift speeds comparable to or greater than the H^+ thermal speed.

Completely general collision terms for the system of moment equations have been derived only for the 5-moment approximation. In this approximation, stress and heat flow are neglected and the behavior of the gas is expressed in terms of the species density, drift velocity and temperature. In the 5-moment approximation, the appropriate collision terms can be expressed in the form

$$\frac{\delta n_s}{\delta t} = 0 \tag{22}$$

$$\frac{\delta \mathbf{M}_s}{\delta t} = \sum_t n_s m_s \nu_{st}(\mathbf{u}_t - \mathbf{u}_s)\Phi_{st} \tag{23}$$

$$\frac{\delta E_s}{\delta t} = \sum_t \frac{n_s m_s \nu_{st}}{m_s + m_t}[3k(T_t - T_s)\Psi_{st} + m_t(\mathbf{u}_s - \mathbf{u}_t)^2\Phi_{st}] \tag{24}$$

where Φ_{st} and Ψ_{st} are velocity-dependent correction factors; values are given in Appendix A for the gas mixture appropriate to the Earth's topside ionosphere. Because heat flow is neglected in the 5-moment approximation, the momentum collision term (Equation 23) does not contain heat flow interference terms; consequently, thermal diffusion effects are not included in the 5-moment approximation.

When heat flow and stress are neglected, the 13-moment system of transport Equations 7a–c, 15, and 16 reduces to just a continuity, momentum, and energy equation for each species:

$$\frac{\partial n_s}{\partial t} + \nabla \cdot (n_s \mathbf{u}_s) = \frac{\delta n_s}{\delta t} \tag{25}$$

$$n_s m_s \frac{D_s \mathbf{u}_s}{Dt} + \nabla p_s - n_s m_s \mathbf{G} - n_s e_s \left(\mathbf{E} + \frac{1}{c}\mathbf{u}_s \times \mathbf{B}\right) = \frac{\delta \mathbf{M}_s}{\delta t} \tag{26}$$

$$\frac{D_s}{Dt}\left(\tfrac{3}{2}p_s\right) + \tfrac{5}{2}p_s(\nabla \cdot \mathbf{u}_s) = \frac{\delta E_s}{\delta t} \tag{27}$$

where use has been made of the fact that the species pressure tensor is isotropic since stress is neglected.

Equations 22 to 27 correspond to the 5-moment approximation for a gas mixture. When collisional-transfer effects are neglected, this system of equations reduces to Euler's classical hydrodynamic equations, with the energy equation reducing to the simple adiabatic energy equation with the ratio of specific heats equal to 5/3.

Diffusive Equilibrium for a Multicomponent Ionosphere

As shown schematically in Figure 3, in the regions far from the Earth there is a certain plasma pressure for which there is a negligibly small ionization flow along geomagnetic field lines. For this situation, the ion density profiles in the topside ionosphere can be obtained from the 13-moment system of transport equations by setting the ion and electron flow velocities equal to zero. For an ionospheric plasma composed of two major ions (denoted by subscripts 1 and 2) and electrons, the steady-state equations for the distribution of ionization along geomagnetic field lines become (Schunk & Walker 1969)

$$\frac{1}{n_1}\frac{dn_1}{dS} = -\frac{m_1 g_{\parallel}}{kT_i} - \gamma_1 \frac{T_e/T_i}{n_e}\frac{dn_e}{dS} - \left[1 + \frac{n_2 \alpha_{12}}{n_1 + n_2}\right]\frac{1}{T_i}\frac{dT_i}{dS} - \frac{(\gamma_1 - D_1)}{T_i}\frac{dT_e}{dS} \tag{28}$$

$$\frac{1}{n_2}\frac{dn_2}{dS} = -\frac{m_2 g_{\parallel}}{kT_i} - \gamma_2 \frac{T_e/T_i}{n_e}\frac{dn_e}{dS} - \left[1 - \frac{n_1 \alpha_{12}}{n_1 + n_2}\right]\frac{1}{T_i}\frac{dT_i}{dS} - \frac{(\gamma_2 + D_2)}{T_i}\frac{dT_e}{dS} \tag{29}$$

where $g_{\parallel}$ is the component of gravitational acceleration along the geomagnetic field, S is the spatial coordinate along $\mathbf{B}$ and α_{12}, D_1, and D_2 are thermal diffusion coefficients. In deriving Equations 28 and 29, it was assumed that both ion species have a common temperature, T_i, which is different from the electron temperature, T_e.

Equations 28 and 29 and the equation for charge neutrality,

$$n_e = n_1\gamma_1 + n_2\gamma_2, \tag{30}$$

describe the distribution of ionization along the geomagnetic field when diffusive equilibrium conditions prevail.

Table 1 Values of $-\alpha_{12}$ as a function of n_{12} and m_{12} for $\gamma_{12} = 1$

n_{12}	$m_{12} =$	16	14	4	7/2	8/7
0.001		2.38	2.34	1.67	1.55	0.18
0.01		2.36	2.32	1.66	1.54	0.18
0.05		2.26	2.23	1.62	1.51	0.18
0.10		2.17	2.14	1.58	1.48	0.18
0.125		2.13	2.10	1.56	1.46	0.18
0.25		1.96	1.94	1.49	1.40	0.18
0.5		1.75	1.74	1.39	1.32	0.18
1.0		1.55	1.54	1.29	1.23	0.18
2.0		1.39	1.39	1.21	1.16	0.18
4.0		1.29	1.29	1.16	1.11	0.17
8.0		1.23	1.23	1.13	1.09	0.17
16.0		1.20	1.20	1.11	1.07	0.17
100.0		1.17	1.17	1.10	1.06	0.17
1000.0		1.16	1.16	1.10	1.06	0.17

Table 2 Values of $-\alpha_{12}$ as a function of n_{12} and m_{12} for $\gamma_{12} = 2$

n_{12}	$m_{12} =$	16	14	4	7/2	8/7	1
0.001		9.55	9.42	7.05	6.63	2.09	1.52
0.01		8.96	8.84	6.70	6.32	2.02	1.48
0.05		7.10	7.02	5.53	5.25	1.80	1.34
0.10		5.73	5.68	4.61	4.40	1.59	1.20
0.125		5.26	5.21	4.28	4.09	1.51	1.14
0.25		3.86	3.83	3.25	3.12	1.24	0.95
0.5		2.76	2.74	2.38	2.30	0.97	0.76
1.0		2.03	2.03	1.79	1.73	0.76	0.61
2.0		1.61	1.61	1.43	1.38	0.63	0.51
4.0		1.38	1.38	1.23	1.19	0.55	0.45
8.0		1.26	1.26	1.13	1.09	0.51	0.42
16.0		1.20	1.20	1.07	1.04	0.49	0.40
100.0		1.15	1.14	1.03	1.00	0.47	0.38
1000.0		1.14	1.14	1.02	1.00	0.47	0.38

Table 3 Values of D_1 and D_2 as functions of n_{12} for $\gamma_1 = 2$ and $\gamma_2 = 1$

n_{12}	D_1	D_2
0.001	1.60	0.00
0.01	1.56	0.02
0.05	1.38	0.07
0.10	1.20	0.12
0.125	1.13	0.14
0.25	0.87	0.22
0.5	0.60	0.30
1.0	0.37	0.37
2.0	0.21	0.41
4.0	0.11	0.44
8.0	0.06	0.46
16.0	0.03	0.47
100.0	0.01	0.47
1000.0	0.00	0.47

The thermal diffusion coefficient α_{12} depends on the ion density ratio, $n_{12} = n_1/n_2$, the ion mass ratio, $m_{12} = m_1/m_2$ and the ion charge ratio, $\gamma_{12} = \gamma_1/\gamma_2$. Values of α_{12} are given in Tables 1 and 2 for values of m_{12} appropriate to ionospheric mixtures of ionized oxygen, nitrogen, helium and hydrogen. The thermal diffusion coefficients D_1 and D_2 depend only on the ratios n_{12} and γ_{12}. When the two ion species have the same charge, $\gamma_1 = \gamma_2$, both D_1 and D_2 are zero. For $\gamma_1 = 2$ and $\gamma_2 = 1$, values of D_1 and D_2 are given in Table 3. The general expressions for α_{12}, D_1, and D_2 are given by Schunk & Walker (1969).

In a given altitude region of the topside ionosphere, there are generally no more than two ions with densities comparable to the electron density. The remaining minor ions (subscript x) are governed by a diffusive equilibrium equation of the form (Schunk & Walker 1969),

$$\frac{1}{n_x}\frac{dn_x}{dS} = -\frac{m_x g_{\parallel}}{kT_i} - \gamma_x \frac{T_e/T_i}{n_e}\frac{dn_e}{dS} - \frac{(1-\beta_x)}{T_i}\frac{dT_i}{dS} - \frac{(\gamma_x - D_x)}{T_i}\frac{dT_e}{dS}, \tag{31}$$

where the minor ions are assumed to have the same temperature as the major ions and where β_x and D_x are thermal diffusion coefficients.

The thermal diffusion coefficient β_x depends on the mass and charge ratios of the minor ion with both major ions as well as on the major ion density ratio. Values of β_x appropriate to the topside ionosphere are given in Table 4. The thermal diffusion coefficient D_x depends on the charges of the three ions as well as on the major ion density ratio. When all three charges are the same, D_x is zero. For multiply charged minor ions in a mixture of singly charged major ions ($\gamma_1 = \gamma_2 = 1$), $D_x = 0.8\gamma_x(\gamma_x - 1)$. The complete expressions for β_x and D_x are given by Schunk & Walker (1969).

Table 4 Values of β_x for various ions in mixtures of O^+ and H^+; $\gamma_{12} = 1$ and $m_{12} = 16$

n_{12}	He^+	N^+	O^{2+}	N^{2+}
0.001	1.67	2.34	9.60	9.47
0.01	1.62	2.29	9.45	9.31
0.05	1.44	2.11	8.82	8.67
0.10	1.25	1.92	8.17	8.01
0.125	1.16	1.84	7.88	7.72
0.25	0.82	1.51	6.76	6.57
0.5	0.38	1.09	5.40	5.16
1.0	−0.07	0.68	4.07	3.77
2.0	−0.45	0.35	3.04	2.67
4.0	−0.72	0.12	2.36	1.93
8.0	−0.89	−0.02	1.97	1.49
16.0	−1.00	−0.09	1.75	1.25
100.0	−1.08	−0.16	1.56	1.03
1000.0	−1.09	−0.17	1.53	1.00

Low Speed Plasma Flow

For low speed plasma flows, where the species flow velocities are small compared to the species thermal speeds, the 13-moment system of transport equations can be simplified by neglecting the inertial and stress terms in the species momentum and energy equations. Further assuming that the plasma is collision-dominated, the species heat flow vectors are simply related to the individual species temperature gradients.

Although these approximations significantly simplify the 13-moment system of transport equations, flow equations that allow for thermal diffusion effects still have to be derived for the H^+–O^+ transition region, where there are two ions with densities comparable to the electron density. However, for polar wind conditions, where H^+ is a minor ion, the appropriate transport equations for the topside ionosphere have been derived by Banks et al (1974) and Raitt et al (1975). For low speed flows, the steady-state equations of continuity, momentum, and energy for H^+, O^+, and electrons become

O^+ continuity

$$\frac{\mathrm{d}}{\mathrm{d}S}[n(O^+)u(O^+)] = P(O^+) - L(O^+)n(O^+), \tag{32}$$

O^+ momentum

$$u(O^+) = \frac{\nu(O^+, H^+)\Phi(O^+, H^+)}{\nu(O^+)}u(H^+) - D(O^+)\left\{\frac{1}{n(O^+)}\frac{\mathrm{d}n(O^+)}{\mathrm{d}S} + \frac{T_e/T(O^+)}{n_e}\frac{\mathrm{d}n_e}{\mathrm{d}S} + \frac{m(O^+)g_{\parallel}}{kT(O^+)} + \frac{1}{T(O^+)}\frac{\mathrm{d}}{\mathrm{d}S}[T(O^+) + T_e]\right\}, \tag{33}$$

O^+ energy

$$-\frac{\mathrm{d}}{\mathrm{d}S}\left(K(\mathrm{O}^+)\frac{\mathrm{d}T(\mathrm{O}^+)}{\mathrm{d}S}\right)=n(\mathrm{O}^+)\nu(\mathrm{O}^+,e)3k[T_e-T(\mathrm{O}^+)]$$

$$+n(\mathrm{O}^+)[\tfrac{4}{11}\nu(\mathrm{O}^+,N_2)+\tfrac{1}{3}\nu(\mathrm{O}^+,\mathrm{O}_2)+\tfrac{1}{2}\nu(\mathrm{O}^+,\mathrm{O})]3k[T_n-T(\mathrm{O}^+)]$$

$$+\frac{n(\mathrm{O}^+)m(\mathrm{O}^+)\nu(\mathrm{O}^+,\mathrm{H}^+)}{m(\mathrm{O}^+)+m(\mathrm{H}^+)}\{3k[T(\mathrm{H}^+)-T(\mathrm{O}^+)]\Psi(\mathrm{O}^+,\mathrm{H}^+)$$

$$+m(\mathrm{H}^+)u^2(\mathrm{H}^+)\Phi(\mathrm{O}^+,\mathrm{H}^+)\}, \qquad (34)$$

H^+ continuity

$$\frac{\mathrm{d}}{\mathrm{d}S}[n(\mathrm{H}^+)u(\mathrm{H}^+)]=P(\mathrm{H}^+)-L(\mathrm{H}^+)n(\mathrm{H}^+), \qquad (35)$$

H^+ momentum

$$u(\mathrm{H}^+)=-D(\mathrm{H}^+)\left\{\frac{1}{n(\mathrm{H}^+)}\frac{\mathrm{d}n(\mathrm{H}^+)}{\mathrm{d}S}+\frac{T_e/T(\mathrm{H}^+)}{n_e}\frac{\mathrm{d}n_e}{\mathrm{d}S}+\frac{m(\mathrm{H}^+)g_{\parallel}}{kT(\mathrm{H}^+)}\right.$$

$$\left.+\frac{1}{T(\mathrm{H}^+)}\frac{\mathrm{d}}{\mathrm{d}S}[T(\mathrm{H}^+)+T_e]-\frac{\alpha(\mathrm{H}^+)}{T(\mathrm{H}^+)}\frac{\mathrm{d}T(\mathrm{H}^+)}{\mathrm{d}S}\right\}, \qquad (36)$$

H^+ energy

$$\tfrac{3}{2}n(\mathrm{H}^+)ku(\mathrm{H}^+)\frac{\mathrm{d}T(\mathrm{H}^+)}{\mathrm{d}S}+\tfrac{3}{2}kT(\mathrm{H}^+)\frac{\mathrm{d}}{\mathrm{d}S}[n(\mathrm{H}^+)u(\mathrm{H}^+)]+n(\mathrm{H}^+)kT(\mathrm{H}^+)\frac{\mathrm{d}u(\mathrm{H}^+)}{\mathrm{d}S}$$

$$-\frac{\mathrm{d}}{\mathrm{d}S}\left(K(\mathrm{H}^+)\frac{\mathrm{d}T(\mathrm{H}^+)}{\mathrm{d}S}\right)=n(\mathrm{H}^+)\nu(\mathrm{H}^+,e)3k[T_e-T(\mathrm{H}^+)]\Psi(\mathrm{H}^+,e)$$

$$+\sum_j\frac{n(\mathrm{H}^+)m(\mathrm{H}^+)\nu(\mathrm{H}^+,j)}{m(\mathrm{H}^+)+m(j)}\{3k[T(j)-T(\mathrm{H}^+)]\Psi(\mathrm{H}^+,j)+m(j)u^2(\mathrm{H}^+)\Phi(\mathrm{H}^+,j)\},$$

$$(37)$$

and electron energy

$$-\frac{\mathrm{d}}{\mathrm{d}S}\left(K_e\frac{\mathrm{d}T_e}{\mathrm{d}S}\right)=\frac{m_e}{m(\mathrm{H}^+)}n_e\nu(e,\mathrm{H}^+)\{3k[T(\mathrm{H}^+)-T_e]+m(\mathrm{H}^+)[u_e-u(\mathrm{H}^+)]^2\}$$

$$+\frac{3m_e}{m(\mathrm{O}^+)}n_e\nu(e,\mathrm{O}^+)k[T(\mathrm{O}^+)-T_e], \qquad (38)$$

where the geomagnetic field is assumed to be vertical and where the electron density and drift velocity are obtained from the requirements of charge neutrality and charge conservation,

$$n_e=n(\mathrm{O}^+)+n(\mathrm{H}^+) \qquad (39)$$

$$n_e u_e=n(\mathrm{O}^+)u(\mathrm{O}^+)+n(\mathrm{H}^+)u(\mathrm{H}^+). \qquad (40)$$

In the O^+ and H^+ continuity equations, production and loss terms have been added in order to correctly describe the ion flow in both the F region and topside ionosphere. Production of O^+ ions results from photoionization of neutral atomic oxygen, and loss is caused by reactions with N_2 and O_2,

$$P(O^+) = 2\times 10^{-7} n(O)\,\text{cm}^{-3}\,\text{sec}^{-1} \tag{41}$$

$$L(O^+) = 1.2\times 10^{-12} n(N_2) + 2\times 10^{-11} n(O_2)\,\text{sec}^{-1}. \tag{42}$$

Production of H^+ ions results from charge exchange between O^+ and H, and loss is caused by charge exchange between H^+ and O (Banks & Kockarts 1973),

$$P(H^+) = 2.5\times 10^{-11} T_n^{1/2} n(H) n(O^+)\,\text{cm}^{-3}\,\text{sec}^{-1} \tag{43}$$

$$L(H^+) = 2.2\times 10^{-11} T^{1/2}(H^+) n(O)\,\text{sec}^{-1}. \tag{44}$$

In the O^+ and H^+ momentum equations, the quantities $D(O^+)$ and $D(H^+)$ are diffusion coefficients (Schunk & Walker 1970b, Schunk 1975),

$$D(O^+) = \frac{kT(O^+)}{m(O^+)\nu(O^+)} \tag{45}$$

$$D(H^+) = \frac{kT(H^+)}{m(H^+)\nu(H^+)} \tag{46}$$

where

$$\nu(O^+) = \nu(O^+, N_2) + \nu(O^+, O_2) + \nu(O^+, O) + \nu(O^+, H^+)\Phi(O^+, H^+)/2.15 \tag{47}$$

$$\nu(H^+) = \nu(H^+, N_2) + \nu(H^+, O_2) + \nu(H^+, O)\Phi(H^+, O) + \nu(H^+, O^+)\Phi(H^+, O^+)/2.15. \tag{48}$$

The momentum transfer collision frequencies, $\nu(i, j)$, and the velocity-dependent correction factors, $\Phi(i, j)$ and $\Psi(i, j)$, are given in Appendix A. For low speed flows, $\Phi(i,j)$ is ~ 1 and $\Psi(i,j)$ is ~ 1.

In the O^+, H^+, and electron energy equations, the quantities $K(O^+)$, $K(H^+)$, and K_e are thermal conductivities (Schunk & Walker 1973, Schunk 1975),

$$K_e = 7.7\times 10^5 T_e^{5/2} \tag{49}$$

$$K(O^+) = 7.75\times 10^3 \frac{n(O^+)}{n_e} T^{5/2}(O^+) \tag{50}$$

$$K(H^+) = 3.1\times 10^4 \frac{n(H^+)}{n_e} T^{5/2}(H^+)\left[1 + \frac{10}{13}\frac{\nu(H^+, N_2)}{\nu(H^+, O^+)} + \frac{10}{13}\frac{\nu(H^+, O_2)}{\nu(H^+, O^+)} + \frac{57}{52}\frac{\nu(H^+, O)}{\nu(H^+, O^+)}\right]^{-1} \tag{51}$$

where the units are electron volts per degree Kelvin per centimeter per second. In deriving $K(O^+)$, O^+ interactions with N_2, O_2, and O have been neglected. These interactions have a negligibly small effect on $K(O^+)$ at the altitudes where O^+

thermal conduction is important (Banks 1967). Likewise, in deriving K_e, electron interactions with N_2, O_2, and O have been neglected. These interactions have a significant effect on K_e only at low altitudes (Banks & Kockarts 1973).

With regard to the ion and electron energy equations, several facts should be noted. First, the summation in the H^+ energy Equation 37 is over the species N_2, O_2, O, and O^+. Next, the electron energy Equation 38 does not contain electron cooling to the neutral gases N_2, O_2, and O. This cooling is important only at low altitudes, and its inclusion does not significantly alter the temperature and density structure of O^+ and H^+ (cf Raitt et al 1975). Finally, the collisional terms appearing in the ion and electron energy equations contain terms that are proportional to the square of the relative flow velocities between O^+, H^+, and electrons. In the low speed flow approximation, these terms are generally neglected (compare Equations 19 and 24). However, for polar wind studies these terms must be retained because they account for frictional heating of the ion and electron gases. For H^+ in particular, this frictional heating can be important, even for relatively low H^+ drift velocities.

It is instructive to consider the behavior of the ion flow at high altitudes where ion production and loss processes can be neglected. Consider first the continuity Equation 32 for the major ion O^+. At the altitudes where O^+ production and loss can be neglected, the O^+ continuity equation indicates that the flux of O^+ is constant,

$$n(O^+)u(O^+) = F_a = \text{constant}, \tag{52}$$

where $u(O^+)$ is given by Equation 33. Substituting Equation 52 into Equation 33 and using the fact that H^+ is a minor ion [$n_e \approx n(O^+)$; H^+–O^+ frictional drag is negligible], the O^+ momentum equation reduces to

$$\frac{1}{n_e}\frac{dn_e}{dS} = -\frac{1}{H_p} - \frac{1}{T(O^+)+T_e}\frac{d}{dS}[T(O^+)+T_e] - \frac{F_a}{n_e D_a}, \tag{53}$$

where the ambipolar diffusion coefficient, D_a, and the plasma scale height, H_p, are given by

$$D_a = D(O^+)\frac{T(O^+)+T_e}{T(O^+)} \tag{54}$$

$$H_p = \frac{k[T(O^+)+T_e]}{m(O^+)g_{\parallel}}. \tag{55}$$

For most ionospheric conditions, the ambipolar flux is too small to affect the electron density at the altitudes where O^+ is the dominant ion, so that the last term in Equation 53 is generally negligible. If we further assume that the O^+ and electron temperatures are constant and neglect the altitude dependence of gravity, Equation 53 can be readily integrated to obtain an expression for the electron density,

$$n_e = n_{er} \exp[-(S-r)/H_p], \tag{56}$$

where r represents the lower reference point. Equation 56 indicates that the electron or O^+ density decreases exponentially with altitude with the plasma scale height (55).

As with the major ion O^+, the H^+ continuity Equation 35 indicates that the H^+ flux at high altitudes is constant,

$$n(H^+)u(H^+) = F_m = \text{constant} \tag{57}$$

where $u(H^+)$ is given by the H^+ momentum Equation 36. Substituting Equations 53 and 57 into Equation 36, the momentum equation for the minor ion H^+ can be expressed in the form

$$\frac{1}{n(H^+)}\frac{dn(H^+)}{dS} = -\frac{1}{H_m} - \frac{1}{T(H^+)}\frac{d}{dS}[T(H^+)+T_e] + \frac{\alpha(H^+)}{T(H^+)}\frac{dT(H^+)}{dS}$$
$$+\frac{T(H^+)}{T_e}\left\{\frac{1}{H_p} + \frac{1}{T(O^+)+T_e}\frac{d}{dS}[T(O^+)+T_e] + \frac{F_a}{n_e D_a}\right\} - \frac{F_m}{n(H^+)D(H^+)} \tag{58}$$

where

$$H_m = \frac{kT(H^+)}{m(H^+)g_{\parallel}}. \tag{59}$$

If we consider the simple case discussed above where the ion and electron temperatures and gravity are constant and the ambipolar flux is small, Equation 58 reduces to

$$\frac{1}{n(H^+)}\frac{dn(H^+)}{dS} = -\frac{1}{H_m} + \frac{T_e/T(H^+)}{H_p} - \frac{F_m}{n(H^+)D(H^+)}. \tag{60}$$

In contrast with the solution for the major ion O^+, the escape flux of H^+ is generally sufficient to affect the H^+ density profile. In order to determine the effect of F_m, it is convenient to express Equation 60 in a slightly different form. Multiplying Equation 60 by $n(H^+)D(H^+)$ and then taking its derivative, the equation becomes

$$\frac{d^2n(H^+)}{dS^2} + \left\{\frac{1}{D(H^+)}\frac{dD(H^+)}{dS} - \left[\frac{T_e/T(H^+)}{H_p} - \frac{1}{H_m}\right]\right\}\frac{dn(H^+)}{dS}$$
$$-\left[\frac{T_e/T(H^+)}{H_p} - \frac{1}{H_m}\right]\frac{1}{D(H^+)}\frac{dD(H^+)}{dS}n(H^+) = 0. \tag{61}$$

Because O^+ is the dominant ion impeding the flow of H^+ at high altitudes, it follows from Equations 46, 48, 56, and the expression for $\nu(H^+, O^+)$ given in Appendix A that

$$\frac{1}{D(H^+)}\frac{dD(H^+)}{dS} = \frac{1}{H_p}. \tag{62}$$

With allowance for Equation 62, the two linearly independent solutions to Equation 61 are

$$n(H^+) = n(H^+)_r \exp\left\{\left[\frac{T_e/T(H^+)}{H_p} - \frac{1}{H_m}\right](S-r)\right\} \tag{63}$$

and

$$n(\mathrm{H}^+) = n(\mathrm{H}^+)_r \exp\{-(S-r)/H_p\} \tag{64}$$

where r represents the lower reference point.

The first solution (63) corresponds to a diffusive equilibrium condition for the minor constituent H^+. Since $T_e/[H_p T(\mathrm{H}^+)] > (1/H_m)$, the H^+ density increases with altitude until H^+ is no longer a minor constituent.

For the second solution (64), the H^+ density decreases with altitude with the scale height of the major ion O^+. When this solution is substituted into the H^+ momentum Equation 36, the H^+ velocity becomes

$$u(\mathrm{H}^+) = D(\mathrm{H}^+)\left[\frac{1+T_e/T(\mathrm{H}^+)}{H_p} - \frac{1}{H_m}\right]. \tag{65}$$

Because the first term in the brackets in Equation 65 is greater than the second, the minor constituent H^+ flows upward with a speed that increases with altitude as $\exp\{(S-r)/H_p\}$ because of the dependence of $D(\mathrm{H}^+)$ on the O^+ density.

The H^+ flux associated with Solution 64 is given by

$$F_m = n(\mathrm{H}^+)u(\mathrm{H}^+) = n(\mathrm{H}^+)_r D(\mathrm{H}^+)_r\left[\frac{1+T_e/T(\mathrm{H}^+)}{H_p} - \frac{1}{H_m}\right] \tag{66}$$

where r is a reference level. This height independent flux represents the maximum rate at which H^+ ions can flow through the frictional medium provided by the dominant O^+ ions. For thermal equilibrium, this limiting or critical H^+ flux becomes (Geisler 1967)

$$F_m = 3.1\times 10^{-3}[n(\mathrm{O}^+)_r^{5/8}/n(\mathrm{O})_r^{3/8}]n(\mathrm{H})_r T^{3/2}\ \mathrm{cm}^{-2}\ \mathrm{sec}^{-1}. \tag{67}$$

High Speed Plasma Flow

For polar wind flow, the H^+ escape flux can approach its limiting value. When this occurs, the upward flow speed of H^+ increases exponentially with altitude in the manner described at the end of the last section. Eventually, above a certain altitude H^+ inertia and stress, which were neglected in the low speed formulation, become more important than H^+–O^+ friction in limiting the rate at which the H^+ velocity can increase with altitude. Consequently, H^+ inertia and stress must be considered in order to correctly describe high speed H^+ outflow.

For high speed polar wind flow, the O^+ and electron densities, drift velocities, and temperatures are still governed by Equations 32 to 34 and 38 to 40. Likewise, the H^+ continuity equation is still given by Equation 35. The H^+ momentum equation, on the other hand, must be altered to allow for H^+ inertia and stress. With allowance for these effects, the H^+ momentum equation for high speed flow becomes (Raitt et al 1975)

$$u(\mathrm{H}^+) + \frac{1}{\nu(\mathrm{H}^+)}\left[u(\mathrm{H}^+)\frac{\mathrm{d}u(\mathrm{H}^+)}{\mathrm{d}S} + \frac{\mathrm{d}\tau(\mathrm{H}^+)/\mathrm{d}S}{n(\mathrm{H}^+)m(\mathrm{H}^+)}\right] = -D(\mathrm{H}^+)\left\{\frac{1}{n(\mathrm{H}^+)}\frac{\mathrm{d}n(\mathrm{H}^+)}{\mathrm{d}S}\right.$$
$$\left. + \frac{T_e/T(\mathrm{H}^+)}{n_e}\frac{\mathrm{d}n_e}{\mathrm{d}S} + \frac{m(\mathrm{H}^+)g_\parallel}{kT(\mathrm{H}^+)} + \frac{1}{T(\mathrm{H}^+)}\frac{\mathrm{d}}{\mathrm{d}S}[T(\mathrm{H}^+)+T_e] - \frac{\alpha(\mathrm{H}^+)}{T(\mathrm{H}^+)}\frac{\mathrm{d}T(\mathrm{H}^+)}{\mathrm{d}S}\right\} \tag{68}$$

where $\tau(H^+) = \tau(H^+)\!:\mathbf{BB}/B^2$ is the parallel component of the H^+ stress tensor.

An expression for $\tau(H^+)$ can be obtained from the 13-moment pressure tensor Equation 15. However, as noted earlier, general collision terms for the 13-moment system of transport equations have been derived only for low speed flow. Rather than neglect H^+ stress effects, we use the low speed pressure tensor collision term (20) and derive the following expression for $\tau(H^+)$ in the limit of a collision-dominated plasma:

$$\tau(H^+) = -\frac{10}{9}\frac{n(H^+)kT(H^+)}{\nu'(H^+)}\frac{du(H^+)}{dS} \tag{69}$$

where

$$\nu'(H^+) = \nu(H^+, H^+) + 1.04\nu(H^+, O^+) + 1.02\nu(H^+, N_2) + 1.02\nu(H^+, O_2) + 0.88\nu(H^+, O). \tag{70}$$

Equation 69 corresponds to the low speed Navier-Stokes approximation for the parallel component of the H^+ stress tensor (cf Schunk 1975).

With regard to the H^+ energy equation for high speed flow, the heating associated with H^+ stress effects is negligible compared with H^+ thermal conduction at the altitudes where the H^+ gas is collision-dominated. At these altitudes, the H^+ energy equation for high speed flow is similar to the low speed Equation 37, except that for high speed flow the velocity dependent correction factors Φ and Ψ are not equal to unity, but instead are functions of the H^+ flow velocity (see Appendix A).

To emphasize the high speed nature of the above system of H^+ flow equations, it is useful to introduce a local H^+ Mach number, M, defined as

$$M = u(H^+)/C(H^+) \tag{71}$$

where $C(H^+) = [kT(H^+)/m(H^+)]^{1/2}$ is the H^+ isothermal sound speed. In terms of the local Mach number, the H^+ momentum Equation 68 becomes

$$\frac{1}{M}\frac{dM}{dS}(M^2-1) + \frac{d\tau(H^+)/dS}{n(H^+)m(H^+)C^2(H^+)} = -\frac{g_\parallel}{C^2(H^+)} - \frac{\nu(H^+)M}{C(H^+)} - \frac{1}{Q(H^+)}\frac{dQ(H^+)}{dS} - \frac{1}{2T(H^+)}\frac{dT(H^+)}{dS}(M^2+1) - \frac{T_e/T(H^+)}{n_e}\frac{dn_e}{dS} - \frac{1}{T(H^+)}\frac{dT_e}{dS} + \frac{\alpha(H^+)}{T(H^+)}\frac{dT(H^+)}{dS} \tag{72}$$

where the H^+ pressure gradient term has been eliminated with the aid of the H^+ continuity equation and where

$$Q(H^+) = \int_r^S [P(H^+) - L(H^+)n(H^+)]\, dS' \tag{73}$$

is the net ion flux flowing through the magnetic field tube.

In the original studies of the polar wind (Banks & Holzer 1968, 1969b; Marubashi 1970), H^+ stress was neglected; as a consequence, when H^+ was a minor ion, the

H^+ momentum Equation 72 possessed singularities at $M = \pm 1$, that is, at the point of transition from subsonic to supersonic flow. In a subsequent study, Strobel & Weber (1972) studied the extent to which the electron density gradient term influences the location of the critical point. This term was thought to be important because $n_e = \Sigma n_i$ and, consequently, the dn_e/dS term yields a term proportional to dM/dS. However, they found the H^+ Mach number at the critical point was approximately 1.1, rather than 1.

The effect of H^+ stress on the density and temperature structure of polar wind flow has been studied recently by Raitt et al (1975). From the mathematical point of view, H^+ stress introduces a term in the H^+ momentum equation that is proportional to d^2M/dS^2, consequently eliminating the singularity that appeared in the original studies of the polar wind. The effect of stress on the H^+ density, drift velocity, and temperature profiles is discussed in the section on models of the topside ionosphere.

The theoretical models of polar wind outflow presented in this and the previous subsection are valid at the altitudes where the H^+ gas is collision-dominated. As a rough guide, the H^+ gas is effectively collision-dominated when

$$u(H^+)/H\nu(H^+, O^+) \ll 1 \tag{74}$$

where H is the H^+ density scale height. Typically, this condition begins to break down at 1000 km and is clearly violated at 2000 km (cf Raitt et al 1975). When the plasma is not collision-dominated, the perpendicular component of the H^+ pressure tensor is important (Holzer et al 1971, Schunk 1975). Furthermore, the H^+ heat flow vector is not simply related to the gradient in the H^+ temperature. Under these conditions, it is more appropriate to obtain the components of the H^+ pressure tensor and heat flow vector directly from flow equations (Schunk 1975). Unfortunately, high speed collision terms for the appropriate system of flow equations are currently not available. Consequently, it is not possible at present to obtain accurate values of the H^+ density, drift velocity, and temperature in the altitude region where the flow changes from collision-dominated to collisionless conditions.

Kinetic Models of the Polar Wind

Although our review is primarily devoted to that region of the topside ionosphere where the H^+ gas is strongly influenced by collisions, in this subsection we briefly discuss kinetic models of the polar wind.

Kinetic polar wind models have been developed by Dessler & Cloutier (1969) and by Lemaire & Scherer (1970, 1971, 1972) in order to describe polar wind outflow at the altitudes where the H^+ gas is collisionless. In these models, collisions are neglected beyond some level h_0, referred to as the exobase or baropause. This exospheric level corresponds to the altitude where the H^+ mean free path becomes equal to the electron density scale height. In reality, the transition layer between the collision-dominated and collisionless domains is of the order of the n_e scale height. In practical computations, however, the exobase is generally taken as a sharp boundary separating the collision-dominated and collisionless regimes.

In the exosphere, which is defined to be the altitude region above the exobase, the

H^+ gas satisfies the Vlasov or collisionless Boltzmann equation (Equation 1 with $\delta f_s/\delta t = 0$). This equation is equivalent to stating that the density of particles in phase space is constant along the particle trajectories. Consequently, if the H^+ distribution function is known at the exobase, it can be readily determined at any altitude in the exosphere. Since the H^+ distribution function obtained in this manner satisfies Vlasov's equation, the physically significant moments of the distribution function such as density, drift velocity, temperature, pressure tensor, and heat flow automatically satisfy the collisionless transport equations.

An important step in the development of any kinetic model is the selection of the H^+ velocity distribution at the exobase, $f_{H^+}(\mathbf{v}_0, h_0)$. Because the geomagnetic field lines at polar latitudes are open, incoming particles must be excluded. Consequently, the distribution function $f_{H^+}(\mathbf{v}_0, h_0)$ selected in kinetic calculations need not be a realistic microscopic representation of the actual velocity distribution at the top of the collision-dominated region. Any parametric function of $\mathbf{v}_0$ can be chosen provided its first γ moments are equal to the corresponding γ moments of the actual velocity distribution (Lemaire & Scherer 1973, 1974).

For a one-parametric Maxwellian distribution function, the H^+ velocity distribution at the exobase is given by (Lemaire 1972).

$$f_{H^+}(\mathbf{v}_0, h_0) = \begin{cases} 0; \quad \text{for} \quad v_{0_\parallel} < 0 \quad \text{and} \quad v_0 > v_{\text{escape}} \\ N(H^+)[m(H^+)/2\pi k T(H^+)]^{3/2} \exp\left[-m(H^+)v_0^2/2kT(H^+)\right]; \\ \qquad\qquad\qquad\qquad \text{otherwise} \end{cases} \tag{75}$$

where the subscript $\parallel$ denotes the component parallel to the magnetic field and v_{escape} is the velocity necessary for H^+ to escape along the open magnetic field lines. The temperature $T(H^+)$ is assumed to be equal to the H^+ temperature at the exobase. The parameter $N(H^+)$ is chosen to match the density (the first moment of f) to its actual value just below the exobase,

$$n(H^+; h_{0+}) = \int f_{H^+}(\mathbf{v}_0, h_0)\, d\mathbf{v}_0 = n(H^+; h_{0-}). \tag{76}$$

Using Equations 75 and 76, one can readily obtain expressions for the H^+ density, $n(H^+; h_{0+})$, and flux, $F(H^+; h_{0+})$, just above the exobase,

$$n(H^+; h_{0+}) = \frac{N(H^+)}{2} \tag{77}$$

$$F(H^+; h_{0+}) = \frac{N(H^+)}{4}\left[\frac{8kT(H^+)}{\pi m(H^+)}\right]^{1/2}. \tag{78}$$

From Equations 77 and 78, one can then obtain the effusion velocity of the H^+ ions,

$$u(H^+; h_{0+}) = \left[\frac{2kT(H^+)}{\pi m(H^+)}\right]^{1/2} = 0.8C(H^+) \tag{79}$$

where $C(H^+) = [kT(H^+)/m(H^+)]^{1/2}$.

In the polar wind model developed by Lemaire (1972), the H^+ density and drift velocity profiles below the exobase are obtained from a solution to Euler's hydrodynamic equations, whereas the H^+ density and drift velocity profiles above the exobase are obtained from the kinetic theory of Lemaire & Scherer (1970, 1971). The solutions are joined at the exobase by requiring that the hydrodynamic drift velocity, $u(H^+; h_{0-})$, and diffusion flux, $F(H^+; h_{0-})$, at the exobase be equal to the effusion velocity (Equation 79) and escape flux (Equation 78), respectively.

It should be noted that kinetic models have also been developed using asymmetric Maxwellian distributions,

$$f_{H^+}(\mathbf{v}_0, h_0) \sim \exp\left\{-\frac{m(H^+)[\mathbf{v}_0 - u(H^+)]^2}{2kT(H^+)}\right\} \tag{80}$$

as well as asymmetric bi-Maxwellian distributions,

$$f_{H^+}(\mathbf{v}_0, h_0) \sim \exp\left\{-\frac{m(H^+)[v_{0_{\parallel}} - u(H^+)]^2}{2kT_{\parallel}(H^+)} - \frac{m(H^+)v_{0\perp}^2}{2kT_{\perp}(H^+)}\right\} \tag{81}$$

for the nonzero part of the H^+ velocity distribution at the exobase. These improvements to the theoretical model do not appear to have a significant effect on the H^+ density and drift velocity profiles (Lemaire 1972).

Finally, it should be noted that Holzer et al (1971) have compared the H^+ density and drift velocity profiles obtained from a kinetic model of the polar wind with those obtained from a set of collisionless transport equiations equivalent to the Chew-Goldberger-Low (CGL) approximation (Chew et al 1956). From this comparison, they have found that for supersonic flows there is only a small difference between corresponding H^+ density and drift velocity profiles obtained from the two methods.

MODELS OF THE TOPSIDE IONOSPHERE

Extensive studies of the behavior of the topside ionosphere have been made using models based upon the ideas described in the preceding section (e.g. Hanson & Patterson 1964, Banks & Holzer 1969b, Banks 1969, Schunk & Walker 1970b, Marubashi 1970, Banks et al 1971, Mayr et al 1972, Banks 1973, Bailey & Moffett 1974, Banks et al 1974, Raitt et al 1975, and others). Primary emphasis in these studies was on the varieties of behavior expected as a consequence of changes in atmospheric and ionospheric composition and temperature, as well as the plasma (proton) pressure at the high altitude upper boundary.

In terms of the global dynamics, the basic problem is to understand the relationship between H^+ and O^+ densities and the H^+ flow speed along the geomagnetic field. At middle latitudes, where $\mathbf{E} \times \mathbf{B}$ plasma drifts are usually small, plasma in the region above the topside ionosphere essentially corotates with the Earth so that over a period of several days (Park 1974) the plasma density along a given magnetic flux tube can rise to a high value ($\sim 10^3$ ions/cm^3 in the equatorial plane), thereby restricting the H^+ flow velocity to small values ($\sim$10–100 m sec^{-1}) corresponding to a diurnal ebb and flow. At high magnetic latitudes, in contrast,

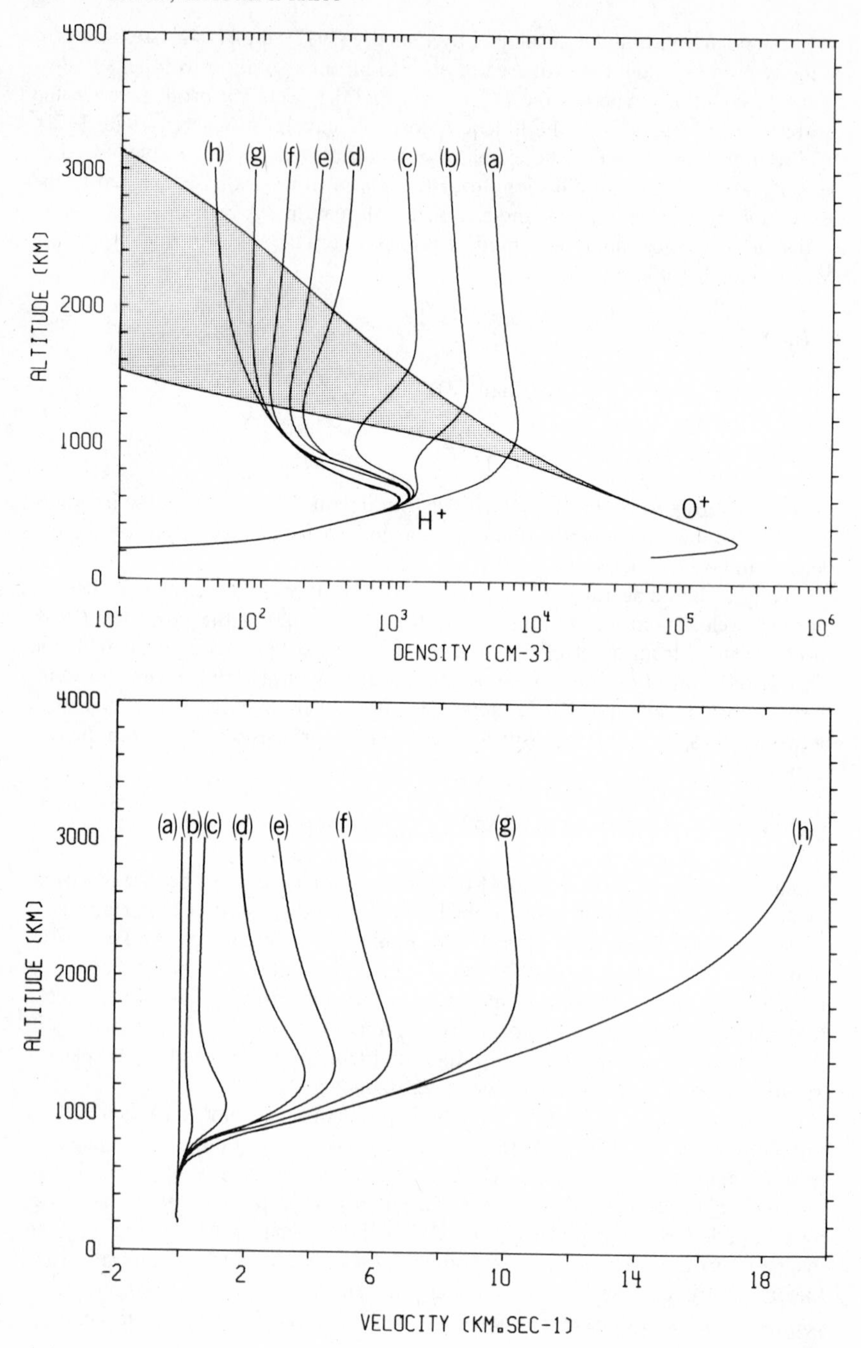

ALTITUDE (KM)
DENSITY (CM-3)
(h) (g) (f) (e) (d) (c) (b) (a)
H+
O+
ALTITUDE (KM)
VELOCITY (KM.SEC-1)
(a) (b) (c) (d) (e) (f) (g) (h)

E × **B** plasma drifts are large (up to 3 km sec^{-1}) and provide a mechanism for transferring plasma from closed to open geomagnetic field lines. As a consequence, the H^+ pressure is low at high altitudes, and strong upward flows of H^+ (the polar wind), which greatly reduce the H^+ density in the topside ionosphere, are present.

To illustrate the effect that the H^+ flow velocity has on the H^+ composition, Figure 5 gives theoretical H^+ and O^+ density profiles for a wide range of outward H^+ velocities at 3000 km. Curve *a* represents near diffusive equilibrium, with H^+ becoming the dominant ion at 900 km. (The O^+ density in this case follows the lower curve of the shaded regime.) As the upper boundary velocity is increased, the H^+ density is progressively diminished with a peak in the H^+ density profile appearing near 600–700 km altitude. Curve *h*, which represents a flow velocity of 20 km sec^{-1} at 3000 km, corresponds to a supersonic flow of H^+ with a flux of 8.5×10^7 ions cm^{-2} sec^{-1}. Such results provide the most direct explanation for the division of the protonosphere into two distinct regions: (*a*) the plasmasphere, the toruslike volume confined to low and middle latitudes, and (*b*) the regions of very low cool plasma density external to the plasmasphere (see Figure 4).

With regard to the H^+ flow velocity, Figure 5 also gives profiles corresponding to the density curves of the upper panel. Near-diffusive equilibrium, curve *a*, has only very low speed flow, whereas curve *h* rapidly climbs to 20 km sec^{-1}. In any real situation, such velocities would correspond to boundary conditions imposed by plasma pressure gradients from above. Owing to the retarding frictional effect of H^+–O^+ collisions, significant H^+ velocities are found only above 700–800 km in the present model (taken from Raitt et al 1975).

The H^+ and O^+ temperatures are both affected by the H^+ flow. Figure 6 gives the respective H^+ and O^+ temperatures for the various flow velocities of Figure 5. For H^+ there are clearly counteracting tendencies, with the peak H^+ temperature being reached for a moderate velocity of 3.0 km sec^{-1} at 3000 km. The relative contributions made to the H^+ thermal balance by convection, advection, thermal conduction, and collisions are shown in Figure 7, corresponding to curve *b* of Figures 5 and 6. Although these results predict separate temperatures for H^+ and O^+, with H^+ being substantially hotter, the actual topside ion temperatures outside the plasmasphere may be substantially higher than indicated. Such a condition may arise as a consequence of high altitude Joule heating connected with plasma convection in directions perpendicular to **B**. Although the rate of ion-neutral collisions is small, such collisions play an important part in heating the ion gases to 800 km, and one can expect ion temperatures of 2500° to 3500°K for moderate electric fields (~30 to 50 mVm^{-1}) found at high latitudes and in the polar caps.

A separate peculiarity of H^+ flow is its flux-limiting character. As the outward flow velocity increases, the H^+ flux rapidly rises to a saturation limit, first described by Hanson & Patterson (1964) and given analytic form by Geisler (1967). Figure 8 illustrates this behavior in terms of the H^+ density at 3000 km. At

← *Figure 5* Profiles of H^+ and O^+ density and velocity for different H^+ velocities at 3000 km: (*a*) 0.06, (*b*) 0.34, (*c*) 0.75, (*d*) 2.0, (*e*) 3.0, (*f*) 5.0, (*g*) 10.0, (*h*) 20.0 km sec^{-1}. From Raitt et al (1975).

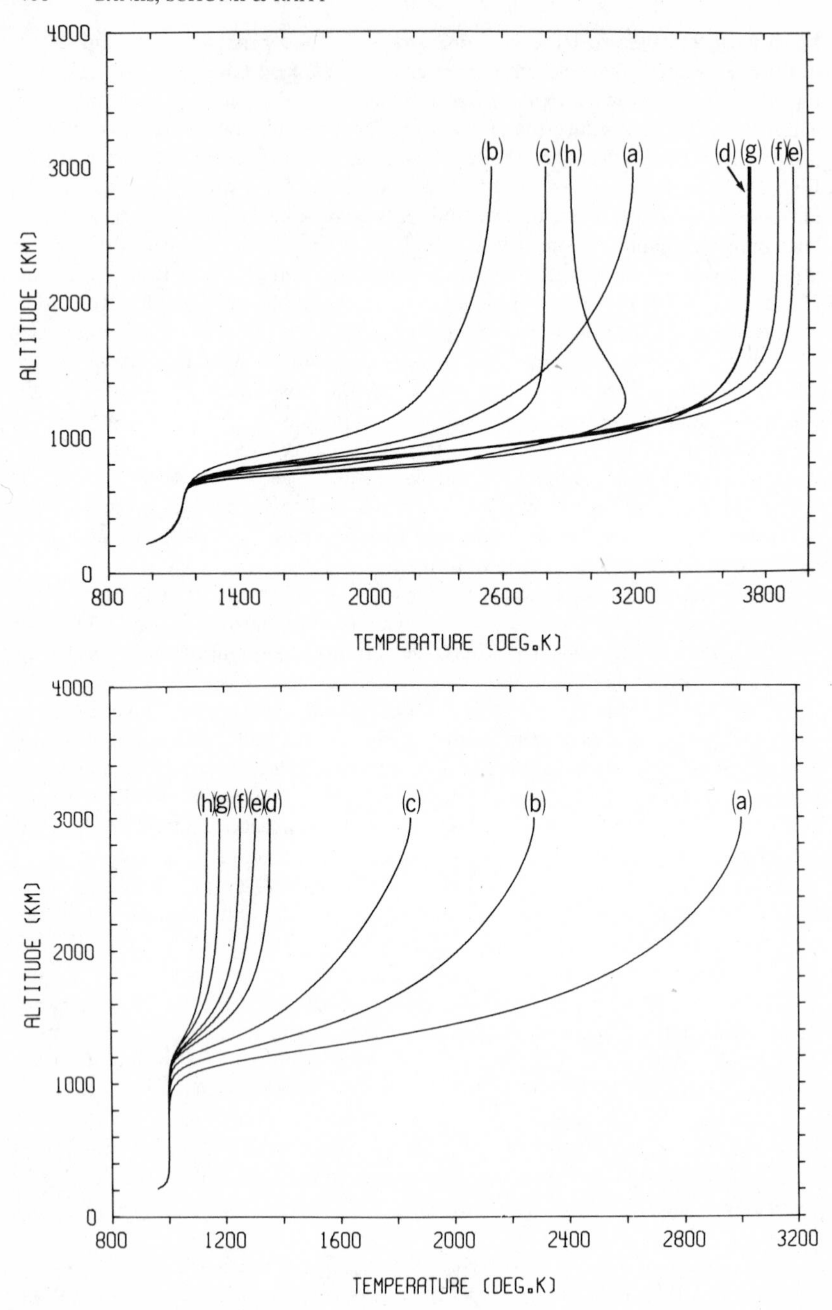

Figure 6 H^+ (*top*) and O^+ (*bottom*) temperature profiles for the model of Figure 5. From Raitt et al (1975).

sufficiently high densities, the flux is inward (negative), but, as the H^+ density is lowered, the outward flux is quickly established and soon saturates in magnitude. Thus, although arbitrarily large plasma inflow can occur in response to high densities within the protonosphere, the outflow is limited by various atmospheric constraints. These include the plasma temperatures, the neutral atomic hydrogen concentration, the O^+ density, and the atmosphere composition. The limiting fluxes for several atomic hydrogen densities are included in Figure 8, showing that the H^+ flux decreases proportionately to the density of H.

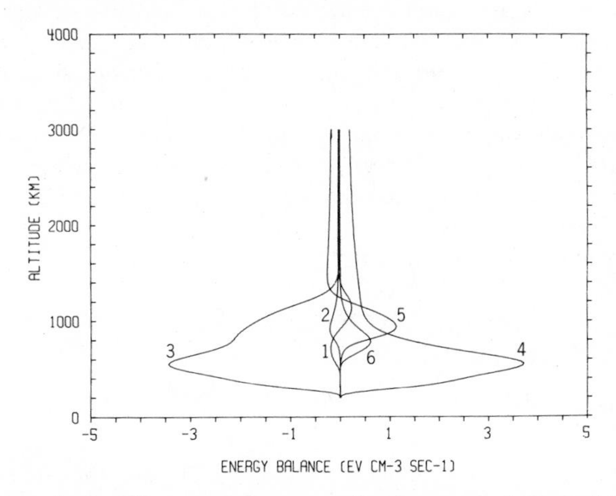

Figure 7 Terms in the H^+ energy balance equation for case *b* of Figure 5. The terms are [1] convection, [2] advection, [3] H^+–O^+ collisional coupling, [4] H^+-*e* collisional coupling, [5] thermal conduction, and [6] Joule heating by H^+–O^+ collisions. From Raitt et al (1975).

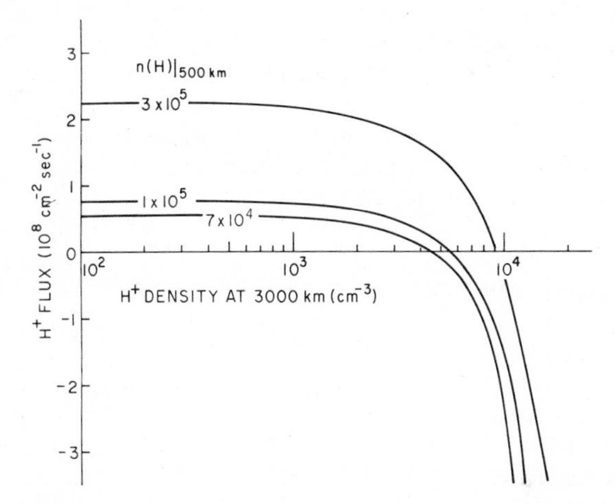

Figure 8 H^+ flux for different H^+ boundary densities and neutral hydrogen densities. From Banks (1972).

One of the newest results for high speed plasma flow shows the effect viscous stress parallel to **B** has upon computed profiles of density, velocity, and temperature. Figure 9, taken from Raitt et al (1975), gives vertical profiles of these quantities for an outflow velocity of 3 km sec^{-1} at 3000 km (see curve *e* of Figure 5). The profiles labeled 1 are obtained with a standard coefficient of viscosity, and those labeled 2 are based upon reduction by a factor of 3. Decreasing viscous stress tends to decrease the H^+ density while increasing the flow speed and temperature.

This brief summary of the extensive studies of the topside environment explains the wide variety of experimental results described in the next section. However, specific models of the global topside ionosphere have not been made, largely because they require time-dependent variation of the magnetospheric convection electric field. Convective motion of plasma relative to the Earth, together with the presence of auroral oval phenomena, sufficiently complicate theoretical models to the extent that no immediate progress can be made. However, as better models of the global electric field pattern become available, it will be possible to treat in a unified fashion the topside as a region of dynamic transition between the underlying F region and the higher protonosphere.

EXPERIMENTAL OBSERVATIONS OF THE TOPSIDE IONOSPHERE

Introduction

The topside ionosphere is a weakly ionized plasma with the ratio of electron to neutral particle densities being about 10^{-4} near the F_2 peak, rising to about 10^{-1} at 3000 km altitude. In order to relate the actual ionosphere to theoretical models of its formation, it is necessary to make measurements of the generation of thermal plasma, the properties of the thermal plasma itself, and also the neutral gas. In addition, it is necessary to study the magnetic and electric force fields acting on the topside plasma. The direct effects of neutral winds are less important at topside altitudes because of the negligible momentum transfer above about 250 km.

In this section, we limit ourselves to a discussion of the measured thermal plasma properties in order not to widen the scope of the paper excessively. We also omit any measurements of plasma instabilities, which can be expected to influence the distribution of ionospheric plasma in more dynamic regions such as the auroral zones and polar caps. Finally, we restrict ourselves to observations in which the ion and electron velocity distributions are considered to be isotropic, and, therefore, we can discuss particle temperatures as measured by a number of techniques.

The main experimental techniques used in gathering measurements of the properties of the topside ionosphere are (*a*) plasma probes carried on rockets and satellites, (*b*) satellite-borne ionosondes (topside sounders), (*c*) ground-based beacon satellite observations, (*d*) ground-based natural wave propagation (whistlers), (*e*) incoherent scatter radar, and (*f*) spectroscopic measurements.

The various techniques have their own advantages and disadvantages. Rocket probes give a detailed vertical profile, but it is a snapshot measurement in position and time. Satellite-borne probes provide data with fine detail in an essentially

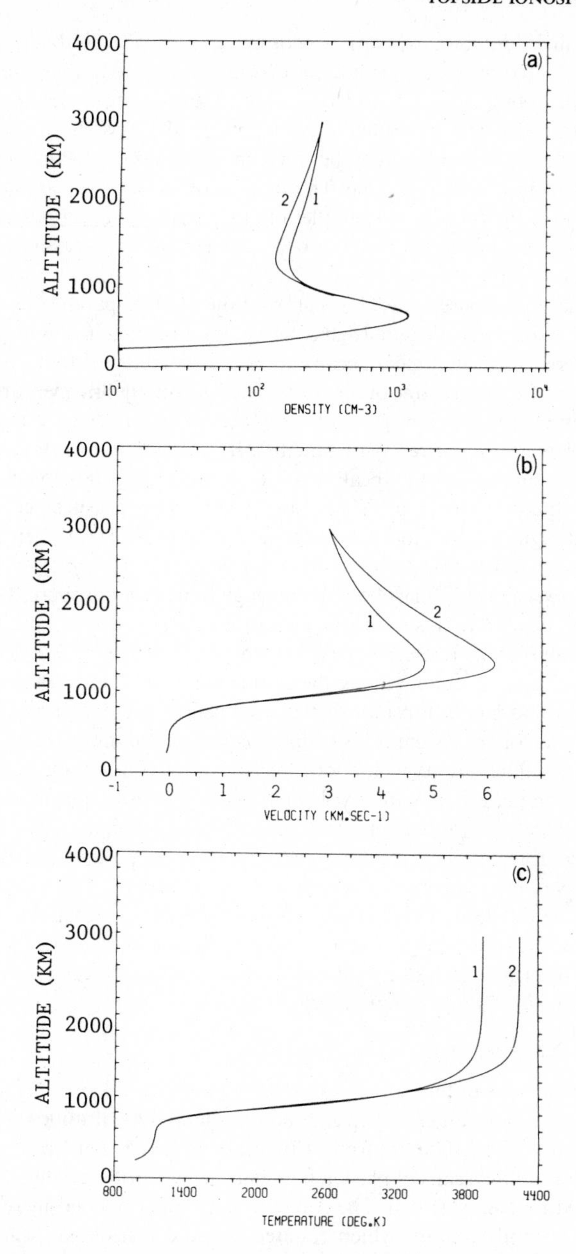

Figure 9 Profiles of H^+ density, velocity and temperature for an outflow velocity of 3 km sec^{-1} at 3000 km. Profile 1 corresponds to a standard viscosity used in results of Figures 5, 6, and 7, and profile 2 is smaller by a factor of 3. From Raitt et al (1975).

horizontal direction, but, by various combinations of orbital eccentricity and inclination, limited scans in altitude and local time are possible. However, such scans generally require a long period to average over, resulting in blurring of temporal features of the measurements. Satellite-borne ionosondes have the advantage that an altitude profile of electron density between the F_2 peak and the satellite can be measured in a short period of time. The time to obtain an ionosonde is normally longer than the interval between samples of in situ measuring probes, resulting in a poorer spatial and temporal resolution of ionospheric parameters from topside sounder data.

Ground-based measurements all suffer from the limited spatial extent over which measurements can be made, although some spatial scanning in the magnetospheric frame of reference can be made because of the asymmetry of that system relative to the Earth's corotating frame of reference. The beacon satellite measurements have a further disadvantage: they give an integrated electron density in the column linking the satellite to the receiving antenna. By combining this data with ground ionosonde measurements of F_2 peak densities, however, it is possible to make more meaningful studies of the topside ionosphere. Whistler measurements enable the more distant regions of the topside ionosphere to be probed, and their applicability to the types of measurements presently being discussed is largely in defining the boundary between the plasmasphere and the high latitude ionosphere. The technique can, however, be used to determine the equatorial electron density and thus provide a measurement directly related to the plasma pressure difference that drives H^+ ions to flow out of the ionosphere along the geomagnetic field lines. Perhaps the most successful ground-based topside ionosphere measurement technique is the incoherent scatter radar system. This technique uses scattering of radio waves from plasma density fluctuations in the ionosphere, and, by analyzing the spectrum of the returned signal, it is possible to deduce electron density, plasma drift velocity, electron temperature, ion temperature, and ion composition.

Spectroscopic techniques have been used both from the ground and from satellites. Observations of emission lines characteristic of recombining ions enables the ion densities to be deduced. The method has the disadvantage that the spectrometer collects photons from a large volume of ionosphere and there is the consequent problem of blurring detail of height structure in vertical observations or latitude/longitude structure in horizontal observations.

Global Distribution of Electron Density

A number of surveys of the global morphology of electron density have been made by satellite-borne instruments. Topside sounder data gives altitude distributions of electron density up to 1000 km from Alouette-1 and 3000 km from Alouette-2. A number of latitudinal scans of plasma frequency in the altitude range 300–1000 km are presented by Nelms (1964). The Alouette data was obtained in real time by a network of ground stations, which resulted in wide meridional scans at only a limited number of longitudes, the best being that in the American continent ($\approx 75°$W). An example of the data obtained from topside sounders, shown in Figure 10, is taken from Nelms & Lockwood (1967), who used the Alouette-2 data giving

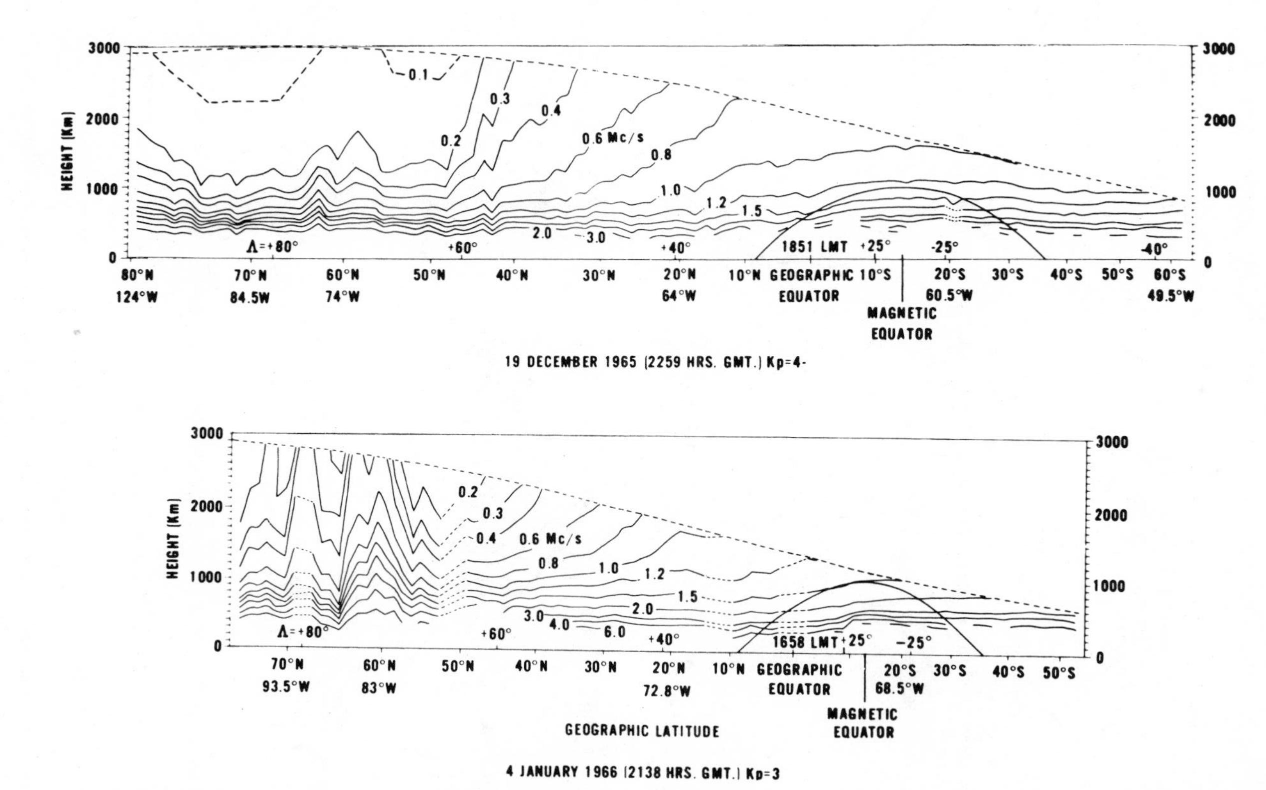

Figure 10 Contours of constant plasma frequency as a function of height and geographic latitude deduced from Alouette-2 topside sounder ionograms. From Nelms & Lockwood (1967).

measurements up to 3000 km. The authors use plasma frequency rather than electron density, but the two parameters are simply related by the expression

$$f_p = 8.98 \times 10^3 n_e^{1/2} H_z$$

where f_p = plasma frequency and n_e = electron density (cm^{-3}). The density map of December 19, 1965, shows a marked difference between the low and midlatitude ionosphere and the high latitude ionosphere, namely, that a clear scale height change from that characteristic of heavier ions (O^+) to light ions (H^+) can be seen as the altitude increases at latitudes equatorward from about 50°N. At higher latitudes, the scale height indicates that O^+ is the dominant ion down to the limit of sensitivity of the ionosonde. The December 19 map also shows a density trough at about 60°N, and this feature is more pronounced in the January 4, 1966, data, again at about 60°N.

Systematic studies of the location and depth of the electron density trough, often referred to as the midlatitude trough, have been made by a number of workers

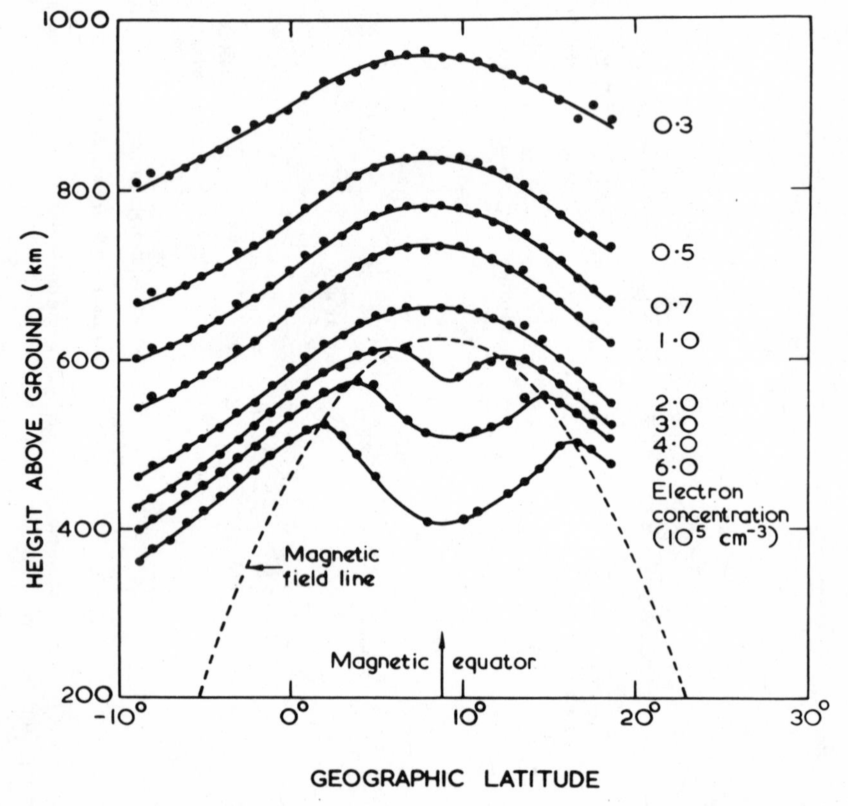

Figure 11 Contours of constant electron density as a function of height and geographic latitude deduced from Alouette-1 topside ionograms. From Eccles & King (1969).

(Muldrew 1965; Thomas & Andrews 1968, 1969; Rycroft & Thomas 1970). There is a close association of the trough with the expected position of the plasmapause, although its width (5°–15° in latitude) make it difficult to be specific on which part of the trough maps to the equatorial plasmapause. The electron density trough is also mainly observed at night because of filling by photoionized O^+ during the day. It is, however, also present in light ions (see section on topside ionospheric plasmapause for further discussion of these results).

Another pronounced feature of the topside studied by topside sounders is the equatorial anomaly (King et al 1964, Eccles & King 1969). In the example of the anomaly shown in Figure 11, the symmetrically placed electron density enhancements on either side of the magnetic equator can be seen at altitudes below about 600 km. Theories of the formation of the anomaly have been reviewed by Goldberg (1969); he summarizes a mechanism postulating electrodynamic lifting of the equatorial ionosphere due to electrostatic fields originating in the E region, followed by a downward flow of ionization along the magnetic field lines, resulting in enhanced electron density centered about dip latitude of 20°. There is apparently pronounced asymmetry with longitude–data taken at 110°E shows the anomaly more strongly than at 75°W.

An alternative method of obtaining the average global distribution of electron density is to use in situ measurements from an eccentric orbit satellite with a high inclination orbit. The precession of the line of apsides and the orbital plane results in all altitudes from apogee to perigee and all local times being covered typically in a period of four to six months. The data can then be fitted to a complex function and reduced to a set of coefficients (Willmore 1965), or it can be collected in cells of latitude, altitude, and local time and a mean density found for each cell (Clark et al 1972). An example of the averaging technique used to map out the electron density distribution in a fixed altitude range is shown in Figure 12. This type of presentation smooths out details of spatial or temporal variations comparable to the cell sizes and so does not show evidence of the midlatitude trough or the equatorial anomaly.

Global Distribution of Ion Composition

We have seen from earlier discussions that the major ions expected to be found in the topside ionosphere are O^+ at lower altitudes and H^+ at higher altitudes. The distribution of these major ions is quite variable and the transition altitude from O^+ dominance to H^+ dominance is strongly affected by dynamic processes occuring in the topside ionosphere and by diurnal and seasonal solar energy control. In addition to the two major ion species, the minor atomic ions, O^{2+}, N^+, and He^+, and the minor molecular ions, O_2^+ and NO^+, are detected in the topside ionosphere by satellite-borne mass spectrometers. The ions are constrained to move along geomagnetic field lines; thus, in addition to diurnal solar effects, further diurnal variations are seen because of the offset between the magnetic and geographic poles.

Most global measurements of ion composition have been made by in situ measurements on satellites. Early measurements of the global distribution of the major ion species were made by the Ariel-1 satellite (Raitt et al 1965), which clearly showed an increasing fraction of O^+ ions towards high latitudes even in the situation where the satellite altitude was increasing from about 500 to 1000 km. More comprehensive measurements have been made by H. A. Taylor (1971) using a Bennett radio frequency spectrometer on the OGO-4 satellite. Figure 13(*a–d*) shows families of pole-to-pole profiles of O^+, N^+, H^+, and He^+ density profiles taken at local noon when the satellite apogee was over the equator, thus giving symmetrical altitude variations in each hemisphere. Seasonal and geomagnetic asymmetries can be seen during the course of the 24-hour period covered by the data. At low latitude and near 180° longitude, another feature is the marked equatorial trough in He^+.

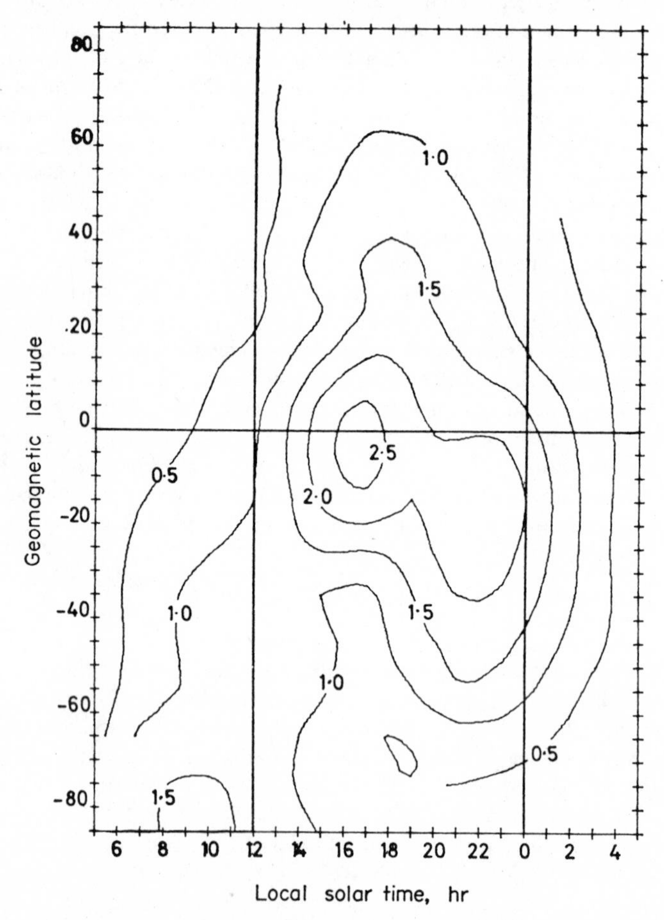

Figure 12 Contours of average electron density ($\times 10^{-5}$ cm^{-3}) for the altitude range 600–700 km, during magnetically quiet times for the period October 1968 to April 1969, computed from ESRO-1 Langmuir probe data. From Clark et al (1972).

This feature can also be seen to a lesser extent in the H^+ densities, but not at all in the O^+ and N^+ densities. This does not conflict with observations of the equatorial anomaly in n_e discussed earlier, because the satellite apogee is above the anomalous region. Recent measurements from ISIS-2 (Hoffman et al 1974) have shown similar global patterns of ion density and, in addition, have identified measurable quantities of O^{2+} ions.

The minor molecular ion distributions have also been studied on a global basis by satellite-borne mass spectrometers. At topside ionospheric altitudes, the molecular ion densities during geomagnetically quiet times are normally very much lower than the major atomic densities as shown in Figure 14 from H. A. Taylor (1974). However, during periods of magnetic disturbance, molecular ion densities in general, and NO^+ densities in particular, can become much enhanced as shown in Figure 15 and Figure 16, again from H. A. Taylor (1974). The high latitude irregularities seen in Figure 15 are thought to be associated with electric field–induced ion temperature increases changing the $N_2+O^+ \rightarrow NO^+ + N$ reaction rate (Schunk et al 1975), although it is likely that the low latitude increases seen in Figure 16 are a consequence of enhanced N_2 density similar to that measured by Prölss & von Zahn (1974). During periods of severe disturbance, such as during the August 1972 magnetic storm, molecular ion densities of about 10^3 cm^{-3} were measured by Hoffman et al (1974) at 1400 km in the high latitude region near dusk.

One of the clearest features seen in pole to pole scans of ion composition is the rapid decline in light ion density poleward of about 60° magnetic latitude. This effect is seen at all local times, although the decline is steeper during the night than during the day. The feature has been named the light ion trough (LIT), and an example is shown in Figure 17 from H. A. Taylor & Walsh (1972). Detailed mapping of gross composition changes from O^+ to H^+ has been inferred from topside sounder records by using the measured scale height of the electron density. Watt (1965) showed the lack of light ions at high latitudes during the night from Alouette-1 ionograms. However, his daytime averages did not show the predominance of O^+ to high altitudes so clearly, probably because of the large local time interval over which the data were averaged. On the other hand, Banks & Doupnik (1974) analyzed a smaller selection of Alouette-2 ionograms and showed that O^+ dominated to 2500 km during the morning sector at invariant latitudes as low as 48°. Watt (1965) interpreted his result in terms of a change in plasma temperature, whereas Banks & Doupnik (1974) invoked an outflow of H^+ ions from the topside F region along magnetic field lines to the outer regions of the plasmasphere. This latter view is generally held to be the reason for the low H^+ densities at high latitudes and is supported by satellite beacon measurements (Titheridge 1974), as well as by numerous incoherent scatter measurements discussed later. Similarly, discussion of the location of the light ion trough and its relation to magnetospheric phenomena are discussed later.

Global Distribution of Plasma Temperature

The global characteristics of electron temperature have been studied more comprehensively than those of the positive ion temperature. Willmore (1965) made a study

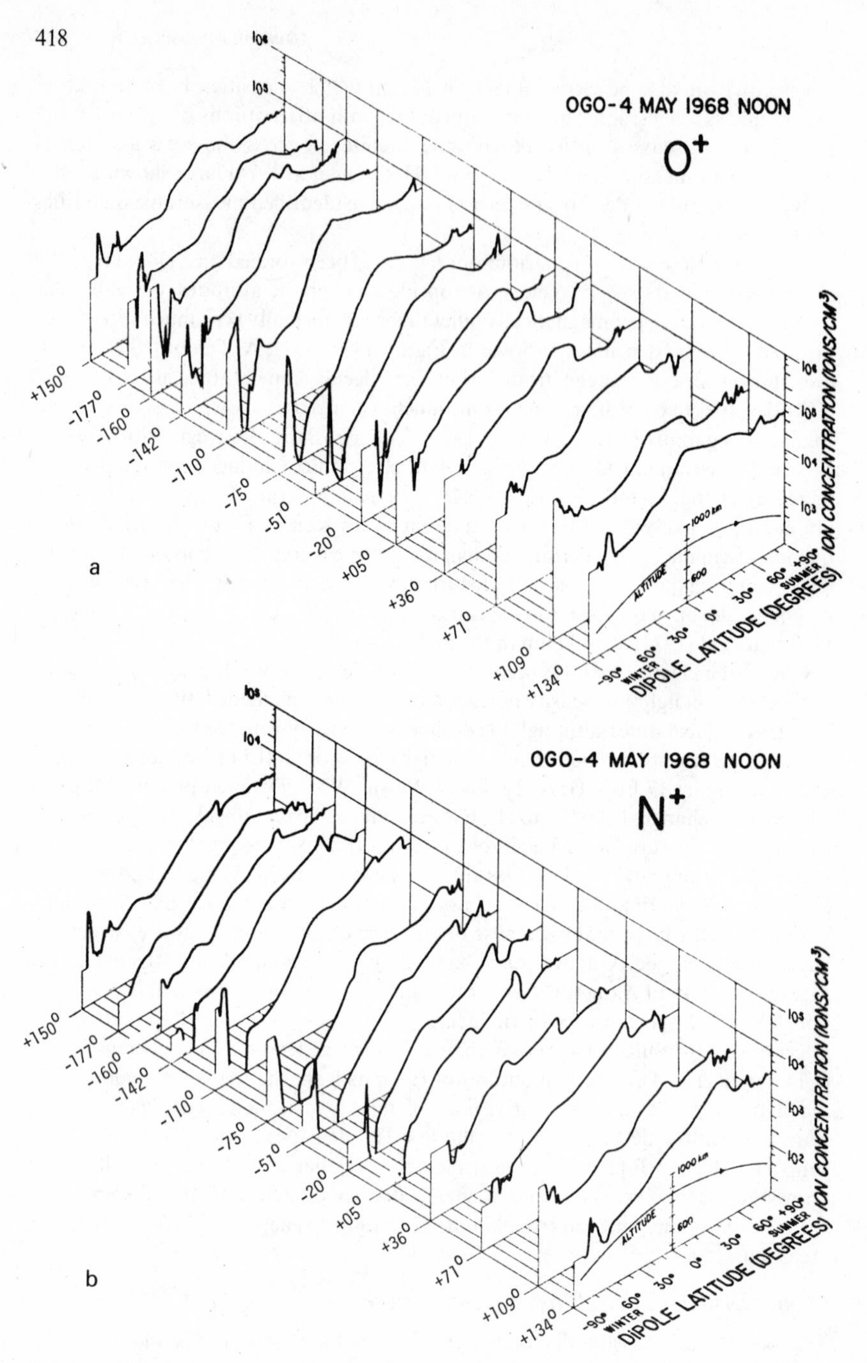

Figure 13 Families of pole-to-pole ion distributions of (*a*) O^+, (*b*) N^+, (*c*) H^+, and (*d*) He^+ for a 24-hour period in May 1968 and near noon local time. The curve to the right of

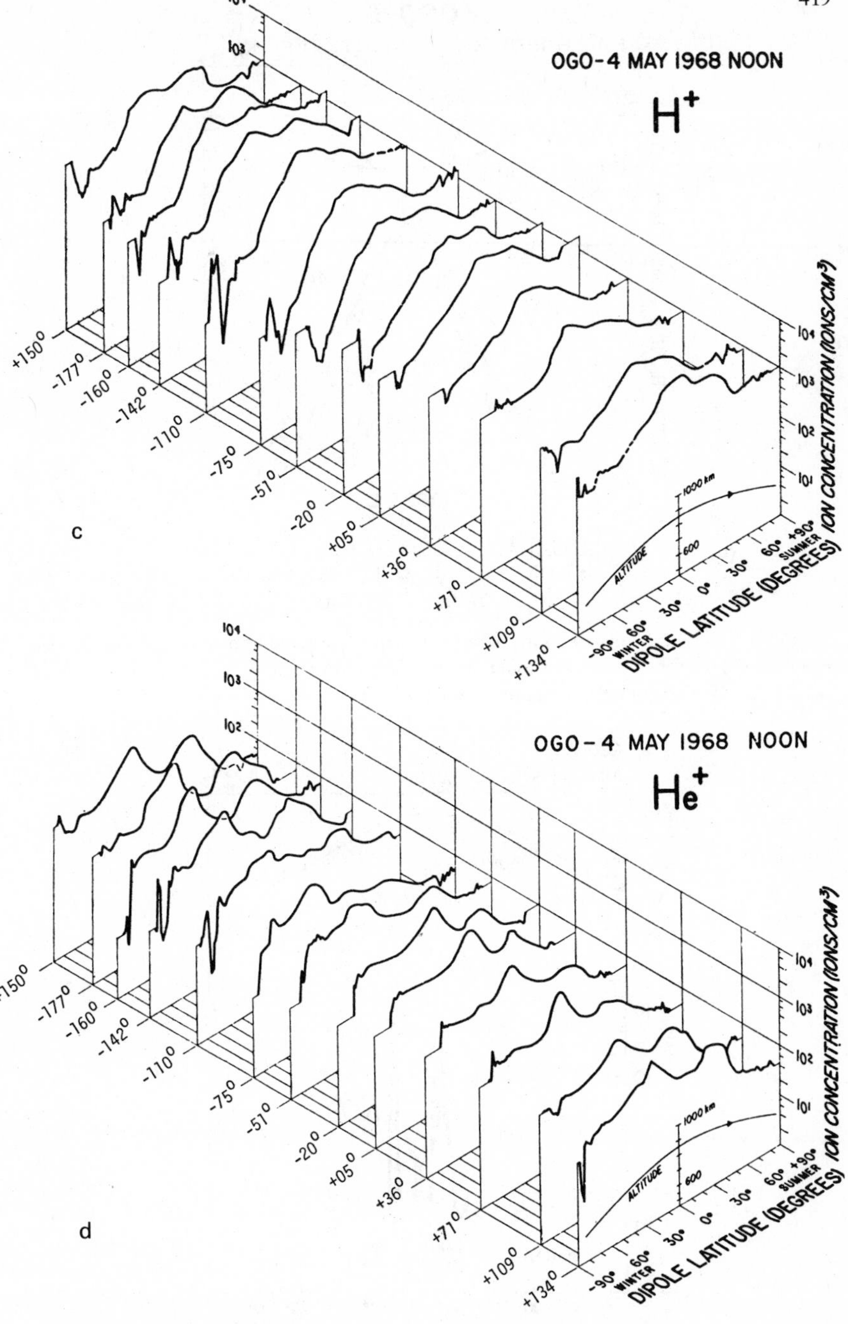

each figure shows the altitude variation for the first profile. Adapted from H. A. Taylor et al (1971). Reproduced by permission of the American Geophysical Union.

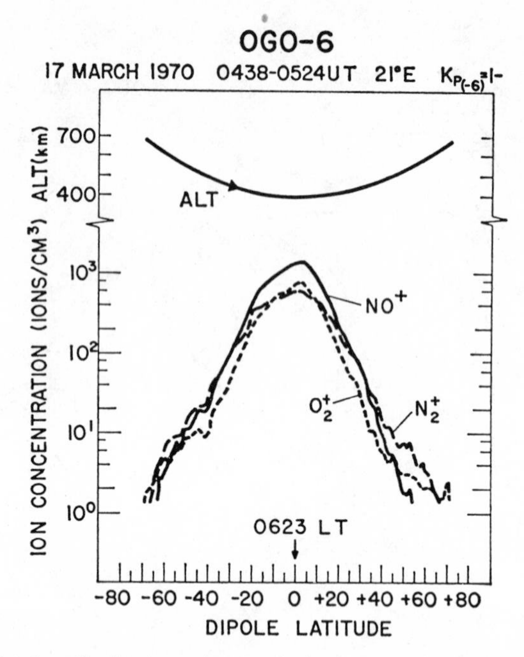

Figure 14 Quiet time distributions of molecular ions observed near dawn local time, near equinox. K_p-6 refers to the maximum value of K_p during the 6-hour interval preceding the measurement. From H. A. Taylor (1974).

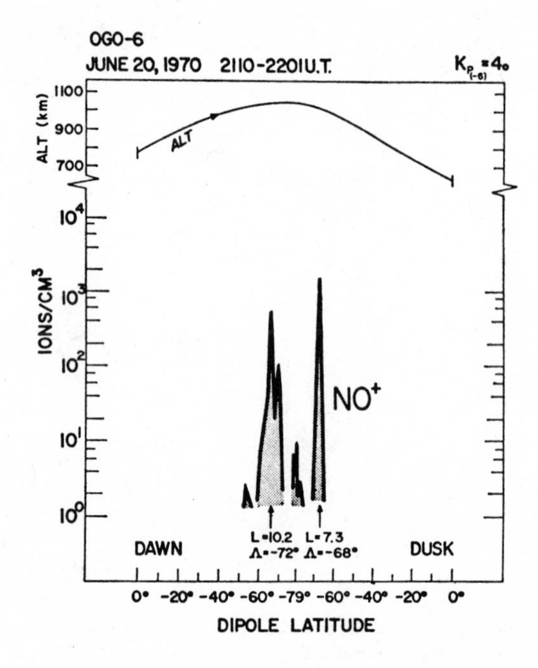

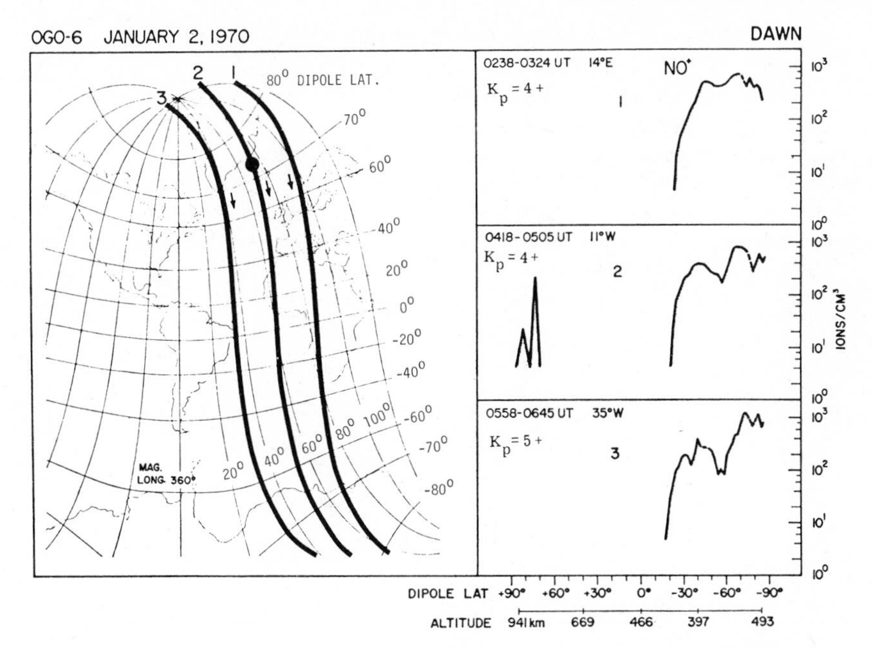

Figure 16 An example of a broad low to midlatitude increase in NO^+ observed in a sequence of three consecutive OGO-6 passes at the geographic position indicated in the left-hand panel. From H. A. Taylor (1974). Reproduced by permission of the American Geophysical Union.

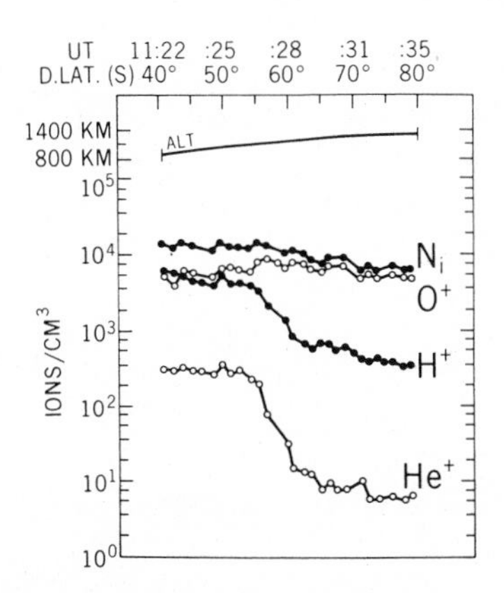

Figure 17 An example of the light ion trough observed in H^+ and He^+ ions; n_e shows no trough effect. Data from OGO-2 (H. A. Taylor & Walsh 1972).

← *Figure 15* An example of pronounced high latitude enhancement of NO^+ seen at high altitudes during a disturbed period. From H. A. Taylor (1974).

of electron temperatures measured by the Ariel-1 satellite and found increases in T_e with both increasing altitude and increasing latitude. Willmore fitted all of the data collected for a six-month period from a Langmuir probe on Ariel-1 and produced a complex empirical function describing the mean global variation of T_e. An example showing both the latitudinal and altitudinal increases is shown in Figure 18 from Willmore (1965). The global mapping of T_e was repeated by Clark et al (1972), using ESRO-1 data and averaging in cells as described earlier for n_e distributions. The results generally agreed with the Ariel-1 results; however, the absolute temperatures differed, and this was attributed to a solar cycle effect with Ariel-1 being active near solar maximum, and ESRO-1 near solar minimum. A map

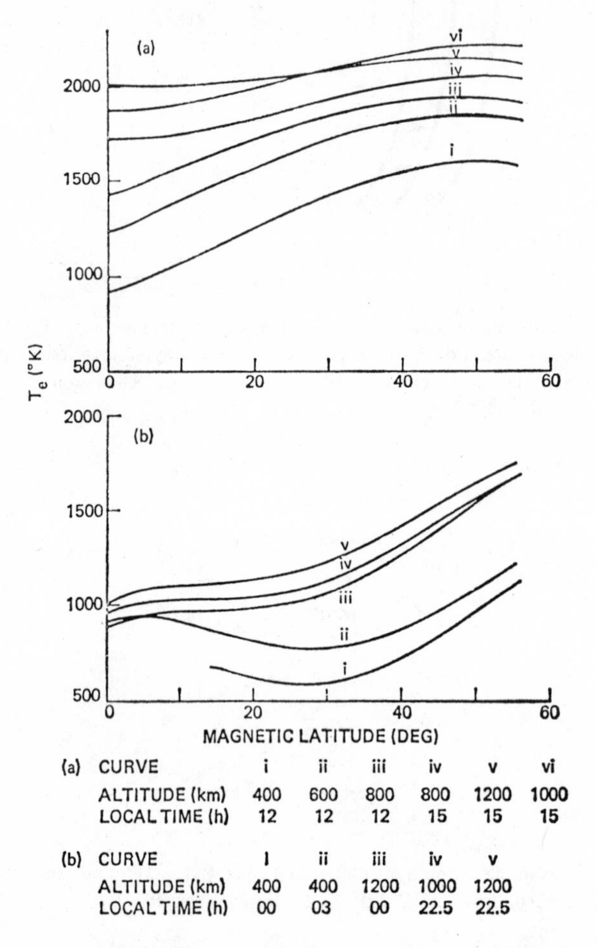

Figure 18 Variations of electron temperature with latitude and altitude from a function fitted to Ariel-1 data. From Willmore (1965).

of the ESRO-1 T_e data is shown in Figure 19; the variation with altitude is shown in Figure 20 by the map of the gradient in T_e, deduced from the mean measured values at 500 km and 1000 km.

The averaged maps or profiles of T_e do not show much of the detail existing in a pole-to-pole section through the ionosphere. An example of this type of structure and its temporal development is shown in Figure 21, after Raitt (1974). This data taken from successive orbits of ESRO-1 before, during, and after a geomagnetic storm shows the buildup of zones of elevated T_e, including the narrow subauroral band responsible for SAR arcs, and a general poleward increase in T_e. A notable feature of the data is the long persistence of the storm effects, which are still apparent 10 days after the storm period.

Most ion temperature data have been obtained from incoherent scatter results

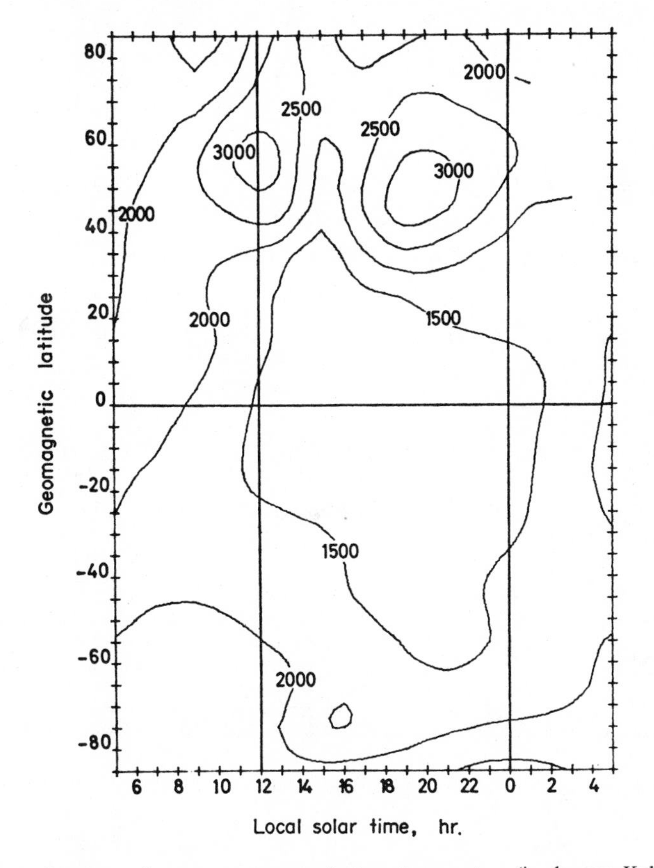

Figure 19 Contours of constant average electron temperature (in degrees Kelvin) as a function of geomagnetic latitude and local time in the altitude range 600–700 km during magnetically quiet times. From Clark et al (1972).

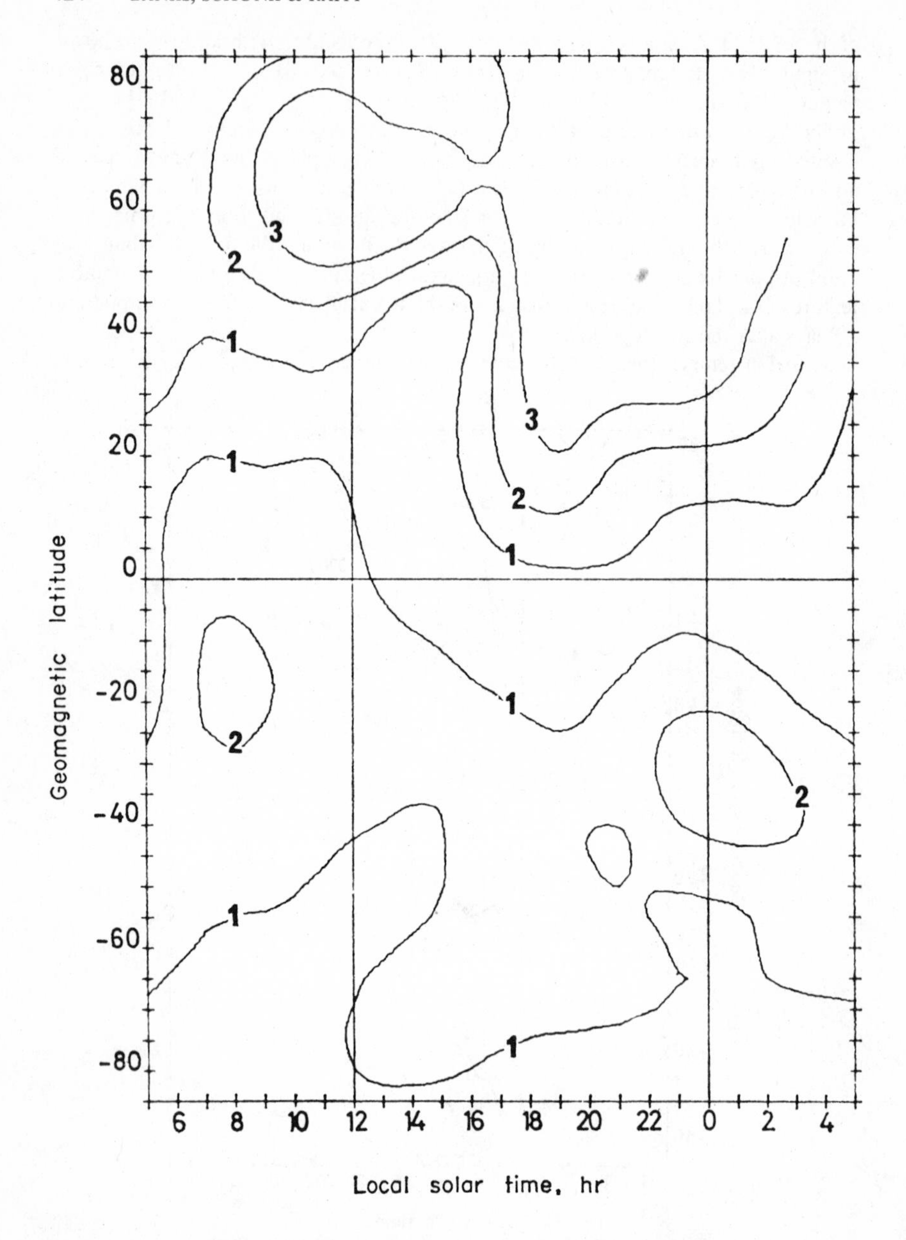

Figure 20 Contours of constant average electron temperature gradient (in degrees Kelvin per kilometer) between 500 km and 1000 km, as a function of geomagnetic latitude and local time during magnetically quiet times. From Clark et al (1972).

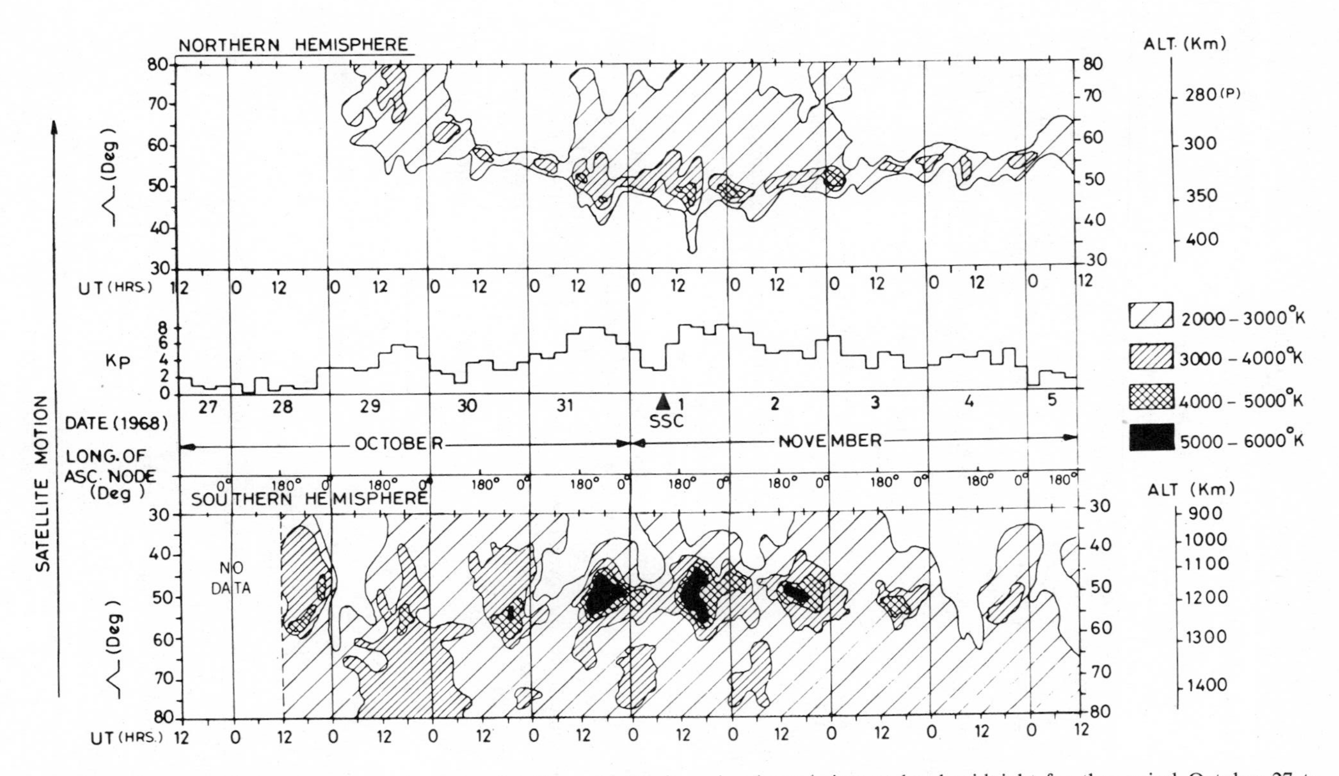

Figure 21 Contours of electron temperature as a function of latitude and universal time at local midnight for the period October 27 to November 5, 1968. The altitude scales are mean variations, and individual orbits may deviate by up to 100 km. From Raitt (1974). Reproduced by permission of the American Geophysical Union.

and are therefore restricted in location. We therefore defer the discussion of ion temperature measurements to the next section. However, some pole to pole profiles of T_i have been measured (cf Harris et al 1967) showing temperatures in the range 400°K–1000°K with large fluctuations and generally higher temperatures in polar regions.

Vertical Sounding of the Topside Ionosphere From Fixed Locations

In contrast to the global measurements of ionospheric parameters obtained from satellite observations (discussed earlier), there have been numerous studies of the topside ionosphere by vertical soundings. The earliest measurements of the topside plasma were made by rocket-borne plasma diagnostic instruments making in situ measurements. An example of a more recent measurement of nighttime ion composition above Wallops Island is shown in Figure 22, from Hoffman (1967). Hanson et al (1969) have studied the vertical plasma temperature profile in conjunction with a coincident Alouette-2 pass to estimate the downward heat flux into the ionosphere. Rocket soundings provide a virtually instantaneous height profile with greater spatial resolution than topside sounders or incoherent scatter radars, but they can tell us very little about latitudinal or temporal changes in the topside ionosphere.

Perhaps the most successful technique of probing the topside ionosphere from the ground has been the incoherent backscatter radar installations at various locations around the world. These equipments have made measurements of electron density and temperature, ion composition and temperature, and plasma drift velocity on regular schedules, enabling a comprehensive picture to be made of the diurnal and seasonal changes in the topside ionosphere in the vicinity of the radar site. Data have been obtained at an equatorial site (Jicamarca, Peru), a low latitude site (Arecibo, Puerto Rico), several midlatitude sites (Millstone Hill, United States; St. Santin, France; and Malvern, United Kingdom) and a high latitude site (Chatanika, Alaska).

An example of the diurnal variation of equatorial profiles of electron density is

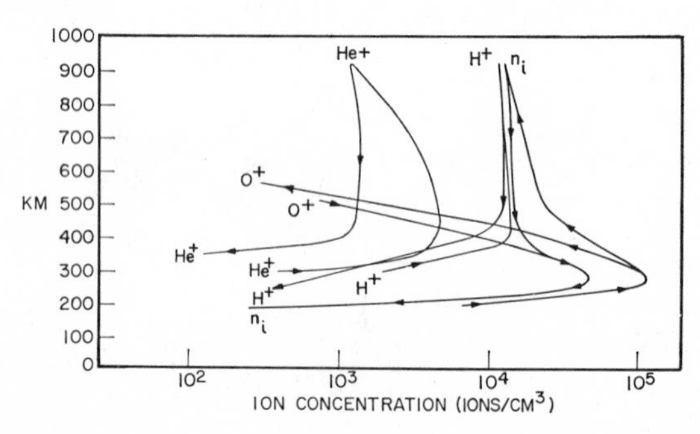

Figure 22 Ion composition measurements from a rocket launch from Wallops Island showing the vertical distribution of O^+, H^+, and He^+. From Hoffman (1967). Reproduced by permission of the American Geophysical Union.

shown in Figure 23, taken from Farley et al (1967), using data from the Jicamarca installation. The plot shows the height profiles of electron density near equinox for a complete cycle of local time interrupted only by the appearance of spread F in the early morning hours. An interesting feature of the equatorial measurements is that the electron and ion temperatures are essentially isothermal between 300 km and 500 km, as shown in Figure 24, again from Farley et al (1967). This behavior contrasts with the measurements at other latitudes showing a strong T_e gradient with altitude due to downward heat conduction along the magnetic field lines. The

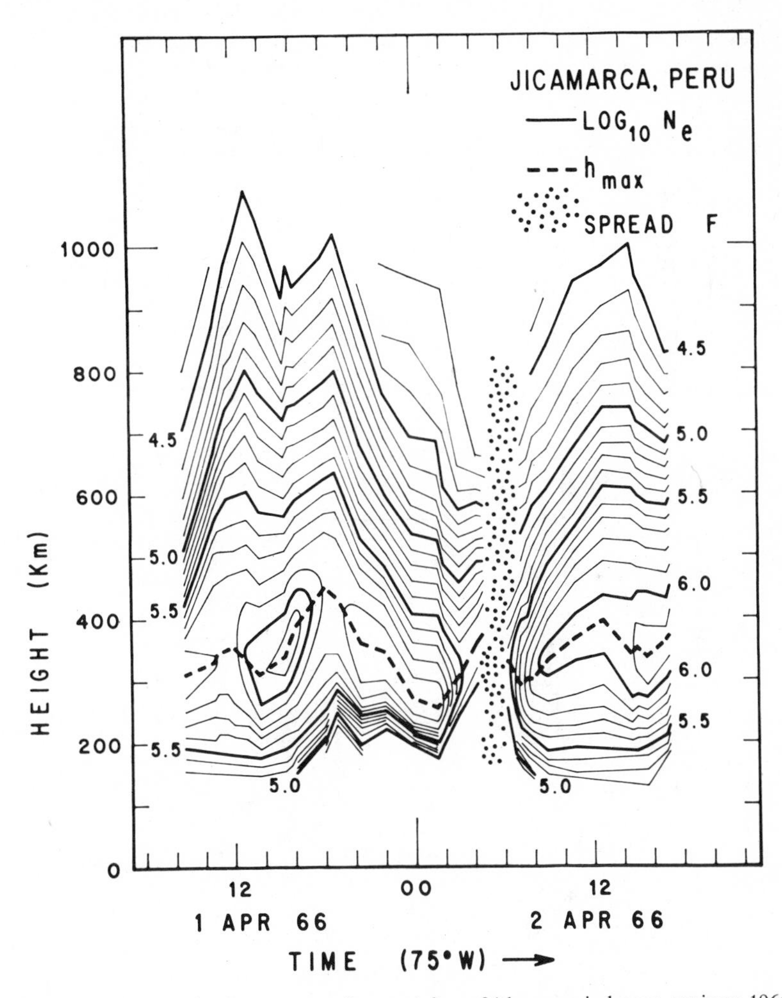

Figure 23 Electron density contours ($\log_{10} n_e$) for a 24-hour period near equinox 1966, taken by the equatorial incoherent scatter installation at Jicamarca, Peru. From Farley et al (1967). Reproduced by permission of the American Geophysical Union.

Arecibo installation has also been used to measure vertical ionospheric drifts (Woodman 1970). The pattern of behavior observed is a vertical drift upwards of about 20 m sec^{-1} by day and a similar magnitude downward drift at night. These drift velocities indicate a perpendicular electric field of about ± 0.5 mV/m and provides experimental confirmation of the existence of the type of electric field invoked to account for the equatorial anomaly in n_e.

Low latitude ionosphere data from the Arecibo ionospheric observatory have been reported by a number of workers including Carlson & Gordon (1966), Prasad (1970), Ho & Moorcroft (1971), and Hagen & Hsu (1974). An example of the capabilities of the incoherent scatter technique to resolve topside ionospheric composition is shown in Figure 25, from Hagen and Hsu (1974). Figure 25*a* shows

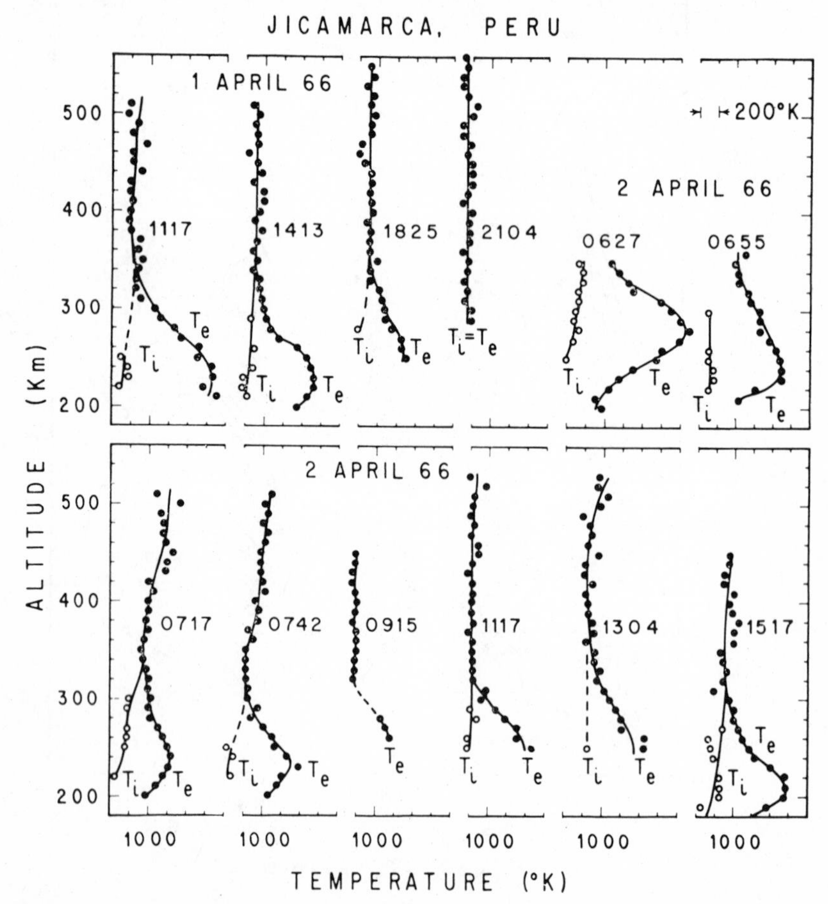

Figure 24 A sequence of electron and ion temperatures from incoherent scatter data in the topside ionosphere above Jicamarca, Peru. These profiles correspond to times within the period covered by the data shown in Figure 23. From Farley et al (1967). Reproduced by permission of the American Geophysical Union.

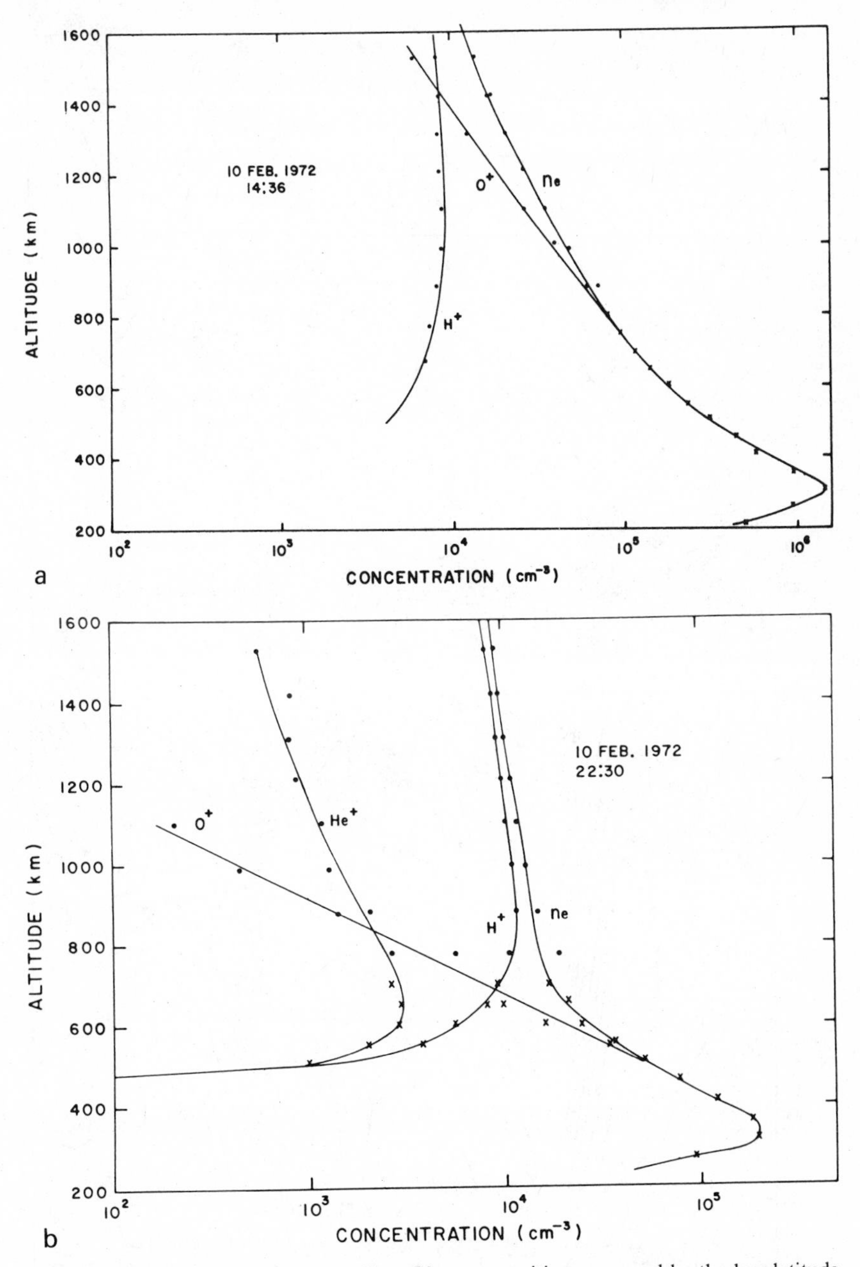

Figure 25 Examples of vertical profiles of ion composition measured by the low latitude incoherent scatter installation at Arecibo, Puerto Rico. The profiles correspond to (*a*) daytime and (*b*) nighttime. From Hagen & Hsu (1974). Reproduced by permission of the American Geophysical Union.

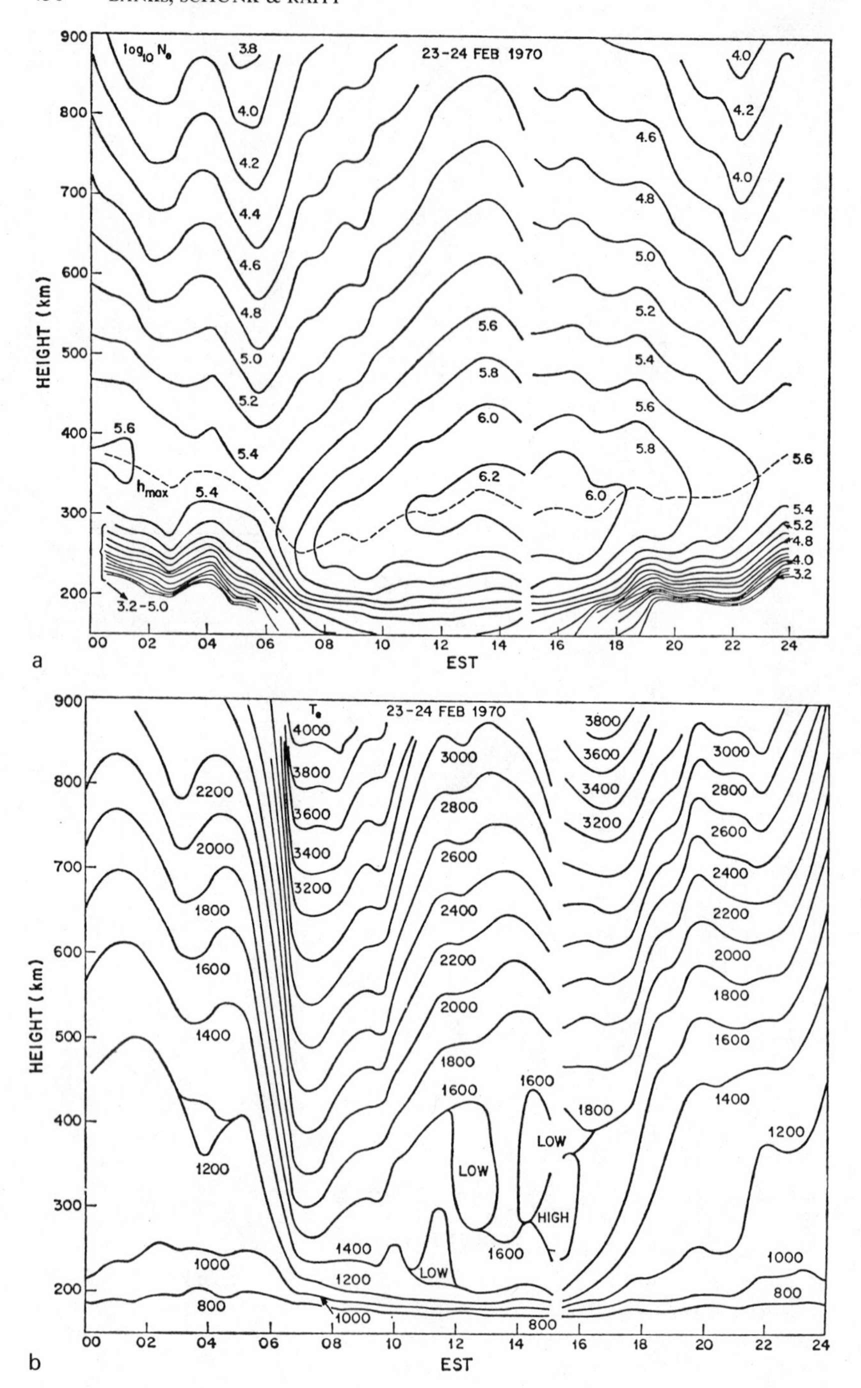

a
log10 Ne
23-24 FEB 1970
HEIGHT (km)
EST
hmax
3.2-5.0
b
Te
23-24 FEB 1970
HEIGHT (km)
EST
LOW
HIGH

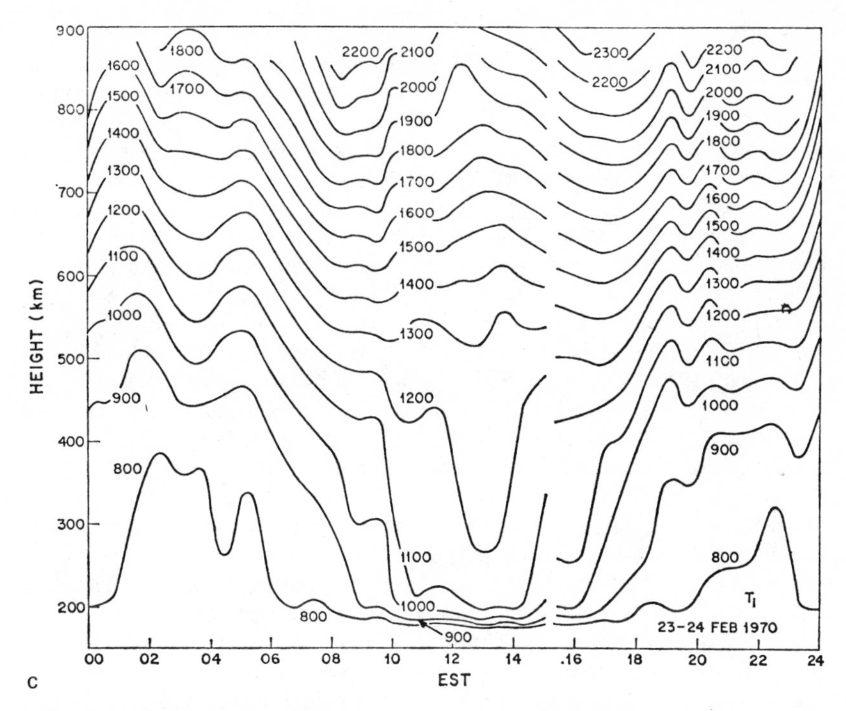

Figure 26 Contours of the principal ionospheric parameters (*a*) $\log_{10} n_e$, (*b*) T_e, and (*c*) T_i, measured in the F region and topside ionosphere above Millstone Hill. The data cover a 24-hour period in February 1970. From Evans (1974).

a daytime profile showing an O^+–H^+ transitional altitude occurring at about 1400 km, and the nightime profile shown in Figure 25*b* shows the transition altitude at 600 km and the appearance of measurable He^+ densities. Hagen & Hsu (1974) have presented vertical plasma velocity measurements above Arecibo showing a diurnal variation from upflow by day to downflow by night, and Ho & Moorcroft (1971) have made similar deductions based on the diurnal variation of the plasma density. The measurements support the concept of an ebb and flow of ionization in the form of an upward and downward flux of H^+ ions, but with flux values far less than the flux-limited condition. This, then, accounts for the near diffusive equilibrium form of both the day and night ion density profiles of Hagen & Hsu (1974) shown in Figure 25.

The incoherent backscatter data taken at the mid- and high-latitude sites have been of particular interest in studying the dynamic processes coupling the ionosphere to the outer regions of the plasmasphere and to the magnetosphere. The Millstone Hill site, at an *L* value of 3.2, has been used for numerous measurements relating to the dynamic coupling between the ionosphere and protonosphere. Evans (1965) made a detailed study of the seasonal and diurnal properties of the F_2 and topside variations. He has also reported on the distribution of vertical plasma velocities and their interpretation in terms of fluxes in and out of the protonosphere (Evans 1971a,b,c). Measurements of horizontal drifts have also been made at Millstone Hill (Evans 1972) which indicate that the drift velocity at this latitude is normally about 25 m sec^{-1} and rotates 360° in direction during a period of 12 hours. The behavior is consistent with electric fields generated by the dynamo effect; however, following a magnetic storm, a much-enhanced drift velocity was observed with a peak velocity of 200 m sec^{-1}. This high velocity is thought to result from penetration of magnetospheric electric fields to the Millstone Hill latitude following contraction of the plasmasphere, a well-known storm effect. Large flow velocities have also been reported at Malvern by G. N. Taylor (1974) and at St. Santin by Vasseur (1969, 1970) and Amayenc & Vasseur (1972). The diurnal variation of the principal topside ionospheric parameters (n_e, T_e, and T_i) above Millstone Hill for a 24-hour period in February 1970 are shown in Figure 26(*a–c*), taken from Evans (1974). The pronounced T_e and T_i gradients with altitude can be seen together with the rapid increase in T_e at dawn, and the less rapid decrease at dusk. The similar, but more sluggish, behavior of n_e and T_i at dawn and dusk can also be seen on the figure.

The more recently installed Chatanika facility at $L = 5.8$ has provided a powerful ground-based instrument for studying the complexities of the electron density distribution in the auroral zone and in the plasma trough. It has also enabled measurements of the plasma drift velocity and hence the perpendicular electric field to be made. The general diurnal pattern of these electric fields (Doupnik et al 1972, Banks & Doupnik 1975) is consistent with theories of magnetospheric convection electric fields penetrating to the high latitude ionosphere (Axford 1969). An investigation made during a geomagnetically disturbed period (Banks et al 1973) showed the appearance of a large westward electric field at the time of a westward travelling surge observed by a network of all sky cameras.

The Topside Ionospheric Plasmapause

One interesting dynamic region of the topside ionosphere, which has received considerable attention, is the projection of the equatorial plasmapause to ionospheric heights. This boundary separates the torus of cool, relatively static plasma, referred to as the plasmasphere, from the much more dynamic region of plasma at high latitudes, whose distribution is often more influenced by magnetospheric electric fields and particle fluxes than by solar electromagnetic radiation.

Recent measurements of the morphology of the plasmapause in the equatorial plane have been summarized by Chappell (1972). The toroidal shape of the plasmasphere was mapped out by Brinton et al (1968), who showed that the average plasmapause was bounded by the $L = 4.5$ magnetic shell. At lower altitudes, the association of the midlatitude electron density trough with the plasmapause was investigated, but the effect of photoionization filling the trough during daytime was a limitation. The use of mass spectrometers enabled the ion species to be studied individually, and it was seen that the latitudinal distribution of the light (H^+ and He^+) ions was a superior indicator of the ionospheric plasmapause.

The light ion trough (LIT) was described earlier, and an example is shown in Figure 17, from H. A. Taylor & Walsh (1972). The reduction in H^+ density with latitude is considered to be associated with increasing upward fluxes of H^+ ions at increasing latitudes. The theoretical H^+ density altitude profiles discussed earlier show that a change from diffusive equilibrium to a modest upward flux of H^+ can result in an order of magnitude decrease in H^+ density at, say, 1000 km. Thus, we can view the LIT observed at a constant altitude as a progression from the H^+ density altitude profile represented by near diffusive equilibrium at the equator to a flux-limited polar wind flow at high latitudes. There have been few direct measurements of H^+ density altitude profiles, but by using time averages of satellite data, Brinton et al (1971) showed the transition between a diffusive equilibrium type of H^+ density profile at low latitudes to a profile characteristic of outflowing H^+ ions at high latitudes.

The location of the band of latitudes representing the transition from the region of diurnal ebb and flow of light ions to the continuous outflow region is probably the most physically meaningful definition of the ionospheric plasmapause. It is difficult to identify this position by just using an arbitrary density cutoff, as was done by Brace & Theis (1974). On the other hand, such a point is related to the boundary and is a parameter readily obtained from satellite measurements. The theoretical studies of this region discussed earlier indicate that the outward flow velocity and the H^+ temperature are also important parameters that show marked changes at the plasmapause. These two parameters are difficult to measure, but there has been published direct measurement of the upward H^+ velocity by Maier & Hoffman (1974), who observed H^+ velocities in the range 3.7 to 10.8 km sec^{-1} outside the plasmasphere. These were, however, spot values and could not be used to identify the plasmapause. There have also been few published results of light ion temperature measurements; however, Maier & Hoffman (1974) have analyzed some ISIS-2 data and found that in the polar wind $T(H^+) > T(O^+)$ is true, as predicted by theoretical calculations. The absolute values of $T(H^+)$ have been measured as high as 9486°K, when the outflow velocity was 6.7 km sec^{-1}. However, sufficient data have not yet been published to establish any clear experimental correlation between $T(H^+)$ and the H^+ outflow velocity.

The location of the plasmapause has also been studied experimentally by ground-based observations, principally by using whistler measurements (Park 1970, 1973, 1974; Carpenter & Park 1973). Although this is essentially a fixed location observation, the direction-finding ability of the whistler detection equipment does enable latitudinal scans of the plasmapause to be made.

A further feature of plasmasphere/plasmapause measurements at ionospheric levels is the high degree of variability of the decline in H^+ density with latitude. Although some of this effect is probably caused by local variations in the ionospheric plasma, other fluctuations in the otherwise monotonic decline in H^+ density with increasing latitude have been attributed to detached plasma regions observed near the equatorial plasmapause (Chappell 1974, Chen & Grebowsky 1974). Irregular structure in the duskside high altitude H^+ density measurements were reported by H. A. Taylor et al (1970) and at lower altitudes by H. A. Taylor et al (1971) and H. A. Taylor & Cordier (1974). A sequence of H^+ LIT measurements are shown in Figure 27, taken from H. A. Taylor & Cordier (1974). The figure shows

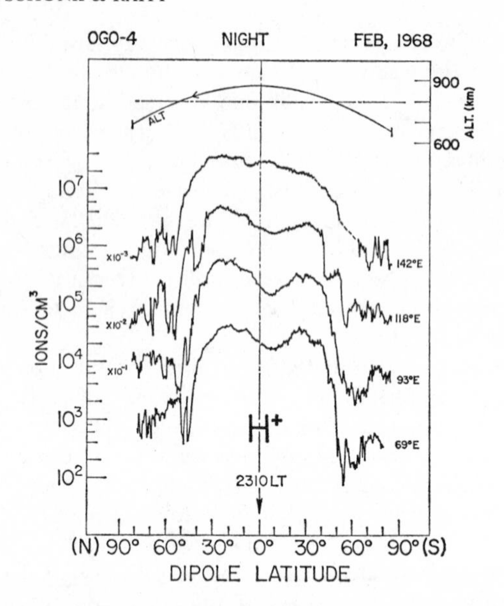

Figure 27 A sequence of pole-to-pole profiles of H^+ density showing the light ion trough and the irregularities observed within the trough. The densities are scaled by the figure on the left-hand side of each plot to make the diagram more legible. From H. A. Taylor & Cordier (1974).

numerous subtroughs and H^+ density recoveries within the main LIT. These enhanced H^+ density features are postulated to represent regions of the topside ionosphere where the H^+ ions have a smaller outflow velocity than the ions at adjoining latitudes because they are linked to regions of enhanced H^+ density in the equatorial plane. These regions of enhanced densities are attributed to tails or detached islands of thermal plasma separating from the plasmasphere at the plasmapause.

CONCLUSION

At present, theoretical models of the topside ionosphere are adequate to explain the broad ranges of variability noted on a global scale. The concept of plasma flow to and from the ionosphere and protonosphere along magnetic field lines in response to changes in ionospheric and magnetospheric plasma pressures and electric fields appears well substantiated. However, even within the relatively quiescent plasmasphere, where nearly simultaneous measurements of the relevant plasma parameters are available, the experiments have not been precise enough to test the importance of a number of predicted effects such as thermal diffusion or thermoelectric transport processes. Even greater uncertainty exists for regions outside the plasmasphere, because no simultaneous measurements of plasma density, flow velocity, temperatures, and atmospheric densities have yet been made. Thus, although high speed flows of light ions are known to occur at high latitudes, their correspondence to the predictions of current theoretical models remains unchecked. This is especially bothersome in the case of supersonic flows owing to the possibility that plasma instabilities, unmentioned in the previous sections, may arise to prevent large relative velocities between the different ion species. In addition, in situ measurements of H^+ production and loss rates remain unverified outside the laboratory, keeping open the question of actual equilibrium times or the magnitudes of H^+ escape fluxes in the polar wind.

The resolution of many of these difficulties is difficult because profiles of the relevant variables along magnetic field lines are required. Proposed spacecraft missions, such as the NASA-sponsored Electrodynamics Explorer, may help to better match experiment with theory. However, the problems are complex, and a well-designed observational program is required before meaningful tests of the theory can be made.

ACKNOWLEDGMENTS

This research was supported in part by the National Aeronautics and Space Administration and the National Science Foundation. W.J.R. would like to acknowledge the support and interest of the Director, Professor R. L. F. Boyd, C.B.E., F.R.S., and other colleagues of the Mullard Space Science Laboratory.

APPENDIX A

The momentum transfer collision frequencies and the velocity dependent correction factors are given by (Schunk & Walker 1970a, 1973; Schunk 1975; Burgers 1969)

$$\nu(O^+, N_2) = 6.82 \times 10^{-10} n(N_2)$$

$$\nu(O^+, O_2) = 6.66 \times 10^{-10} n(O_2)$$

$$\nu(O^+, O) = 3.42 \times 10^{-11} n(O) T_r^{1/2} [1.08 - 0.139 \log_{10} T_r + 4.51 \times 10^{-3} (\log_{10} T_r)^2]$$

$$\nu(O^+, H^+) = 0.084 \frac{n(H^+)}{T^{3/2}(H^+)}$$

$$\nu(O^+, e) = 1.87 \times 10^{-3} \frac{n_e}{T_e^{3/2}}$$

$$\nu(H^+, N_2) = 3.36 \times 10^{-9} n(N_2)$$

$$\nu(H^+, O_2) = 3.20 \times 10^{-9} n(O_2)$$

$$\nu(H^+, O) = 6.61 \times 10^{-11} n(O) [1 - 0.047 \log_{10} T(H^+)]^2 T^{1/2}(H^+)$$

$$\nu(H^+, O^+) = 1.35 \frac{n(O^+)}{T^{3/2}(H^+)}$$

$$\nu(H^+, H^+) = 0.9 \frac{n(H^+)}{T^{3/2}(H^+)}$$

$$\nu(\mathrm{H}^+, e) = 0.03\,\frac{n_e}{T_e^{3/2}}$$

$$\Phi(\mathrm{H}^+,\mathrm{O}^+) = \frac{3\sqrt{\pi}}{4}\frac{\operatorname{erf}[\varepsilon(\mathrm{H}^+,\mathrm{O}^+)]}{\varepsilon^3(\mathrm{H}^+,\mathrm{O}^+)} - \frac{3}{2}\frac{\exp[-\varepsilon^2(\mathrm{H}^+,\mathrm{O}^+)]}{\varepsilon^2(\mathrm{H}^+,\mathrm{O}^+)}$$

$$\Phi(\mathrm{H}^+,\mathrm{O}) = \frac{3\sqrt{\pi}}{8}\left[\varepsilon(\mathrm{H}^+,\mathrm{O}) + \frac{1}{\varepsilon(\mathrm{H}^+,\mathrm{O})} - \frac{1}{4\varepsilon^3(\mathrm{H}^+,\mathrm{O})}\right]\operatorname{erf}[\varepsilon(\mathrm{H}^+,\mathrm{O})] + \frac{3}{8}\left[1 + \frac{1}{2\varepsilon^2(\mathrm{H}^+,\mathrm{O})}\right]\exp[-\varepsilon^2(\mathrm{H}^+,\mathrm{O})]$$

$$\Phi(\mathrm{H}^+,\mathrm{N}_2) = \Phi(\mathrm{H}^+,\mathrm{O}_2) = 1$$

$$\Psi(\mathrm{H}^+,\mathrm{O}^+) = \exp[-\varepsilon^2(\mathrm{H}^+,\mathrm{O}^+)]$$

$$\Psi(\mathrm{H}^+,\mathrm{O}) = \frac{\sqrt{\pi}}{2}\left[\varepsilon(\mathrm{H}^+,\mathrm{O}) + \frac{1}{2\varepsilon(\mathrm{H}^+,\mathrm{O})}\right]\operatorname{erf}[\varepsilon(\mathrm{H}^+,\mathrm{O})] + \tfrac{1}{2}\exp[-\varepsilon^2(\mathrm{H}^+,\mathrm{O})]$$

$$\Psi(\mathrm{H}^+,e) = \exp[-\varepsilon^2(\mathrm{H}^+,e)]$$

$$\Psi(\mathrm{H}^+,\mathrm{N}_2) = \Psi(\mathrm{H}^+,\mathrm{O}_2) = 1$$

where

$$\varepsilon(\mathrm{H}^+,\mathrm{O}^+) \approx \frac{|u(\mathrm{H}^+)|}{[2kT(\mathrm{H}^+)/m(\mathrm{H}^+)]^{1/2}}$$

$$\varepsilon(\mathrm{H}^+,\mathrm{O}) \approx \frac{|u(\mathrm{H}^+)|}{[2kT(\mathrm{H}^+)/m(\mathrm{H}^+)]^{1/2}}$$

$$\varepsilon(\mathrm{H}^+,e) \approx \frac{|u_e - u(\mathrm{H}^+)|}{(2kT_e/m_e)^{1/2}}$$

and

$$T_r = \frac{T(\mathrm{O}^+) + T_n}{2}.$$

The velocity dependent correction factors are symmetric so that $\Phi(i, j) = \Phi(j, i)$ and $\Psi(i, j) = \Psi(j, i)$. The momentum transfer collision frequencies satisfy the relation $n(i)m(i)\nu(i, j) = n(j)m(j)\nu(j, i)$.

Appendix B

The quantities z_{st}, z'_{st}, and z''_{st} depend only on the form of the interaction potential, whereas $D^{(1)}_{st}$ and $D^{(4)}_{st}$ depend on the particle masses in addition to the interaction potential. General expressions for z_{st}, z'_{st}, z''_{st}, $D^{(1)}_{st}$, and $D^{(4)}_{st}$ are given by Burgers (1969) and Schunk (1975). For ionospheric studies, the relevant interaction potentials are coulomb

$$z_{st} = \tfrac{3}{5}, \quad z'_{st} = \tfrac{13}{10}, \quad z''_{st} = 2$$

$$D_{st}^{(1)} = (3m_s^2 + \tfrac{1}{10}m_s m_t - \tfrac{1}{5}m_t^2)/(m_s + m_t)^2$$

$$D_{st}^{(4)} = (\tfrac{6}{5}m_t^2 - \tfrac{3}{2}m_s m_t)/(m_s + m_t)^2$$

hard sphere

$$z_{st} = -\tfrac{1}{5}, \quad z'_{st} = \tfrac{13}{10}, \quad z''_{st} = 2$$

$$D_{st}^{(1)} = (3m_s^2 + \tfrac{21}{10}m_s m_t + \tfrac{9}{5}m_t^2)/(m_s + m_t)^2$$

$$D_{st}^{(4)} = (\tfrac{16}{5}m_t^2 + \tfrac{1}{2}m_s m_t)/(m_s + m_t)^2$$

Maxwell molecule

$$z_{st} = 0, \quad z'_{st} = 1, \quad z''_{st} = 2$$

$$D_{st}^{(1)} = (3m_s^2 + m_t^2 + \tfrac{8}{5}m_s m_t)/(m_s + m_t)^2$$

$$D_{st}^{(4)} = \tfrac{12}{5}m_t^2/(m_s + m_t)^2$$

Literature Cited

Amayenc, P., Vasseur, G. 1972. Neutral winds deduced from incoherent scatter observations and their theoretical interpretation. *J. Atmos. Terr. Phys.* 34:351–64

Axford, W. I. 1969. Magnetospheric convection. *Rev. Geophys. Space Phys.* 7: 421–59

Bailey, G. J., Moffett, R. J. 1974. Temperatures in the polar wind. *Planet. Space Sci.* 22:1193–99

Banks, P. M. 1967. Ion temperature in the upper atmosphere. *J. Geophys. Res.* 72: 3365–85

Banks, P. M. 1969. Plasma transport in the topside polar ionosphere. In *The Polar Ionosphere and Magnetospheric Processes,* ed. G. Skovli, pp. 193–208. New York: Gordon & Breach

Banks, P. M. 1972. Behavior of thermal plasma in the magnetosphere and topside ionosphere. In *Critical Problems of Magnetospheric Physics,* ed. E. R. Dyer, pp. 157–78. Washington DC: Inter-Union Comm. Solar Terr. Phys., N.A.S.

Banks, P. M. 1973. Ion heating in thermal plasma flows. *J. Geophys. Res.* 78:3186

Banks, P. M., Doupnik, J. R., Akasofu, S.-I. 1973. Electric field observations by incoherent scatter radar in the auroral zone. *J. Geophys. Res.* 78:6607–22

Banks, P. M., Doupnik, J. R. 1974. Thermal proton flow in the plasmasphere: the morning sector. *Planet. Space Sci.* 22: 79–94

Banks, P. M., Doupnik, J. R. 1975. A review of auroral zone electrodynamics deduced from incoherent scatter radar observations. *J. Atmos. Terr. Phys.* 37:951–72

Banks, P. M., Holzer, T. E. 1968. The polar wind. *J. Geophys. Res.* 73:6846–54

Banks, P. M., Holzer, T. E. 1969a. Features of plasma transport in the upper atmosphere. *J. Geophys. Res.* 74:6304–16

Banks, P. M., Holzer, T. E. 1969b. High-latitude plasma transport: the polar wind. *J. Geophys. Res.* 74:6317–32

Banks, P. M., Kockarts, G. 1973. *Aeronomy.* New York: Academic

Banks, P. M., Nagy, A. F., Axford, W. I. 1971. Dynamical behavior of thermal protons in the mid-latitude ionosphere and magnetosphere. *Planet. Space Sci.* 19: 1053–67

Banks, P. M., Schunk, R. W., Raitt, W. J. 1974. Temperature and density structure of thermal proton flows. *J. Geophys. Res.* 79:4691–4702

Bauer, S. J. 1966. The structure of the topside ionosphere. In *Electron Density Profiles in the Ionosphere and Exosphere,* ed. J. Frihagen. Amsterdam: North-Holland

Brace, L. H., Theis, R. F. 1974. The behavior of the plasmapause at mid-

latitudes: ISIS-1 Langmuir probe measurements. *J. Geophys. Res.* 79: 1871–84
Bradbury, N. E. 1938. Ionization, negative-ion formation and recombination in the ionosphere. *Terr. Magn. Atmos. Electr.* 43: 55–66
Brinton, H. C., Grebowsky, J. M., Mayr, H. G. 1971. Altitude variation of ion composition in the mid-latitude trough region: evidence for upward plasma flow. *J. Geophys. Res.* 76: 3738–45
Brinton, H. C., Rickett, R. A., Taylor, H. A. 1968. Thermal ion structure of the plasmasphere. *Planet. Space Sci.* 16: 899–910
Burgers, J. M. 1969. *Flow Equations for Composite Gases.* New York: Academic
Carlson, H. C., Gordon, W. E. 1966. Radar spectrographic estimates of ionic composition from 225 to 1400 km for solar minimum winter and summer conditions. *J. Geophys. Res.* 71: 5573–78
Carpenter, D. L., Park, C. G. 1973. On what ionospheric workers should know about the plasmapause-plasmasphere. *Rev. Geophys. Space Phys.* 11: 133–54
Chapman, S., Cowling, T. G. 1970. *The Mathematical Theory of Non-Uniform Gases.* Cambridge, Engl: Cambridge Univ. Press
Chappell, C. R. 1972. Recent satellite measurements of the morphology and dynamics of the plasmasphere. *Rev. Geophys. Space Phys.* 10: 951–79
Chappell, C. R. 1974. Detached plasma regions in the magnetosphere. *J. Geophys. Res.* 79: 1861–70
Chen, A. J., Grebowsky, J. M. 1974. Plasma tail interpretations of pronounced detached plasma regions measured by OGO-5. *J. Geophys. Res.* 79: 3851–55
Chew, G. F., Goldberger, M. L., Low, F. E. 1956. The Boltzmann equation and the one-fluid hydromagnetic equations in the absence of particle collisions. *Proc. R. Soc. London, Ser. A* 236: 112–18
Clark, D. H., Raitt, W. J., Willmore, A. P. 1972. The global morphology of electron temperature in the topside ionosphere as measured by an A.C. Langmuir probe. *J. Atmos. Terr. Phys.* 34: 1865–80
Dessler, A. J., Cloutier, P. A. 1969. Discussion of the letter by Peter M. Banks and Thomas E. Holzer, "The polar wind." *J. Geophys. Res.* 74: 3730–33
Doupnik, J. R., Banks, P. M., Baron, M. J., Rino, C. L., Petriceks, J. 1972. Direct measurements of plasma drift velocities at high magnetic latitudes. *J. Geophys. Res.* 77: 4268–71
Dungey, J. W. 1955. *The Physics of the Ionosphere.* Phys. Soc. London
Eccles, D., King, J. W. 1969. A review of topside sounder studies of the equatorial ionosphere. *Proc. IEEE* 57: 1012–18
Evans, J. V. 1965. Cause of the mid-latitude evening increase in foF_2. *J. Geophys. Res.* 70: 1175–86
Evans, J. V. 1971a. Observations of F-region vertical velocities at Millstone Hill: I. Evidence for drifts due to expansion, contraction and winds. *Radio Sci.* 6: 609–26
Evans, J. V. 1971b. Observations of F-region vertical velocities at Millstone Hill: II. Evidence for fluxes into and out of the protonosphere. *Radio Sci.* 6: 843–54
Evans, J. V. 1971c. Observations of F-region vertical velocities at Millstone Hill: III. Determination of altitude distribution of H^+. *Radio Sci.* 6: 855
Evans, J. V. 1972. Measurements of horizontal drifts in the E- and F-regions at Millstone Hill. *J. Geophys. Res.* 77: 2341–52
Evans, J. V. 1974. Some post-war developments in ground based radiowave sounding of the ionosphere. *J. Atmos. Terr. Phys.* 36: 2183–2234
Farley, D. T., McClure, J. P., Sterling, D. L., Green, J. L. 1967. Temperature and composition of the equatorial ionosphere. *J. Geophys. Res.* 72: 5837–52
Ferraro, V. C. A. 1945. Diffusion of ions in the ionosphere. *Terr. Magn. Atmos. Electr.* 50: 215–22
Geisler, J. E. 1967. On the limiting daytime flux of ionization into the protonosphere. *J. Geophys. Res.* 72: 81–85
Geisler, J. E., Bowhill, S. A. 1965. Exchange of energy between the ionosphere and the protonosphere. *J. Atmos. Terr. Phys.* 27: 1119–46
Goldberg, R. A. 1969. A review of theories concerning the equatorial F_2 region ionosphere. *Proc. IEEE* 57: 1119–26
Grad, H. 1949. On the kinetic theory of rarefied gases. *Commun. Pure Appl. Math.* 2: 331–407
Grad, H. 1958. Principles of the kinetic theory of gases. *Handb. Phys.* 12: 205–94
Hagen, J. B., Hsu, P. Y.-S. 1974. The structure of the protonosphere above Arecibo. *J. Geophys. Res.* 79: 4269–75
Hanson, W. B., Ortenburger, I. B. 1961. The coupling between the protonosphere and the normal F-region. *J. Geophys. Res.* 66: 1425–35
Hanson, W. B., Patterson, T. N. L. 1964. The maintenance of the night-time F-layer. *Planet. Space Sci.* 12: 979–97
Hanson, W. B., Sanatani, S., Brace, L. H., Findlay, J. A. 1969. Thermal structure of

an Alouette-2 topside profile as deduced from rocket measurements. *J. Geophys. Res.* 74:2229–39
Harris, K. K., Sharp, G. W., Knudsen, W. C. 1967. Ion temperature and relative ion composition measurements from a low-altitude, polar-orbiting satellite. *J. Geophys. Res.* 72:5939–48
Ho, M. C., Moorcroft, D. R. 1971. Hydrogen density and proton flux in the topside ionosphere over Arecibo, Puerto Rico, from incoherent scatter observations. *Planet. Space Sci.* 19:1441–56
Hoffman, J. H. 1967. A mass spectrometric determination of the composition of the nighttime topside ionosphere. *J. Geophys. Res.* 72:1883–88
Hoffman, J. H., Dodson, W. H., Lippincott, C. R., Hammack, H. D. 1974. Initial ion composition results from the ISIS-2 satellite. *J. Geophys. Res.* 79:4246–51
Holzer, T. E., Fedder, J. A., Banks, P. M. 1971. A comparison of kinetic and hydrodynamic models of an expanding ion-exosphere. *J. Geophys. Res.* 76:2453–68
Johnson, F. S. 1960. The ion distribution above the F_2-maximum. *J. Geophys. Res.* 65:577
King, J. W., Smith, P. A., Eccles, D., Fooks, G. F., Helm, H. 1964. Preliminary investigation of the structure of the upper ionosphere as observed by the topside sounder satellite, Alouette. *Proc. R. Soc. London, Ser. A* 281:464–87
Kockarts, G., Nicolet, M. 1963. *Ann. Geophys.* 19:370
Lemaire, J. 1972. O^+, H^+ and He^+ ion distributions in a new polar wind model. *J. Atmos. Terr. Phys.* 34:1647–58
Lemaire, J., Scherer, M. 1970. Model of the polar ion-exosphere. *Planet. Space Sci.* 18:103–20
Lemaire, J., Scherer, M. 1971. Simple model for an ion-exosphere in an open magnetic field. *Phys. Fluids* 14:1683–94
Lemaire, J., Scherer, M. 1972. Ion-exosphere with asymmetric velocity distribution. *Phys. Fluids* 15:760–66
Lemaire, J., Scherer, M. 1973. Kinetic models of the solar and polar winds. *Rev. Geophys. Space Phys.* 11:427–468
Lemaire, J., Scherer, M. 1974. Exospheric models of the topside ionosphere. *Space Sci. Rev.* 15:591–640
Maier, E. J., Hoffman, J. H. 1974. Observation of a two-temperature ion energy distribution in regions of polar wind flow. *J. Geophys. Res.* 79:2444–47
Mange, P. 1960. The distribution of minor ions in electrostatic equilibrium in the high atmosphere. *J. Geophys. Res.* 65:3833–34
Marubashi, K. 1970. Escape of the polar-ionospheric plasma into the magnetospheric tail. *Rep. Ionos. Space Res. Jpn.* 24:322–46
Mayr, H. G., Fontheim, E. G., Brace, L. H., Brinton, H. C., Taylor, H. A. 1972. A theoretical model of the ionosphere dynamics with interhemispheric coupling. *J. Atmos. Terr. Phys.* 34:1659–80
Muldrew, D. B. 1965. F-layer ionization troughs deduced from Alouette data. *J. Geophys. Res.* 70:2635–50
Nelms, G. L. 1964. Ionospheric results from the topside sounder satellite Alouette. *Space Res.* 4:437–48
Nelms, G. L., Lockwood, G. E. K. 1967. Early results from the topside sounder in the Alouette-2 satellite. *Space Res.* 7:604–23
Park, C. G. 1970. Whistler observations of the interchange of ionization between the ionosphere and the protonosphere. *J. Geophys. Res.* 75:4249–60
Park, C. G. 1973. Whistler observations of the depletion of the plasmasphere during a magnetospheric substorm. *J. Geophys. Res.* 78:672–83
Park, C. G. 1974. Some features of plasma distribution in the plasmasphere deduced from Antarctic whistlers. *J. Geophys. Res.* 79:169–73
Prasad, S. S. 1970. Ionic composition and temperature over Arecibo 2. *J. Geophys. Res.* 75:1911–18
Prölss, G. W., von Zahn, U. 1974. ESRO-4 gas analyzer results 2. Direct measurements of changes in the neutral composition during an ionospheric storm. *J. Geophys. Res.* 79:2535–39
Raitt, W. J. 1974. The temporal and spatial development of mid-latitude thermospheric electron temperature enhancements during a geomagnetic storm. *J. Geophys. Res.* 79:4703–8
Raitt, W. J., Laflin, S., Boyd, R. L. F. 1965. A synoptic view of ionic constitution above the F-layer maximum. *Space Res.* 5:629–34
Raitt, W. J., Schunk, R. W., Banks, P. M. 1975. A comparison of the temperature and density structure in high and low speed thermal proton flows. *Planet. Space Sci.* 23:1103–17
Rees, M. H., Jones, R. A., Walker, J. C. G. 1971. The influence of field-aligned currents on auroral electron temperatures. *Planet. Space Sci.* 19:313–25
Rycroft, M. J., Thomas, J. O. 1970. The magnetospheric plasmapause and the

electron density trough at the Alouette-1 orbit. *Planet. Space Sci.* 18:65–80
Schunk, R. W. 1975. Transport equations for aeronomy. *Planet. Space Sci.* 23:437–85
Schunk, R. W., Raitt, W. J., Banks, P. M. 1975. Effect of electric fields on the day-time high-latitude E- and F-regions. *J. Geophys. Res.* 80:3121–30
Schunk, R. W., Walker, J. C. G. 1969. Thermal diffusion in the topside ionosphere for mixtures which include multiply-charged ions. *Planet. Space Sci.* 17:853–68
Schunk, R. W., Walker, J. C. G. 1970a. Thermal diffusion in the F_2-region of the ionosphere. *Planet. Space Sci.* 18:535–57
Schunk, R. W., Walker, J. C. G. 1970b. Minor ion diffusion in the F_2-region of the ionosphere. *Planet. Space Sci.* 18:1319–34
Schunk, R. W., Walker, J. C. G. 1970c. Transport properties of the ionospheric electron gas. *Planet. Space Sci.* 18:1535–50
Schunk, R. W., Walker, J. C. G. 1973. The theory of charged particle temperatures in the upper atmosphere. In *Progress in High Temperature Physics and Chemistry,* ed. C. A. Rouse, 5:1–62. New York: Pergamon
Storey, L. R. O. 1953. An investigation of whistling atmospherics. *Philos. Trans. R. Soc. London, Ser. A* 246:113–41
Strobel, D. F., Weber, E. J. 1972. Mathematical model of the polar wind. *J. Geophys. Res.* 77:6864–69
Taylor, G. N. 1974. Meridional F_2-region plasma drifts at Malvern. *J. Atmos. Terr. Phys.* 36:267–86
Taylor, H. A. 1971. Evidence of solar geomagnetic seasonal control of the topside ionosphere. *Planet. Space Sci.* 19:77–94
Taylor, H. A. 1974. High latitude minor ion enhancements: a clue for studies of magnetosphere-atmosphere coupling. *J. Atmos. Terr. Phys.* 36:1815–23
Taylor, H. A., Brinton, H. C., Deshmukh, A. R. 1970. Observations of irregular structure in thermal ion distributions on the dusk side magnetosphere. *J. Geophys. Res.* 75:2481–89
Taylor, H. A., Cordier, G. R. 1974. In-situ observations of irregular ionospheric structure associated with the plasmapause. *Planet. Space Sci.* 22:1289–96
Taylor, H. A., Grebowsky, J. M., Walsh, W. J. 1971. Structural variations of the plasmapause: evidence of a co-rotating plasma tail. *J. Geophys. Res.* 76:6806–14
Taylor, H. A., Walsh, W. J. 1972. The light-ion trough, the main trough, and the plasmapause. *J. Geophys. Res.* 77:6716–23
Thomas, J. O., Andrews, M. K. 1968. Transpolar exospheric plasma: 1. Plasmasphere termination. *J. Geophys. Res.* 73:7407–17
Thomas, J. O., Andrews, M. K. 1969. The transpolar exospheric plasma: 3. A unified picture. *Planet. Space Sci.* 17:433–46
Titheridge, J. E. 1974. Exospheric temperature and composition from satellite beacon measurements. *Planet. Space Sci.* 22:209–22
Vasseur, G. 1969. Dynamics of the F-region observed with Thomson scatter-I. Atmospheric circulation and neutral winds. *J. Atmos. Terr. Phys.* 31:397–420
Vasseur, G. 1970. Dynamics of the F-region observed with Thomson scatter-II. Influence of neutral winds on the ionospheric F-region. *J. Atmos. Terr. Phys.* 32:775–89
Walker, J. C. G. 1967. Thermal diffusion in the topside ionosphere. *Planet. Space Sci.* 15:1151–56
Watt, T. M. 1965. Ion distribution and temperature in the topside ionosphere obtained from the Alouette satellite. *J. Geophys. Res.* 70:5849–59
Willmore, A. P. 1965. Geographical and solar activity variations in the electron temperature of the upper F-region. *Proc. R. Soc. London, Ser. A* 286:537–58
Woodman, R. F. 1970. Vertical drift velocities and east-west electric fields at the magnetic equator. *J. Geophys. Res.* 75:6249–59

INTERACTIONS OF CH_4 AND CO IN THE EARTH'S ATMOSPHERE

Steven C. Wofsy
Center for Earth and Planetary Physics, Harvard University,
Cambridge, Massachusetts 02138

INTRODUCTION

The presence of CH_4 and CO in the Earth's atmosphere was unknown until the 1940s, when Migeotte (1948, 1950) identified their infrared absorption lines in the solar spectrum. Using Migeotte's fairly accurate column abundances, Bates & Witherspoon (1952) made the first study of the photochemistry and atmospheric cycles of CO and CH_4. They correctly deduced many of the basic features of the atmospheric processes involving these gases, and their ideas were accepted with little innovation for nearly 20 years.

Bates & Witherspoon (1952) recognized that CH_4 must have a sizeable natural source, and they singled out anaerobic fermentation for special attention. They discounted fermentation, however, because an appreciable fraction ($\sim 1\%$) of all biologically fixed carbon would have to be released as CH_4, which in fact appears to be the case (Koyama 1964, McElroy 1975). They suspected that formaldehyde would be an intermediate product when CH_4 is photooxidized in the atmosphere and they identified reactions involving OH and HO_2 as likely sinks.

Bates & Witherspoon (1952) recognized combustion of fossil fuels as an important source of CO, but could not identify any others. They realized that there should be a reasonably efficient atmospheric sink, based on the anthropogenic source and the observed concentration. They studied reactions with OH and HO_2; however, they calculated inadequate removal rates because these radicals were thought to be formed only by H_2O photolysis, and therefore confined to the upper atmosphere.

In 1969, understanding of atmospheric CO and CH_4 began to change considerably. Weinstock (1969) obtained an estimate of 0.1 yr for the CO mean atmospheric residence time, based on the calculated source of cosmogenic $C^{14}O$ and measurements of the C^{14} content of CO. This value conflicted with estimates of one or two years (cf Robinson & Robbins 1968; Jaffe 1970, 1973) based on the observed atmospheric abundance of CO and the assumed anthropogenic sources. Furthermore, Seiler & Junge (1970), Junge, Seiler & Warneck (1971), and Robinson &

Table 1 Chemical reactions[a]

Reaction number (R)	Reaction	Rate expression	Reference
1.	$CO + OH \rightarrow CO_2 + H$	$2.1 \times 10^{-13} \exp(-115/T)$	Greiner (1969)
2.	$CH_4 + OH \rightarrow CH_3 + H_2O$	$5.5 \times 10^{-12} \exp(-1186/T)$	Greiner (1970)
3a.	$O_3 + h\nu \rightarrow O_2(^1\Delta g) + O(^1D)$		Griggs (1968)
3b.	$O_3 + h\nu \rightarrow O_2 + O(^3P)$		Lin & Demore (1973)
4.	$O(^1D) + N_2 \rightarrow O(^3P) + N_2$	$1.2 \times 10^{-11} \exp(253/T)$	Streit et al (1976)[b]
5.	$O(^1D) + O_2 \rightarrow O(^3P) + O_2$	$2.1 \times 10^{-11} \exp(+193/T)$	Streit et al (1976)[b]
6.	$O(^1D) + H_2O \rightarrow 2OH$	2.1×10^{-10}	Davidson et al (1976)[b]
7.	$H + O_2 + M \rightarrow HO_2 + M$	$1.83 \times 10^{-32} \exp(342/T)$	Wong & Davis (1974)
8.	$HO_2 + HO_2 \rightarrow H_2O_2 + O_2$	9.5×10^{-12}; (3×10^{-12})[b]	Hochanadel et al (1972)
9.	$HO_2 + O_3 \rightarrow OH + 2O_2$	$1 \times 10^{-12} \exp(-1875/T)$	Demore (1973), Demore & Tschuikow-Roux (1974)
10.	$HO_2 + NO \rightarrow OH + NO_2$	$2 \times 10^{-11} \exp(-606/T)$	Payne et al (1973), Hack et al (1975)
11.	$HO_2 + O \rightarrow OH + O_2$	6×10^{-11}	Estimated value;
			Hochanadel et al (1972), McConnell & McElroy (1973)
12.	$H_2O_2 + h\nu \rightarrow OH + OH$		Cross sections from Schumb et al (1955)
13.	$H_2O_2 + OH \rightarrow H_2O + HO_2$	$4.1 \times 10^{-13} T^{\frac{1}{2}} \exp(-600/T)$	Greiner (1968, 1969)
14.	$OH + HO_2 \rightarrow H_2O + O_2$	2×10^{-10}	Hochanadel et al (1972)
		(5×10^{-11})[c]	Demore & Tschuikow-Roux (1974)
15.	$OH + HNO_3 \rightarrow H_2O + NO_3$	8.9×10^{-14}	Margitan et al (1975)
16.	$OH + CH_3OOH \rightarrow H_2O + CH_3O_2$	$1/2\ k_{13}$	Estimate by Wofsy et al (1972)
17.	$O(^1D) + H_2 \rightarrow OH + H$	2×10^{-10}	Scott & Cvetanovic (1971)
			Paraskevopoulos & Cvetanovic (1971)
18.	$OH + H_2 \rightarrow H_2O + H$	$1.8 \times 10^{-11} \exp(-2342/T)$	Smith & Zellner (1974)
19.	$CH_3 + O_2 + M \rightarrow CH_3O_2 + M$	1.3×10^{-31}	Basco et al (1971)
20.	$CH_3O_2 + NO \rightarrow CH_3O + NO_2$	0	(See text)

21.	$CH_3O + O_2 \rightarrow H_2CO + HO_2$	$1.7 \times 10^{-13} \exp(-3200/T)$	Heicklen (1968)
22.	$H_2CO + h\nu \rightarrow H_2 + CO$		Cross sections from Calvert et al (1972)
23.	$H_2CO + h\nu \rightarrow H + HCO$		
24.	$H_2CO + OH \rightarrow H_2O + HCO$	1.4×10^{-11}	Morris & Niki (1971)
25.	$HCO + O_2 \rightarrow HO_2 + CO$	1.5×10^{-11}	Washida et al (1974)
26.	$CH_3O_2 + HO_2 \rightarrow CH_3OOH + O_2$	5×10^{-12}	Estimate by Heicklen (1968)
27.	$CH_3OOH + h\nu \rightarrow CH_3O + OH$	$2/3\ J_{12}$	Estimate by Wofsy et al (1972)
28.	$CH_3O_2 + CH_3O \rightarrow CH_3OOH + H_2CO$	0	Assumed negligible
29.	$CH_3O_2 + CH_3O_2 \rightarrow 2CH_3O + O_2$	$5.3 \times 10^{-13} \exp(-3000/T)$	Benson & Shaw (1971)
30.	$O(^1D) + CH_4 \rightarrow OH + CH_3$	1.3×10^{-10}	Davidson et al (1976)[b]
31a.	$O(^1D) + N_2O \rightarrow N_2 + O_2$	$2.8 \times 10^{-10} \exp(-430/T)$	Our estimates based on Streit et al (1976)[b]
31b.	$O(^1D) + N_2O \rightarrow NO + NO$	$3.05 \times 10^{-10} \exp(-430/T)$	
32.	$CO_2 + h\nu \rightarrow CO + O$		Cross sections from Shemansky (1972)
33.	$CO + O + M \rightarrow CO_2 + M$	2.2×10^{-36}	Stuhl & Niki (1971)
34.	$H_2O + h\nu \rightarrow H + OH$		Cross sections from Thompson et al (1963)
35.	$HO_2 + h\nu \rightarrow OH + O$		Estimated cross sections from McConnell & McElroy (1973)
36.	$O + OH \rightarrow O_2 + H$	5.0×10^{-11}	Kaufman (1969)
37.	$H + O_3 \rightarrow OH + O_2$	2.6×10^{-11}	Kaufman (1969)
38.	$OH + O_3 \rightarrow HO_2 + O_2$	$1.3 \times 10^{-12} \exp(-956/T)$	Anderson & Kaufman (1973)
39a.	$O + H_2O_2 \rightarrow O_2 + HO_2$	$1.5 \times 10^{-12} \exp(-2222/T)$	Davis et al (1976)
39b.	$O + H_2O_2 \rightarrow OH + HO_2$	$1.5 \times 10^{-12} \exp(-2222/T)$	
40a.	$H + H_2O_2 \rightarrow H_2 + HO_2$	$5.0 \times 10^{-12} \exp(-2100/T)$	Davis, Wong & Schiff (1974)
40b.	$H + H_2O_2 \rightarrow OH + H_2O$	$5.0 \times 10^{-12} \exp(-2100/T)$	
41a.	$H + HO_2 \rightarrow H_2 + O_2$	$3 \times 10^{-11} \exp(-333/T)$	McConnell & McElroy (1973)
41b.	$H + HO_2 \rightarrow OH + OH$	$1 \times 10^{-10} \exp(-333/T)$	Hampson & Garvin (1975)
41c.	$H + HO_2 \rightarrow H_2O + O$	0	
42.	$OH + OH \rightarrow H_2O + O$	2.6×10^{-12}	Dixon-Lewis et al (1966), Kaufman (1969)
43.	$O + HO_2 \rightarrow O_2 + OH$	6×10^{-11}	Estimate
44.	$O + H_2 \rightarrow OH + H$	$5.3 \times 10^{-11} \exp(-5100/T)$	Schofield (1967)
45.	$O + NO_2 \rightarrow O_2 + NO$	9.12×10^{-12}	Davis, Herron & Huie (1973); Davis et al (1973)
46.	$O_3 + NO \rightarrow NO_2 + O_2$	$1.5 \times 10^{-12} \exp(-1330/T)$	Baulch et al (1973) review

Table 1 (Continued)

Reaction number (R)	Reaction	Rate expression	Reference
47.	$H + NO_2 \rightarrow OH + NO$	$1.2 \times 10^{-9} \exp(-965/T)$	Kaufman (1969)
48.	$O_3 + NO_2 \rightarrow NO_3 + O_2$	$1.31 \times 10^{-13} \exp(-2475/T)$	Graham & Johnston (1974); cf Davis et al (1974)
49.	$NO + NO_3 \rightarrow 2NO_2$	8.7×10^{-12}	Harker & Johnston (1973)
50.	$NO_2 + NO_3 \rightarrow NO + O_2 + NO_2$	$2.2 \times 10^{-13} \exp(-1850/T)$	McConnell & McElroy (1973)
51.	$NO_3 + NO_3 \rightarrow 2NO_2 + O_2$	$5 \times 10^{-15} \exp(-3000/T)$	Baulch et al (1973)
52.	$OH + NO_2 + M \rightarrow HNO_3 + M$		Based on Tsang (1973); function of z
53.	$HNO_3 + h\nu \rightarrow NO_2 + OH$		Cross sections from Johnston & Graham (1973); cf Johnston et al (1974) for quantum yield
54.	$NO_2 + NO_3 + M \rightarrow N_2O_5 + M$		See expression adopted by McConnell & McElroy (1973)
55.	$N_2O_5 + M \rightarrow NO_2 + NO_3$		
56.	$NO_2 + h\nu \rightarrow NO + O$		Cross sections from Hall & Blacet (1952)
57a.	$NO_3 + h\nu \rightarrow NO_2 + O$		Cross section from Johnston (private communication, 1975)
57b.	$NO_3 + h\nu \rightarrow NO + O_2$	0	
58.	$O + HNO_3 \rightarrow OH + NO_3$	$1 \times 10^{-11} \exp(-1860/T)$	Upper limit; see McConnell & McElroy (1973)
59.	$O + O_2 + M \rightarrow O_3 + M$	$1.05 \times 10^{-34} \exp(520/T)$	Huie, Herron & Davis (1972)
60.	$O + O_3 \rightarrow 2O_2$	$1.32 \times 10^{-11} \exp(-2140/T)$	McCrumb & Kaufman (1972)
61.	$O + O + M \rightarrow O_2 + M$		See Campbell & Gray (1973)
62.	$2NO + O_2 \rightarrow 2NO_2$	$1.0 \times 10^{-38} \exp(478/T)$	Baulch et al (1973)
63.	Soluble gas → precipitation scavenging {HCl, H_2O_2, H_2CO, HNO_3, CH_3OOH}	2.3×10^{-6} $0 < z < 4$ km 1.15×10^{-6} $4 < z < 6$ km 5.78×10^{-7} $6 < z < 8$ km 2.89×10^{-7} $8 < z < 10$ km 0 10 km $< z$	See text. At each latitude and season, these rates corrected for precipitation amount.
64.	$OH + HCl \rightarrow H_2O + Cl$	$2 \times 10^{-12} \exp(-313/T)$	Zahniser et al (1974)
65.	$Cl + H_2 \rightarrow HCl + H$	$5.7 \times 10^{-11} \exp(-2400/T)$	Davis and Watson (1975, private communication)

66.	$H + HCl \rightarrow Cl + H_2$	$1 \times 10^{-11} \exp(-1605/T)$	Benson & Shaw (1971)
67.	$O + HCl \rightarrow OH + Cl$	$2.5 \times 10^{-12} \exp(-2980/T)$	Brown & Smith (1975)
68.	$Cl + O_3 \rightarrow ClO + O_2$	$2.35 \times 10^{-11} \exp(-170/T)$	Anderson et al (1975)
69.	$Cl + Cl + M \rightarrow Cl_2 + M$	$1.6 \times 10^{-33} \exp(800/T)$	
70.	$Cl + O_2 + M \rightarrow ClOO + M$	1.7×10^{-33}	See references in Wofsy & McElroy (1974)
71.	$ClOO + M \rightarrow Cl + O_2 + M$	$1.66 \times 10^{-8} \exp(-3424/T)$	
72.	$ClO + ClO \rightarrow Cl + ClOO$	$1.3 \times 10^{-12} \exp(-1150/T)$	
73.	$Cl + ClOO \rightarrow ClO + ClO$	1.44×10^{-12}	See Wofsy & McElroy (1974)
74.	$Cl + ClOO \rightarrow Cl_2 + O_2$	1.5×10^{-10}	
75.	$ClO + NO \rightarrow Cl + NO_2$	1.7×10^{-11}	Watson (1974)
76.	$O + ClO \rightarrow Cl + O_2$	5.3×10^{-11}	Watson (1974)
77.	$ClO + ClO + M \rightarrow Cl_2 + O_2$	5×10^{-32}	See Wofsy & McElroy (1974)
78.	$O + Cl_2 \rightarrow ClO + C$	$1.36 \times 10^{-11} \exp(-1560/T)$	See Wofsy & McElroy (1974)
79.	$ClO + O_3 \rightarrow$ products	0	Watson (1974)
80.	$Cl + H_2O_2 \rightarrow HCl + HO_2$	6×10^{-13}	Davis & Watson (1975, private communication)
81.	$Cl + HO_2 \rightarrow HCl + O_2$	1×10^{-11}	Assumed
82.	$ClOO + O_3 \rightarrow ClO + 2O_2$	0	
83.	$ClOO + O \rightarrow ClO + O_2$	0	
84.	$Cl + CH_4 \rightarrow HCl + CH_3$	$5.4 \times 10^{-12} \exp(-1133/T)$	Davis & Watson (1975, private communication)
85.	$CH_4 + h\nu \rightarrow$ products		See Strobel (1975)
86.	$Cl_2 + h\nu \rightarrow 2Cl$		See Wofsy & McElroy (1974)
87.	$O_2 + h\nu \rightarrow 2O$		See McElroy et al (1974), McConnell & McElroy (1973)
88.	$HCl + h\nu \rightarrow H + Cl$		Cross sections from Wofsy & McElroy (1974)
89.	$N_2O_5 + h\nu \rightarrow NO_2 + NO_3$		Cross section from Johnston (1974, private communication)
90.	$N_2O + h\nu \rightarrow N_2 + O$		See Bates & Hays (1967)
91.	$ClO + h\nu \rightarrow Cl + O$		See Wofsy & McElroy (1974) and Nicholls (1975)

[a] Units for rate constants are $cm^3\ sec^{-1}$ for bimolecular reactions, $cm^6\ sec^{-1}$ termolecular reactions. Photodissociation rates (per second) are calculated from solar flux (Vernazza 1975) and relevant cross-section data.

[b] Both values give similar results. The atmospheric rate may be the faster value. See Hamilton 1975).

[c] Both values of K_{14} were used, with very small differences between the results. The slower rate is indicated by the work of Hack et al (1975).

Robbins (1969) had shown that CO is present in the southern hemisphere. These results suggested that significant natural sources of CO must exist and that the unidentified sink of CO must be even more efficient than previously assumed. The new estimates for the CO lifetime were inconsistent with the proposal that the stratosphere was the principal sink (Pressman & Warneck 1970).

One key step to resolving these problems was provided by Levy (1971). He extended stratospheric models by Hunt (1966) and Nicolet (1970) into the troposphere and predicted that the OH radical should have an appreciable abundance near the ground. Because low temperature reaction rates (reaction numbers at right are keyed to Table 1) were available for

$$CO + OH \rightarrow CO_2 + H \qquad \text{(R1)}$$

and

$$CH_4 + OH \rightarrow CH_3 + H_2O, \qquad \text{(R2)}$$

Levy (1971) postulated that tropospheric OH could play an important role in the atmospheric photochemistry of CH_4 and CO.

Finally, McConnell et al (1971) proposed that formaldehyde (H_2CO) formed from methane photooxidation should in turn be photooxidized principally to CO, and therefore that CH_4 oxidation initiated by Reaction 2 (R2) should be the missing natural source of CO. The same authors subsequently presented detailed calculations on the global budgets for CO and CH_4 (Wofsy et al 1972) in which CH_4 oxidation was estimated to provide two thirds of the global source of CO. A number of similar calculations that followed reached the same general conclusions but, for various reasons, tended to estimate shorter CO lifetimes and a larger percentage of CO derived from CH_4 oxidation (Levy 1972, 1973; Stevens et al 1972; Kummler & Baurer 1973).

These developments stimulated much interest in CH_4 and CO, which resulted in a great expansion of the available data on distributions of these gases in the atmosphere. The new measurements have been variously interpreted as supporting or contradicting the proposed role of CH_4 in the CO budget. In Part I of this review, we examine this question in detail. In Part II, we discuss CO and CH_4 in the upper atmosphere.

PART I: GLOBAL DISTRIBUTIONS, SOURCES, AND SINKS OF CO AND CH_4

Elements of the Theory

If one knows the steady-state concentration of a gas such as CH_4 and can quantify the atmospheric sources and sinks for the gas (e.g. for CH_4, there are no sources and sinks are equal to R2 and R30), then one can directly calculate the net flux of the gas across the Earth's surface from the continuity equation. At steady state, all excess molecules created (lost) in the atmosphere must flow into (out of) the Earth's surface. A simple mathematical formulation is given in the appendix (Equations 2–4).

The calculations of production and loss in the atmosphere provide estimates for sources, sinks, and fluxes independent of any assumptions about industrial or biological processes. These estimates can be calibrated by comparison with other gases that are influenced by similar atmospheric reactions, and frequently the theory can be tested by examining the observed distributions in time and space. We adopt this approach in the present work.

In the calculations described below, we assume that each molecule of CH_4 oxidized in the atmosphere is subsequently converted to a molecule of CO, except for minor losses of H_2CO or CH_3OOH in rain. The appendix reviews chemical evidence that this conversion approaches unit efficiency by the pathway summarized in Figure 1, but the precise chemical mechanism is unimportant. Throughout the present discussion, the source of CO from methane oxidation is very nearly equal to the methane oxidation rate, R2 and R30 (see Table 1).

Under this assumption, global and seasonal distributions of OH and $O(^1D)$ are the principal requirements needed to specify the atmospheric sources and sinks of CO and CH_4. Table 1 lists the complete set of chemical reactions used to calculate these distributions, and Figure 2*a* summarizes the chemistry of HO_x radicals. This chemistry has been discussed in detail elsewhere (Wofsy et al 1972, Levy 1974), so that only a few highlights are discussed below (see also appendix).

The troposphere accounts for more than 90% of the global sources and sinks of CO and CH_4, and the lowest scale height accounts for nearly 80%. The chemistry of OH in the lower troposphere is much simpler than that shown in Figure 2*a*, as illustrated by Figure 2*b*. The double lines trace the most important path, which leads from OH directly to precipitation scavenging, by way of reaction with CO or CH_4. The OH concentration in the lower troposphere is controlled by balance between production from water (Hampson 1964),

$$O(^1D) + H_2O \rightarrow 2OH, \tag{R6}$$

and loss due to reactions with CO and CH_4 (R1 and R2). Regeneration of OH from HO_2 or H_2O_2 increases the OH concentration by less than 50%; therefore,

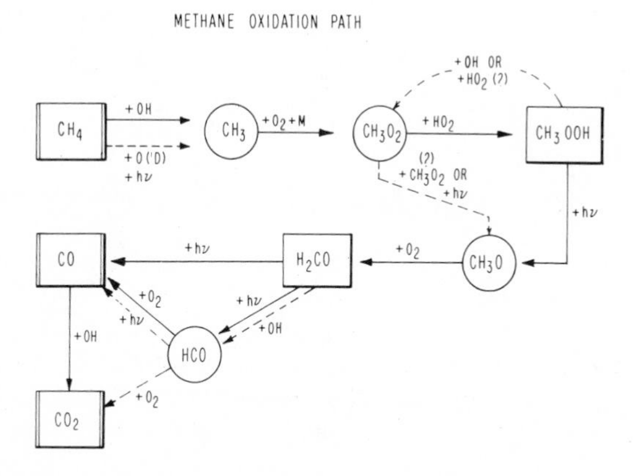

Figure 1 The reaction pathway between CH_4 and CO is shown, from Wofsy et al (1972).

to a rough approximation, each OH molecule produced by R6 consumes one molecule of either CO or CH_4. The relative probability of reaction with CO or CH_4 depends on their relative concentrations and on temperature, because of the significant activation energy of R2.

Because atmospheric consumption of CO and CH_4 is proportional to the

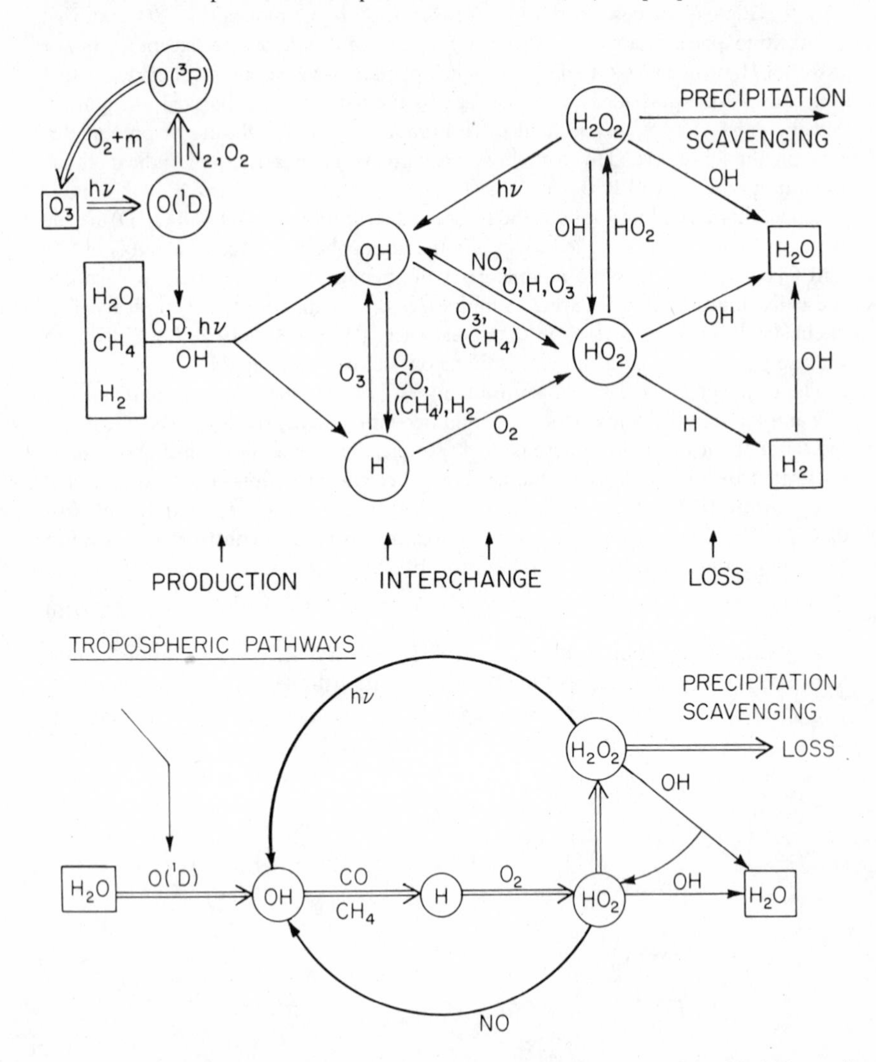

Figure 2 (*a*) The chemical reactions affecting atmospheric HO_x (OH, H, HO_2 and H_2O_2) are illustrated schematically. (*b*) The chemical reactions which control OH in the lower troposphere are isolated from Figure 2*a*. The double lines denote the principal pathway.

concentrations of $O(^1D)$ and H_2O, the crucial parameters in accurate calculation are the distributions of ozone, temperature, relative humidity, and insolation. The complex chemistry of Table 1, although important in the stratosphere, is secondary in the calculation of global sources and sinks. Likewise, the washout profile given in Table 1 does not strongly affect the calculation, provided only that the washout lifetime of H_2O_2 is shorter than, or comparable to, the photolysis lifetime (i.e. 9 days).[1] Perhaps the major uncertainty in the present calculations is the neglect of scattering and reflection of UV radiation, which influences the calculated photolysis rates.

The differences between Wofsy et al (1972) and subsequent calculations (e.g. Levy 1973, Kummler & Baurer 1973) result largely from inclusion of precipitation scavenging by Wofsy et al (1972). Of course, the adopted scavenging rates are intended to simulate mean conditions, and the higher radical concentrations obtained by Levy (1973) and others might well occur at locations where precipitation is ineffective.

Results of the Global Calculations

Figure 3 shows the global distribution of OH from the complete calculation. The CO and CH_4 distributions were adopted from the observations by Seiler & Schmidt (1974) and Wilkniss et al (1973) in the troposphere and were calculated as described below in the stratosphere. Distributions of other trace gases, temperature, pressure, etc were obtained as discussed in the appendix.

The calculations predict 30–40% more OH at southern midlatitudes than at corresponding northern locations. This asymmetry is mostly due to anthropogenic CO, which is concentrated in the north (see below), and to a lesser extent to excess CH_4 ($\sim 10\%$) found in the northern hemisphere (Wilkniss et al 1973).[2] This situation is somewhat disturbing, because the hydroxyl radical plays a crucial role in many tropospheric self-cleansing processes, such as removal of various organic contaminents (cf Yung et al 1975). Present emissions of CO are apparently large enough to perturb OH chemistry on the global scale.

The chemical lifetimes of CO and CH_4 are shown in Figures 4 and 5. The annual mean lifetimes for CO are less than three months, except at polar latitudes, whereas CH_4 lifetimes are greater than two years. The available global data (Ehhalt 1974, Wilkniss et al 1973) indicate that tropospheric CH_4 is well mixed vertically and horizontally, with a small north/south hemispheric asymmetry. The chemical lifetimes shown in Figure 5 appear to be quite consistent with this distribution and

[1] If "washout" were omitted, regeneration of OH would remain inefficient. Concentrations of H_2O_2, CH_3OOH_3 would become sufficiently large so that reaction of OH with these species would terminate the radical chains.

[2] The asymmetry of CH_4 could also be influenced by human activities. Man-related sources of CH_4 include rice paddies (Koyama 1969), enteric fermentation in ruminants, coal mining, sewage processing, and natural gas operations. Filling of wetlands should reduce the natural sources of CH_4. According to Ehhalt (1974), the man-related sources account for as much as 50% of the total. They should have a northern bias and could account for the hemispheric differences.

estimates for interhemispheric transport times (Newell et al 1969, Czeplak & Junge 1974).

The tropospheric distribution of CO is considerably more complex than that for CH_4. Figure 6*a* shows the latitude/altitude distribution of CO inferred by Seiler & Schmidt (1974) from a number of measurements. Curve *A* of Figure 6*b* gives the observed latitudinal distribution of CO, vertically averaged, curve *B* shows the distribution in the upper troposphere (~8 km), and curve *C* shows the distribution at the surface, also according to Seiler & Schmidt (1974). These data clearly indicate interhemispheric transport of anthropogenic CO in the upper troposphere (cf curves *B* and *C*), where the CO lifetime is longer than one year. Indeed, direct observations

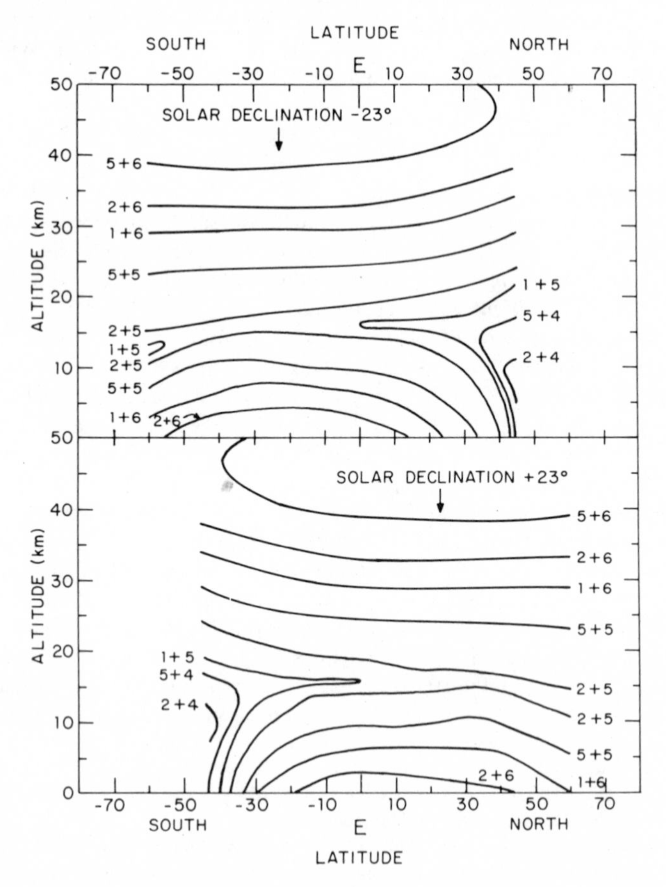

Figure 3 The global distribution of OH (in cubic centimeters, 24-hour average) is shown for January 1 in the upper panel and for July 1 in the lower panel. Note the reduction of OH in the lower northern troposphere, as compared to the southern troposphere, due to anthropogenic CO emitted in the north.

by Seiler & Schmidt (1974) show nearly uniform CO across the equator at ~8 km, even when concurrent data show a sharp break at the surface.

The CO surface data in the southern hemisphere can be explained quantitatively by local balance between production from CH_4 oxidation and loss by reaction with OH. Curve *D* in Figure 6*b* shows the annual average of the expression

$$f_{CO} = \frac{\int_0^{z_{mix}} k_2[OH][n]\,dz}{\int_0^{z_{mix}} k_1[OH][n]\,dz} f_{CH_4}$$

where f_i is the mixing ratio (v/v) of species i, $[n]$ is the atmospheric number density, and the integration is over altitude, z, between the ground and the top of the

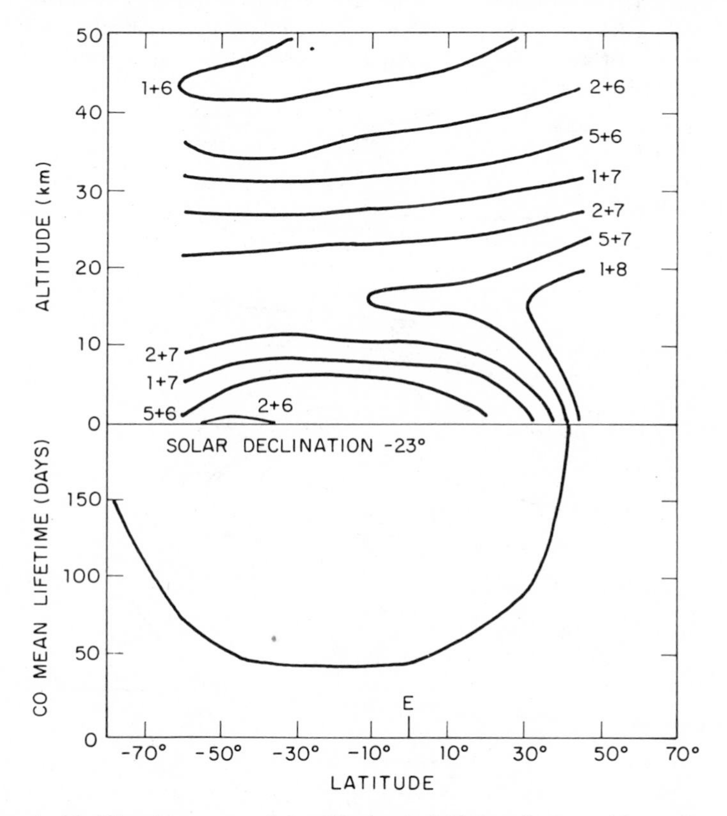

Figure 4 Meridional contours of the CO chemical lifetime (in seconds) are shown for January 1 in the upper panel, and the density-weighted atmospheric mean is presented in the lower panel. The chemical lifetime is defined at [CO]/*L*, where [CO] is the concentration (in cubic centimeters) of CO and *L* is the chemical loss rate, approximately equal to the rate of Reaction R1. The notation "1+6" should be read as 1×10^6.

well-mixed layer, denoted z_{mix}. The integration over z is important because the scale height for R2 (2.5 km) is considerably shorter than that of R1 (4 km) [see Figures 4 and 5], due to the temperature dependence of R2. The level z_{mix} should be higher than the top of the planetary boundary layer (2–3 km), which is rapidly stirred by convection. Curve *D* corresponds to the somewhat arbitrary value of 4 km for z_{mix}; if convection were neglected ($z_{mix} = 0$), curve *D* would be about 30% higher, whereas vertical mixing of CO through the whole troposphere ($z_{mix} \simeq 10$ km) would lower curve *D* by about 25%.

Comparison of curves *C* and *D* of Figure 6*b* reveals that the surface values for [CO] follow almost exactly the latitude distribution predicted by Equation 1 in the Southern Hemisphere. The southern tropics are enriched in CO, principally because the rate of R2 is enhanced at warmer temperatures. In the northern tropics, chemistry is not quite able to establish a local balance in the face of the huge concentrations of anthropogenic CO at midlatitudes, but consumption of CO in the tropics is fast enough to maintain quite a steep latitudinal gradient of CO in the Northern Hemisphere.

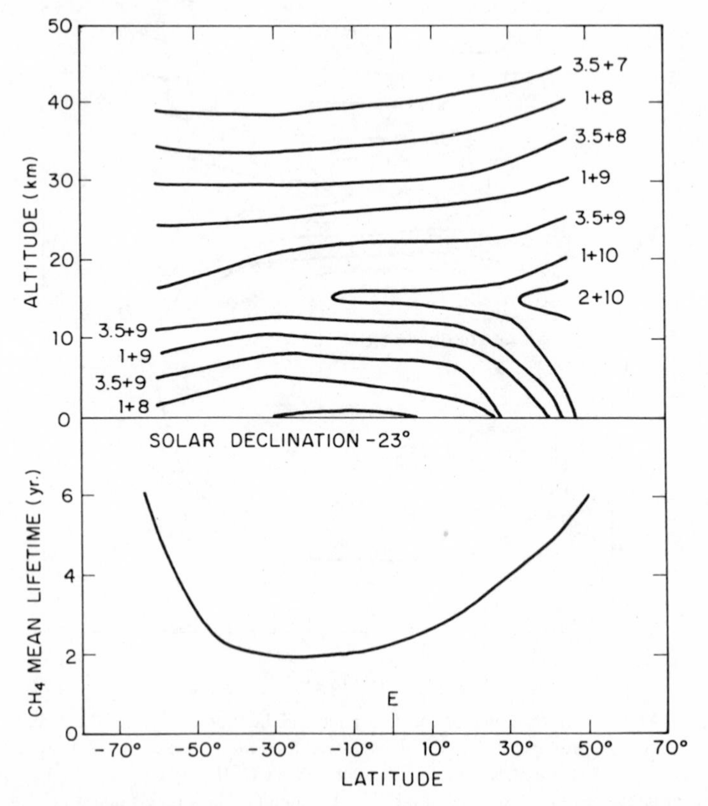

Figure 5 Chemical lifetimes for CH_4 are given, in the same arrangement as Figure 4, with Reaction R2 being the loss process.

Global Budgets for CO and CH_4

The vertically integrated rates for production and loss of CO and CH_4 are presented in Figure 7. Curve *A* gives the vertically integrated rate of CH_4 oxidation via R2, curve *C* shows the loss of CO via R1, and curve *B* gives the difference between *C* and *A*, i.e. the excess atmospheric loss of CO. Where more CO is consumed via R1 than is produced via R2 (curve $B > 0$), other sources of CO are required. We can draw a number of important conclusions from Figure 7.

1. Nowhere in the atmosphere is *C-A* significantly negative. Negative values for *C-A* would imply that more CO is put into the atmosphere from CH_4 oxidation than is lost by R1. Hence the calculation does not support the suggestions of an additional important sink for CO, e.g. soil microorganisms (Seiler 1974).

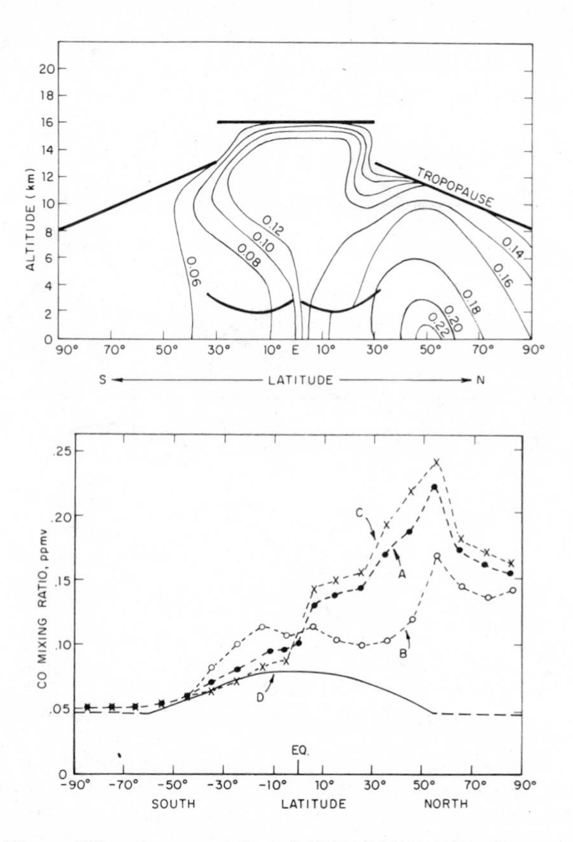

Figure 6 (*a*) The meridional cross section of CO mixing ratio is shown, according to Seiler & Schmidt (1974). The units are parts per million by volume. (*b*) The latitudinal variation of CO mixing ratio is shown for the troposphere: *A* = density weighted mean value; *B* = high troposphere (8–10 km); *C* = near-surface values; *D* = calculated near-surface values from Equation 1. Curves *A, B,* and *C* from Seiler & Schmidt (1974).

2. The Southern Hemisphere appears to be nearly in balance, except near the equator where anthropogenic CO perturbs the higher altitudes.

Points 1 and 2 lend strong support to the theory. Curve *B* of Figure 7 is derived from the differences between larger quantities, which makes fortuitious balance unlikely.

3. CO and especially CH_4 are rather inert at high latitudes and altitudes. Therefore estimates of global sources and sinks could be in error by a factor of 2 or more if variations of insolation, temperature, and OH were ignored. For example, if surface conditions at 30°N in summer were adopted as typical, the CH_4 oxidation rate would be overestimated by a factor of 4.

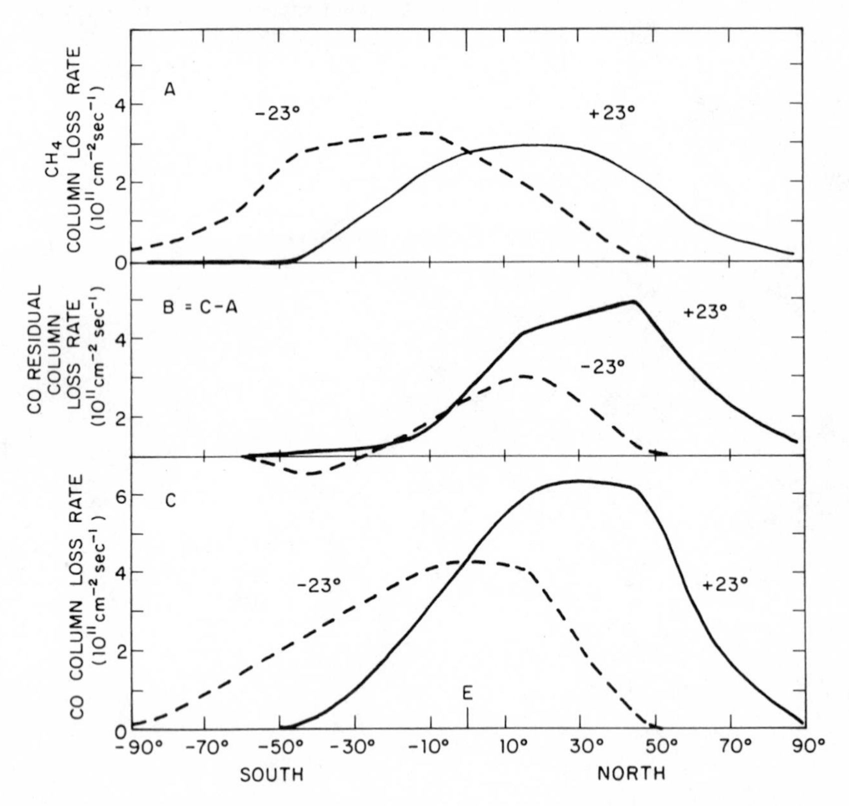

Figure 7 The vertically integrated atmospheric loss rates are shown as functions of latitude for solar declinations of $\pm 23°$. *A* = methane; *C* = carbon monoxide; *B* = excess loss of CO in the atmosphere, (*C*-*A*), after account is taken of the methane oxidation source. Units are square centimeters per second. The quantity plotted is

$$F_\theta = \frac{1}{2\pi}\int_0^{2\pi} d\phi \int_0^{\infty} dz\{P_i - L_i\}$$

(see appendix), where P_i and L_i are the chemical production and loss terms for species *i*.

4. We can integrate the results of Figure 7 over time and space to obtain our best estimates for the global budgets of CO and CH_4, given in Table 2.

The results in Table 2 demonstrate that CO in the Southern Hemisphere is derived almost entirely (~87%) from CH_4. In the Northern Hemisphere, oxidation of CH_4 provides about half (~46%) the input of CO into the atmosphere. The oxidation rate of CH_4 is lower in the Northern Hemisphere even though there is more CH_4 there—a result of [OH] reduction due to anthropogenic CO. As a global average, CH_4 oxidation accounts for 65% of the CO source in the atmosphere.

Anthropogenic CO is emitted mostly at latitudes of higher than 35°N where the CO lifetime becomes rather long. Therefore, anthropogenic CO accounts for a

Table 2 Global budget for CH_4 and CO[a]

Mean flux,[b] R1 and R2:	CH_4	CO (total)	CO (net)
Southern Hemisphere			
summer	2.7	2.9	0.24
winter	1.3	1.7	0.37
annual	2.0	2.3	0.31
Northern Hemisphere			
summer	2.3	5.2	2.9
winter	1.2	2.4	1.2
annual	1.8	3.8	2.0
Global mean	1.9	3.0	1.1

Processes other than R1 and R2 that generate or consume CO:	Flux[b]	Reference
Sinks for CO:		
possible loss of CH_3OOH	0.3	Shearer (1969), Dhar & Ram (1939)
Sources for CO:		
oceans	0.13	Seiler & Schmidt (1974)
industrial		
north	1.5	Seiler & Schmidt (1974)
south	0.26	
Residuals, including oceans and industry and omitting soils:		
north	0.4	
south	0.1	
Soil (net)		
north	0.8	Seiler (1974)
south	0.4	
global	0.6	
Residuals, including soils		
north	−0.4	
south	−0.3	

[a] Values are vertically integrated to 50 km. The troposphere accounts for ~90% of these totals.
[b] Flux is calculated per square centimeter per second times 10^{11}.

somewhat larger fraction of the CO molecules in the atmosphere than we might expect on the basis of the relative atmospheric source strengths in Table 2. Indeed, we find that only about 55% of the global atmospheric inventory is derived from CH_4. This factor must be remembered when isotope data are analyzed: for example, according to Figure 6*b*, nearly 75% of the atmospheric CO is anthropogenic at 50°N.

Table 2 shows unexpectedly good agreement between our estimates for other sources of CO and estimates derived from studies of the sources themselves. Seiler & Schmidt (1974) recently reviewed the available data and arrived at the oceanic and industrial source strengths given in part B of Table 2. When we include other sources, we obtain balance in both the Northern and Southern Hemispheres to better than 10%, denoted residuals in the table. No doubt this high accuracy is fortuitious, but the general agreement gives considerable confidence that the important processes are included in the theory.

Seiler (1974) and Seiler & Schmidt (1974) postulated a significant sink for CO in the soil, based on unpublished data. According to Seiler (1974), soil organisms can either produce or consume CO, depending on soil conditions and, presumably, on the particulars of the microbial ecology. Unfortunately, accurate estimates of the importance of soil in the CO budget are necessarily very difficult to obtain because of the impossibility of experimentally simulating global conditions. If Seiler's (1974) estimates for the soil sinks are included in Table 2, global balance to 10% is preserved, indicating that soil microorganisms do not affect the basic balance of the global CO budget. Nevertheless, the contribution of microbial consumption could be moderately significant and should be the subject of further research.

To summarize, we have studied the global distributions of CO and CH_4 and the sources and sinks for these gases. Oxidation of CH_4 is the largest source of atmospheric CO, about 5/8 of the total, in agreement with Wofsy et al (1972). In the Southern Hemisphere, CH_4 oxidation is nearly an order of magnitude greater than all other CO sources, but in the Northern Hemisphere natural and anthropogenic sources are comparable.

This situation is somewhat ironic. When Bates & Witherspoon (1952) proposed combustion as the principal source of CO, methane dominated combustion in both hemispheres. But by the time McConnell et al (1971) identified CH_4 oxidation as the main source, anthropogenic emissions had overtaken natural sources in the Northern Hemisphere.

PART II: CO AND CH_4 IN THE UPPER ATMOSPHERE

Carbon Monoxide

Observations are inadequate to define the global distribution of CO or CH_4 in the stratosphere and mesosphere. Consequently, we attempt to understand the existing observations by calculating theoretical distributions and comparing them to the measurements.

Stratospheric observations of CO have been reported by Seiler & Warneck (1972), who measured vertical profiles over the North Sea, and by Farmer (1974), who reported latitudinal measurements at 16 km. The vertical profiles reported by Seiler & Warneck (1972) appeared to provide some support for oxidation of atmospheric CH_4 to CO, as well as some puzzles. The CO concentration at 45°N dropped sharply above the tropopause, which was at ~8.5 km. Above 10 km, however, the CO mixing ratio assumed a constant value, up to the highest level (~12 km) accessible to their aircraft. The constant mixing ratio implied a high altitude source of CO, and the value, 0.04 ppmv, was only slightly higher than the predictions of McConnell et al (1971) and Wofsy et al (1972). However, the global mean calculations predicted a gradient at the tropopause much less steep than that observed by Seiler & Warneck (1972).

The observed vertical profiles of CO, as well as Farmer's (1974) observations of latitudinal distribution, can be understood using a simple transport model. Hering (1965), Danielson (1968), List & Telegadas (1969), and Johnston et al (1975) have demonstrated that various long-lived tracers distribute themselves meridionally along identical preferred mixing surfaces in the lower stratosphere. These surfaces are quasi-horizontal; between equator and pole, they follow approximately isentropic trajectories which descend toward the pole. The poleward slope is steeper in winter than in summer. Because CH_4 and CO are long-lived tracers in the lower stratosphere (see Figures 4 and 5), they should be similarly distributed.

In the spirit of the earlier work (List & Telegadas 1969 and others), two-dimensional distributions were derived from a simplified model, based on the following two postulates:

1. Transport along the "preferred mixing surfaces" is more rapid than vertical mixing across them and is fast enough to smooth out variations of the CO or CH_4 mixing ratios along these surfaces (cf Warneck et al 1973). Thus the mixing surfaces become isopleths of the time-averaged CO and CH_4 mixing ratios. Below 20 km, they coincide with the isopleths of ozone mixing ratio (Hering 1975) and above 20 km they are adopted from C^{14} and potential vorticity isopleths (Johnston et al 1975, Danielson 1968). A graph showing the isopleths has been presented elsewhere [McElroy et al (1976)].

2. Vertical transport can be simulated using the one-dimensional eddy-diffusion equation (cf Wofsy et al 1972, McElroy et al 1974) with the transport coefficient (K_{zz}) adjusted to the observed CH_4 profile (see below). The one-dimensional calculation represents a global average, but it should be closest to a 30° profile because the CH_4 observations were made at 30°N and conditions at 30°N approximate mean stratospheric values for temperature, insolation, etc. The calculated one-dimensional profile is therefore imposed at 30°N latitude, and extended to two dimensions using the isopleths described above.

Curve *A* of Figure 8 shows the global average profile of CO, calculated by solving the one-dimensional eddy transport equation, with the eddy-diffusion coefficient adjusted to the observed CH_4 profile (cf Wofsy & McElroy 1973). At the lower boundary, the mixing ratio of CO was set equal to its global mean value (~0.11 ppm) from Figure 6*b*. Curve *B* of Figure 8 shows the calculated CO profile at 45°N, in winter. Above the tropopause, curve *B* was derived directly by imposing the global mean profile of curve *A* at 30°N and assuming a constant mixing ratio along each preferred mixing surface. Near the tropopause this profile was joined to the observed values in the troposphere (from Figure 6).

This procedure derives from the basic postulates of the model, which imply that stratospheric distributions are necessarily influenced by transport to or from the troposphere at distant latitudes. Consequently, latitudinal variations in the troposphere are largely independent of those in the stratosphere. In order to simulate the profile at a given location in the Northern Hemisphere, the observed, anthropogenically perturbed CO profile in the troposphere should be joined to the calculated stratospheric profile.

The stratospheric CO observed at 45°N, say at 16 km, was not transported vertically at 45°N. Instead, most of it entered the stratosphere at lower latitudes (see Danielson 1968 and Newell et al 1969), or was created from CH_4 at various points on the appropriate mixing surface, and it then flowed along the mixing surface to 45°N. Contributions from vertical transport were effectively averaged over all latitudes and longitudes, on a given mixing surface. By contrast, anthropogenic CO in the northern troposphere was mostly generated near the latitude where it was observed, because of the geographical concentration of sources. Comparison of

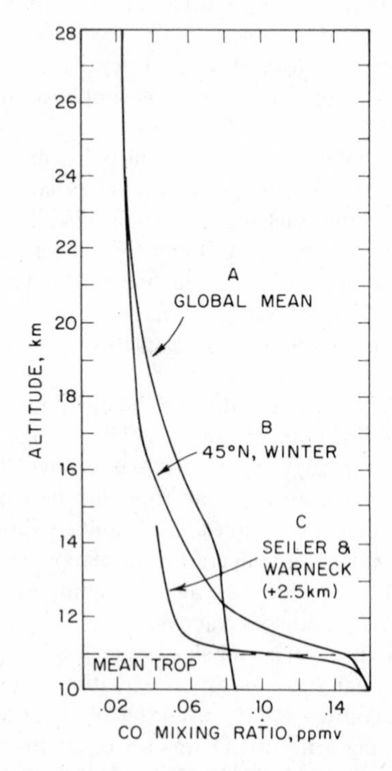

Figure 8 The vertical distribution of CO in the stratosphere is shown. A = calculated global mean; B = calculated profile at 45°N, winter; C = winter data of Seiler & Warneck (1972) over the North Sea, normalized to a nominal tropopause height of 11 km.

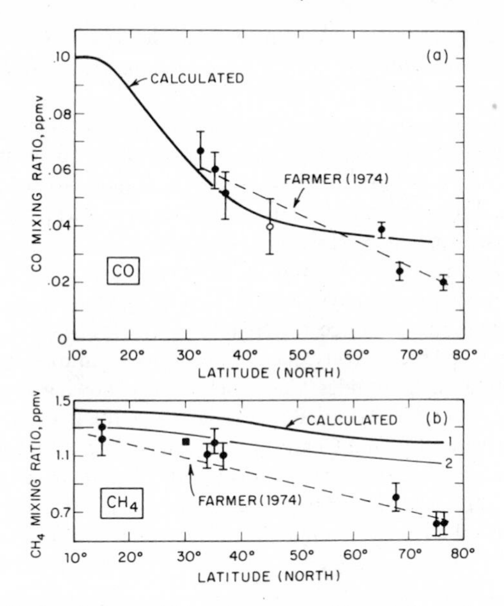

Figure 9 Latitudinal variations of (*a*) carbon monoxide and (*b*) methane are presented at 16 km. The calculated curves are discussed in the text. The experimental points are ● (Farmer 1974), ■ (Ehhalt & Heidt 1973), ○ (Seiler & Warneck 1972).

curve *B* in Figure 8 with the observations shown in curve *C*,[3] reveals reasonably good agreement. Between 18 and 12 km, curve *B* is displaced downward from curve *A* because the mixing surfaces descend at high latitudes. Near the tropopause there is a sharp break caused by the weak coupling between tropospheric and stratospheric CO at 45°N. Either distortion of the mean CO profile due to the peculiar meteorological conditions[4] or shortcomings in our procedure for calculating curve *B* could account for the remaining discrepancies.

Figure 9*a* shows the latitudinal distribution of stratospheric CO reported by Farmer (1974). We note that the north/south trend is opposite from the tropospheric CO trend in Figure 6*b*, that is, there is less CO at high latitudes. The calculated curve was obtained from the global mean profile and preferred mixing surfaces as described above. The agreement certainly supports the basic ideas underlying the model. The latitudinal variation of CO at 16 km is evidently not controlled by CO

[3] Curve *C* shows the data of Seiler & Warneck (1972) displaced by 2.5 km, so that the observed tropopause coincides with the mean tropopause, assumed in the present calculations at 11 km.

[4] For a tracer moving down, such as O_3, transport from stratosphere to troposphere might have been more efficient under the meteorological conditions sought by Seiler & Warneck (1972). Strong winter storms with a low tropopause are likely to induce downward transport (Danielson 1968). Seiler & Warneck's deductions (1972) based on O_3 profiles need to be evaluated in this light.

transported vertically from, say, 8 km. Instead, the decrease of CO at high latitudes reflects the vertical distribution of the gas in the chemically active regions at lower latitudes.

The only discrepancies appear over the north polar region, where the data suggest the influence of downwelling from higher altitudes. The winter polar vortex could induce significant downward transport, which we have not been able to include in our empirical mixing surfaces because of inadequate tracer data and expected deviations from isentropic trajectories near the pole.

The latitudinal distributions of a number of different gases can be understood by reference to the mixing surface hypothesis. The mixing ratio of a tracer that flows down from a stratospheric source region (NO_x, O_3, HCl) should increase with height above the tropopause; consequently, at a given altitude, the tracer should be more abundant at high latitudes. Conversely, tracers flowing up from the troposphere (CO, CH_4, N_2O, $CFCl_3$) should decrease above the tropopause and at high latitudes. Gases such as CO and $CFCl_3$, which have strong, geographically concentrated anthropogenic sources, may exhibit contradictory trends with latitude in the troposphere as compared to the stratosphere. A chemical lifetime of one year or more in the lower stratosphere appears to be the only requirement for this rule to apply.

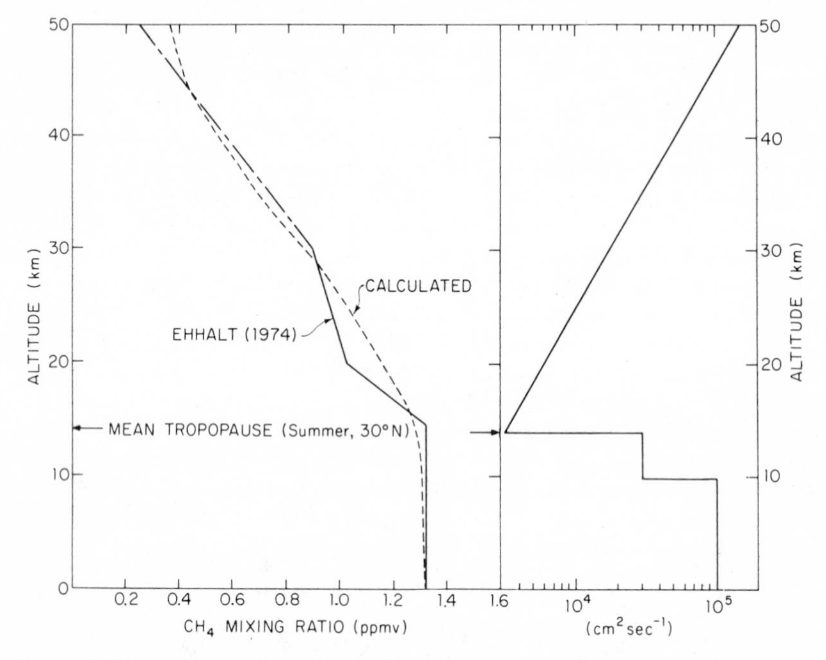

Figure 10 The left panel shows the vertical profile of CH_4 according to measurements summarized by Ehhalt (1974). Also shown is our calculated profile, using the one-dimensional eddy-diffusion equation and the chemistry of Table 1. The eddy-diffusion parameter, K_{zz}, is shown in the right-hand panel.

Methane

The vertical distribution of CH_4 has been measured by Bainbridge & Heidt (1966), Kyle et al (1969), Ackerman & Muller (1973), and Ehhalt and co-workers (1972, 1973, 1974). These data extend up to 30 km, usually near 30°N latitude, with a few observations near 50 km. Data on the latitudinal variation were recently summarized by Farmer (1974).

Figure 10 shows the average vertical distribution of CH_4 at 30°N in summer, according to Ehhalt (1974). Between the tropopause and 30 km, two distinct regions appear that have a gradient significantly steeper below 20 km than above that level. Above 30 km, two rocket observations provide the average mixing ratio between 44 and 60 km, which we have connected to the mixing ratio data at 30 km by a dashed line.

Wofsy & McElroy (1973) pointed out that the observation of CH_4 near the stratopause implied substantially faster vertical mixing above 30 km than below. As indicated by Figure 5, the CH_4 chemical lifetime becomes rather short (less than one year) in the upper stratosphere, and earlier calculations (Wofsy et al 1972; Whitten, Sims & Turco 1973) predicted significantly less CH_4 near 50 km than observed. Wofsy & McElroy (1973) proposed values for the vertical mixing coefficient, K_{zz}, which were quite small in the lower stratosphere but increased roughly as $[n]^{-\frac{1}{2}}$ in the upper stratosphere. Small values of K_{zz} were required in the lower stratosphere to obtain the long observed residence times for stratospheric tracers, whereas the increase of K_{zz} above 30 km was implied by the CH_4 data. The form of K_{zz} was chosen to concur with a suggestion by Lindzen (1971) that vertically propagating disturbances may be responsible for tracer transport in the upper stratosphere and mesosphere.

Subsequently, Hunten (1975) presented a simplified form for K_{zz} that adequately fit the data. Figure 10 shows the CH_4 profile calculated with a modification of Hunten's (1975) coefficient K_{zz}, which is also shown in the figure. Evidently, the K_{zz} derived in this manner depends on the profile of OH, which is somewhat different in the present work than that of Wofsy & McElroy (1973) or Hunten (1975).

The calculated profile reproduces the decay of CH_4 in the upper stratosphere fairly well. However, it does not exhibit the two distinct gradients below 30 km, and vertical transport appears too fast near the stratopause. Unfortunately, further refinements of K_{zz} are not possible without new tracer observations. The present calculation of K_{zz} depends on [OH], which has never been measured below 40 km, and it assumes that the limited data define the global-average CH_4 profile, which is not assured. Nevertheless, the CH_4 profile has provided very concrete evidence that vertical transport is faster in the upper stratosphere than in the lower stratosphere.

Figure 9*b* shows the latitudinal variation of CH_4 at 16 km (Farmer 1974). The calculated curves were taken from two-dimensional distributions derived using the mixing surfaces as described above. Curve 1 used the calculated global average profile of 30°N, and Curve 2 used Ehhalt's (1974) data. The poleward slope of the calculated curves are slightly smaller than Farmer's (1974) observations. Ehhalt's

(1974) observed CH_4 profile at 30° predicts the CH_4 distribution in nearly exact agreement with Farmer's (1974) points at latitudes lower than 60°K. The CH_4 data, like the CO data, show anomalously low concentrations in the polar region.

CONCLUSION

The availability of global data for CO and CH_4 permits a number of tests for the present theory of tropospheric and stratospheric photochemistry. The theory is apparently consistent with the observations down to a level of rather fine detail, which strongly supports the proposed role of OH in atmospheric photochemistry and the calculated abundance of OH in the lower troposphere. The conversion of CH_4 to CO is also strongly implied by the present results.

Global pollution of the atmosphere by CO emissions has reached disturbing levels, although the system is not yet seriously overloaded. The source of CO from methane oxidation constitutes an appreciable fraction of biogenically fixed carbon (cf McElroy 1975), and the CH_4-CO-CO_2 cycle is crucial to the self-cleansing mechanisms of the troposphere, because of interactions with OH and other radicals. Anthropogenic emissions of CO are beginning to perturb these very fundamental atmospheric processes.

This situation has some peculiar aspects. Carbon monoxide sources are concentrated in urban areas, where large quantities of shorter-lived pollutants such as NO_x and reactive hydrocarbons are also emitted. The urban air pollutants are localized, and can generate high radical concentrations through efficient chain branching and propagating reactions; the short-lived pollutants tend to be rapidly consumed as they are converted into the noxious elements of smog. However, CO is long-lived enough to become a global pollutant. Chain propagation is inefficient in clean air, so the effect of CO on the global scale is to suppress the concentration of OH and to inhibit the capability of the atmosphere to remove a large number of other long-lived pollutants. Pollution sources have evidently reached a magnitude that can significantly stress the atmosphere on regional and global scales, as well as the more familiar local scales.

ACKNOWLEDGMENTS

The author is indebted to J. Davidson, G. Streit, C. Howard, Y. L. Yung, and J. E. Penner for making available their unpublished results. Helpful discussions with R. Lindzen, M. B. McElroy, E. Sarachik, and C. B. Farmer are gratefully acknowledged. This work was supported by a grant from the National Aeronautics and Space Administration, NASA NSG 2031, to Harvard University.

APPENDIX

A few aspects of the chemistry of OH, CH_4, CO, and intermediate species may need clarification, and a brief technical description of the present calculations is also appropriate.

One important feature of the present model is the significant role attributed to rainout and washout of soluble gases. In the lower troposphere, the chemical lifetimes for H_2O_2 and CH_3OOH are between 1 and 2 weeks, and for HNO_3 the lifetime is longer than 10 weeks. Wofsy et al (1975b) utilized observations of long-lived soluble gases such as HNO_3 and HCl (Lazrus & Gandrud 1974, Duce 1969, Duce et al 1969) and data on scavenging of radon daughters (see Junge 1963) to construct the loss profile shown in Table 1 (R63). This loss profile corresponds to a mean atmospheric lifetime of 7 days for soluble gases, and it removes H_2O_2, CH_3OOH, and HNO_3 more rapidly than chemical reactions in the lower troposphere.

Matsui (1949) (cf Junge 1963, p. 316) claimed to observe H_2O_2 in rain at concentrations which, if globally distributed, would account for loss of $\sim 3 \times 10^{10}$ molecules cm^{-2} sec^{-1}. Washout of CH_3OOH (Wofsy et al 1972) and HNO_3 (Warneck 1974) could each provide a similar flux. These fluxes are the same order as the vertically integrated rate for OH production (i.e. approximately twice the rate of R6). The present calculations predict a heterogeneous loss rate of 5×10^{10} cm^{-2} sec^{-1} of H_2O_2 and $\sim 3 \times 10^{10}$ cm^{-2} sec^{-1} of CH_3OOH and HNO_3 using the loss profiles of Table 1. The agreement with Matsui (1949) indicates that precipitation scavenging is indeed faster than chemical loss for H_2O_2, although it does not provide a quantitative measure of the washout rate.

Warneck (1974) showed that a significant fraction of atmospheric HO_2 could be lost by collision with the surface of aerosol particles, if the HO_2 were destroyed on every collision. This requirement seems unlikely, however, as even the best catalytic surfaces rarely approach unit efficiency.

Another key tenet of the theory is efficient conversion of CH_4 to CO by atmospheric photooxidation. Levy (1971) proposed that atmospheric CH_4 should be oxidized to H_2CO by the following steps:

$$CH_4 + OH \rightarrow CH_3 + H \tag{R2}$$

$$CH_3 + O_2 + M \rightarrow CH_3O_2 + M \tag{R19}$$

$$CH_3O_2 + NO \rightarrow CH_3O + NO_2 \tag{R20}$$

$$CH_3O + O_2 \rightarrow H_2CO + HO_2 \tag{R21}$$

The fate of CH_3 radicals (Reaction R19) appears well established from laboratory data (Parkes et al 1973). No evidence exists for R20, and there is only indirect evidence for R21.

The fate of H_2CO in the atmosphere is almost certainly conversion to CO via

$$H_2CO + h\nu \rightarrow H_2 + CO \tag{R22}$$

$$H_2CO + h\nu \rightarrow H + HCO \tag{R23}$$

and

$$H_2CO + OH \rightarrow H_2O + HCO, \tag{R24}$$

followed by

$$HCO + O_2 \rightarrow HO_2 + CO. \tag{R25}$$

The photolysis rates are fairly well determined (Calvert et al 1972), as is the rate for R24 (Morris & Niki 1971). Reaction 25 has also been observed (Radford et al 1974) as the principal path for HCO in the presence of O_2 and the rate has been measured by Washida et al (1974). Buffalini & Brubaker (1971) reported smog-chamber experiments confirming that CO is the major product of H_2CO photo-oxidation.

The uncertainties lie mainly in the pathway between CH_4 and H_2CO. Wofsy et al (1972) rejected R20 as unlikely and proposed a different scheme in which methyl hydroperoxide (CH_3OOH) is an intermediate:

$$CH_3O_2 + HO_2 \rightarrow CH_3OOH + O_2 \qquad \text{(R26)}$$

$$CH_3OOH + h\nu \rightarrow CH_3O + OH \qquad \text{(R27)}$$

$$CH_3O + O_2 \rightarrow H_2CO + HO_2. \qquad \text{(R21)}$$

CH_3OOH was observed in virtually all the laboratory systems devised to study the low temperature oxidation of CH_3 radicals (Kleimenov & Nalbandian 1960, Hummel & Hearne 1966, Fisher & Tipper 1963, Subbaratnam & Calvert 1962, Dever & Calvert 1962). Rection 26 is supported by the experiments of Hummel & Hearne (1966), who found CH_3OOH as the major product when mixtures of CH_4 and O_2 were subjected to very low doses of γ-rays. Because of the extremely small conversions in their system and the very low concentrations expected for CH_3O, they considered R26 a more likely source of CH_3OOH than

$$CH_3O_2 + CH_3O \rightarrow CH_3OOH + H_2CO, \qquad \text{(R28)}$$

as proposed by Subbaratnam & Calvert (1962).

In all the reported experiments, CH_3OOH interacts strongly with the vessel walls. To further complicate the matter, experiments that included NO or NO_2 in the reaction mixture (Heicklen 1973) produced mainly methyl peroxynitrate (CH_3OONO_2), quite possibly on the walls. The laboratory data are clearly inadequate to establish the atmospheric fate of CH_3 radicals. However, observations of formaldehyde in rain provide some support for the proposed atmospheric pathway (Wofsy et al 1972) involving CH_3OOH.

Finally, the basic atmospheric model should be described briefly. The reactions in Table 1 were combined into a set of photochemical equations, which were then solved using a modification of Newton's method. In addition to OH, we calculated H, H_2O_2, HO_2, NO, NO_2, NO_3, N_2O_5, HNO_3, H_2CO, HCl, Cl, ClO, $O(^3P)$, $O(^1D)$, and O_3 above 30 km (cf McElroy et al 1974). Solar fluxes were taken from the recent compilation by Vernazza (1975), and photodissociation rates averaged over a 24-hour day appropriate to the latitude and solar declination. Total chlorine, H_2 and NO_x were calculated by a one-dimensional model in the manner of McElroy et al (1974) and Wofsy et al (1975a,b) and extended to two dimensions by assuming uniform horizontal distribution in the troposphere[5] and uniform distr-bution in the stratosphere along the empirical mixing surfaces described in Part II.

[5] A north/south asymmetry was adopted for H_2 from Seiler & Schmidt (1974).

Temperatures and pressures were taken from the U.S. Standard Atmosphere Supplements (1966), and tropospheric water vapor calculated from these temperatures and relative humidities from London (quoted by Lorenz 1967). Stratospheric water vapor was taken as 4.2 ppmv. Ozone data below 30 km were adopted from Hering & Borden (1966) and Hering (1975).

The flux across the Earth's surface is derived from the chemical reaction rates and the mass continuity equation, assuming steady-state conditions:

$$\nabla \cdot \Phi = P - L. \tag{2}$$

Here Φ is the flux ($cm^{-2}\ sec^{-1}$), and P and L are chemical production and loss terms ($cm^{-3}\ sec^{-1}$) for a given molecular species. The mean flux across the Earth's surface, F, is given by

$$F = \frac{1}{4\pi}\int_0^{2\pi} d\phi \int_{-\pi/2}^{\pi/2} d\theta\, \Phi_z(z=0)\cos\theta = \frac{1}{4\pi}\int_0^{2\pi} d\phi \int_{-\pi/2}^{\pi/2} d\theta \cos\theta \int_0^{\infty} dz\{P-L\}$$

$$\equiv \frac{1}{2}\int_{-\pi/2}^{\pi/2} \cos\theta F_\theta\, d\theta \tag{3}$$

where ϕ, θ, and z are longitude, latitude, and altitude, respectively, and the flux at the top of the atmosphere is taken as zero. In order to help identify the source regions, we have introduced F_θ, the zonally averaged contribution to F at latitude θ,

$$F_\theta = \frac{1}{2\pi}\int_0^{2\pi} d\phi \int_0^{\infty} dz\{P-L\} \tag{4}$$

F_θ is plotted against latitude in Figure 7 for CO and CH_4.

Literature Cited

Ackerman, M., Muller, C. 1973. Stratospheric methane and nitrogen dioxide from infrared spectra. *Pure Appl. Geophys.* 106–8:1325–35

Anderson, J. G., Kaufman, F. 1973. Kinetics of the reaction $OH(V=O)+O_3 \rightarrow HO_2+O_2$. *Chem. Phys. Lett.* 19:483–86

Anderson, J. G., Margitan, J. J., Zahniser, M. S., Kaufman, F. 1976. Laboratory studies of stratospheric OH and Cl reactions. In press

Bainbridge, A. E., Heidt, L. E. 1966. Measurements of methane in the troposphere and lower stratosphere. *Tellus* 18: 221–25

Basco, L., James, D., James, F. 1971. A kinetic study of the combination reaction: CH_3+O_2. *Chem. Phys. Lett.* 8:265–66

Bates, D. R., Hays, P. B. 1967. Atmospheric nitrous oxide. *Planet. Space Sci.* 15:189–97

Bates, D. R., Witherspoon, A. E. 1952. The photochemistry of some minor constituents of the Earth's atmosphere (CO_2, CO, CH_4, N_2O). *Mon. Not. R. Astron. Soc.* 112:101–24

Baulch, D. L., Drysdale, D. D., Horne, D. G. 1973. *Chemical Kinetics Data Survey V*, pp. 49–115. Washington DC: Nat. Bur. Stand.

Benson. S. W., Shaw, R. 1971. Thermochemistry of organic peroxides, hydroperoxides, polyoxides, and their radicals. *Organic Peroxides*, ed. D. Swern, 1:105–39. New York: Interscience

Brown, R., Smith, I. W. M. 1975. Absolute rate constants for the reactions of $O(^3P)$ atoms with HCl and HBr. *Int. J. Chem. Kinet.* 7:301–15

Buffalini, J. J., Brubaker, K. L. 1971. The photochemistry of formaldehyde at low partial pressures. *Chemical Reactions in Urban Atmospheres*, ed. C. S. Tuesday, pp. 225–38. New York: Elsevier

Calvert, J. G., Kerr, J. A., Dermerjian, K. L., McQuigg, R. D. 1972. Photolysis of formaldehyde as a hydrogen atom source in the lower atmosphere. *Science* 175:751–52

Campbell, I. M., Gray, C. N. 1973. Rate constants for $O(^3P)$ recombination and association with $N(^4S)$. *Chem. Phys. Lett.* 18:607–9

Czeplak, G., Junge, C. 1974. Studies of interhemispheric exchange in the troposphere by a diffusion model. *Adv. Geophys.* 18B: 57–72

Danielson, E. F. 1968. Stratospheric-tropospheric exchange based on radioactivity, ozone and potential vorticity. *J. Atmos. Sci.* 25:502–18

Davidson, J. A., Sadowski, C. M., Schiff, H. I., Streit, G. E., Howard, C. J., Jennings, D. A., Schmeltekopf, A. L. 1976. Absolute rate constant determinations for the deactivation of $O(^1D)$ by time-resolved decay of $O(^1D) \rightarrow O(^3P)$ emission. *J. Chem. Phys.* 64:57–62

Davis, D. D., Herron, J. T., Huie, R. E. 1973. Absolute rate constants for $O(^3O) + NO_2 \rightarrow NO + O_2$. *J. Chem. Phys.* 58: 530–35

Davis, D. D., Kim, P., Pusaczyk, J., Dwyer, M. 1974. *J. Phys. Chem.* 78:1775–81

Davis, D. D., Wong, W., Schiff, R. 1974. *J. Phys. Chem.* 78:463–71

Demore, W. B. 1973. Rate constants for the reactions of hydroxyl and hydroperoxyl with ozone. *Science* 180:735–37

Demore, W. B., Tschuikow-Roux, E. 1974. Temperature dependence of the reactions of OH and HO_2 with O_3. *J. Phys. Chem.* 78:1447–51

Dever, D. G., Calvert, J. G. 1962. Rate studies of the oxidation of methyl radicals in oxygen-rich media at 25°. *J. Am. Chem. Soc.* 84:1362–68

Dixon-Lewis, G., Wilson, W. E., Westenberg, A. A. 1966. Studies of OH kinetics by quantitative ESR. *J. Chem. Phys.* 44: 2877–84

Duce, R. A. 1969. On the source of gaseous chlorine in the marine atmosphere. *J. Geophys. Res.* 74:4597–99

Duce, R. A., Seto, Y. B., Moyers, J. L. 1969. Variation of sodium and chloride concentrations with rainfall intensity in Hawaiian Trade wind showers. *Pac. Sci.* 23:483–95

Ehhalt, D. H. 1974. The atmospheric cycle of methane. *Tellus* 26:58–70

Ehhalt, D. H., Heidt, L. E., Martell, E. A. 1972. The concentration of atmospheric methane between 44 and 62 km altitude. *J. Geophys. Res.* 77:3293–96

Ehhalt, D. H., Heidt, L. E. 1973. The concentration of molecular H_2 and CH_4 in the stratosphere. *Pure Appl. Geophys.* 106–8:1352–60

Farmer, C. B. 1974. Infrared measurements of stratospheric composition. *Can. J. Chem.* 52:1544–59

Fisher, I. P., Tipper, C. F. H. 1963. Oxidation of methane at about 400°C. *Trans. Faraday Soc.* 59:1163–88

Graham, R. A., Johnston, H. S. 1974. Kinetics of the gas-phase reaction between O_3 and NO_2. *J. Chem. Phys.* 60:4628–29

Greiner, N. R. 1968. Hydroxyl radical kinetics by kinetic spectroscopy, 3, reactions with H_2O_2 in the range 300–458°K. *J. Phys. Chem.* 72:406–10

Greiner, N. R. 1969. Hydroxyl radical kinetics by kinetic spectroscopy, 4, reactions with H_2 and CO in the range 300–500°K. *J. Chem. Phys.* 51:5049–51

Greiner, N. R. 1970. Hydroxyl radical kinetics by kinetic spectroscopy, 6, reactions with alkanes in the range 300–500°K. *J. Chem. Phys.* 53:1070–76

Griggs, M. 1968. Absorption coefficients of ozone in the ultraviolet and visible regions. *J. Chem. Phys.* 49:857–59

Hack, W., Hoyermann, K., Wagner, H. G. 1975. The reaction $NO + HO_2 \rightarrow NO_2 + OH$ with $OH + H_2O \rightarrow H_2O + HO_2$ as an HO_2 source. *Int. J. Chem. Kin. Symp.* 1: 329–41

Hall, T. C., Blacet, F. E. 1952. Separation of the absorption of NO_2 and N_2O_4 in the range 2400–5000Å. *J. Chem. Phys.* 20:1745–49

Hamilton, E. J. 1975. Water vapor dependence of the kinetics of the self-reaction of HO_2 in the gas phase. *J. Chem. Phys.* 63:3682–83

Hampson, J. 1964. *Photochemical Behavior of the Ozone Layer.* Quebec: C.A.R.D.E. 280 pp.

Harker, A., Johnston, H. S. 1973. Photolysis of NO_2 to produce transient O, NO_3 and N_2O_5. *J. Phys. Chem.* 77:1153–56

Heicklen, J. 1968. Gas phase reactions of alkyl peroxy and alkoxy radicals. *Adv. Chem. Ser.* 76:23–39

Heicklen, J. 1973. Photochemical and rate data for methyl nitrite, methoxy and methyl peroxy. *Chemical Kinetics Data Survey V, NBSIR73-206,* ed. D. Garvin, pp. 43–48. Washington DC: Nat. Bur. Stand.

Hering, W. S. 1975. Ozone and atmospheric transport processes. *Tellus* 23:229–336

Hering, W. S., Borden, T. R. 1966. *Ozonesonde Observations Over North America. Environ. Res. Paper No. 279,* Air Force Cambridge Res. Lab. 365 pp.

Hochanadel, C. J., Ghormley, J. A., Ogren, P. J. 1972. Absorption spectrum and reaction kinetics of the HO_2 radical in the gas phase. *J. Chem. Phys.* 56:4426–32

Huie, R. E., Herron, J. J., Davis, D. D. 1972. Absolute rate constants for the reaction $O+O_2+M \rightarrow O_3+M$ over the temperature range 200–346°K. *J. Phys. Chem.* 76:2653–58

Hummel, R. W., Hearne, J. A. 1966. The radiolytic oxidation of methane. *Radiat. Chem.* 23:353–64

Hunt, B. G. 1966. Photochemistry of ozone in a moist atmosphere. *J. Geophys. Res.* 71:1385–90

Hunten, D. M. 1975. *The Philosophy of One-Dimensional Modelling.* Paper presented at Conf. Climatic Impact Assessment Progr., 4th, US Dep. Transport., Cambridge, Mass.

Jaffe, L. S. 1970. Sources, characteristics and fate of atmospheric carbon monoxide. *Ann. N.Y. Acad. Sci.* 174:76–88

Jaffe, L. S. 1973. Carbon monoxide in the biosphere. *J. Geophys. Res.* 78:5293–5505

Johnston, H. S., Graham, R. 1973. Gas phase ultraviolet absorption spectrum of nitric acid vapor. *J. Phys. Chem.* 77:62–63

Johnston, H. S., Chang, S.-G., Whitten, G. 1974. Photolysis of HNO_3 vapor. *J. Phys. Chem.* 78:1–7

Johnston, H. S., Kaltenhorn, D., Whitten, G. 1975. *Use of Excess Carbon-14 Data to Calibrate Models of Stratospheric Ozone Depletion by Supersonic Transports. Lawrence Livermore Lab. LBL-3548*

Junge, C. E. 1963. *Air Chemistry and Radioactivity.* New York: Academic

Junge, C. E., Seiler, W., Warneck, P. 1971. The atmospheric ^{12}CO and ^{14}CO budget. *J. Geophys. Res.* 76:2866–79

Kaufman, F. 1969. Neutral reactions involving hydrogen and other minor constituents. *Can. J. Chem.* 47:1917–24

Kleimenov, N. A., Nalbandyan, A. B. 1960. Oxidarea metanuli la temperaturi joase. *Rev. Chim. (Bucharest)* 4:391–401

Koyama, T. 1969. Biogeochemical studies on lake sediments and paddy soils and the production of atmospheric methane and hydrogen. In *Recent Researches in the Fields of Hydrosphere, Atmosphere, and Nuclear Geochemistry,* ed. Y. Miyake, T. Koyama. Water Res. Lab., Nagoya Univ., Nagoya, Japan

Kummler, R. H., Bauer, T. 1973. Atmospheric carbon-hydrogen chemistry. *J. Geophys. Res.* 78:5306–16

Kyle, T. G., Murcray, D. G., Murcray, F. H., Williams, W. J. 1969. Abundance of methane in the atmosphere above 20 km. *J. Geophys. Res.* 74:3421–25

Lazrus, A. L., Gandrud, B. W. 1974. Distribution of stratospheric nitric acid vapor. *J. Atmos. Sci.* 31:1102–8

Levy, H. 1971. Normal atmosphere: large radical and formaldehyde concentrations predicted. *Science* 173:141–43

Levy, H. Jr. 1972. Photochemistry of the lower troposphere. *Planet. Space Sci.* 20: 919–31

Levy, H. Jr. 1973. Tropospheric budgets for methane, carbon monoxide, and related species. *J. Geophys. Res.* 78:5325–32

Levy, H. Jr. 1974. Photochemistry of the troposphere. *Adv. Photochem.* 9:369–524

Lin, C.-L., Demore, W. B. 1973. $O(^1D)$ production in ozone photolysis near 3100 Å. *J. Photochem.* 2:161–64

Lindzen, R. S. 1971. Tides and gravity waves in the upper atmosphere. *Mesospheric Models and Related Experiments. Proc. Esrin-Esrolab Conf., 4th Frascatti.* Dordrecht: Reidel

List, R. J., Telegadas, K. 1969. Using radioactive tracers to develop a model of the circulation of the stratosphere. *J. Atmos. Sci.* 26:112–36

Lorenz, E. N. 1967. The nature and theory of the general circulation of the atmosphere. *WMO Tech. Note.* 161 pp.

Margitan, J. J., Kaufman, F., Anderson, J. G. 1975. *Int. J. Chem. Kin. Symp.* 1:281–89

McConnell, J. C., McElroy, M. B. 1973. Odd nitrogen in the atmosphere. *J. Atmos. Sci.* 30:1465–80

McConnell, J. C., McElroy, M. B., Wofsy, S. C. 1971. Natural sources of atmospheric CO. *Nature* 233:187–88

McCrumb, J. L., Kaufman, F. 1972. Kinetics of the $O+O_3$ reaction. *J. Chem. Phys.* 57: 1270–76

McElroy, M. B. 1976. Chemical processes in the solar system: a kinetic perspective. In *MTP International Review of Science,* ed. D. R. Herschbach. Baltimore: Univ. Park Press. In press

McElroy, M. B., McConnell, J. C. 1971. Nitrous oxide: a natural source of stratospheric NO. *J. Atmos. Sci.* 28:1095–98

McElroy, M. B., Elkins, J. W., Wofsy, S. C., Yung, Y. L. 1976. Sources and sinks for atmospheric N_2O. *Rev. Geophys. Space Phys.* In press

McElroy, M. B., Wofsy, S. C., Penner, J. E., McConnell, J. C. 1974. Atmospheric ozone: possible impact of stratospheric avaiation. *J. Atmos. Sci.* 31:287–303

Matsui, H. 1949. On the content of hydrogen peroxide of atmospheric precipitate. *J. Meteorol. Soc. Jpn* 27:380–81

Migeotte, M. V. 1948. Lines of methane at 7.7μ in the solar spectrum. *Phys. Rev.* 74:112–13

Migeotte, M. V. 1948. The fundamental bond of carbon monoxide at 4.7μ in the solar spectrum. *Phys. Rev.* 75:1108–9

Migeotte, M. V. 1948. Methane in the Earth's

atmosphere. *Ap. J.* 107:400–3

Migeotte, M. V., Nevin, L. 1950. Détection du monoxyde de carbone dans l'atmosphere terrestre, à 3580 mètres d'altitude. *Physica* 16:423–27

Morris, E. D., Niki, H. 1971. Mass spectrometric study of the reaction of OH with H_2CO. *J. Chem. Phys.* 55:1991–92

Newell, R. E., Vincent, D. G., Kidson, J. W. 1969. Interhemispheric mass exchange from meteorological and trace substance observation. *Tellus* 21:641–47

Nicholls, R. W. 1975. Spectroscopic implications of stratospheric ClO. *J. Atmos. Sci.* 32:856–60

Nicolet, M. 1970. Aeronomic reactions of hydrogen and ozone. *Aeronomica* A79: 1–100

Paraskevopoulos, G., Cvetanovic. R. J. 1971. Relative rate of reaction of $O(^1D_2)$ with H_2O. *Chem. Phys. Lett.* 9:603–5

Parkes, D. A., Paul, D. M., Quinn, C. P., Robson, R. C. 1973. The ultraviolet absorption by alkylperoxy radicals and their mutual reactions. *Chem. Phys. Lett.* 33:425–29

Payne, W. A., Stief, L. J., Davis, D. D. 1973. A kinetics study of the reaction of HO_2 with SO_2 and NO. *J. Am. Chem. Soc.* 95: 7614–19

Pressman, J., Warneck, P. 1970. The stratosphere as a chemical sink for carbon monoxide. *J. Atmos. Sci.* 27:155–63

Radford, H. E., Evenson, K. M., Howard, C. J. 1974. HO_2 detected by laser magnetic resonance. *J. Chem. Phys.* 60:3178–83

Robinson, E., Robbins, R. C. 1968. *Sources, Abundance and Fate of Gaseous Atmospheric Pollutants. Res. Proj. PR-6755,* Stanford Univ., Stanford, Calif.

Robinson, E., Robbins, R. C. 1969. Atmospheric CO concentrations on the Greenland ice cap. *J. Geophys. Res.* 74:1968–73

Schofield, K. 1967. An evaluation of kinetic rate data of atmospheric interest. *Planet. Space Sci.* 15:643–70

Schumb, W. C., Satterfield, C. N., Wentworth, R. L. 1955. *Hydrogen Peroxide.* New York: Reinhold

Scott, P. M., Cvetanovic, R. J. 1971. Relative rate constants for reactions of $O(^1D)$ atoms generated by flash photolysis of ozone. *J. Chem. Phys.* 54:1440–45

Seiler, W. 1974. The cycle of atmospheric CO. *Tellus* 26:116–35

Seiler, W., Junge, C. E. 1969. Decrease of carbon monoxide mixing ratio above the polar tropopause. *Tellus* 21:447–49

Seiler, W., Junge, C. E. 1970. Carbon monoxide in the atmosphere. *J. Geophys. Res.* 75:2217–26

Seiler, W., Schmidt, U. 1974. New aspects on CO and H_2 cycles in the atmosphere. *Proc. Int. Conf. Structure, Composition, General Circulation in the Upper and Lower Atmospheres and Possible Anthropogenic Perturbations,* pp. 192–222. Toronto: IAMAP

Seiler, W., Warneck, P. 1972. Decrease of the carbon monoxide mixing ratio at the tropopause. *J. Geophys. Res.* 77:3204–14

Shemansky, D. E. 1972. CO_2 extinction coefficient 1700–3000Å. *J. Chem. Phys.* 56:1582–87

Smith, I. W. M., Zellner, R. 1974. Rate measurements of reactions of OH by resonance absorption. Part III. Reactions of OH with H_2, D_2 and hydrogen and deuterium halides. *J. Chem. Soc. Faraday Trans. II* 70:1045–56

Stevens, C. M., Krout, L., Walling, D., Venters, A., Engelkemeir, A., Ross, L. E. 1972. The isotopic composition of atmospheric CO. *Earth Planet. Sci. Lett.* 16: 147–65

Streit, G. E., Howard, C. J., Schmeltekopf, A. L., Davidson, J. A., Schiff, H. I. 1976. Temperature dependence of $O(^7D)$ rate constants for reactions with O_2, N_2, O_3, and N_2O. *J. Chem. Phvs.* 64: In press

Strobel, D. F. 1975. Aeronomy of the major planets. *Rev. Geophys. Space Phys.* 13: 372–82

Stuhl, F., Niki, H. 1971. Measurements of rate constants for termolecular reactions of $O(^3P)$ with NO, O_2, CO, N_2, and CO_2 using a pulsed vacuum UV photolysis chemiluminescent method. *J. Chem. Phys.* 55:3943–53

Subbaratnam, N. R., Calvert, J. G. 1962. The mechanism of methyl hydroperoxide formation in the photooxidation of azomethane at 25°. *J. Am. Chem. Soc.* 84: 1113–18

Tsang, W. 1973. Comparison between experimental and calculated rate constants for dissociation and combination reactions involving small polyatomic molecules. *Int. J. Chem. Kinet* 5:947–63

U.S. Standard Atmosphere Supplement 1966. Prepared under sponsorship of ESSA, NASA, and USAF. Washington DC: GPO. 289 pp.

Vernazza, J. E. 1975. Unpublished work. Cited in Vernazza, J. E., Averett, E. H., Loeser, R. 1976. Structure of the solar chromosphere II. The underlying photosphere and temperature minimum region. *Ap. J.* In press

Warneck, P. 1974. On the role of OH and HO_2 radicals in the troposphere. *Tellus* 26:39–46

Warneck, P., Junge, C. E., Seiler, W. 1973. OH radical concentration in the stratosphere. *Pure Appl. Geophys.* 106–8:1417–30

Washida, N., Martinez, R. I., Bayes, K. D. 1974. The oxidation of formyl radicals. *Z. Naturforsch.* 29A:251–44

Watson, R. T. 1974. *Chemical Kinetics Data Survey VIII, NBSIR74-516.* Washington DC: Nat. Bur. Stand.

Weinstock, B. 1969. The residence time of carbon monoxide in the atmosphere. *Science* 166:224–25

Whitten, R. C., Sims, J. S., Turco, R. P. 1973. Model of carbon compounds in the stratosphere and mesosphere. *J. Geophys. Res.* 78:5362–74

Wilkniss, P. E., Lamontagne, R. A., Larson, R. E., Swinnerton, J. W., Dickson, C. R., Thompson, T. 1973. Atmospheric trace gases in the southern hemisphere. *Nature* 245:45–57

Wofsy, S. C., McConnell, J. C., McElroy, M. B. 1972. Atmospheric CH_4, CO and CO_2. *J. Geophys. Res.* 77:4477–93

Wofsy, S. C., McElroy, M. B. 1973. On vertical mixing in the stratosphere and lower mesosphere. *J. Geophys. Res.* 78:2619–24

Wofsy, S. C., McElroy, M. B. 1974. HO_x, NO_x and ClO_x: their role in atmospheric photochemistry. *Can. J. Chem.* 52:1582–91

Wofsy, S. C., McElroy, M. B., Sze, N. D. 1975a. Freon consumption: implications for atmospheric ozone. *Science* 187:535–38

Wofsy, S. C. McElroy, M. B., Yung, Y. L. 1975b. The chemistry of atmospheric bromine. *Geophys. Res. Lett.* 2:215–18

Wong, W., Davis, D. D. 1974. A flash photolysis-resonance fluorescence study of the reaction of atomic hydrogen with molecular oxygen: $H + O_2M \rightarrow HO_2 + M$. *Int. J. Chem. Kinet* 6:401–16

Yung, Y. L., McElroy, M. B., Wofsy, S. C. 1975. Atmospheric halocarbons: a discussion with emphasis on chloroform. *Geophys. Res. Lett.* 2:397–99

Zahniser, M. S., Kaufman, F., Anderson, J. G. 1974. Kinetics of the reaction of OH with HCl. *Chem. Phys. Lett.* 27:507–10

AUTHOR INDEX

N

O

P

T

CUMULATIVE INDEXES

CONTRIBUTING AUTHORS VOLUMES 1-4

CHAPTER TITLES VOLUMES 1-4